# Recent Advances in HTLV Research 2015

## Special Issue Editor
Louis M. Mansky

**MDPI**

*Guest Editor*
Louis M. Mansky
University of Minnesota
Minneapolis, MN
USA

*Editorial Office*
MDPI AG
St. Alban-Anlage 66
Basel, Switzerland

This edition is a reprint of the Special Issue published online in the open access journal *Viruses* (ISSN 1999-4915) from 2015–2016 (available at: http://www.mdpi.com/journal/viruses/special_issues/HTLV_research).

For citation purposes, cite each article independently as indicated on the article page online and as indicated below:

Author 1; Author 2; Author 3 etc. Article title. *Journalname*. **Year**. Article number/page range.

ISBN 978-3-03842-376-8 (Pbk)
ISBN 978-3-03842-377-5 (PDF)

# Table of Contents

## Section 1: Review Articles

## Section 2: Research Articles

# About the Guest Editor

**Louis Mansky**, Ph.D., is currently a Professor and Director of the Institute for Molecular Virology at the University of Minnesota Twin Cities. The Mansky research group has an intense interest in the use of interdisciplinary approaches (spanning disciplines including molecular and cell biology, pharmacology, biophysics, medicinal chemistry) to the study of HIV-1 and HTLV-1 replication; particularly aspects relating to: (1) viral mutagenesis, genetic variation, and evolution; and (2) virus particle assembly, release and maturation. Dr. Mansky also serves as Director of the Institute for Molecular Virology Training Program, which seeks to train the next generation of scientists who will study virus replication and pathogenesis as well as help discover new antiviral drugs and vaccines.

# Preface to "Recent Advances in HTLV Research 2015"

Human T-cell leukemia virus (HTLV) has been foundational in our understanding of the molecular pathology of virus-induced cancers. The study of adult T-cell leukemia (ATL) and associated neurological pathologies, including HTLV-associated-myelopathy/tropical spastic paraparesis (HAM/TSP) continues to enhance our understanding regarding how viruses can cause cancer and associated pathologies. This volume presents the latest advancements in knowledge in the study of HTLV replication and pathology, and outlines prospects for future advancement in the field.

Louis M. Mansky
*Guest Editor*

# Section 1:
# Review Articles

*viruses*

MDPI

*Review*

# Molecular Studies of HTLV-1 Replication: An Update

Jessica L. Martin [1], José O. Maldonado [2], Joachim D. Mueller [3], Wei Zhang [4] and Louis M. Mansky [5,*]

[1]   Institute for Molecular Virology, Pharmacoimmunology Training Program & Pharmacology Graduate Program, University of Minnesota, 18-242 Moos Tower, 515 Delaware Street SE, Minneapolis, MN 55455, USA; mart3243@umn.edu

[2]   Institute for Molecular Virology & DDS-PhD Dual Degree Program, University of Minnesota, 18-242 Moos Tower, 515 Delaware Street SE, Minneapolis, MN 55455, USA; jmaldo@umn.edu

[3]   Institute for Molecular Virology & School of Physics and Astronomy, University of Minnesota, 18-242 Moos Tower, 515 Delaware Street SE, Minneapolis, MN 55455, USA; mueller@physics.umn.edu

[4]   Institute for Molecular Virology, School of Dentistry & Characterization Facility, University of Minnesota, 18-242 Moos Tower, 515 Delaware Street SE, Minneapolis, MN 55455, USA; zhangwei@umn.edu

[5]   Institute for Molecular Virology, School of Dentistry & Pharmacology Graduate Program, University of Minnesota, 18-242 Moos Tower, 515 Delaware Street SE, Minneapolis, MN 55455, USA

*   Correspondence: mansky@umn.edu; Tel.: +1-612-626-5525; Fax: +1-612-626-5515

Academic Editor: Eric O. Freed
Received: 25 November 2015; Accepted: 18 January 2016; Published: 27 January 2016

**Abstract:** Human T-cell leukemia virus type 1 (HTLV-1) was the first human retrovirus discovered. Studies on HTLV-1 have been instrumental for our understanding of the molecular pathology of virus-induced cancers. HTLV-1 is the etiological agent of an adult T-cell leukemia (ATL) and can lead to a variety of neurological pathologies, including HTLV-1-associated-myelopathy/tropical spastic paraparesis (HAM/TSP). The ability to treat the aggressive ATL subtypes remains inadequate. HTLV-1 replicates by (1) an infectious cycle involving virus budding and infection of new permissive target cells and (2) mitotic division of cells harboring an integrated provirus. Virus replication initiates host antiviral immunity and the checkpoint control of cell proliferation, but HTLV-1 has evolved elegant strategies to counteract these host defense mechanisms to allow for virus persistence. The study of the molecular biology of HTLV-1 replication has provided crucial information for understanding HTLV-1 replication as well as aspects of viral replication that are shared between HTLV-1 and human immunodeficiency virus type 1 (HIV-1). Here in this review, we discuss the various stages of the virus replication cycle—both foundational knowledge as well as current updates of ongoing research that is important for understanding HTLV-1 molecular pathogenesis as well as in developing novel therapeutic strategies.

**Keywords:** deltaretrovirus; antiretroviral; lentivirus

## 1. Introduction

Human T-cell leukemia virus type 1 (HTLV-1) was independently discovered in 1980 by two research groups and identified as the etiological agent of an adult T-cell leukemia (ATL) [1,2]. As the first human retrovirus discovered, research on HTLV-1 laid the foundational framework for subsequent studies of human immunodeficiency virus type 1 (HIV-1), infectious causes of cancer, and the molecular mechanisms of leukemogenesis [3].

Shortly after the discovery of HTLV-1, another human retrovirus was discovered—human T-cell leukemia virus type 2, HTLV-2—which closely resembled HTLV-1 in genome structure and nucleotide sequence [4]. Unlike HTLV-1, HTLV-2 has not been convincingly associated with human pathology. Nevertheless, both HTLV-1 and HTLV-2 are included in worldwide prevalence estimates. Historically, it has been estimated that 15–20 million people are infected worldwide [5,6]. A more recent study has estimated the number closer to 5–10 million, with the majority of these individuals residing in Japan and the Caribbean Basin [7]. A third and fourth type of HTLV, human T-cell leukemia virus type 3 (HTLV-3) and human T-cell leukemia virus type 4 (HTLV-4), have been discovered in central Africa in the past decade; both are closely related to HTLV-1, and likely share similarities in replication, pathogenesis and transmission [8,9].

HTLV-1 is the etiological agent of ATL as well as a variety of neurological pathologies, primarily HTLV-1-associated-myelopathy/tropical spastic paraparesis (HAM/TSP) [10]. Both ATL and HAM/TSP have a low incidence among HTLV-1 carriers. It is thought that approximately 2%–6% of patients infected with HTLV-1 will acquire either pathology [11,12]. ATL generally presents after a long latency in patients infected during childhood. This is in contrast to HAM/TSP, which is associated with infection later in life [13].

ATL is an aggressive malignancy of the peripheral T-cells and can be divided into four subtypes—acute, lymphomatous, chronic, or smoldering. Patients with the acute form of ATL have a prognosis of approximately 6 months—an estimate that has not significantly changed since the discovery of the disease, despite advances in treatments [14]. Current recommended therapies for ATL include chemotherapy, monoclonal antibodies, allogeneic bone marrow transplants, and a combination of interferon-α (IFN-α) and azidothymidine (AZT) [15–18]. Interestingly, the mechanism of action of the combination of IFN-α and AZT appears to correlate with an induction of cell apoptosis by phosphorylation of p53 [19].

HAM/TSP is characterized by spasticity and weakness of the legs along with urinary disturbances [19]. The primary pathology of HAM/TSP is associated with HTLV-1 infection in the spinal cord leading to inflammation. Unlike ATL, which appears to have a complex and multi-faceted pathology, the incidence of HAM/TSP has been shown to correlate with HTLV-1 proviral loads as well as the site of proviral integration [20,21]. Treatment of HAM/TSP is symptom-based and includes antispasmodic and anti-inflammatory medications [22].

Research into the HTLV-1 life cycle to date has been essential in the discovery and development of better therapeutic strategies. Here in this review, we highlight what is currently known as well as recent advances in the study of HTLV-1 replication. The recent advances help to provide further reason for hope in effective therapeutic options for HTLV-1-infected individuals.

## 2. HTLV-1 Infectious Replication Cycle

### 2.1. Attachment and Fusion

HTLV-1 primarily infects CD4[+] T-cells but has the potential to infect a wide variety of cells, including CD8[+] T-cells, B-lymphocytes, endothelial cells, myeloid cells, fibroblasts, as well as other mammalian cells [23–27]. This wide variety of target cells is due in part to the ability of the surface subunit (SU) of the HTLV-1 envelope glycoprotein (Env) to interact with three widely distributed cellular surface receptors including the glucose transporter (GLUT1) [28], heparin sulfate proteoglycan (HSPG) [29], and the VEGF-165 receptor neuropilin-1 (NRP-1) [30]. Once HTLV-1 has attached to the cell, the membrane fusion process occurs by a series of proposed sequential events between SU and the target cell receptor proteins (Figure 1A,B) [30,31]. Briefly, the HTLV-1 Env interacts with HSPG first followed by NRP-1, which results in the formation of a complex. Following this event, GLUT1 associates with the HSPG/NRP-1 complex to initiate the fusion process, through interactions with the HTLV-1 Env transmembrane (TM) protein, which allows for the HTLV-1 capsid (CA) core containing the viral genome and viral proteins to be released into the cytoplasm of the permissive target cell (Figure 1B).

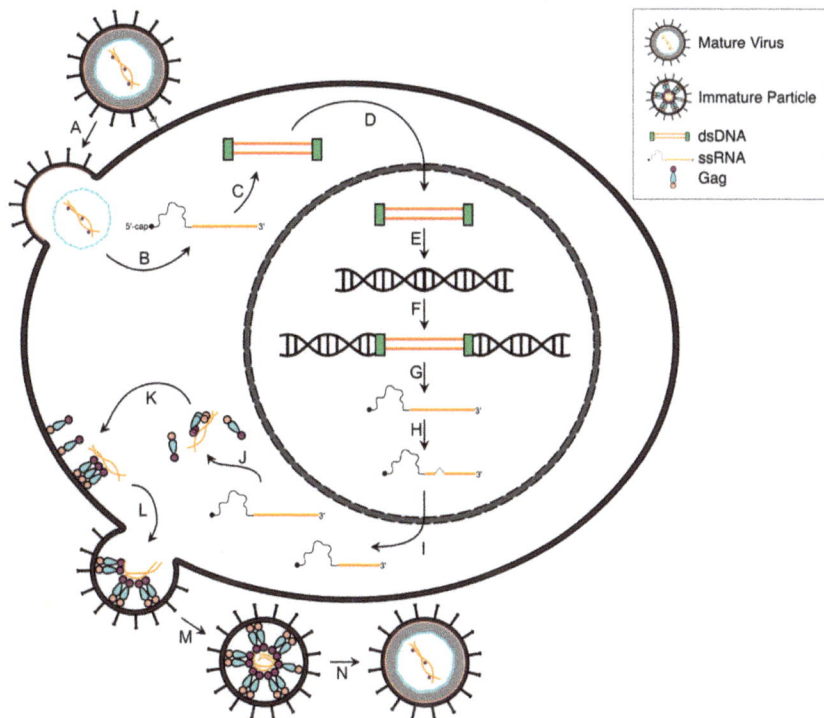

**Figure 1.** HTLV-1 life cycle. The major steps in the life cycle of HTLV-1 are shown. A mature, infectious HTLV-1 virion attaches and fuses to the target cell membrane through interaction with the target cell surface receptors GLUT1/HSPG/NRP-1 via the HTLV-1 envelope surface and transmembrane domains of the envelope (Env) protein (**A**). Following fusion, the viral core containing the viral genomic RNA (gRNA) is delivered into the cytoplasm (**B**), and during and/or following entry the gRNA genome undergoes reverse transcription to convert the gRNA into double stranded DNA (dsDNA) (**C**). The dsDNA is then transported into the nucleus (**D**), and it is integrated into the host genome; (**E,F**). The provirus is then transcribed by cellular RNA polymerase II (**G**), as well as post-transcriptionally modified (**H**). Both full-length and spliced viral mRNAs are exported from the nucleus to the cytoplasm (**I**). The viral proteins are then translated by the host cell translation machinery (**J**), and the Gag, Gag-Pol and Env proteins transported to the plasma membrane (PM) along with two copies of the gRNA genome (**K**). These viral proteins and gRNA assemble at a virus budding site along the PM to form an immature virus particle (**L**). The budding particle releases from the cell surface (**M**), and undergoes a maturation process through the action of the viral protease, which cleaves the viral polyproteins to form an infectious, mature virus particle (**N**).

It has been demonstrated that GLUT1 plays a key role in both the binding of the SU and the infection of $CD4^+$ cells [32]. Paradoxically, other retroviruses have mechanisms that decrease surface expression of their receptors, such as HIV-1 Nef and Vpu [33,34]. The decrease of receptor expression on the cell surface is thought to prevent both superinfection and intracellular Env-receptor interactions, which can inhibit proper proteolytic processing of the Env precursor polyprotein. HTLV-1 does not encode for an accessory protein that reduces surface expression of GLUT1, and it is therefore unclear how HTLV-1 modulates plasma membrane receptor expression. However, it has been recently shown that HTLV-1-based virus-like particles (VLPs) produced in cells with high levels of GLUT1 were better able to fuse with target cells than those produced from cells with low levels of GLUT1 [35]. In 293T cells,

HTLV-1 Env avoids interaction with GLUT1 through the separate intracellular localization of GLUT1 and Env [35]. This recent observation is important because it suggests that separate intracellular localization of GLUT1 and HTLV-1 Env is required for proper fusion activity of the HTLV-1 Env. This study may also have implications for HTLV-1 cellular tropism, as CD4$^+$ regulatory T-cells, the primary viral reservoir for HTLV-1-infected individuals, express GLUT1 at low levels as compared to other types of CD4$^+$ T-cells [36].

### 2.2. Reverse Transcription, Nuclear Transport and Integration

The HTLV-1 CA core enters the infected cell and contains two copies of the viral genomic RNA (gRNA) along with reverse transcriptase (RT), integrase (IN), and the viral protease (PR). Reverse transcription of HTLV-1 RNA to double-stranded DNA (dsDNA) has not been extensively studied but likely occurs after virus entry (Figure 1C) [37,38]. It is thought that HIV-1 reverse transcription is linked to intracellular uncoating of the CA core [39]. Additionally, HIV-1 RT and IN interactions have been shown to be necessary for production of early reverse transcription products [40]. Complementary studies with HTLV-1 have yet to be done, so it is unclear whether HTLV-1 CA uncoating correlates with reverse transcription or if RT-IN interactions occur during early reverse transcription. Recombination can occur during reverse transcription, and recent evidence from phylogenetic analyses strongly suggests that recombination played a distinct role in emergence of HTLV-1 in the human population approximately 4000 years ago [41].

Unlike HIV-1, which is highly sensitive to the effects of the APOBEC family of cytidine deaminases, HTLV-1 appears less sensitive APOBECs. There is some evidence that APOBEC3G may lead to G-to-A hypermutation in some HTLV-1 sequences *in vivo* [42,43], but the overall effect on HTLV-1 sequence diversity appears to be negligible—perhaps due to the propensity of HTLV-1 to be propagated by clonal expansion of infected cells rather than replication via reverse transcription. HTLV-1 has been previously shown to prevent APOBEC3G packaging through an element at the C-terminal nucleocapsid (NC) region of Gag [44].

The partially disassembled core containing the reverse transcription complex (preintegration complex) is translocated to the nucleus (Figure 1D) where integration into the host cell chromosome occurs to form the provirus (Figure 1E,F). It has been found that HTLV-1 integrates into the genome in the absence of preferred sites [45–50]. Such studies have analyzed hundreds of thousands of HTLV-1 integration sites [51,52] and have not been able to identify HTLV-1 proviral integration site hotspots. Interestingly, in HTLV-1-induced disease states, the integration sites of HTLV-1 become non-random. For example, it was recently demonstrated that the clinical diagnosis of HAM/TSP correlates with proviral integration into transcriptionally active regions [53].

### 2.3. Viral Gene Transcription

The long terminal repeats (LTRs) of the HTLV-1 provirus contain the necessary promoter and enhancer elements to initiate RNA transcription (Figure 1G), with the polyadenylation signal located in the 3′LTR [1]. Tax, a non-structural protein and the main driver of viral transcription, potently activates viral transcription during the early phase of infection by recruiting multiple cellular transcription factors [54]. Three conserved 21-bp repeat elements, known as the Tax-responsive element 1 (TRE-1), bind the cyclic AMP response element binding protein (CREB) at the TRE-1 site through its N-terminus (NTD) [55–61], while the C-terminal domain (CTD) of Tax is believed to promote the transcriptional initiation and RNA polymerase elongation by directly interacting with the TATA binding protein [5,62]. The Tax-CREB promoter complex recruits the multifunctional cellular coactivators CREB binding protein (CBP), p300, and the p300/CBP-associated factor to the LTR [63–68].

Recently, several host factors that directly interfere with HTLV-1 viral transcription have been identified. TCF1 and LEF1 are transcription factors specifically found in T-cells. They antagonize Tax activity through physical association with Tax, preventing transcription of the viral proteins. In most HTLV-1-infected cell lines, however, TCF1 and LEF1 expression is low due to downregulation via

STAT5a, which is activated by Tax [69]. The host protein SIRT1 deacetylase has also been shown to downregulate HTLV-1 viral transcription by inhibiting Tax. Unlike TCF1 and LEF1, SIRT1 appears to inhibit Tax-CREB interactions. Interestingly, the well-known SIRT1 activator resveratrol significantly decreases the transmission of HTLV-1 produced from MT2 cells [70,71]. This suggests that resveratrol may be a potential therapeutic option for patients infected with HTLV-1 or a prophylactic option to prevent virus transmission. In addition to these cellular host factors, the facilitate chromatin transcription (FACT)proteins SUPT16H and SSRP1 have been shown to inhibit both HTLV-1 and HIV-1 transcription by preventing interaction of HTLV-1 Tax and HIV-1 Tat with their respective viral LTRs [72].

### 2.4. Post-Transcriptional Regulation

Rex is a positive post-transcriptional regulator essential for splicing and transport of HTLV-1 mRNA (Figure 1H,I). Rex specifically interacts with the U3 and R regions of the HTLV-1 gRNA known as the Rex-responsive element (RexRE). During the early stages of viral gene transcription, suboptimal levels of Rex are present [73], which results in the exclusive export of doubly spliced (*tax*, *rex*, *p30II*, *p12*, *p13*, and *hbz*) viral mRNAs to the cytoplasm (Figure 1I) [74]. Once Rex accumulates in the nucleus, Rex reduces splicing of viral mRNA and the singly spliced (*env*) and unspliced (*gag-pro-pol*) mRNAs are then exported from the nucleus to the cytoplasm leading to the production of enzymatic and structural proteins (Figure 1J) [74]. Rex binds to the RexRE through a highly basic RNA-binding NTD, while the CTD is important for protein oligomerization [75,76]. Rex also contains an activation domain containing the nuclear export signal, which targets Rex to the nuclear pore complex in order for Rex to move between the nucleus and cytoplasm [77,78].

Despite the presence of host cell mechanisms to export doubly spliced RNA, all HTLV-1 mRNA transcripts, including those that are doubly spliced, have RexREs present. A recent study has shown that Rex may have a CRM1-dependent role in nuclear export of all HTLV-1 mRNAs, even the doubly spliced mRNAs that should be exported via host cell mechanisms [79]. These observations suggest that viral mRNA export is under a more complex regulation than previously thought. Furthermore, another recent study has suggested that there are three alternatively spliced HTLV-1 transcripts that encode for novel Rex isoforms, which may also contribute to the regulation of HTLV-1 protein expression levels [80].

### 2.5. Viral Protein Translation

As soon as HTLV-1 mRNAs are exported to the cytoplasm, the host protein-synthesis machinery translates the viral proteins. Presumably, the full-length viral gRNA is either translated or trafficked to the plasma membrane, where it can dimerize, interact with the Gag polyprotein, and be packaged into assembling particles (Figure 1K,L) [81]. The doubly spliced and unspliced mRNAs are translated by free ribosomes to express the enzymatic and structural proteins, respectively, while the singly spliced mRNA is translated by membrane-bound ribosomes to express Env [45].

Many RNA viruses use a cap-independent mechanism to recruit the 40S ribosomal subunit to an internal ribosome entry segment (IRES) within the 5′UTR of the mRNA, which allows for ribosomal scanning and protein translation to occur [82–84]. It was thought that HTLV-1 mRNA contains an IRES element [85] used for the translation of the Gag protein, but another study suggests that a 5′ proximal post-transcriptional control element modulates post-transcriptional HTLV-1 gene expression by interacting with the host RNA helicase A instead of an IRES element, implying that the translation of the HTLV-1 mRNA is cap-dependent [86]. Interestingly, a recent study has demonstrated that HTLV-1 translation is inhibited by the drug edeine, a cap-independent translation inhibitor, suggesting that an IRES element in the 5′ UTR recruits the ribosome to the mRNA [87]. Obviously, more research is needed to firmly establish the mechanism(s) used by HTLV-1 to translate its viral proteins.

## 2.6. Gag and Viral RNA Trafficking

Viral particle formation occurs after Gag traffics from the cytoplasm to the plasma membrane (PM) (Figure 1K). How HTLV-1 Gag translocates from the site of translation to the membrane is poorly understood. However, it is known that monomeric forms of HTLV-1 Gag exist in the cytoplasm and are detected at the membrane shortly after the initiation of viral protein translation [88]. This is in contrast to HIV-1 Gag, where low ordered oligomers are observed in the cytoplasm until micromolar concentrations are reached prior to detecting oligomeric Gag at the plasma membrane [88]. HIV-1 Gag interacts with many cellular proteins, including cytoskeleton-associated proteins, though their relationship to HIV-1 Gag trafficking is unclear [89]. HTLV-1-infected cells regulate cytoskeletal polarization [90], though it is unclear if this is related to Gag trafficking to the plasma membrane.

HTLV-1 Gag nucleocapsid (NC) protein binds to HTLV-1 RNA relatively weakly as compared to that of other retroviral NC proteins, due in part to the anionic carboxy-terminal domain (CTD) of the HTLV-1 NC [91]. The HTLV-1 MA has been recently reported to bind RNA, and it was found that HTLV-2 MA binds RNA at much higher affinity than HTLV-2 NC [92]. This is in direct contrast to HIV-1, in which NC binds to RNA more strongly than HIV-1 MA [92]. These recent findings highlight the importance of both the MA and NC domains in viral RNA interactions that are likely critically important for viral gRNA recognition and gRNA packaging. How HTLV-1 RNA traffics through the cytoplasm in order to get to the plasma membrane (and to virus budding sites) is poorly understood, but a recent study with HIV-1 gRNA suggests that the viral gRNA diffuses through the cytoplasm to the membrane [93]. It is formally possible that HTLV-1 gRNA also diffuses through the cytoplasm to reach the membrane, but it could also bind to Gag before reaching the membrane (Figure 1I–K). There is a significant need for future studies in order to better understand these aspects of HTLV-1 replication.

## 2.7. Assembly, Budding and Maturation

Gag-gRNA, Gag-Gag and Gag-membrane interactions are all required for the assembly and budding of virus particles (Figure 1L) [94]. Gag forms higher order oligomers by oligomerizing with other Gag molecules through interactions primarily involving the CA domain and to some extent the NC domain [95–100]. Once at the PM, virus budding sites are identified and are characterized by the interaction of HIV-1 MA with lipid-rich [101] assembly sites known as lipid rafts [102–104]. Membrane binding of HIV-1 Gag is dependent upon interaction of MA with phosphatidylinositol-(4,5)-bisphosphate PI(4,5)P$_2$ [105]. HTLV-1 Gag has been shown to not have a preference for binding to PI(4,5)P$_2$, which has implications for how HTLV-1 Gag targets the PM and identifies virus budding sites [105]. Cellular factors are also recruited to the virus budding sites, resulting in budding and subsequent release of immature virus particles (Figure 1L,M) [100,106,107]. The viral protease (PR) cleaves the Gag and Pol polyproteins during and shortly after the release of immature virus particles (Figure 1N) [108]. MA remains closely associated with the PM; CA forms a capsid shell that contains reverse transcriptase, integrase and the NC-coated gRNA. The mature virus particle, if infectious, is capable of infecting a permissive target cell (Figure 1N) [109].

## 3. HTLV-1 Transmission

### 3.1. Inter-Host Transmission

There are generally three modes of inter-host HTLV-1 transmission described: (1) blood and blood products, (2) vertical or (3) sexual transmission [110], but the main mode of transmission is thought to be vertical, *i.e.* from mother-to-child through breastfeeding [111]. Mother-to-child transmission rates vary from 5% to 27% for children nursed by infected mothers and correlate with the duration of breastfeeding [112,113]. While it is not clear precisely how infection occurs through the mucosal and epithelial barriers of the gastrointestinal tract, it is thought that infected lymphocytes in breast milk carry the virus into the gut [114]. Once in the gut, either cell-free virus or cells carrying the virus must pass through the epithelium. A recent study demonstrated *in vitro* that cell-free HTLV-1 may cross the epithelial barrier via transcytosis before infecting subepithelial dendritic cells [115]. The precise mechanism of transcytosis for HTLV-1 remains unclear. However, studies with HIV-1 have shown that transcytosis across vaginal epithelial cells occurs via the endocytic recycling pathway [116]. It is plausible that other mechanism(s) are involved in HTLV-1 infection across the gut epithelial barrier due to the low infectivity of cell-free virus. While cell-free HIV-1 is generally thought to be much more infectious than cell-free HTLV-1, it has been suggested that HIV-1-infected lymphocytes more efficiently infect target cells in the gut than cell-free virus – possibly through the formation of a viral synapse that induces transcytosis [117]. The role of the virological synapse in these transmission events has not been carefully studied. It is also not known whether HTLV-1 infected lymphocytes can transmigrate as a whole cell across the epithelial barrier and infect subepithelial immune cells.

Zoonotic transmission events of simian T-cell leukemia virus type 1 (STLV-1) to humans after contact with nonhuman primates through bites or bushmeat slaughtering still occur in Africa, establishing the emergence of new HTLV-1 infections in humans. A recent study found that more than 8% of individuals bitten by nonhuman primates in Africa are infected with HTLV-1, and virus transmission cannot be attributed to mother-to-child transmission [118]. The strains of HTLV-1 found in those infected closely resembled the subtypes of STLV-1 commonly found in the primate species from which they were bitten [118,119]. In fact, it is likely that the emergence of HTLV-3 and HTLV-4 may be attributable to recent STLV zoonotic transmission events, as STLV-4 is known to be endemic in African gorillas, and phylogenetic analyses have shown that HTLV-4 is not an ancient human virus but recently emerged in the human population [120]. While these findings highlight the potential ongoing role of nonhuman primates as virus reservoirs, they also highlight interest in the virus-host interactions that facilitate cross-species transmission as well as potential risks in transmission and emergence of more highly pathogenic types of HTLV. While monkeys in Japan also harbor STLV-1 strains [121], those strains are more highly divergent from the HTLV-1 strains in Japanese patients, indicating that zoonotic transmission of HTLV-1 may not be a major public health threat in regions outside of Africa [122].

### 3.2. Cell-to-Cell Transmission

In general, there are two distinct methods of virus transmission between cells: virus infection of cells in the absence of cell-to-cell contacts and virus infection involving cell-to-cell contacts. Most retroviruses can efficiently infect target cells in the absence of cell-to-cell contacts—in which the virus buds from the cell and infects a target cell through diffusion. HTLV-1 is notorious for being poorly infectious in the absence of direct cell-to-cell transmission, and co-cultivation of permissive target cells with virus-producing cells are the most effective means of virus transmission [123].

### 3.3. Virological Synapses

Immunofluorescence and confocal microscopy were used previously to demonstrate that Gag and Env proteins are more evenly distributed in isolated T-cells, but once the cell comes into contact with another cell, cell polarization occurs—impacting the localization of HTLV-1 Gag, Env and the genomic RNA towards the cell-cell junction. This cell-to-cell junction, termed the virological synapse (VS), shares many features with the previously described immunological synapse, which includes features such as ordered talin domains and microtubule organizing center (MTOC) polarization [124]. Cryoelectron tomography studies of HTLV-1 associated VS structures suggest that there is no fusion of the cell membranes [125]. To the contrary, HTLV-1 transmission occurs via rapid budding and fusion of the HTLV-1 virus across the VS from the infected to uninfected cell (Figure 2).

It has been reported that the formation of the VS is triggered by HTLV-1 infection and is not dependent on signaling through the T-cell receptor as is seen in immunological synapses [124]. The VS forms when the surface adhesion molecule intercellular adhesion molecule-1 (ICAM-1) is engaged by its ligand lymphocyte function-associated antigen 1 (LFA-1) [90,126]. ICAM-1 then activates the MEK/ERK pathway, which contributes to MTOC relocation. The HTLV-1 Tax protein, the key virus transcription accessory protein, works in synergy with ICAM-1 to facilitate MTOC polarization. While it is primarily a nuclear protein, Tax can be found in the cytoplasm near the MTOC as well as in the cell-cell contact region [127]. Tax activates the CREB-signaling pathway during the formation of the HTLV-1 VS [126]. The CREB pathway increases expression of Gem, a small GTP-binding protein in the RAS superfamily, which is involved in cytoskeleton remodeling and cell migration [128]. Tax also appears to upregulate ICAM-1 in HTLV-1-infected cells, indicating that ICAM-1 and Tax appear to have synergistic roles in HTLV-1 VS formation [129].

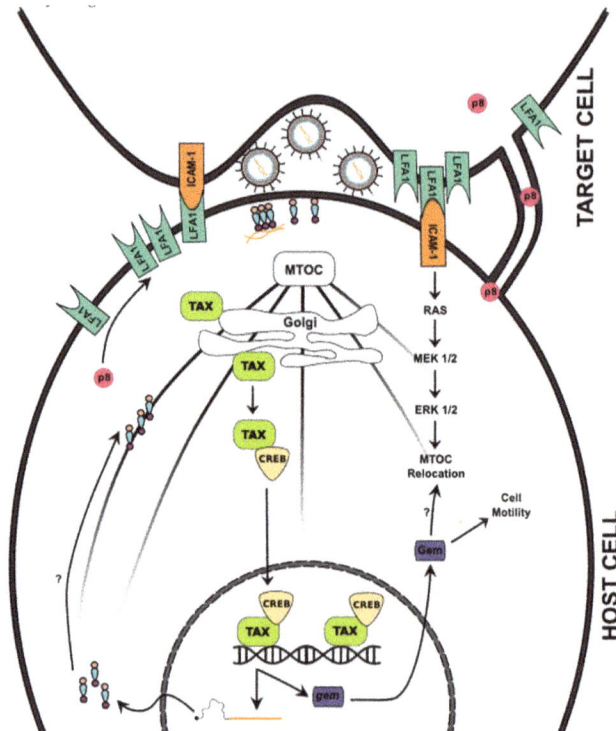

**Figure 2.** HTLV-1 Cell-to-Cell Transmission. Shown in the diagram is a host cell (**bottom**) that has anchored itself to a permissive target cell (**top**) using ICAM-1 and LFA1. The HTLV-1 accessory protein p8 has been shown to increase the expression of the LFA1 receptors as well as increase the number of cell-cell synapses, which p8 then traffics through to increase LFA1 in the target cell. Once ICAM-1 is bound, it triggers a signaling cascade that leads to the relocation of the MTOC. Additionally, HTLV-1 Tax is found bound to Golgi bodies that are attached to the MTOC. This initiates a variety of cell signaling cascades. For example, one established pathway increases expression of Gem, a protein that increases cell motility and possibly MTOC relocation. As HTLV-1 Gag is produced, it is found concentrated at the MTOC. Subsequently, HTLV-1 Gag is found at the sites of cell-cell contacts where the virus particles bud into the viral synapse formed by the cellular adhesion points. Virus particles produced can bind receptors for entry into the permissive target cell. The question marks along the lines with arrows indicate mechanisms of transport or activation that are not well understood.

When comparing HTLV-1 cell-cell transmission to HIV-1 cell-cell transmission, HTLV-1 is more dependent on cell contact for infectivity. Coculturing of virus-producing cells with permissive target cells significantly enhances HTLV-1 transmission by several thousand-fold, but only increases HIV-1 transmission 10–100 hundred-fold [130,131]. Furthermore, disruption of actin and tubulin polymerization inhibits HTLV-1 spread to a greater extent than that of HIV-1 spread [131]. Nonetheless, HIV-1 cell-cell transmission remains a highly relevant form of virus spread *in vivo*. A recent study helps to highlight this through the observation that granulocytes (e.g., basophils) may actually capture HIV-1 and contribute to virus transmission [132].

*3.4. Viral Biofilms*

In addition to transmission via the VS, HTLV-1 particles have been reported to have the ability to form a biofilm-like, carbohydrate-rich extracellular structure on the surface of cells. These structures are composed of collagen, agrin, tetherin, and galectin-3 and may function as a way to concentrate HTLV-1 particles in a single location to increase the likelihood of infection of a permissive target cell [133]. A recent report described the creation of antiviral biofilm monoclonal antibodies using purified biofilms from MT2 cells. It was found that in addition to the structural proteins previously identified, the antigens CD4, CD150, CD25, CD70, and CD80 were also identified in these viral biofilms. Tax expression was found to modulate the level of these antigens in the viral biofilm [134].

While it has been observed that dendritic cells may be infected by cell-free HTLV-1 *in vitro* [31], a recent study analyzed the infectivity of chronically infected C91PL cell culture supernatant as compared to purified biofilms in primary human monocyte-derived dendritic cells (MDDCs) and T lymphocytes. It was found that while MDDCs were more easily infected than were T lymphocytes, both cell types achieved significantly higher proviral loads when exposed to viral biofilms as opposed to supernatant-derived virus [135].

## 4. Monoclonal Expansion of HTLV-1 Infected Cells and Leukemogenesis

While HTLV-1 cell-cell transmission is likely a critical determinant of virus transmission from an infected individual to a susceptible individual, many studies have established that the primary route of replication for HTLV-1 *in vivo* is through mitotic division of host cells and subsequent propagation of the provirus by clonal expansion. DNA analysis of HTLV-1 sequences from patients in geographically distinct locations shows very little genetic variation among HTLV-1 isolates [136]. This is likely because of a low evolutionary rate of $7.06 \times 10^{-7}$–$1.38 \times 10^{-5}$ substitutions per site per year in the *LTR* and *env* regions [137]. While the HTLV-1 reverse transcriptase has a reduced mutation rate as compared to HIV-1 reverse transcriptase ($7 \times 10^{-6}$ mutations per target base pair per replication cycle for HTLV-1 as compared to $3.4 \times 10^{-5}$ for HIV-1), the fourfold difference is likely not sufficient alone to explain the relative difference in genetic diversity between the two viruses [138,139]. Also, administration of reverse transcriptase inhibitors was found to not reduce proviral loads, even when administered shortly after infection [140,141].

The most compelling evidence in favor of a clonal expansion of HTLV-1 proviruses is the clonality of T-cells in infected individuals. HTLV-1 integrates randomly into the genome [142,143]. When the HTLV-1 proviral sites are amplified by PCR, it has been consistently found that the T-cells are clonal in both symptomatic and asymptomatic carriers [144,145]. In some HTLV-1 infected individuals, more than one in every 1500 peripheral blood mononuclear cells (PBMCs) were found to be clonal [146].

Since HTLV-1 persistence depends to some extent on the clonal expansion of infected T-cells, it is not surprising that HTLV-1 encodes for gene products that have been found to increase cell proliferation. The oncogenic potential of HTLV-1 is likely a byproduct of the induction of these cellular proliferative pathways. In fact, expression of the HTLV-1 basic leucine zipper (HBZ) and Tax proteins in double transgenic mice was recently shown to be sufficient for the development of lymphoma in the absence of any other viral genes [147]. Since similar pathways influence both cell immortalization and cellular transformation, these two processes are difficult to uncouple.

*4.1. Tax*

Tax has been established to have many roles in the HTLV-1 proliferative cycle, including essential roles in mediating transcription of HTLV-1 genes as well as the recruitment of the MTOC to sites of cell-cell contact. Tax can also interfere with host cell cycle regulation, apoptotic pathways, and proliferative pathways through a variety of mechanisms. The most commonly studied pathway through which Tax increases cellular proliferation is via the NF-κB/Rel family of proteins.

*4.2. Tax and Canonical NF-κB Signaling*

The NF-κB/Rel proteins are transcription factors that share a Rel homology domain responsible for DNA binding and dimerization. There are five proteins in this family: RelA (p65), RelB, c-Rel, p105 (processed to p50), and p100 (processed to p52). When the NF-κB pathways are turned off, NF-κB dimers are bound to the inhibitor of κB (IκB) proteins. Activation of NF-κB can occur through a canonical pathway in which nuclear localization of NF-κB is induced by inflammatory stimuli such as tumor necrosis factor (TNF) or through a non-canonical pathway reviewed in [148].

NF-κB is constitutively active in HTLV-1 infected cells, an abnormality due to the tight regulation of NF-κB by its inhibitor, IκB [149]. IκB kinase (IKK) phosphorylates IκB molecules, leading to the eventual ubiquitination and degradation of IκB and freeing NF-κB dimers to traffic to the nucleus. HTLV-1 Tax increases the activity of IKK by regulating a variety of upstream effectors. Tak1 is a kinase that can phosphorylate and activate IKK. Tax binds directly to Tak1 to form an IKK-Tak1-Tax complex that appears to increase the efficiency of IKK [150]. Tax can also increase the activity of Tak1 by binding to Tak1-binding protein 2 (TAB2); the physiological function of this remains unclear [151]. Tax can also bind directly to TRAF6 and can stimulate its function, leading to the activation of NF-κB through IKK phosphorylation [152]. Finally, Tax can also stimulate MEKK1, which is an upstream regulator of TAK1 [153].

In addition to its upstream effects on IKK, Tax also binds directly to an IKK subunit, IKKγ (referred to as NEMO) [154,155]. IKKγ is a regulatory subunit that modulates IKK activity through an unknown mechanism [156]. Protein phosphatase 2A (PP2A) is a positive regulator of IKKγ, and PP2A binding to helix 2 (HLX2) is necessary for proper IKK function [157]. Tax binds IKKγ in a region two heptads downstream of HLX2 termed coiled-coil region 2 (CCR2). When Tax binds this region, it exposes the HLX2 domain for PP2A binding and simultaneously inactivates PP2A, preventing it from dissociating, resulting in IKK being constitutively active [158,159].

Tax also has the ability to form complexes with NF-κB monomers in the cytoplasm. Tax can bind directly to cytoplasmic RelA (p65) and can induce translocation into the nucleus, leading to high transcriptional activity [151,160–162]. This particular protein-protein interaction appears to also rely on the CREB-binding protein (CBP or p300), which facilitates the transcriptional activity of RelA.

A role for Tax that has only recently been described is the recognition and inactivation of the ubiquitin-editing enzyme A20. This enzyme is a negative regulator of the canonical NF-κB pathway. It binds with the regulatory protein TAX1BP1 and E3 ligase Itch to form a complex that inactivates essential NF-κB upstream regulators. Tax binds to TAX1BP1, preventing the interaction of TAX1BP1, A20, and Itch [163,164].

*4.3. Tax and Non-Canonical NF-κB Signaling*

Tax is known to facilitate the non-canonical pathway of NF-κB as well as the canonical pathway traditionally induced by MEKK1 and Tak1. This is an important distinction, as the NF-κB dimers are distinct for each pathway and increase expression of different gene products. For example, when the non-canonical pathway is blocked in the presence of Tax, tumorigenesis is significantly delayed in Tax transgenic mice [165]. IKKα phosphorylates p100, which leads to its ubiquitination and cleavage to p52. Tax facilitates this process by recruiting IKKα to p100 and inducing cleavage [166,167].

*4.4. Tax and Other Cell Proliferative Pathways*

In addition to its interaction with NF-κB, HTLV-1 Tax has roles in a variety of other cell proliferation pathways. In particular, Tax interacts directly or indirectly with cyclin-dependent kinases (CDK) [168,169], phosphoinositide 3-kinase (PI3K) [170], transforming growth factor β-1

*Viruses* **2016**, *8*, 31

(TGFβ-1) [171,172], and p53 [173–175]. Through a series of complex interactions, Tax ensures that HTLV-1 replication occurs.

*4.5. Tax Downregulation*

Tax plays a key role in the proliferation of T-cells but is known to be a highly immunogenic protein, and Tax-expressing cells are targeted by cytotoxic T-cells [176,177]. That Tax is expressed only at early time points helps to avoid immune surveillance. In patients with ATL, *tax* mRNA transcripts were only found in 34% of all cases [178], indicating that there are downregulation mechanisms in place to prevent HTLV-1-infected cells from being destroyed by the immune system.

Several mechanisms regulate the downregulation of Tax expression. Many *tax* genes become mutated so that non-functional transcripts are produced [178–180]. Cells containing these *tax* gene sequences are more likely to replicate due to decreased immune surveillance. Additionally, the *tax* gene is often methylated, leading to gene silencing [178,181]. Finally, the HTLV-1 bZIP factor (HBZ) works to downregulate transcription of many HTLV-1 genes, including Tax [182].

*4.6. HBZ*

The *HBZ* gene is an important HTLV-1 gene that is consistently associated with ATL. *HBZ* mRNA transcripts have been found in virtually all ATL cells, and it was shown that HBZ plays a key role in the proliferation of T-cells [183]. Interestingly, the *HBZ* mRNA appears to play a different role in T-cell proliferation than the HBZ protein [183]. In mouse T-cells, an HBZ start codon mutant that distinguishes between RNA function and protein function was found to increase T-cell proliferation by inhibiting apoptosis and promoting S-phase entry [184]. *HBZ* RNA attenuates apoptosis by promoting the transcription of Survivin, a caspase inhibitor that prevents apoptosis. While HBZ protein promotes S-phase entry, it can also promote apoptosis through its pro-inflammatory effects [184].

HBZ has been reported to directly or indirectly interact with the following proteins that are essential in CREB-dependent cellular proliferation: CREB, CBP, ATF-1, and ATF-3 [185–188]. Additionally, HBZ interacts with the Jun family of transcription factors, JunB, JunD, and c-Jun [189–191]. The combined activities of Tax and HBZ are essential for cellular proliferation. Taken together, the activities of both HBZ and Tax are essential for the transformation of T-cells.

## 5. Conclusions and Future Directions

As the first human retrovirus discovered in the early 1980s, HTLV-1 has been studied extensively, yet there is still no treatment or vaccine for HTLV-1 infection. Additionally, ATL and HAM/TSP treatments are symptom-based and do not directly treat the viral infection. Continued research on the molecular aspects of HTLV-1 replication will enhance opportunities for the discovery of potential antiretroviral targets that can be exploited for the development of effective therapeutic strategies. For example, a recent candidate peptide ($HBZ_{157--176}$) has been described for use in vaccine development [192]. Such observations help to enhance the likelihood for specific forms of therapeutic intervention for the treatment of HTLV-1 infection.

**Acknowledgments:** This work is supported by National Institutes of Health grant R01 GM098550. Jessica L. Martin has been supported by NIH grant T32 DA007097. José O. Maldonado has been supported from NIH grants T32 AI083196 (Institute for Molecular Virology Training Program) and F30 DE022286.

**Author Contributions:** Jessica L. Martin and José O. Maldonado drafted the manuscript, and Joachim D. Mueller, Wei Zhang, and Louis M. Mansky edited the text to create the final submitted version.

**Conflicts of Interest:** The authors declare no conflict of interest.

## References

1. Poiesz, B.J.; Ruscetti, F.W.; Gazdar, A.F.; Bunn, P.A.; Minna, J.D.; Gallo, R.C. Detection and isolation of type C retrovirus particles from fresh and cultured lymphocytes of a patient with cutaneous T-cell lymphoma. *Proc. Natl. Acad. Sci. USA* **1980**, *77*, 7415–7419. [CrossRef] [PubMed]

2. Yoshida, M.; Miyoshi, I.; Hinuma, Y. Isolation and characterization of retrovirus from cell lines of human adult t-cell leukemia and its implication in the disease. *Proc. Natl. Acad. Sci. USA* **1982**, *79*, 2031–2035. [CrossRef] [PubMed]

3. Gallo, R.C. History of the discoveries of the first human retroviruses: HTLV-1 and HTLV-2. *Oncogene* **2005**, *24*, 5926–5930. [CrossRef] [PubMed]

4. Kalyanaraman, V.S.; Sarngadharan, M.G.; Robert-Guroff, M.; Miyoshi, I.; Golde, D.; Gallo, R.C. A new subtype of human T-cell leukemia virus (HTLV-II) associated with a T-cell variant of hairy cell leukemia. *Science* **1982**, *218*, 571–573. [CrossRef] [PubMed]

5. Caron, C.; Rousset, R.; Beraud, C.; Moncollin, V.; Egly, J.M.; Jalinot, P. Functional and biochemical interaction of the HTLV-I Tax1 transactivator with tbp. *EMBO J.* **1993**, *12*, 4269–4278. [PubMed]

6. Proietti, F.A.; Carneiro-Proietti, A.B.; Catalan-Soares, B.C.; Murphy, E.L. Global epidemiology of HTLV-I infection and associated diseases. *Oncogene* **2005**, *24*, 6058–6068. [CrossRef] [PubMed]

7. Gessain, A.; Cassar, O. Epidemiological aspects and world distribution of HTLV-1 infection. *Front. Microbiol.* **2012**, *3*. [CrossRef] [PubMed]

8. Calattini, S.; Chevalier, S.A.; Duprez, R.; Bassot, S.; Froment, A.; Mahieux, R.; Gessain, A. Discovery of a new human T-cell lymphotropic virus (HTLV-3) in central africa. *Retrovirology* **2005**, *2*. [CrossRef] [PubMed]

9. Wolfe, N.D.; Heneine, W.; Carr, J.K.; Garcia, A.D.; Shanmugam, V.; Tamoufe, U.; Torimiro, J.N.; Prosser, A.T.; Lebreton, M.; Mpoudi-Ngole, E.; *et al.* Emergence of unique primate T-lymphotropic viruses among central african bushmeat hunters. *Proc. Natl. Acad. Sci. USA* **2005**, *102*, 7994–7999. [CrossRef] [PubMed]

10. Yoshida, M.; Seiki, M.; Yamaguchi, K.; Takatsuki, K. Monoclonal integration of human T-cell leukemia provirus in all primary tumors of adult T-cell leukemia suggests causative role of human T-cell leukemia virus in the disease. *Proc. Natl. Acad. Sci. USA* **1984**, *81*, 2534–2537. [CrossRef] [PubMed]

11. Murphy, E.L.; Figueroa, J.P.; Gibbs, W.N.; Brathwaite, A.; Holding-Cobham, M.; Waters, D.; Cranston, B.; Hanchard, B.; Blattner, W.A. Sexual transmission of human T-lymphotropic virus type I (HTLV-I). *Ann. Int. Med.* **1989**, *111*, 555–560. [CrossRef] [PubMed]

12. Kaplan, J.E.; Osame, M.; Kubota, H.; Igata, A.; Nishitani, H.; Maeda, Y.; Khabbaz, R.F.; Janssen, R.S. The risk of development of HTLV-I-associated myelopathy/tropical spastic paraparesis among persons infected with HTLV-I. *J. Acquir. Immune Defic. Syndr.* **1990**, *3*, 1096–1101. [PubMed]

13. Hisada, M.; Stuver, S.O.; Okayama, A.; Li, H.C.; Sawada, T.; Hanchard, B.; Mueller, N.E. Persistent paradox of natural history of human T lymphotropic virus type I: Parallel analyses of Japanese and Jamaican carriers. *J. Infect. Dis.* **2004**, *190*, 1605–1609. [CrossRef] [PubMed]

14. Tsukasaki, K.; Hermine, O.; Bazarbachi, A.; Ratner, L.; Ramos, J.C.; Harrington, W., Jr.; O'Mahony, D.; Janik, J.E.; Bittencourt, A.L.; Taylor, G.P.; *et al.* Definition, prognostic factors, treatment, and response criteria of adult T-cell leukemia-lymphoma: A proposal from an international consensus meeting. *J. Clin. Oncol.* **2009**, *27*, 453–459. [CrossRef] [PubMed]

15. Tanosaki, R.; Uike, N.; Utsunomiya, A.; Saburi, Y.; Masuda, M.; Tomonaga, M.; Eto, T.; Hidaka, M.; Harada, M.; Choi, I.; *et al.* Allogeneic hematopoietic stem cell transplantation using reduced-intensity conditioning for adult T cell leukemia/lymphoma: Impact of antithymocyte globulin on clinical outcome. *J. Am. Soc. Blood Marrow Transplant.* **2008**, *14*, 702–708. [CrossRef] [PubMed]

16. Tsukasaki, K.; Utsunomiya, A.; Fukuda, H.; Shibata, T.; Fukushima, T.; Takatsuka, Y.; Ikeda, S.; Masuda, M.; Nagoshi, H.; Ueda, R.; *et al.* VCAP-AMP-VECP compared with biweekly CHOP for adult T-cell leukemia-lymphoma: Japan clinical oncology group study JCOG9801. *J. Clin. Oncol.* **2007**, *25*, 5458–5464. [CrossRef] [PubMed]

17. Yamamoto, K.; Utsunomiya, A.; Tobinai, K.; Tsukasaki, K.; Uike, N.; Uozumi, K.; Yamaguchi, K.; Yamada, Y.; Hanada, S.; Tamura, K. Phase I study of KW-0761, a defucosylated humanized anti-CCR4 antibody, in relapsed patients with adult T-cell leukemia-lymphoma and peripheral T-cell lymphoma. *J. Clin. Oncol.* **2010**, *28*, 1591–1598. [CrossRef] [PubMed]

18. Bazarbachi, A.; Plumelle, Y.; Carlos Ramos, J.; Tortevoye, P.; Otrock, Z.; Taylor, G.; Gessain, A.; Harrington, W.; Panelatti, G.; Hermine, O. Meta-analysis on the use of zidovudine and interferon-alfa in adult T-cell leukemia/lymphoma showing improved survival in the leukemic subtypes. *J. Clin. Oncol.* **2010**, *28*, 4177–4183. [CrossRef] [PubMed]

19. Kinpara, S.; Kijiyama, M.; Takamori, A.; Hasegawa, A.; Sasada, A.; Masuda, T.; Tanaka, Y.; Utsunomiya, A.; Kannagi, M. Interferon-α (IFN-α) suppresses HTLV-1 gene expression and cell cycling, while IFN-α combined with zidovudine induces p53 signaling and apoptosis in HTLV-1-infected cells. *Retrovirology* **2013**, *10*. [CrossRef] [PubMed]

20. Meekings, K.N.; Leipzig, J.; Bushman, F.D.; Taylor, G.P.; Bangham, C.R. HTLV-1 integration into transcriptionally active genomic regions is associated with proviral expression and with HAM/TSP. *PLoS Pathog.* **2008**, *4*. [CrossRef] [PubMed]

21. Nagai, M.; Usuku, K.; Matsumoto, W.; Kodama, D.; Takenouchi, N.; Moritoyo, T.; Hashiguchi, S.; Ichinose, M.; Bangham, C.R.; Izumo, S.; et al. Analysis of HTLV-I proviral load in 202 HAM/TSP patients and 243 asymptomatic HTLV-I carriers: High proviral load strongly predisposes to HAM/TSP. *J. Neurovirol.* **1998**, *4*, 586–593. [CrossRef] [PubMed]

22. Oh, U.; Jacobson, S. Treatment of HTLV-I-associated myelopathy/tropical spastic paraparesis: Toward rational targeted therapy. *Neurol. Clin.* **2008**, *26*, 781–797. [CrossRef] [PubMed]

23. Yamamoto, N.; Matsumoto, T.; Koyanagi, Y.; Tanaka, Y.; Hinuma, Y. Unique cell lines harbouring both epstein-barr virus and adult T-cel lleukaemia virus, established from leukaemia patients. *Nature* **1982**, *299*, 367–369. [CrossRef] [PubMed]

24. Ho, D.D.; Rota, T.R.; Hirsch, M.S. Infection of human endothelial cells by human T-lymphotropic virus type I. *Proc. Natl. Acad. Sci. USA* **1984**, *81*, 7588–7590. [CrossRef] [PubMed]

25. Longo, D.L.; Gelmann, E.P.; Cossman, J.; Young, R.A.; Gallo, R.C.; O'Brien, S.J.; Matis, L.A. Isolation of HTLV-transformed B-lymphocyte clone from a patient with HTLV-associated adult T-cell leukaemia. *Nature* **1984**, *310*, 505–506. [CrossRef] [PubMed]

26. Yoshikura, H.; Nishida, J.; Yoshida, M.; Kitamura, Y.; Takaku, F.; Ikeda, S. Isolation of HTLV derived from Japanese adult T-cell leukemia patients in human diploid fibroblast strain IMR90 and the biological characters of the infected cells. *Int. J. Cancer* **1984**, *33*, 745–749. [CrossRef] [PubMed]

27. Koyanagi, Y.; Itoyama, Y.; Nakamura, N.; Takamatsu, K.; Kira, J.; Iwamasa, T.; Goto, I.; Yamamoto, N. *In vivo* infection of human T-cell leukemia virus type I in non-T cells. *Virology* **1993**, *196*, 25–33. [CrossRef] [PubMed]

28. Manel, N.; Kim, F.J.; Kinet, S.; Taylor, N.; Sitbon, M.; Battini, J.L. The ubiquitous glucose transporter GLUT-1 is a receptor for HTLV. *Cell* **2003**, *115*, 449–459. [CrossRef]

29. Jones, K.S.; Petrow-Sadowski, C.; Bertolette, D.C.; Huang, Y.; Ruscetti, F.W. Heparan sulfate proteoglycans mediate attachment and entry of human T-cell leukemia virus type 1 virions into CD4+ T cells. *J. Virol.* **2005**, *79*, 12692–12702. [CrossRef] [PubMed]

30. Ghez, D.; Lepelletier, Y.; Lambert, S.; Fourneau, J.M.; Blot, V.; Janvier, S.; Arnulf, B.; van Endert, P.M.; Heveker, N.; Pique, C.; et al. Neuropilin-1 is involved in human T-cell lymphotropic virus type 1 entry. *J. Virol.* **2006**, *80*, 6844–6854. [CrossRef] [PubMed]

31. Jones, K.S.; Petrow-Sadowski, C.; Huang, Y.K.; Bertolette, D.C.; Ruscetti, F.W. Cell-free HTLV-1 infects dendritic cells leading to transmission and transformation of CD4+ T cells. *Nat. Med.* **2008**, *14*, 429–436. [CrossRef] [PubMed]

32. Jin, Q.; Agrawal, L.; VanHorn-Ali, Z.; Alkhatib, G. Infection of CD4+ T lymphocytes by the human T cell leukemia virus type 1 is mediated by the glucose transporter glut-1: Evidence using antibodies specific to the receptor's large extracellular domain. *Virology* **2006**, *349*, 184–196. [CrossRef] [PubMed]

33. Aiken, C.; Konner, J.; Landau, N.R.; Lenburg, M.E.; Trono, D. Nef induces CD4 endocytosis: Requirement for a critical dileucine motif in the membrane-proximal CD4 cytoplasmic domain. *Cell* **1994**, *76*, 853–864. [CrossRef]

34. Margottin, F.; Bour, S.P.; Durand, H.; Selig, L.; Benichou, S.; Richard, V.; Thomas, D.; Strebel, K.; Benarous, R. A novel human WD protein, H-β TRCP, that interacts with HIV-1 VPU connects CD4 to the ER degradation pathway through an F-box motif. *Mol. Cell* **1998**, *1*, 565–574. [CrossRef]

35. Maeda, Y.; Terasawa, H.; Tanaka, Y.; Mitsuura, C.; Nakashima, K.; Yusa, K.; Harada, S. Separate cellular localizations of human T-lymphotropic virus 1 (HTLV-1) Env and glucose transporter type 1 (glut1) are required for HTLV-1 Env-mediated fusion and infection. *J. Virol.* **2015**, *89*, 502–511. [CrossRef] [PubMed]

36. Macintyre, A.N.; Gerriets, V.A.; Nichols, A.G.; Michalek, R.D.; Rudolph, M.C.; Deoliveira, D.; Anderson, S.M.; Abel, E.D.; Chen, B.J.; Hale, L.P.; *et al.* The glucose transporter glut1 is selectively essential for CD4 T cell activation and effector function. *Cell Metab.* **2014**, *20*, 61–72. [CrossRef] [PubMed]

37. Temin, H.M.; Mizutani, S. RNA-dependent DNA polymerase in virions of Rous sarcoma virus. *Nature* **1970**, *226*, 1211–1213. [CrossRef] [PubMed]

38. Baltimore, D. RNA-dependent DNA polymerase in virions of RNA tumour viruses. *Nature* **1970**, *226*, 1209–1211. [CrossRef] [PubMed]

39. Hulme, A.E.; Perez, O.; Hope, T.J. Complementary assays reveal a relationship between HIV-1 uncoating and reverse transcription. *Proc. Natl. Acad. Sci. USA* **2011**, *108*, 9975–9980. [CrossRef] [PubMed]

40. Tekeste, S.S.; Wilkinson, T.A.; Weiner, E.M.; Xu, X.; Miller, J.T.; Le Grice, S.F.; Clubb, R.T.; Chow, S.A. Interaction between reverse transcriptase and integrase is required for reverse transcription during HIV-1 replication. *J. Virol.* **2015**. [CrossRef] [PubMed]

41. Desrames, A.; Cassar, O.; Gout, O.; Hermine, O.; Taylor, G.P.; Afonso, P.V.; Gessain, A. Northern African strains of human T-lymphotropic virus type 1 arose from a recombination event. *J. Virol.* **2014**, *88*, 9782–9788. [CrossRef] [PubMed]

42. Ooms, M.; Krikoni, A.; Kress, A.K.; Simon, V.; Munk, C. APOBEC3A, APOBEC3B, and APOBEC3H haplotype 2 restrict human T-lymphotropic virus type 1. *J. Virol.* **2012**, *86*, 6097–6108. [CrossRef] [PubMed]

43. Fan, J.; Ma, G.; Nosaka, K.; Tanabe, J.; Satou, Y.; Koito, A.; Wain-Hobson, S.; Vartanian, J.P.; Matsuoka, M. APOBEC3G generates nonsense mutations in human T-cell leukemia virus type 1 proviral genomes *in vivo*. *J. Virol.* **2010**, *84*, 7278–7287. [CrossRef] [PubMed]

44. Derse, D.; Hill, S.A.; Princler, G.; Lloyd, P.; Heidecker, G. Resistance of human T cell leukemia virus type 1 to APOBEC3G restriction is mediated by elements in nucleocapsid. *Proc. Natl. Acad. Sci. USA* **2007**, *104*, 2915–2920. [CrossRef] [PubMed]

45. Coffin, J.M.; Hughes, S.H.; Varmus, H.E. The interactions of retroviruses and their hosts. In *Retroviruses*; Cold Spring Harbor: NY, USA, 1997.

46. Kitamura, Y.; Lee, Y.M.; Coffin, J.M. Nonrandom integration of retroviral DNA *in vitro*: Effect of CPG methylation. *Proc. Natl. Acad. Sci. USA* **1992**, *89*, 5532–5536. [CrossRef] [PubMed]

47. Doi, K.; Wu, X.; Taniguchi, Y.; Yasunaga, J.; Satou, Y.; Okayama, A.; Nosaka, K.; Matsuoka, M. Preferential selection of human T-cell leukemia virus type I provirus integration sites in leukemic *versus* carrier states. *Blood* **2005**, *106*, 1048–1053. [CrossRef] [PubMed]

48. Holman, A.G.; Coffin, J.M. Symmetrical base preferences surrounding HIV-1, avian sarcoma/leukosis virus, and murine leukemia virus integration sites. *Proc. Natl. Acad. Sci. USA* **2005**, *102*, 6103–6107. [CrossRef] [PubMed]

49. Kang, Y.; Moressi, C.J.; Scheetz, T.E.; Xie, L.; Tran, D.T.; Casavant, T.L.; Ak, P.; Benham, C.J.; Davidson, B.L.; McCray, P.B., Jr. Integration site choice of a feline immunodeficiency virus vector. *J. Virol.* **2006**, *80*, 8820–8823. [CrossRef] [PubMed]

50. Derse, D.; Crise, B.; Li, Y.; Princler, G.; Lum, N.; Stewart, C.; McGrath, C.F.; Hughes, S.H.; Munroe, D.J.; Wu, X. Human T-cell leukemia virus type 1 integration target sites in the human genome: Comparison with those of other retroviruses. *J. Virol.* **2007**, *81*, 6731–6741. [CrossRef] [PubMed]

51. Gillet, N.A.; Cook, L.; Laydon, D.J.; Hlela, C.; Verdonck, K.; Alvarez, C.; Gotuzzo, E.; Clark, D.; Farre, L.; Bittencourt, A.; *et al.* Strongyloidiasis and infective dermatitis alter human T lymphotropic virus-1 clonality *in vivo*. *PLoS Pathog.* **2013**, *9*. [CrossRef] [PubMed]

52. Cook, L.B.; Melamed, A.; Niederer, H.; Valganon, M.; Laydon, D.; Foroni, L.; Taylor, G.P.; Matsuoka, M.; Bangham, C.R. The role of HTLV-1 clonality, proviral structure, and genomic integration site in adult T-cell leukemia/lymphoma. *Blood* **2014**, *123*, 3925–3931. [CrossRef] [PubMed]

53. Niederer, H.A.; Laydon, D.J.; Melamed, A.; Elemans, M.; Asquith, B.; Matsuoka, M.; Bangham, C.R. HTLV-1 proviral integration sites differ between asymptomatic carriers and patients with HAM/TSP. *J. Virol.* **2014**, *11*. [CrossRef] [PubMed]

54. Kashanchi, F.; Brady, J.N. Transcriptional and post-transcriptional gene regulation of HTLV-1. *Oncogene* **2005**, *24*, 5938–5951. [CrossRef] [PubMed]

55. Jeang, K.T.; Boros, I.; Brady, J.; Radonovich, M.; Khoury, G. Characterization of cellular factors that interact with the human T-cell leukemia virus type I p40x-responsive 21-base-pair sequence. *J. Virol.* **1988**, *62*, 4499–4509. [PubMed]

56. Baranger, A.M.; Palmer, C.R.; Hamm, M.K.; Giebler, H.A.; Brauweiler, A.; Nyborg, J.K.; Schepartz, A. Mechanism of DNA-binding enhancement by the human T-cell leukaemia virus transactivator Tax. *Nature* **1995**, *376*, 606–608. [CrossRef] [PubMed]

57. Adya, N.; Giam, C.Z. Distinct regions in human T-cell lymphotropic virus type I Tax mediate interactions with activator protein CREB and basal transcription factors. *J. Virol.* **1995**, *69*, 1834–1841. [PubMed]

58. Adya, N.; Zhao, L.J.; Huang, W.; Boros, I.; Giam, C.Z. Expansion of CREB'S DNA recognition specificity by Tax results from interaction with Ala-Ala-Arg at positions 282–284 near the conserved DNA-binding domain of CREB. *Proc. Natl. Acad. Sci. USA* **1994**, *91*, 5642–5646. [CrossRef] [PubMed]

59. Goren, I.; Semmes, O.J.; Jeang, K.T.; Moelling, K. The amino terminus of Tax is required for interaction with the cyclic AMP response element binding protein. *J. Virol.* **1995**, *69*, 5806–5811. [PubMed]

60. Tie, F.; Adya, N.; Greene, W.C.; Giam, C.Z. Interaction of the human T-lymphotropic virus type 1 Tax dimer with CREB and the viral 21-base-pair repeat. *J. Virol.* **1996**, *70*, 8368–8374. [PubMed]

61. Zhao, L.J.; Giam, C.Z. Human T-cell lymphotropic virus type I (HTLV-I) transcriptional activator, Tax, enhances CREB binding to HTLV-I 21-base-pair repeats by protein-protein interaction. *Proc. Natl. Acad. Sci. USA* **1992**, *89*, 7070–7074. [CrossRef] [PubMed]

62. Ching, Y.P.; Chun, A.C.; Chin, K.T.; Zhang, Z.Q.; Jeang, K.T.; Jin, D.Y. Specific TATAA and bZIP requirements suggest that HTLV-I Tax has transcriptional activity subsequent to the assembly of an initiation complex. *Retrovirology* **2004**, *1*. [CrossRef] [PubMed]

63. Seiki, M.; Inoue, J.; Takeda, T.; Yoshida, M. Direct evidence that p40x of human T-cell leukemia virus type I is a trans-acting transcriptional activator. *EMBO J.* **1986**, *5*, 561–565. [PubMed]

64. Giebler, H.A.; Loring, J.E.; van Orden, K.; Colgin, M.A.; Garrus, J.E.; Escudero, K.W.; Brauweiler, A.; Nyborg, J.K. Anchoring of CREB binding protein to the human T-cell leukemia virus type 1 promoter: A molecular mechanism of Tax transactivation. *Mol. Cell. Biol.* **1997**, *17*, 5156–5164. [CrossRef] [PubMed]

65. Kwok, R.P.; Laurance, M.E.; Lundblad, J.R.; Goldman, P.S.; Shih, H.; Connor, L.M.; Marriott, S.J.; Goodman, R.H. Control of CAMP-regulated enhancers by the viral transactivator Tax through CREB and the co-activator CBP. *Nature* **1996**, *380*, 642–646. [CrossRef] [PubMed]

66. Jiang, H.; Lu, H.; Schiltz, R.L.; Pise-Masison, C.A.; Ogryzko, V.V.; Nakatani, Y.; Brady, J.N. PCAF interacts with Tax and stimulates Tax transactivation in a histone acetyltransferase-independent manner. *Mol. Cell. Biol.* **1999**, *19*, 8136–8145. [CrossRef]

67. Harrod, R.; Kuo, Y.L.; Tang, Y.; Yao, Y.; Vassilev, A.; Nakatani, Y.; Giam, C.Z. P300 and p300/CAMP-responsive element-binding protein associated factor interact with human T-cell lymphotropic virus type-1 Tax in a multi-histone acetyltransferase/activator-enhancer complex. *J. Biol. Chem.* **2000**, *275*, 11852–11857. [CrossRef] [PubMed]

68. Bex, F.; Yin, M.J.; Burny, A.; Gaynor, R.B. Differential transcriptional activation by human T-cell leukemia virus type 1 Tax mutants is mediated by distinct interactions with CREB binding protein and p300. *Mol. Cell. Biol.* **1998**, *18*, 2392–2405. [CrossRef] [PubMed]

69. Ma, G.; Yasunaga, J.; Akari, H.; Matsuoka, M. Tcf1 and LEF1 act as T-cell intrinsic HTLV-1 antagonists by targeting Tax. *Proc. Natl. Acad. Sci. USA* **2015**, *112*, 2216–2221. [CrossRef] [PubMed]

70. Zou, T.; Yang, Y.; Xia, F.; Huang, A.; Gao, X.; Fang, D.; Xiong, S.; Zhang, J. Resveratrol inhibits CD4$^+$ T cell activation by enhancing the expression and activity of Sirt1. *PloS ONE* **2013**, *8*, e75139. [CrossRef] [PubMed]

71. Tang, H.M.; Gao, W.W.; Chan, C.P.; Cheng, Y.; Deng, J.J.; Yuen, K.S.; Iha, H.; Jin, D.Y. Sirt1 suppresses human T-cell leukemia virus type 1 transcription. *J. Virol.* **2015**, *89*, 8623–8631. [CrossRef] [PubMed]

72. Huang, H.; Santoso, N.; Power, D.; Simpson, S.; Dieringer, M.; Miao, H.; Gurova, K.; Giam, C.Z.; Elledge, S.; Zhu, J. Fact proteins, SUPT16H and SSRP1, are transcriptional suppressors of HIV-1 and HTLV-1 that facilitate viral latency. *J. Biol. Chem.* **2015**. [CrossRef] [PubMed]

73. Green, P.L.; Chen, I.S. Regulation of human T cell leukemia virus expression. *FASEB J.* **1990**, *4*, 169–175. [PubMed]

74. Hidaka, M.; Inoue, J.; Yoshida, M.; Seiki, M. Post-transcriptional regulator (Rex) of HTLV-1 initiates expression of viral structural proteins but suppresses expression of regulatory proteins. *EMBO J.* **1988**, *7*, 519–523. [PubMed]

75. Bogerd, H.P.; Huckaby, G.L.; Ahmed, Y.F.; Hanly, S.M.; Greene, W.C. The type I human T-cell leukemia virus (HTLV-I) Rex trans-activator binds directly to the HTLV-I Rex and the type 1 human immunodeficiency virus Rev RNA response elements. *Proc. Natl. Acad. Sci. USA* **1991**, *88*, 5704–5708. [CrossRef] [PubMed]

76. Nosaka, T.; Siomi, H.; Adachi, Y.; Ishibashi, M.; Kubota, S.; Maki, M.; Hatanaka, M. Nucleolar targeting signal of human T-cell leukemia virus type I Rex-encoded protein is essential for cytoplasmic accumulation of unspliced viral mrna. *Proc. Natl. Acad. Sci. USA* **1989**, *86*, 9798–9802. [CrossRef] [PubMed]

77. Rehberger, S.; Gounari, F.; DucDodon, M.; Chlichlia, K.; Gazzolo, L.; Schirrmacher, V.; Khazaie, K. The activation domain of a hormone inducible HTLV-1 Rex protein determines colocalization with the nuclear pore. *Exp. Cell Res.* **1997**, *233*, 363–371. [CrossRef] [PubMed]

78. Palmeri, D.; Malim, M.H. The human T-cell leukemia virus type 1 posttranscriptional trans-activator Rex contains a nuclear export signal. *J. Virol.* **1996**, *70*, 6442–6445. [PubMed]

79. Bai, X.T.; Sinha-Datta, U.; Ko, N.L.; Bellon, M.; Nicot, C. Nuclear export and expression of human T-cell leukemia virus type 1 Tax/Rex mRNA are Rxre/Rex dependent. *J. Virol.* **2012**, *86*, 4559–4565. [CrossRef] [PubMed]

80. Rende, F.; Cavallari, I.; Andresen, V.; Valeri, V.W.; D'Agostino, D.M.; Franchini, G.; Ciminale, V. Identification of novel monocistronic HTLV-1 mrnas encoding functional Rex isoforms. *Retrovirology* **2015**, *12*, 58. [CrossRef] [PubMed]

81. Butsch, M.; Boris-Lawrie, K. Destiny of unspliced retroviral RNA: Ribosome and/or virion? *J. Virol.* **2002**, *76*, 3089–3094. [CrossRef] [PubMed]

82. Hellen, C.U.; Sarnow, P. Internal ribosome entry sites in eukaryotic mRNA molecules. *Genes Dev.* **2001**, *15*, 1593–1612. [CrossRef] [PubMed]

83. Sarnow, P.; Cevallos, R.C.; Jan, E. Takeover of host ribosomes by divergent IRES elements. *Biochem. Soc. Trans.* **2005**, *33*, 1479–1482. [CrossRef] [PubMed]

84. Jackson, R.J. Alternative mechanisms of initiating translation of mammalian mrnas. *Biochem. Soc. Trans.* **2005**, *33*, 1231–1241. [CrossRef] [PubMed]

85. Attal, J.; Theron, M.C.; Taboit, F.; Cajero-Juarez, M.; Kann, G.; Bolifraud, P.; Houdebine, L.M. The RU5 ("R") region from human leukaemia viruses (HTLV-1) contains an internal ribosome entry site (IRES)-like sequence. *FEBS Let.* **1996**, *392*, 220–224. [CrossRef]

86. Bolinger, C.; Yilmaz, A.; Hartman, T.R.; Kovacic, M.B.; Fernandez, S.; Ye, J.; Forget, M.; Green, P.L.; Boris-Lawrie, K. RNA helicase a interacts with divergent lymphotropic retroviruses and promotes translation of human T-cell leukemia virus type 1. *Nucleic Acids Res.* **2007**, *35*, 2629–2642. [CrossRef] [PubMed]

87. Olivares, E.; Landry, D.M.; Caceres, C.J.; Pino, K.; Rossi, F.; Navarrete, C.; Huidobro-Toro, J.P.; Thompson, S.R.; Lopez-Lastra, M. The 5' untranslated region of the human T-cell lymphotropic virus type 1 mRNA enables cap-independent translation initiation. *J. Virol.* **2014**, *88*, 5936–5955. [CrossRef] [PubMed]

88. Fogarty, K.H.; Chen, Y.; Grigsby, I.F.; Macdonald, P.J.; Smith, E.M.; Johnson, J.L.; Rawson, J.M.; Mansky, L.M.; Mueller, J.D. Characterization of cytoplasmic Gag-gag interactions by dual-color z-scan fluorescence fluctuation spectroscopy. *Biophys. J.* **2011**, *100*, 1587–1595. [CrossRef] [PubMed]

89. Ritchie, C.; Cylinder, I.; Platt, E.J.; Barklis, E. Analysis of HIV-1 Gag protein interactions via biotin ligase tagging. *J. Virol.* **2015**, *89*, 3988–4001. [CrossRef] [PubMed]

90. Barnard, A.L.; Igakura, T.; Tanaka, Y.; Taylor, G.P.; Bangham, C.R. Engagement of specific T-cellsurface molecules regulates cytoskeletal polarization in HTLV-1-infected lymphocytes. *Blood* **2005**, *106*, 988–995. [CrossRef] [PubMed]

91. Qualley, D.F.; Stewart-Maynard, K.M.; Wang, F.; Mitra, M.; Gorelick, R.J.; Rouzina, I.; Williams, M.C.; Musier-Forsyth, K. C-terminal domain modulates the nucleic acid chaperone activity of human T-cell leukemia virus type 1 nucleocapsid protein via an electrostatic mechanism. *J. Biol. Chem.* **2010**, *285*, 295–307. [CrossRef] [PubMed]

92. Sun, M.; Grigsby, I.F.; Gorelick, R.J.; Mansky, L.M.; Musier-Forsyth, K. Retrovirus-specific differences in matrix and nucleocapsid protein-nucleic acid interactions: Implications for genomic RNA packaging. *J. Virol.* **2014**, *88*, 1271–1280. [CrossRef] [PubMed]

93. Chen, J.; Grunwald, D.; Sardo, L.; Galli, A.; Plisov, S.; Nikolaitchik, O.A.; Chen, D.; Lockett, S.; Larson, D.R.; Pathak, V.K.; *et al.* Cytoplasmic HIV-1 RNA is mainly transported by diffusion in the presence or absence of gag protein. *Proc. Natl. Acad. Sci. USA* **2014**, *111*, E5205–E5213. [CrossRef] [PubMed]

94. Zhang, W.; Cao, S.; Martin, J.L.; Mueller, J.D.; Mansky, L.M. Morphology and ultrastructure of retrovirus particles. *AIMS Biophys.* **2015**, *2*, 343–369. [CrossRef] [PubMed]

95. Gamble, T.R.; Yoo, S.; Vajdos, F.F.; von Schwedler, U.K.; Worthylake, D.K.; Wang, H.; McCutcheon, J.P.; Sundquist, W.I.; Hill, C.P. Structure of the carboxyl-terminal dimerization domain of the HIV-1 capsid protein. *Science* **1997**, *278*, 849–853. [CrossRef] [PubMed]

96. Rayne, F.; Bouamr, F.; Lalanne, J.; Mamoun, R.Z. The NH2-terminal domain of the human T-cell leukemia virus type 1 capsid protein is involved in particle formation. *J. Virol.* **2001**, *75*, 5277–5287. [CrossRef] [PubMed]

97. Ganser-Pornillos, B.K.; von Schwedler, U.K.; Stray, K.M.; Aiken, C.; Sundquist, W.I. Assembly properties of the human immunodeficiency virus type 1 CA protein. *J. Virol.* **2004**, *78*, 2545–2552. [CrossRef] [PubMed]

98. Ako-Adjei, D.; Johnson, M.C.; Vogt, V.M. The retroviral capsid domain dictates virion size, morphology, and coassembly of gag into virus-like particles. *J. Virol.* **2005**, *79*, 13463–13472. [CrossRef] [PubMed]

99. Hogue, I.B.; Hoppe, A.; Ono, A. Quantitative fluorescence resonance energy transfer microscopy analysis of the human immunodeficiency virus type 1 Gag-gag interaction: Relative contributions of the CA and NC domains and membrane binding. *J. Virol.* **2009**, *83*, 7322–7336. [CrossRef] [PubMed]

100. Lingappa, J.R.; Reed, J.C.; Tanaka, M.; Chutiraka, K.; Robinson, B.A. How HIV-1 Gag assembles in cells: Putting together pieces of the puzzle. *Virus research* **2014**, *193*, 89–107. [CrossRef] [PubMed]

101. Maldonado, J.O.; Martin, J.L.; Mueller, J.D.; Zhang, W.; Mansky, L.M. New insights into retroviral Gag-gag and Gag-membrane interactions. *Frontiers Microbiol.* **2014**, *5*. [CrossRef] [PubMed]

102. Lingwood, D.; Kaiser, H.J.; Levental, I.; Simons, K. Lipid rafts as functional heterogeneity in cell membranes. *Biochem. Soc. Trans.* **2009**, *37*, 955–960. [CrossRef] [PubMed]

103. Ono, A. HIV-1 assembly at the plasma membrane: Gag trafficking and localization. *Futur. Virol.* **2009**, *4*, 241–257. [CrossRef] [PubMed]

104. Sonnino, S.; Prinetti, A. Membrane domains and the "lipid raft" concept. *Curr. Med. Chem.* **2013**, *20*, 4–21. [PubMed]

105. Inlora, J.; Collins, D.R.; Trubin, M.E.; Chung, J.Y.; Ono, A. Membrane binding and subcellular localization of retroviral Gag proteins are differentially regulated by MA interactions with phosphatidylinositol-(4,5)-bisphosphate and RNA. *mBio* **2014**, *5*, e02202. [CrossRef] [PubMed]

106. Demirov, D.G.; Freed, E.O. Retrovirus budding. *Virus Res.* **2004**, *106*, 87–102. [CrossRef] [PubMed]

107. Morita, E.; Sundquist, W.I. Retrovirus budding. *Ann. Rev. Cell Dev. Biol.* **2004**, *20*, 395–425. [CrossRef] [PubMed]

108. Le Blanc, I.; Grange, M.P.; Delamarre, L.; Rosenberg, A.R.; Blot, V.; Pique, C.; Dokhelar, M.C. HTLV-1 structural proteins. *Virus Res.* **2001**, *78*, 5–16. [CrossRef]

109. Konvalinka, J.; Krausslich, H.G.; Muller, B. Retroviral proteases and their roles in virion maturation. *Virology* **2015**, *479–480*, 403–417. [CrossRef] [PubMed]

110. Goncalves, D.U.; Proietti, F.A.; Ribas, J.G.; Araujo, M.G.; Pinheiro, S.R.; Guedes, A.C.; Carneiro-Proietti, A.B. Epidemiology, treatment, and prevention of human T-cell leukemia virus type 1-associated diseases. *Clin. Microbiol. Rev.* **2010**, *23*, 577–589. [CrossRef] [PubMed]

111. Wiktor, S.Z.; Pate, E.J.; Murphy, E.L.; Palker, T.J.; Champegnie, E.; Ramlal, A.; Cranston, B.; Hanchard, B.; Blattner, W.A. Mother-to-child transmission of human T-cell lymphotropic virus type I (HTLV-I) in Jamaica: Association with antibodies to envelope glycoprotein (gp46) epitopes. *J. Acquir. Immune Defic. Syndr.* **1993**, *6*, 1162–1167. [PubMed]

112. Takahashi, K.; Takezaki, T.; Oki, T.; Kawakami, K.; Yashiki, S.; Fujiyoshi, T.; Usuku, K.; Mueller, N.; Osame, M.; Miyata, K.; *et al.* Inhibitory effect of maternal antibody on mother-to-child transmission of human T-lymphotropic virus type I. The mother-to-child transmission study group. *Int. J. Cancer* **1991**, *49*, 673–677. [CrossRef] [PubMed]

113. Nyambi, P.N.; Ville, Y.; Louwagie, J.; Bedjabaga, I.; Glowaczower, E.; Peeters, M.; Kerouedan, D.; Dazza, M.; Larouze, B.; van der Groen, G.; *et al.* Mother-to-child transmission of human T-cell lymphotropic virus types I and II (HTLV-I/II) in Gabon: A prospective follow-up of 4 years. *J. Acquir. Immune Defic. Syndr.* **1996**, *12*, 187–192. [CrossRef]

114. Li, H.C.; Biggar, R.J.; Miley, W.J.; Maloney, E.M.; Cranston, B.; Hanchard, B.; Hisada, M. Provirus load in breast milk and risk of mother-to-child transmission of human T lymphotropic virus type I. *J. Infec. Dis.* **2004**, *190*, 1275–1278. [CrossRef] [PubMed]

115. Martin-Latil, S.; Gnadig, N.F.; Mallet, A.; Desdouits, M.; Guivel-Benhassine, F.; Jeannin, P.; Prevost, M.C.; Schwartz, O.; Gessain, A.; Ozden, S.; *et al.* Transcytosis of HTLV-1 across a tight human epithelial barrier and infection of subepithelial dendritic cells. *Blood* **2012**, *120*, 572–580. [CrossRef] [PubMed]

116. Kinlock, B.L.; Wang, Y.; Turner, T.M.; Wang, C.; Liu, B. Transcytosis of HIV-1 through vaginal epithelial cells is dependent on trafficking to the endocytic recycling pathway. *PloS ONE* **2014**, *9*, e96760. [CrossRef] [PubMed]

117. Alfsen, A.; Yu, H.; Magerus-Chatinet, A.; Schmitt, A.; Bomsel, M. HIV-1-infected blood mononuclear cells form an integrin- and agrin-dependent viral synapse to induce efficient HIV-1 transcytosis across epithelial cell monolayer. *Mol. Biol. Cell* **2005**, *16*, 4267–4279. [CrossRef] [PubMed]

118. Filippone, C.; Betsem, E.; Tortevoye, P.; Cassar, O.; Bassot, S.; Froment, A.; Fontanet, A.; Gessain, A. A severe bite from a nonhuman primate is a major risk factor for HTLV-1 infection in hunters from central Africa. *Clin. Infect. Dis.* **2015**, *60*, 1667–1676. [CrossRef] [PubMed]

119. Kazanji, M.; Mouinga-Ondeme, A.; Lekana-Douki-Etenna, S.; Caron, M.; Makuwa, M.; Mahieux, R.; Gessain, A. Origin of HTLV-1 in hunters of nonhuman primates in central Africa. *J. Infect. Dis.* **2015**, *211*, 361–365. [CrossRef] [PubMed]

120. LeBreton, M.; Switzer, W.M.; Djoko, C.F.; Gillis, A.; Jia, H.; Sturgeon, M.M.; Shankar, A.; Zheng, H.; Nkeunen, G.; Tamoufe, U.; *et al.* A gorilla reservoir for human T-lymphotropic virus type 4. *Emerg. Microbes Infect.* **2014**, *3*. [CrossRef] [PubMed]

121. Miura, M.; Yasunaga, J.; Tanabe, J.; Sugata, K.; Zhao, T.; Ma, G.; Miyazato, P.; Ohshima, K.; Kaneko, A.; Watanabe, A.; *et al.* Characterization of simian T-cell leukemia virus type 1 in naturally infected Japanese macaques as a model of HTLV-1 infection. *Retrovirology* **2013**, *10*. [CrossRef] [PubMed]

122. Song, K.J.; Nerurkar, V.R.; Saitou, N.; Lazo, A.; Blakeslee, J.R.; Miyoshi, I.; Yanagihara, R. Genetic analysis and molecular phylogeny of simian T-cell lymphotropic virus type I: Evidence for independent virus evolution in Asia and Africa. *Virology* **1994**, *199*, 56–66. [CrossRef] [PubMed]

123. Fan, N.; Gavalchin, J.; Paul, B.; Wells, K.H.; Lane, M.J.; Poiesz, B.J. Infection of peripheral blood mononuclear cells and cell lines by cell-free human T-cell lymphoma/leukemia virus type I. *J. Clin. Microbiol.* **1992**, *30*, 905–910. [PubMed]

124. Igakura, T.; Stinchcombe, J.C.; Goon, P.K.; Taylor, G.P.; Weber, J.N.; Griffiths, G.M.; Tanaka, Y.; Osame, M.; Bangham, C.R. Spread of HTLV-I between lymphocytes by virus-induced polarization of the cytoskeleton. *Science* **2003**, *299*, 1713–1716. [CrossRef] [PubMed]

125. Majorovits, E.; Nejmeddine, M.; Tanaka, Y.; Taylor, G.P.; Fuller, S.D.; Bangham, C.R. Human T-lymphotropic virus-1 visualized at the virological synapse by electron tomography. *PloS ONE* **2008**, *3*, e2251. [CrossRef] [PubMed]

126. Nejmeddine, M.; Negi, V.S.; Mukherjee, S.; Tanaka, Y.; Orth, K.; Taylor, G.P.; Bangham, C.R. HTLV-1-Tax and ICAM-1 act on T-cellsignal pathways to polarize the microtubule-organizing center at the virological synapse. *Blood* **2009**, *114*, 1016–1025. [CrossRef] [PubMed]

127. Nejmeddine, M.; Barnard, A.L.; Tanaka, Y.; Taylor, G.P.; Bangham, C.R. Human T-lymphotropic virus, type 1, Tax protein triggers microtubule reorientation in the virological synapse. *J. Biol. Chem.* **2005**, *280*, 29653–29660. [CrossRef] [PubMed]

128. Chevalier, S.A.; Turpin, J.; Cachat, A.; Afonso, P.V.; Gessain, A.; Brady, J.N.; Pise-Masison, C.A.; Mahieux, R. Gem-induced cytoskeleton remodeling increases cellular migration of HTLV-1-infected cells, formation of infected-to-target T-cellconjugates and viral transmission. *PLoS Pathog.* **2014**, *10*, e1003917. [CrossRef] [PubMed]

129. Fukudome, K.; Furuse, M.; Fukuhara, N.; Orita, S.; Imai, T.; Takagi, S.; Nagira, M.; Hinuma, Y.; Yoshie, O. Strong induction of ICAM-1 in human T cells transformed by human T-cell-leukemia virus type 1 and depression of ICAM-1 or LFA-1 in adult T-cell-leukemia-derived cell lines. *Int. J. Cancer.* **1992**, *52*, 418–427. [CrossRef] [PubMed]

130. Dimitrov, D.S.; Willey, R.L.; Sato, H.; Chang, L.J.; Blumenthal, R.; Martin, M.A. Quantitation of human immunodeficiency virus type 1 infection kinetics. *J. Virol.* **1993**, *67*, 2182–2190. [PubMed]

131. Mazurov, D.; Ilinskaya, A.; Heidecker, G.; Lloyd, P.; Derse, D. Quantitative comparison of HTLV-1 and HIV-1 cell-to-cell infection with new replication dependent vectors. *PLoS Pathog.* **2010**, *6*, e1000788. [CrossRef] [PubMed]

132. Jiang, A.P.; Jiang, J.F.; Guo, M.G.; Jin, Y.M.; Li, Y.Y.; Wang, J.H. Human blood-circulating basophils capture HIV-1 and mediate viral trans-infection of CD4$^+$ T cells. *J. Virol.* **2015**, *89*, 8050–8062. [CrossRef] [PubMed]

133. Pais-Correia, A.M.; Sachse, M.; Guadagnini, S.; Robbiati, V.; Lasserre, R.; Gessain, A.; Gout, O.; Alcover, A.; Thoulouze, M.I. Biofilm-like extracellular viral assemblies mediate HTLV-1 cell-to-cell transmission at virological synapses. *Nat. Med.* **2010**, *16*, 83–89. [CrossRef] [PubMed]

134. Tarasevich, A.; Filatov, A.; Pichugin, A.; Mazurov, D. Monoclonal antibody profiling of cell surface proteins associated with the viral biofilms on HTLV-1 transformed cells. *Acta Virol.* **2015**, *59*, 247–256. [CrossRef] [PubMed]

135. Alais, S.; Mahieux, R.; Dutartre, H. Viral source-independent high susceptibility of dendritic cells to human T-cell leukemia virus type 1 infection compared to that of T lymphocytes. *J. Virol.* **2015**, *89*, 10580–10590. [CrossRef] [PubMed]

136. Gessain, A.; Gallo, R.C.; Franchini, G. Low degree of human T-cell leukemia/lymphoma virus type I genetic drift *in vivo* as a means of monitoring viral transmission and movement of ancient human populations. *J. Virol.* **1992**, *66*, 2288–2295. [PubMed]

137. Van Dooren, S.; Pybus, O.G.; Salemi, M.; Liu, H.F.; Goubau, P.; Remondegui, C.; Talarmin, A.; Gotuzzo, E.; Alcantara, L.C.; Galvao-Castro, B.; *et al.* The low evolutionary rate of human T-cell lymphotropic virus type-1 confirmed by analysis of vertical transmission chains. *Mol. Biol. Evolut.* **2004**, *21*, 603–611. [CrossRef] [PubMed]

138. Mansky, L.M. *In vivo* analysis of human T-cell leukemia virus type 1 reverse transcription accuracy. *J. Virol.* **2000**, *74*, 9525–9531. [CrossRef] [PubMed]

139. Mansky, L.M.; Temin, H.M. Lower *in vivo* mutation rate of human immunodeficiency virus type 1 than that predicted from the fidelity of purified reverse transcriptase. *J. Virol.* **1995**, *69*, 5087–5094. [PubMed]

140. Taylor, G.P.; Goon, P.; Furukawa, Y.; Green, H.; Barfield, A.; Mosley, A.; Nose, H.; Babiker, A.; Rudge, P.; Usuku, K.; *et al.* Zidovudine plus lamivudine in human T-lymphotropic virus type-I-associated myelopathy: A randomised trial. *Retrovirology* **2006**, *3*, 63. [CrossRef] [PubMed]

141. Miyazato, P.; Yasunaga, J.; Taniguchi, Y.; Koyanagi, Y.; Mitsuya, H.; Matsuoka, M. *De novo* human T-cell leukemia virus type 1 infection of human lymphocytes in NOD-SCID, common γ-chain knockout mice. *J. Virol.* **2006**, *80*, 10683–10691. [CrossRef] [PubMed]

142. Seiki, M.; Eddy, R.; Shows, T.B.; Yoshida, M. Nonspecific integration of the HTLV provirus genome into adult T-cell leukaemia cells. *Nature* **1984**, *309*, 640–642. [CrossRef] [PubMed]

143. Ohshima, K.; Ohgami, A.; Matsuoka, M.; Etoh, K.; Utsunomiya, A.; Makino, T.; Ishiguro, M.; Suzumiya, J.; Kikuchi, M. Random integration of HTLV-1 provirus: Increasing chromosomal instability. *Cancer Lett.* **1998**, *132*, 203–212. [CrossRef]

144. Wattel, E.; Vartanian, J.P.; Pannetier, C.; Wain-Hobson, S. Clonal expansion of human T-cell leukemia virus type I-infected cells in asymptomatic and symptomatic carriers without malignancy. *J. Virol.* **1995**, *69*, 2863–2868. [PubMed]

145. Etoh, K.; Tamiya, S.; Yamaguchi, K.; Okayama, A.; Tsubouchi, H.; Ideta, T.; Mueller, N.; Takatsuki, K.; Matsuoka, M. Persistent clonal proliferation of human T-lymphotropic virus type I-infected cells *in vivo*. *Cancer Res.* **1997**, *57*, 4862–4867. [PubMed]

146. Cavrois, M.; Gessain, A.; Wain-Hobson, S.; Wattel, E. Proliferation of HTLV-1 infected circulating cells *in vivo* in all asymptomatic carriers and patients with TSP/HAM. *Oncogene* **1996**, *12*, 2419–2423. [PubMed]

147. Zhao, T.; Satou, Y.; Matsuoka, M. Development of t cell lymphoma in HTLV-1 bZIP factor and Tax double transgenic mice. *Archives Virol.* **2014**, *159*, 1849–1856. [CrossRef] [PubMed]

148. Oeckinghaus, A.; Hayden, M.S.; Ghosh, S. Crosstalk in NF-κB signaling pathways. *Nat. Immunol.* **2011**, *12*, 695–708. [CrossRef] [PubMed]

149. Peloponese, J.M., Jr.; Jeang, K.T. Role for Akt/protein kinase B and activator protein-1 in cellular proliferation induced by the human T-cell leukemia virus type 1 Tax oncoprotein. *J. Biol. Chem.* **2006**, *281*, 8927–8938. [CrossRef] [PubMed]

150. Wu, X.; Sun, S.C. Retroviral oncoprotein Tax deregulates NF-κB by activating Tak1 and mediating the physical association of tak1-IKK. *EMBO Rep.* **2007**, *8*, 510–515. [CrossRef] [PubMed]

151. Avesani, F.; Romanelli, M.G.; Turci, M.; Di Gennaro, G.; Sampaio, C.; Bidoia, C.; Bertazzoni, U.; Bex, F. Association of HTLV Tax proteins with Tak1-binding protein 2 and RelA in calreticulin-containing cytoplasmic structures participates in Tax-mediated NF-κB activation. *Virology* **2010**, *408*, 39–48. [CrossRef] [PubMed]

152. Choi, Y.B.; Harhaj, E.W. HTLV-1 Tax stabilizes mcl-1 via traf6-dependent k63-linked polyubiquitination to promote cell survival and transformation. *PLoS Pathog.* **2014**, *10*, e1004458. [CrossRef] [PubMed]

153. Yin, M.J.; Christerson, L.B.; Yamamoto, Y.; Kwak, Y.T.; Xu, S.; Mercurio, F.; Barbosa, M.; Cobb, M.H.; Gaynor, R.B. HTLV-I Tax protein binds to mekk1 to stimulate IκB kinase activity and NF-κB activation. *Cell* **1998**, *93*, 875–884. [CrossRef]

154. Harhaj, E.W.; Sun, S.C. Ikkγ serves as a docking subunit of the IκB kinase (Ikk) and mediates interaction of IKK with the human T-cell leukemia virus Tax protein. *J. Biol. Chem.* **1999**, *274*, 22911–22914. [CrossRef] [PubMed]

155. Shembade, N.; Harhaj, N.S.; Yamamoto, M.; Akira, S.; Harhaj, E.W. The human T-cell leukemia virus type 1 Tax oncoprotein requires the ubiquitin-conjugating enzyme UBC13 for NF-κB activation. *J. Virol.* **2007**, *81*, 13735–13742. [CrossRef] [PubMed]

156. Israel, A. The IKK complex, a central regulator of NF-κB activation. *Cold Spring Harb. Perspect. Biol.* **2010**, *2*, a000158. [CrossRef] [PubMed]

157. Kray, A.E.; Carter, R.S.; Pennington, K.N.; Gomez, R.J.; Sanders, L.E.; Llanes, J.M.; Khan, W.N.; Ballard, D.W.; Wadzinski, B.E. Positive regulation of IκB kinase signaling by protein serine/threonine phosphatase 2A. *J. Biol. Chem.* **2005**, *280*, 35974–35982. [CrossRef] [PubMed]

158. Fu, D.X.; Kuo, Y.L.; Liu, B.Y.; Jeang, K.T.; Giam, C.Z. Human T-lymphotropic virus type I Tax activates I-κB kinase by inhibiting I-κB kinase-associated serine/threonine protein phosphatase 2A. *J. Biol. Chem.* **2003**, *278*, 1487–1493. [CrossRef] [PubMed]

159. Hong, S.; Wang, L.C.; Gao, X.; Kuo, Y.L.; Liu, B.; Merling, R.; Kung, H.J.; Shih, H.M.; Giam, C.Z. Heptad repeats regulate protein phosphatase 2A recruitment to I-κB kinase γ/NF-κB essential modulator and are targeted by human T-lymphotropic virus type 1 Tax. *J. Biol. Chem.* **2007**, *282*, 12119–12126. [CrossRef] [PubMed]

160. Azran, I.; Jeang, K.T.; Aboud, M. High levels of cytoplasmic HTLV-1 Tax mutant proteins retain a Tax-NF-κB-CBP ternary complex in the cytoplasm. *Oncogene* **2005**, *24*, 4521–4530. [CrossRef] [PubMed]

161. Lamsoul, I.; Lodewick, J.; Lebrun, S.; Brasseur, R.; Burny, A.; Gaynor, R.B.; Bex, F. Exclusive ubiquitination and sumoylation on overlapping lysine residues mediate NF-κB activation by the human T-cell leukemia virus Tax oncoprotein. *Mol. Cell. Biol.* **2005**, *25*, 10391–10406. [CrossRef] [PubMed]

162. Petropoulos, L.; Lin, R.; Hiscott, J. Human T cell leukemia virus type 1 Tax protein increases NF-κB dimer formation and antagonizes the inhibitory activity of the IκBα regulatory protein. *Virology* **1996**, *225*, 52–64. [CrossRef] [PubMed]

163. Shembade, N.; Harhaj, N.S.; Parvatiyar, K.; Copeland, N.G.; Jenkins, N.A.; Matesic, L.E.; Harhaj, E.W. The E3 ligase ITCH negatively regulates inflammatory signaling pathways by controlling the function of the ubiquitin-editing enzyme A20. *Nat. Immunol.* **2008**, *9*, 254–262. [CrossRef] [PubMed]

164. Pujari, R.; Hunte, R.; Thomas, R.; van der Weyden, L.; Rauch, D.; Ratner, L.; Nyborg, J.K.; Ramos, J.C.; Takai, Y.; Shembade, N. Human T-cell leukemia virus type 1 (HTLV-1) Tax requires CADM1/TSLC1 for inactivation of the NF-κB inhibitor A20 and constitutive NF-κB signaling. *PLoS Pathog.* **2015**, *11*, e1004721. [CrossRef] [PubMed]

165. Fu, J.; Qu, Z.; Yan, P.; Ishikawa, C.; Aqeilan, R.I.; Rabson, A.B.; Xiao, G. The tumor suppressor gene *WWOX* links the canonical and noncanonical NF-κB pathways in HTLV-I Tax-mediated tumorigenesis. *Blood* **2011**, *117*, 1652–1661. [CrossRef] [PubMed]

166. Higuchi, M.; Tsubata, C.; Kondo, R.; Yoshida, S.; Takahashi, M.; Oie, M.; Tanaka, Y.; Mahieux, R.; Matsuoka, M.; Fujii, M. Cooperation of NF-κB2/p100 activation and the PDZ domain binding motif signal in human T-cell leukemia virus type 1 (HTLV-1) Tax1 but not HTLV-2 Tax2 is crucial for interleukin-2-independent growth transformation of a T-cell line. *J. Virol.* **2007**, *81*, 11900–11907. [CrossRef] [PubMed]

167. Xiao, G.; Cvijic, M.E.; Fong, A.; Harhaj, E.W.; Uhlik, M.T.; Waterfield, M.; Sun, S.C. Retroviral oncoprotein Tax induces processing of NF-κB2/p100 in T cells: Evidence for the involvement of IKKα. *EMBO J.* **2001**, *20*, 6805–6815. [CrossRef] [PubMed]

168. Fraedrich, K.; Muller, B.; Grassmann, R. The HTLV-1 Tax protein binding domain of cyclin-dependent kinase 4 (CDK4) includes the regulatory pstaire helix. *Retrovirology* **2005**, *2*, 54. [CrossRef] [PubMed]

169. Suzuki, T.; Yoshida, M. HTLV-1 Tax protein interacts with cyclin-dependent kinase inhibitor p16INK4A and counteracts its inhibitory activity to CDK4. *Leukemia* **1997**, *11* (Suppl. 3), 14–16. [PubMed]

170. Liu, Y.; Wang, Y.; Yamakuchi, M.; Masuda, S.; Tokioka, T.; Yamaoka, S.; Maruyama, I.; Kitajima, I. Phosphoinositide-3 kinase-PKB/Akt pathway activation is involved in fibroblast Rat-1 transformation by human T-cell leukemia virus type I Tax. *Oncogene* **2001**, *20*, 2514–2526. [CrossRef] [PubMed]

171. Kim, S.J.; Kehrl, J.H.; Burton, J.; Tendler, C.L.; Jeang, K.T.; Danielpour, D.; Thevenin, C.; Kim, K.Y.; Sporn, M.B.; Roberts, A.B. Transactivation of the transforming growth factor β 1 (TGF-β1) gene by human T lymphotropic virus type 1 Tax: A potential mechanism for the increased production of TGF-β 1 in adult T cell leukemia. *J. Exp. Med.* **1990**, *172*, 121–129. [CrossRef] [PubMed]

172. Moriuchi, M.; Moriuchi, H. Transforming growth factor-β enhances human T-cell leukemia virus type I infection. *J. Med. Virol.* **2002**, *67*, 427–430. [CrossRef] [PubMed]

173. Ariumi, Y.; Kaida, A.; Lin, J.Y.; Hirota, M.; Masui, O.; Yamaoka, S.; Taya, Y.; Shimotohno, K. HTLV-1 Tax oncoprotein represses the p53-mediated trans-activation function through coactivator CBP sequestration. *Oncogene* **2000**, *19*, 1491–1499. [CrossRef] [PubMed]

174. Suzuki, T.; Uchida-Toita, M.; Yoshida, M. Tax protein of HTLV-1 inhibits CBP/p300-mediated transcription by interfering with recruitment of CBP/p300 onto DNA element of E-box or p53 binding site. *Oncogene* **1999**, *18*, 4137–4143. [CrossRef] [PubMed]

175. Zane, L.; Yasunaga, J.; Mitagami, Y.; Yedavalli, V.; Tang, S.W.; Chen, C.Y.; Ratner, L.; Lu, X.; Jeang, K.T. WIP1 and p53 contribute to HTLV-1 Tax-induced tumorigenesis. *Retrovirology* **2012**, *9*, 114. [CrossRef] [PubMed]

176. Harashima, N.; Kurihara, K.; Utsunomiya, A.; Tanosaki, R.; Hanabuchi, S.; Masuda, M.; Ohashi, T.; Fukui, F.; Hasegawa, A.; Masuda, T.; *et al.* Graft-versus-Tax response in adult T-cell leukemia patients after hematopoietic stem cell transplantation. *Cancer Res.* **2004**, *64*, 391–399. [CrossRef] [PubMed]

177. Rowan, A.G.; Suemori, K.; Fujiwara, H.; Yasukawa, M.; Tanaka, Y.; Taylor, G.P.; Bangham, C.R. Cytotoxic T lymphocyte lysis of HTLV-1 infected cells is limited by weak HBZ protein expression, but non-specifically enhanced on induction of Tax expression. *Retrovirology* **2014**, *11*. [CrossRef] [PubMed]

178. Takeda, S.; Maeda, M.; Morikawa, S.; Taniguchi, Y.; Yasunaga, J.; Nosaka, K.; Tanaka, Y.; Matsuoka, M. Genetic and epigenetic inactivation of Tax gene in adult T-cell leukemia cells. *Int. J. Cancer* **2004**, *109*, 559–567. [CrossRef] [PubMed]

179. Furukawa, Y.; Kubota, R.; Tara, M.; Izumo, S.; Osame, M. Existence of escape mutant in HTLV-I Tax during the development of adult T-cell leukemia. *Blood* **2001**, *97*, 987–993. [CrossRef] [PubMed]

180. Tamiya, S.; Matsuoka, M.; Etoh, K.; Watanabe, T.; Kamihira, S.; Yamaguchi, K.; Takatsuki, K. Two types of defective human T-lymphotropic virus type I provirus in adult T-cell leukemia. *Blood* **1996**, *88*, 3065–3073. [PubMed]

181. Koiwa, T.; Hamano-Usami, A.; Ishida, T.; Okayama, A.; Yamaguchi, K.; Kamihira, S.; Watanabe, T. 5'-long terminal repeat-selective CPG methylation of latent human T-cell leukemia virus type 1 provirus *in vitro* and *in vivo*. *J. Virol.* **2002**, *76*, 9389–9397. [CrossRef] [PubMed]

182. Gaudray, G.; Gachon, F.; Basbous, J.; Biard-Piechaczyk, M.; Devaux, C.; Mesnard, J.M. The complementary strand of the human T-cell leukemia virus type 1 RNA genome encodes a bZIP transcription factor that down-regulates viral transcription. *J. Virol.* **2002**, *76*, 12813–12822. [CrossRef] [PubMed]

183. Satou, Y.; Yasunaga, J.; Yoshida, M.; Matsuoka, M. HTLV-I basic leucine zipper factor gene mRNA supports proliferation of adult T cell leukemia cells. *Proc. Natl. Acad. Sci. USA* **2006**, *103*, 720–725. [CrossRef] [PubMed]

184. Mitobe, Y.; Yasunaga, J.; Furuta, R.; Matsuoka, M. HTLV-1 bZIP factor rna and protein impart distinct functions on T-cellproliferation and survival. *Cancer Res.* **2015**, *75*, 4143–4152. [CrossRef] [PubMed]

185. Lemasson, I.; Lewis, M.R.; Polakowski, N.; Hivin, P.; Cavanagh, M.H.; Thebault, S.; Barbeau, B.; Nyborg, J.K.; Mesnard, J.M. Human T-cell leukemia virus type 1 (HTLV-1) bZIP protein interacts with the cellular transcription factor CREB to inhibit HTLV-1 transcription. *J. Virol.* **2007**, *81*, 1543–1553. [CrossRef] [PubMed]

186. Hagiya, K.; Yasunaga, J.; Satou, Y.; Ohshima, K.; Matsuoka, M. ATF3, an HTLV-1 bZip factor binding protein, promotes proliferation of adult T-cell leukemia cells. *Retrovirology* **2011**, *8*, 19. [CrossRef] [PubMed]

187. Ma, Y.; Zheng, S.; Wang, Y.; Zang, W.; Li, M.; Wang, N.; Li, P.; Jin, J.; Dong, Z.; Zhao, G. The HTLV-1 HBZ protein inhibits cyclin D1 expression through interacting with the cellular transcription factor CREB. *Mol. Biol. Rep.* **2013**, *40*, 5967–5975. [CrossRef] [PubMed]

188. Wurm, T.; Wright, D.G.; Polakowski, N.; Mesnard, J.M.; Lemasson, I. The HTLV-1-encoded protein HBZ directly inhibits the acetyl transferase activity of p300/CBP. *Nucleic Acids Res.* **2012**, *40*, 5910–5925. [CrossRef] [PubMed]

189. Borowiak, M.; Kuhlmann, A.S.; Girard, S.; Gazzolo, L.; Mesnard, J.M.; Jalinot, P.; Dodon, M.D. HTLV-1 bZIP factor impedes the menin tumor suppressor and upregulates JunD-mediated transcription of the *Htert* gene. *Carcinogenesis* **2013**, *34*, 2664–2672. [CrossRef] [PubMed]

190. Hivin, P.; Basbous, J.; Raymond, F.; Henaff, D.; Arpin-Andre, C.; Robert-Hebmann, V.; Barbeau, B.; Mesnard, J.M. The HBZ-SP1 isoform of human T-cell leukemia virus type I represses JunB activity by sequestration into nuclear bodies. *Retrovirology* **2007**, *4*, 14. [CrossRef] [PubMed]

191. Thebault, S.; Basbous, J.; Hivin, P.; Devaux, C.; Mesnard, J.M. HBZ interacts with JunD and stimulates its transcriptional activity. *FEBS Lett.* **2004**, *562*, 165–170. [CrossRef]

192. Sugata, K.; Yasunaga, J.I.; Mitobe, Y.; Miura, M.; Miyazato, P.; Kohara, M.; Matsuoka, M. Protective effect of cytotoxic T lymphocytes targeting HTLV-1 bZIP factor. *Blood* **2015**, *126*, 1095–1105. [CrossRef] [PubMed]

*Review*

# Recent Advances in BLV Research

Pierre-Yves Barez [1,†], Alix de Brogniez [1,†], Alexandre Carpentier [1,†], Hélène Gazon [1,†], Nicolas Gillet [1,†], Gerónimo Gutiérrez [2,†], Malik Hamaidia [1,†], Jean-Rock Jacques [1,†], Srikanth Perike [1,†], Sathya Neelature Sriramareddy [1,†], Nathalie Renotte [1,†], Bernard Staumont [1,†], Michal Reichert [3], Karina Trono [2] and Luc Willems [1,*]

1  Molecular and Cellular Epigenetics (GIGA) and Molecular Biology (Gembloux Agro-Bio Tech), University of Liège (ULg), Liège 4000, Belgium; epy.barez@doct.ulg.ac.be (P.-Y.B.); alix.debrogniez@ulg.ac.be (A.B.); a.carpentier@doct.ulg.ac.be (A.C.); helene.gazon@ulg.ac.be (H.G.); n.gillet@ulg.ac.be (N.G.); mhamaidia@ulg.ac.be (M.H.); jacques.jeanrock@gmail.com (J.-R. J); Srikanthperike@gmail.com (S.P.); sathy.ns@gmail.com (S.N.S.); nrenotte@ulg.ac.be (N.R.); b.staumont@doct.ulg.ac.be (B.S.)
2  Instituto de Virología, Centro de Investigaciones en Ciencias Veterinarias y Agronómicas, INTA, Castelar C.C. 1712, Argentina; gutierrez.geronimo@inta.gob.ar (G.G.); trono.karina@inta.gob.ar (K.T.)
3  Department of Pathology, National Veterinary Research Institute, Pulawy 24-110, Poland; reichert@piwet.pulawy.pl
*  Correspondence: luc.willems@ulg.ac.be; Tel.: +32-4-3664925 or +32-81-622157
†  These authors contributed equally to this work.

Academic Editor: Louis Mansky
Received: 28 September 2015; Accepted: 19 November 2015; Published: 24 November 2015

**Abstract:** Different animal models have been proposed to investigate the mechanisms of Human T-lymphotropic Virus (HTLV)-induced pathogenesis: rats, transgenic and NOD-SCID/γcnull (NOG) mice, rabbits, squirrel monkeys, baboons and macaques. These systems indeed provide useful information but have intrinsic limitations such as lack of disease relevance, species specificity or inadequate immune response. Another strategy based on a comparative virology approach is to characterize a related pathogen and to speculate on possible shared mechanisms. In this perspective, bovine leukemia virus (BLV), another member of the deltaretrovirus genus, is evolutionary related to HTLV-1. BLV induces lymphoproliferative disorders in ruminants providing useful information on the mechanisms of viral persistence, genetic determinants of pathogenesis and potential novel therapies.

**Keywords:** BLV; HTLV-1; Tax; microRNA; vaccine; HDAC

## 1. Introduction

BLV naturally infects cattle, zebu and water buffalo but can also be experimentally transmitted to sheep, goats or alpaca (Vicugna pacos) [1–3]. In cattle, the most prevalent clinical manifestation is a benign accumulation of infected B-lymphocytes called persistent lymphocytosis (PL) affecting about one-third of infected animals [4,5]. In a minority of cases (about 5%–10%), BLV infection can progress to fatal leukemia/lympoma whose most spectacular consequence is spleen disruption consecutive to tumor formation [6]. BLV typically persists in less than 1% of peripheral blood cells, leading to an asymptomatic infection in the majority of infected animals. BLV is transmitted horizontally by direct contact, iatrogenic procedures or insect bites upon transfer of infected cells from milk, blood and body fluids from heavily infected dams [7,8].

Among experimental hosts, sheep provide a useful model to address specific questions pertaining to immunity, viral persistence and pathogenesis. In particular, reverse genetics permitted the development of a life-attenuated vaccine and a novel therapeutic approach. Main advantages of the sheep model include a high frequency of leukemia/lymphoma (close to 100%) and a shorter latency period (typically 2–4 years).

BLV-associated pathogenesis thus shares a series of features with HTLV-1-induced Adult T-cell Leukemia (ATLL) but does apparently not include neurodegenerative diseases such as HTLV-Associated Myelopathy/Tropical Spastic Paraparesis (HAM/TSP) [9]. It is assumed that consumption of raw milk from BLV-infected cattle is not associated with an increased risk of cancer in human, although the link cannot be formally excluded [10].

The goal of this review is to outline interesting observations in the BLV model that are of interest to understand HTLV-1 replication and pathogenesis.

## 2. Viral Oncogenes Drive Proliferation

As deltaretrovirus, BLV carries the classical genes (*gag, pro pol* and *env*) that are required to complete the viral cycle: genesis and budding of a virion, infection of a target cell, reverse transcription and integration into the host cell chromosome. The BLV provirus also encodes a series of additional accessory genes as well as microRNAs that modulate viral and/or cellular gene expression (Figure 1) [11,12]. Among these, Tax and G4 are oncogenes able to promote transformation of primary rat embryo fibroblasts [13,14]. Tax activates transcription by acting on a triplicate 21 bp enhancer motif in the 5′ LTR promoter via the CREB/ATF signaling pathway [15,16]. Although, the mechanisms of cell transformation remain to be further characterized, it is interesting to note that BLV and HTLV-1 Tax share cellular targets. Both transactivators indeed bind to tristetraprolin (TTP), a post-transcriptional modulator of TNFα expression [17]. The Tax proteins promote nuclear accumulation of TTP and restore TNFα expression by inhibiting TTP.

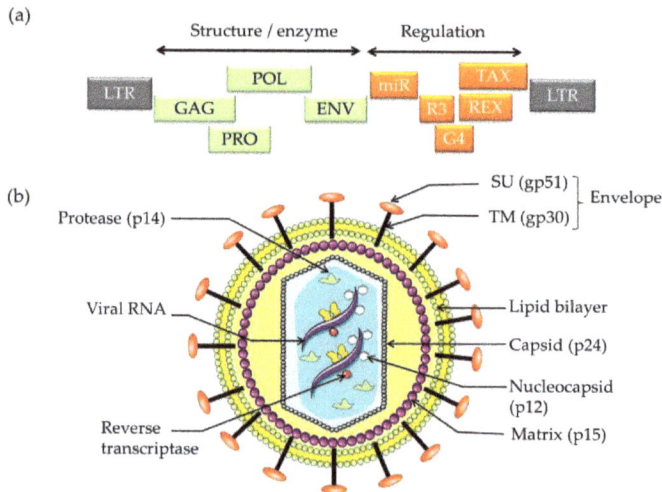

**Figure 1.** Schematic structure of (**a**) the bovine leukemia virus (BLV) genome and (**b**) the viral particle.

Another cellular protein concomitantly targeted by BLV G4 and its ortholog in HTLV-1 (p13) is farnesyl pyrophosphate synthase (FPPS), an enzyme involved in the mevalonate/squalene pathway and in synthesis of FPP, a substrate required for prenylation of Ras [18]. In addition FPPS is involved in synthesis of isoprenoids modulating the membrane fluidity and stability of lipid rafts [19]. Being localized in the nuclear compartment and in mitochondria, G4 and p13 thus exert evolutionary conserved functions.

Recently, a proviral region located 3′ of the env gene was shown to express microRNAs under the control of RNA polymerase III promoters (miR on Figure 1) [20]. The BLV microRNAs associate with Argonaute and mimic cellular analogs (e.g., BLV-miR-B4 for miR-29). BLV microRNAs are transcribed from a region dispensable for *in vivo* infectivity but are abundantly expressed in leukemic B cells (about 40%) [21]. This evidence thus contradicts the dogma that naturally occurring RNA viruses will not encode miRNAs to avoid unproductive cleavage of their genomes. HTLV-1 does not encode microRNAs as indicated by deep sequencing. The role of the BLV microRNAs in viral replication, persistence and disease remains to be further characterized.

BLV thus encodes transformation drivers (Tax, G4) and viral microRNAs likely important in pathogenesis.

## 3. Reverse Genetics Reveals the Significance of Viral Sequences in Infection and Replication

Reverse genetics using a cloned BLV provirus has allowed the screening of regions required for infection, replication and pathogenesis. As expected, large deletions within the *gag*, *pol* or *env* genes destroy infectivity *in vivo*. Discrete regions of the viral genome, such as the ITAM motifs of the envelope transmembrane protein (TM), are particularly important for infection [22]. Contradicting the concept that retroviral genomes are highly condensed, sequences located between the env gene and the Tax/Rex boundary are dispensable for infection [23]. In particular, deletion of R3 and G4 preserves infectivity but affects replication efficiency [24]. Similarly, deletion of HTLV p12$^I$ or p13$^{II}$/p30$^{II}$, the orthologs of BLV R3 and G4, impairs replication in macaques. In contrast, the mutations do not affect viral replication in rabbits, emphasizing the importance of relevant animal models [25]. R3 and G4 are nevertheless dispensable for pathogenesis, although their integrity contributes to disease frequency and latency [26].

Reverse genetics also generated unexpected observations, such as replication at wild-type levels of proviruses expressing fusion-deficient envelope proteins (TM A60V and A64S) [27]. These mutants are thus in principle unable to undergo an infectious cycle and may replicate preferentially through mitotic division of the host cell.

Mutations within the LTR further revealed that the viral promoter contained sub-optimal enhancers (AGACGTCA, TGACGGCA, TGACCTCA) that are essential for viral replication. As expected, site directed mutagenesis of the enhancers into consensus cyclic-AMP responsive elements (TGACGTCA) increases promoter efficiency but strongly impairs viral replication [28,29]. The presence of suboptimal enhancers in all BLV and HTLV-1 isolates suggests an evolutionary conserved mechanism that may reduce basal transcription and facilitate the escape from immune response.

Collectively, these observations thus emphasize the dichotomy between conclusions drawn from *in vitro* experimentations and their relevance in the animal model.

## 4. A Mutation that Increases Pathogenicity: Potential Hyperpathogenic Strain

Until recently, all mutations introduced in the BLV provirus were silent or at most reduced replication and pathogenesis *in vivo*. We recently reported that mutation of a N-linked glycosylation site (N230) affects the stability of the SU envelope protein and increases cell-to-cell transmission suggesting that this site restricts infectivity and viral replication [30]. A mutant carrying the N230 mutation replicates faster and is more pathogenic compared to the isogenic wild-type BLV strain. This observation thus suggests a mechanism of co-evolution restricting excessive pathogenicity that would indirectly impair mutual persistence of the virus and its host. Occurrence of this type of mutation may thus represent a potential threat associated with emergence of hyperpathogenic BLV strains and possibly also of new HTLV variants.

## 5. *In Vivo* Kinetics Indicates that BLV-Infected Cells Undergo High Turnover during Chronic Infection of Sheep

The BLV model has been instrumental to understand the dynamics of cell turnover *in vivo*. In principle, lymphocyte homeostasis is the result of a critical balance between cell proliferation and death. Initial experiments using intravenous injection of bromodeoxyuridine (BrdU) demonstrated that B-lymphocytes are proliferating significantly faster in BLV-positive asymptomatic and persistently lymphocytotic sheep than in uninfected controls. In fact, an excess of 0.9% cells are produced by proliferation each day and, during leukemia, these rates even rise by up to tenfold [31]. Excess of cell proliferation was also reported in HTLV-induced HAM/TSP using a similar strategy based on incorporation of deuterated glucose [32]. In contrast, persistent lymphocytosis in BLV-infected cattle is characterized by a decreased B cell turnover resulting from a reduction of cell death and an overall impairment of proliferation, as observed in human chronic lymphocytic leukemia (CLL) [33,34].

Cell dynamics can also be estimated by intravenous injection of carboxyfluorescein diacetate succinimidyl ester (CFSE) [35]. Since CFSE labels proteins via their $NH_2$ terminal ends, halving of fluorescence indicates that a cell has undergone cell division. Fitting cell numbers and fluorescence intensities revealed massive destruction of B-lymphocytes during chronic infection of sheep. In contrast, lymphocyte trafficking to and from lymphoid organs was unaffected.

Collectively, quantification of the dynamic parameters deduced from BrdU and CFSE kinetics shows that the excess of proliferation in lymphoid organs is compensated by increased death in peripheral blood [36]. Ablative surgery demonstrated that the spleen is a major lymphoid tissue massively destroying BLV-infected cells [37].

BLV chronic infection is thus characterized by a very dynamic equilibrium between a virus attempting to proliferate under a tight control exerted by the immune response.

## 6. Massive Depletion of Clones Located in Genomic Transcriptionally Active Sites during Infection

As retroviruses, BLV and HTLV-1 replicate via an infectious cycle upon expression of progeny virions as well as by mitotic division of provirus-carrying cells (clonal expansion). High throughput sequencing of proviral integration sites revealed the relative importance of these two cycles in viral replication varies during infection [38,39]. The majority of infected clones are created early before the onset of an efficient immune response. Two months from inoculation, the main replication route is mitotic expansion of pre-existing infected clones. Initially, BLV proviral integration significantly favors transcribed regions of the genome. Negative selection then eliminates 97% of the clones detected at seroconversion and disfavors BLV-infected cells carrying a provirus located close to a promoter or a gene. Nevertheless, among the surviving proviruses, clone abundance positively correlates with proximity of the provirus to a transcribed region. Two opposite forces thus operate during primary infection and dictate the fate of long term clonal composition: (1) initial integration inside genes or promoters and (2) host negative selection disfavoring proviruses located next to transcribed regions.

## 7. Tight Control of Virus-Positive Cells by the Immune Response

BLV infection is thus characterized by a massive depletion of provirus-carrying cell clones at early stages and a very dynamic turnover during chronic infection (Section 5). If the host immune response tightly controls viral replication, it is predicted that cells expressing viral antigens would be shorter lived. This question was addressed by comparing the survival rates of two cell pools isolated from the same donor and labeled with different fluorochromes depending on the absence or the presence of viral proteins induced *ex vivo* [40]. As predicted, transient viral expression significantly reduced the lifespan of BLV-infected lymphocytes. Cyclosporine treatment further supported the concept that an efficient immune response is required to control virus-expressing cells. This evidence is consistent with the presence of suboptimal LTR promoters that restrict viral reactivation (see Section 3) enabling escape from immune mediated destruction.

## 8. A Therapy Based on Activation of Viral Expression

BLV persistence is thus a very dynamic process characterized by a virus that continuously attempts to replicate and an active control exerts by the host immune response. As outlined in Section 2, viral proteins promote infectious and mitotic cycles but also expose the infected cell to immune control. Evidence for a very strong immune response is supported by the presence of virus-specific cytotoxic T cells and by high titers of neutralizing antibodies. Persistence of infected cells is thus possible

*Viruses* **2015**, *7*, 6080–6088

providing that viral proteins are not expressed, perhaps under the control of viral microRNAs. In this context, we evaluated the therapeutic effectiveness of a strategy based on the induction of viral gene expression using valproic acid (VPA), a lysine deacetylase inhibitor [41,42]. VPA efficiently induced viral expression in primary cultures and reduced the number of leukemic cells in sheep. This strategy was then translated to another B cell neoplasm [43] and to HTLV-1 infected patients with HAM/TSP [44]. The treatment appeared to be safe but unable to permanently reduce proviral loads over the long term [45]. Instead, combination of VPA and other lysine deacetylase inhibitors with a standard regimen of ATL (AZT + IFN) is promising in ongoing clinical trials [46,47].

### 9. Towards an Efficient Vaccine

Except in the European Union, the herd prevalence of BLV worldwide ranges between 30% and 90% [48]. Major economic losses result from leukemia/lymphoma-induced death, reduction in milk production and custom restrictions. Thus, there is an urgent need for an efficient, safe and cost-effective vaccine against BLV. Previous vaccine candidates faced problems of efficacy (*i.e.*, only a fraction of animals were protected), persistence (*i.e.*, rapid decrease of immune protection), cost (e.g., production of purified proteins) or safety (e.g., genetically modified hybrid viruses). Therefore, we designed another approach based on a life-attenuated BLV strain harboring multiple deletions and mutations. The rationale was to delete pathogenic genes (*i.e.*, the oncogenic drivers) while maintaining a low level of infectivity. After a series of failures, we have identified a deleted BLV provirus that is infectious in cattle but replicates at very low levels. Inoculation of this vaccine elicits a vigorous anti-BLV immune response comparable to that of a wild-type infection. The vaccine does not spread to uninfected sentinels maintained during 7 years in the same herd and could not be detected in colostrum and milk from experimentally infected cows. Passive antibodies are transmitted to the newborn calves via the maternal colostrum. This anti-viral passive immunity persists during several months in the calves. However, the BLV mutant fails to transmit from cows to calves as assessed by nested PCR. In contrast to HIV, there is no significant sequence variation during infection [49,50]. Finally, vaccinated animals but not uninfected controls resist challenge by a wild type BLV virus. Trials are currently ongoing to evaluate the efficacy and safety of the vaccine in large herds in Argentina.

### 10. Conclusions

Understanding the mechanisms of BLV infection has provided valuable information on viral transmission, persistence and pathogenesis. In particular, reverse genetics yielded conclusions that could not be predicted from experiments performed *in vitro* as exemplified by a provirus containing an optimized promoter that was nevertheless attenuated. BLV persistence is characterized by a very dynamic cell turnover, which is rather unusual for a chronic infection. Host immunity is essential to control viral replication as indicated by surgical spleen ablation. Disruption of viral latency with epigenetic modulators has therapeutic value in BLV leukemia and may be useful for treatment of ATL. Finally, availability of an efficient anti-BLV vaccine is informative to develop preventive and curative measures in HTLV.

**Acknowledgments:** This work was supported by the "Fonds National de la Recherche Scientifique" (FNRS), the Télévie, the Interuniversity Attraction Poles (IAP) Program "Virus-host interplay at the early phases of infection" BELVIR initiated by the Belgian Science Policy Office, the Belgian Foundation against Cancer (FBC), the Sixth Research Framework Programme of the European Union (project "The role of infections in cancer" INCA LSHC-CT-2005-018704), the "Neoangio" excellence program and the "Partenariat Public Privé", PPP INCA, of the "Direction générale des Technologies, de la Recherche et de l'Energie/DG06" of the Walloon government, the "Action de Recherche Concertée Glyvir" (ARC) of the "Communauté française de Belgique", the "Centre anticancéreux près ULg" (CAC), the "Subside Fédéral de Soutien à la Recherche Synbiofor and Agricultureislife" projects of Gembloux Agrobiotech (GxABT), the "ULg Fonds Spéciaux pour la Recherche", the "Plan Cancer" of the "Service Public Fédéral". Katrina Trono and Gerónimo Gutiérrez are researchers from "Instituto Nacional de Tecnología Agropecuaria" (INTA) and "Consejo Nacional de Investigaciones Científicas y Técnicas" (CONICET). Alexandre Carpentier, Srikanth Perike, Nicolas Gillet and Alix de Brogniez are supported by grants of the Télévie. Malik Hamaidia is a research fellow of the "Agriculture is life" project of GxABT. Alexandre Carpentier (Télévie) and Pierre-Yves Barez (FNRS research fellow) received a grant from the Fonds Léon Fredericq, Hélène Gazon (post-doctoral researcher) and Luc Willems (Research Director) are members of the FNRS.

**Author Contributions:** L.W. drafted the manuscript. All authors corrected, edited and approved the text.

**Conflicts of Interest:** The authors declare no conflict of interest.

## References

1. Lee, L.C.; Scarratt, W.K.; Buehring, G.C.; Saunders, G.K. Bovine leukemia virus infection in a juvenile alpaca with multicentric lymphoma. *Can. Vet. J.* **2012**, *53*, 283–286. [PubMed]

2. Meas, S.; Ohashi, K.; Tum, S.; Chhin, M.; Te, K.; Miura, K.; Sugimoto, C.; Onuma, M. Seroprevalence of bovine immunodeficiency virus and bovine leukemia virus in draught animals in Cambodia. *J. Vet. Med. Sci.* **2000**, *62*, 779–781. [CrossRef] [PubMed]

3. Rodriguez, S.M.; Florins, A.; Gillet, N.; de Brogniez, A.; Sanchez-Alcaraz, M.T.; Boxus, M.; Boulanger, F.; Gutierrez, G.; Trono, K.; Alvarez, I.; *et al.* Preventive and therapeutic strategies for bovine leukemia virus: Lessons for HTLV. *Viruses* **2011**, *3*, 1210–1248. [CrossRef] [PubMed]

4. Gillet, N.; Florins, A.; Boxus, M.; Burteau, C.; Nigro, A.; Vandermeers, F.; Balon, H.; Bouzar, A.B.; Defoiche, J.; Burny, A.; *et al.* Mechanisms of leukemogenesis induced by bovine leukemia virus: Prospects for novel anti-retroviral therapies in human. *Retrovirology* **2007**, *4*. [CrossRef] [PubMed]

5. Lairmore, M.D. Animal models of bovine leukemia virus and human T-lymphotrophic virus type-1: Insights in transmission and pathogenesis. *Annu. Rev. Anim. Biosci.* **2014**, *2*, 189–208. [CrossRef] [PubMed]

6. Bartlett, P.C.; Norby, B.; Byrem, T.M.; Parmelee, A.; Ledergerber, J.T.; Erskine, R.J. Bovine leukemia virus and cow longevity in Michigan dairy herds. *J. Dairy Sci.* **2013**, *96*, 1591–1597. [CrossRef] [PubMed]

7. Gutierrez, G.; Rodriguez, S.M.; de Brogniez, A.; Gillet, N.; Golime, R.; Burny, A.; Jaworski, J.P.; Alvarez, I.; Vagnoni, L.; Trono, K.; *et al.* Vaccination against delta-retroviruses: The bovine leukemia virus paradigm. *Viruses* **2014**, *6*, 2416–2427. [CrossRef] [PubMed]

8. Kobayashi, S.; Tsutsui, T.; Yamamoto, T.; Hayama, Y.; Muroga, N.; Konishi, M.; Kameyama, K.; Murakami, K. The role of neighboring infected cattle in bovine leukemia virus transmission risk. *J. Vet. Med. Sci.* **2015**, *77*, 861–863. [CrossRef] [PubMed]

9. Boxus, M.; Willems, L. Mechanisms of HTLV-1 persistence and transformation. *Br. J. Cancer* **2009**, *101*, 1497–1501. [CrossRef] [PubMed]

10. Buehring, G.C.; Shen, H.M.; Jensen, H.M.; Jin, D.L.; Hudes, M.; Block, G. Exposure to bovine leukemia virus is associated with breast cancer: A case-control study. *PLoS ONE* **2015**, *10*, e0134304. [CrossRef] [PubMed]

11. Derse, D. Bovine leukemia virus transcription is controlled by a virus-encoded *trans*-acting factor and by *cis*-acting response elements. *J. Virol.* **1987**, *61*, 2462–2471. [PubMed]

12. Derse, D. trans-acting regulation of bovine leukemia virus mRNA processing. *J. Virol.* **1988**, *62*, 1115–1119. [PubMed]

13. Kerkhofs, P.; Heremans, H.; Burny, A.; Kettmann, R.; Willems, L. *In vitro* and *in vivo* oncogenic potential of bovine leukemia virus G4 protein. *J. Virol.* **1998**, *72*, 2554–2559. [PubMed]

14. Willems, L.; Grimonpont, C.; Heremans, H.; Rebeyrotte, N.; Chen, G.; Portetelle, D.; Burny, A.; Kettmann, R. Mutations in the bovine leukemia virus Tax protein can abrogate the long terminal repeat-directed transactivating activity without concomitant loss of transforming potential. *Proc. Natl. Acad. Sci. USA* **1992**, *89*, 3957–3961. [CrossRef] [PubMed]

15. Adam, E.; Kerkhofs, P.; Mammerickx, M.; Burny, A.; Kettmann, R.; Willems, L. The CREB, ATF-1, and ATF-2 transcription factors from bovine leukemia virus-infected B lymphocytes activate viral expression. *J. Virol.* **1996**, *70*, 1990–1999. [PubMed]

16. Willems, L.; Kettmann, R.; Chen, G.; Portetelle, D.; Burny, A.; Derse, D. A cyclic AMP-responsive DNA-binding protein (CREB2) is a cellular transactivator of the bovine leukemia virus long terminal repeat. *J. Virol.* **1992**, *66*, 766–772. [PubMed]

17. Twizere, J.C.; Kruys, V.; Lefebvre, L.; Vanderplasschen, A.; Collete, D.; Debacq, C.; Lai, W.S.; Jauniaux, J.C.; Bernstein, L.R.; Semmes, O.J.; *et al.* Interaction of retroviral Tax oncoproteins with tristetraprolin and regulation of tumor necrosis factor-alpha expression. *J. Natl. Cancer Inst.* **2003**, *95*, 1846–1859. [CrossRef] [PubMed]

18. Lefebvre, L.; Vanderplasschen, A.; Ciminale, V.; Heremans, H.; Dangoisse, O.; Jauniaux, J.C.; Toussaint, J.F.; Zelnik, V.; Burny, A.; Kettmann, R.; *et al.* Oncoviral bovine leukemia virus G4 and human T-cell leukemia virus type 1 p13(II) accessory proteins interact with farnesyl pyrophosphate synthetase. *J. Virol.* **2002**, *76*, 1400–1414. [CrossRef] [PubMed]

19. Wang, X.; Hinson, E.R.; Cresswell, P. The interferon-inducible protein viperin inhibits influenza virus release by perturbing lipid rafts. *Cell Host Microbe* **2007**, *2*, 96–105. [CrossRef] [PubMed]

20. Kincaid, R.P.; Burke, J.M.; Sullivan, C.S. RNA virus microRNA that mimics a B-cell oncomiR. *Proc. Natl. Acad. Sci. USA* **2012**, *109*, 3077–3082. [CrossRef] [PubMed]

21. Rosewick, N.; Momont, M.; Durkin, K.; Takeda, H.; Caiment, F.; Cleuter, Y.; Vernin, C.; Mortreux, F.; Wattel, E.; Burny, A.; *et al.* Deep sequencing reveals abundant noncanonical retroviral microRNAs in B-cell leukemia/lymphoma. *Proc. Natl. Acad. Sci. USA* **2013**, *110*, 2306–2311. [CrossRef] [PubMed]

22. Willems, L.; Gatot, J.S.; Mammerickx, M.; Portetelle, D.; Burny, A.; Kerkhofs, P.; Kettmann, R. The YXXL signalling motifs of the bovine leukemia virus transmembrane protein are required for *in vivo* infection and maintenance of high viral loads. *J. Virol.* **1995**, *69*, 4137–4141. [PubMed]

23. Willems, L.; Burny, A.; Collete, D.; Dangoisse, O.; Dequiedt, F.; Gatot, J.S.; Kerkhofs, P.; Lefebvre, L.; Merezak, C.; Peremans, T.; *et al.* Genetic determinants of bovine leukemia virus pathogenesis. *AIDS Res. Hum. Retrovir.* **2000**, *16*, 1787–1795. [CrossRef] [PubMed]

24. Willems, L.; Kerkhofs, P.; Dequiedt, F.; Portetelle, D.; Mammerickx, M.; Burny, A.; Kettmann, R. Attenuation of bovine leukemia virus by deletion of R3 and G4 open reading frames. *Proc. Natl. Acad. Sci. USA* **1994**, *91*, 11532–11536. [CrossRef] [PubMed]

25. Valeri, V.W.; Hryniewicz, A.; Andresen, V.; Jones, K.; Fenizia, C.; Bialuk, I.; Chung, H.K.; Fukumoto, R.; Parks, R.W.; Ferrari, M.G.; *et al.* Requirement of the human T-cell leukemia virus p12 and p30 products for infectivity of human dendritic cells and macaques but not rabbits. *Blood* **2010**, *116*, 3809–3817. [CrossRef] [PubMed]

26. Florins, A.; Gillet, N.; Boxus, M.; Kerkhofs, P.; Kettmann, R.; Willems, L. Even attenuated bovine leukemia virus proviruses can be pathogenic in sheep. *J. Virol.* **2007**, *81*, 10195–10200. [CrossRef] [PubMed]

27. Gatot, J.S.; Callebaut, I.; Mornon, J.P.; Portetelle, D.; Burny, A.; Kerkhofs, P.; Kettmann, R.; Willems, L. Conservative mutations in the immunosuppressive region of the bovine leukemia virus transmembrane protein affect fusion but not infectivity *in vivo*. *J. Biol. Chem.* **1998**, *273*, 12870–12880. [CrossRef] [PubMed]

28. Debacq, C.; Sanchez Alcaraz, M.T.; Mortreux, F.; Kerkhofs, P.; Kettmann, R.; Willems, L. Reduced proviral loads during primo-infection of sheep by Bovine Leukemia virus attenuated mutants. *Retrovirology* **2004**, *1*. [CrossRef] [PubMed]

29. Merezak, C.; Pierreux, C.; Adam, E.; Lemaigre, F.; Rousseau, G.G.; Calomme, C.; van Lint, C.; Christophe, D.; Kerkhofs, P.; Burny, A.; *et al.* Suboptimal enhancer sequences are required for efficient bovine leukemia virus propagation *in vivo*: Implications for viral latency. *J. Virol.* **2001**, *75*, 6977–6988. [CrossRef] [PubMed]

30. De Brogniez, A.; Bouzar, A.B.; Jacques, J.R.; Cosse, J.P.; Gillet, N.; Callebaut, I.; Reichert, M.; Willems, L. Mutation of a single envelope N-linked glycosylation site enhances the pathogenicity of bovine leukemia virus. *J. Virol.* **2015**, *89*, 8945–8956. [CrossRef] [PubMed]

31. Debacq, C.; Asquith, B.; Kerkhofs, P.; Portetelle, D.; Burny, A.; Kettmann, R.; Willems, L. Increased cell proliferation, but not reduced cell death, induces lymphocytosis in bovine leukemia virus-infected sheep. *Proc. Natl. Acad. Sci. USA* **2002**, *99*, 10048–10053. [CrossRef] [PubMed]

32. Asquith, B.; Zhang, Y.; Mosley, A.J.; de Lara, C.M.; Wallace, D.L.; Worth, A.; Kaftantzi, L.; Meekings, K.; Griffin, G.E.; Tanaka, Y.; *et al.* In vivo T lymphocyte dynamics in humans and the impact of human T-lymphotropic virus 1 infection. *Proc. Natl. Acad. Sci. USA* **2007**, *104*, 8035–8040. [CrossRef] [PubMed]

33. Debacq, C.; Asquith, B.; Reichert, M.; Burny, A.; Kettmann, R.; Willems, L. Reduced cell turnover in bovine leukemia virus-infected, persistently lymphocytotic cattle. *J. Virol.* **2003**, *77*, 13073–13083. [CrossRef] [PubMed]

34. Defoiche, J.; Debacq, C.; Asquith, B.; Zhang, Y.; Burny, A.; Bron, D.; Lagneaux, L.; Macallan, D.; Willems, L. Reduction of B cell turnover in chronic lymphocytic leukaemia. *Br. J. Haematol.* **2008**, *143*, 240–247. [CrossRef] [PubMed]

35. Asquith, B.; Debacq, C.; Florins, A.; Gillet, N.; Sanchez-Alcaraz, T.; Mosley, A.; Willems, L. Quantifying lymphocyte kinetics *in vivo* using carboxyfluorescein diacetate succinimidyl ester (CFSE). *Proc. Biol. Sci.* **2006**, *273*, 1165–1171. [CrossRef] [PubMed]

36. Florins, A.; Gillet, N.; Asquith, B.; Boxus, M.; Burteau, C.; Twizere, J.C.; Urbain, P.; Vandermeers, F.; Debacq, C.; Sanchez-Alcaraz, M.T.; *et al.* Cell dynamics and immune response to BLV infection: A unifying model. *Front. Biosci.* **2007**, *12*, 1520–1531. [CrossRef] [PubMed]

37. Florins, A.; Reichert, M.; Asquith, B.; Bouzar, A.B.; Jean, G.; Francois, C.; Jasik, A.; Burny, A.; Kettmann, R.; Willems, L. Earlier onset of δ-retrovirus-induced leukemia after splenectomy. *PLoS ONE* **2009**, *4*, e6943. [CrossRef] [PubMed]

38. Gillet, N.A.; Gutierrez, G.; Rodriguez, S.M.; de Brogniez, A.; Renotte, N.; Alvarez, I.; Trono, K.; Willems, L. Massive depletion of bovi ne leukemia virus proviral clones located in genomic transcriptionally active sites during primary infection. *PLoS Pathog.* **2013**, *9*, e1003687. [CrossRef] [PubMed]

39. Gillet, N.A.; Malani, N.; Melamed, A.; Gormley, N.; Carter, R.; Bentley, D.; Berry, C.; Bushman, F.D.; Taylor, G.P.; Bangham, C.R. The host genomic environment of the provirus determines the abundance of HTLV-1-infected T-cell clones. *Blood* **2011**, *117*, 3113–3122. [CrossRef] [PubMed]

40. Florins, A.; de Brogniez, A.; Elemans, M.; Bouzar, A.B.; Francois, C.; Reichert, M.; Asquith, B.; Willems, L. Viral expression directs the fate of B cells in bovine leukemia virus-infected sheep. *J. Virol.* **2012**, *86*, 621–624. [CrossRef] [PubMed]

41. Achachi, A.; Florins, A.; Gillet, N.; Debacq, C.; Urbain, P.; Foutsop, G.M.; Vandermeers, F.; Jasik, A.; Reichert, M.; Kerkhofs, P.; *et al.* Valproate activates bovine leukemia virus gene expression, triggers apoptosis, and induces leukemia/lymphoma regression *in vivo*. *Proc. Natl. Acad. Sci. USA* **2005**, *102*, 10309–10314. [CrossRef] [PubMed]

42. Bouzar, A.B.; Boxus, M.; Defoiche, J.; Berchem, G.; Macallan, D.; Pettengell, R.; Willis, F.; Burny, A.; Lagneaux, L.; Bron, D.; *et al.* Valproate synergizes with purine nucleoside analogues to induce apoptosis of B-chronic lymphocytic leukaemia cells. *Br. J. Haematol.* **2009**, *144*, 41–52. [CrossRef] [PubMed]

43. Lagneaux, L.; Gillet, N.; Stamatopoulos, B.; Delforge, A.; Dejeneffe, M.; Massy, M.; Meuleman, N.; Kentos, A.; Martiat, P.; Willems, L.; *et al.* Valproic acid induces apoptosis in chronic lymphocytic leukemia cells through activation of the death receptor pathway and potentiates TRAIL response. *Exp. Hematol.* **2007**, *35*, 1527–1537. [CrossRef] [PubMed]

44. Lezin, A.; Gillet, N.; Olindo, S.; Signate, A.; Grandvaux, N.; Verlaeten, O.; Belrose, G.; Hiscott, J.; de Carvalho Bittencourt, M.; Asquith, B.; *et al.* Histone deacetylase mediated transcriptional activation reduces proviral loads in HTLV-1 associated myelopathy/tropical spastic paraparesis patients. *Blood* **2007**, *110*, 3722–3728. [CrossRef] [PubMed]

45. Olindo, S.; Belrose, G.; Gillet, N.; Rodriguez, S.; Boxus, M.; Verlaeten, O.; Asquith, B.; Bangham, C.; Signate, A.; Smadja, D.; *et al.* Safety of long-term treatment of HAM/TSP patients with valproic acid. *Blood* **2011**, *118*, 6306–6309. [CrossRef] [PubMed]

46. Afonso, P.V.; Mekaouche, M.; Mortreux, F.; Toulza, F.; Moriceau, A.; Wattel, E.; Gessain, A.; Bangham, C.R.; Dubreuil, G.; Plumelle, Y.; *et al.* Highly active antiretroviral treatment against STLV-1 infection combining reverse transcriptase and HDAC inhibitors. *Blood* **2010**, *116*, 3802–3808. [CrossRef] [PubMed]

47. Toomey, N.; Barber, G.; Ramos, J.C. Preclinical efficacy of belinostat in combination with zidovudine in adult T-cell leukemia-lymphoma. *Retrovirology* **2015**, *12*. [CrossRef]

48. European Food Safety Authority. Response to scientific and technical information provided by an NGO on Xylella fastidiosa. *EFSA J.* **2015**, *13*. [CrossRef]

49. Mansky, L.M.; Temin, H.M. Lower mutation rate of bovine leukemia virus relative to that of spleen necrosis virus. *J. Virol.* **1994**, *68*, 494–499. [PubMed]

50. Willems, L.; Thienpont, E.; Kerkhofs, P.; Burny, A.; Mammerickx, M.; Kettmann, R. Bovine leukemia virus, an animal model for the study of intrastrain variability. *J. Virol.* **1993**, *67*, 1086–1089. [PubMed]

*viruses*

MDPI

Review

# From Immunodeficiency to Humanization: The Contribution of Mouse Models to Explore HTLV-1 Leukemogenesis

Eléonore Pérès [1,2], Eugénie Bagdassarian [1,2,3], Sébastien This [1,2,3], Julien Villaudy [4,5], Dominique Rigal [6], Louis Gazzolo [1,2] and Madeleine Duc Dodon [1,2,*]

[1] Laboratoire de Biologie Moléculaire de la Cellule, Unité Mixte de Recherche 5239, Centre National de la Recherche Scientifique, Ecole Normale Supérieure de Lyon, 69364 Lyon Cedex 7, France; eleonore.peres@ens-lyon.fr (E.P.); eugenie.bagdassarian@gmail.com (E.B.); sebastien.this@ens-lyon.fr (S.T.); louis.gazzolo@ens-lyon.fr (L.G.)

[2] SFR UMS3444 BioSciences Lyon-Gerland-Lyon Sud (UMS3444), 69366 Lyon Cedex 7, France

[3] Master BioSciences, Département de Biologie, ENS Lyon, 69366 Lyon Cedex 7, France

[4] AIMM Therapeutics, Meibergdreef 59, 1105 BA Amsterdam Zuidoost, The Netherlands; jvillaudy@aimmtherapeutics.com

[5] Department of Medical Microbiology, Academic Medical Center, University of Amsterdam, Meibergdreef 9, 1105 BA Amsterdam Zuidoost, The Netherlands

[6] Etablissement français du sang, 69007 Lyon, France; dominique.rigal@efs-sante.fr

* Correspondence: mducdodo@ens-lyon.fr; Tel.: +33-047-272-8962; Fax: +33-047-272-8674

Academic Editor: Louis M. Mansky
Received: 30 September 2015; Accepted: 30 November 2015; Published: 7 December 2015

**Abstract:** The first discovered human retrovirus, Human T-Lymphotropic Virus type 1 (HTLV-1), is responsible for an aggressive form of T cell leukemia/lymphoma. Mouse models recapitulating the leukemogenesis process have been helpful for understanding the mechanisms underlying the pathogenesis of this retroviral-induced disease. This review will focus on the recent advances in the generation of immunodeficient and human hemato-lymphoid system mice with a particular emphasis on the development of mouse models for HTLV-1-mediated pathogenesis, their present limitations and the challenges yet to be addressed.

**Keywords:** adult T cell leukemia/lymphoma; HTLV-1; humanized mouse models; oncogenesis

## 1. Introduction

Previously known as RNA tumor viruses upon the identification of numerous avian and murine leukemia/sarcoma viruses, retroviruses were thus termed after the discovery of the viral reverse transcriptase in 1970 allowing these viruses to replicate through a DNA intermediate [1,2]. After the description of retroviruses in non-human primates, the long search of human retroviruses ended with the identification of human T-lymphotropic virus type 1 (HTLV-1) and human immunodeficiency virus type 1 (HIV-1) in 1980 and 1983, respectively [3,4].

The description of retroviruses in many species has underlined their broad diversity and revealed their association with numerous diseases encompassing malignant processes, inflammatory disorders and immune dysfunctions. Importantly, retroviruses have participated in the discovery of new cellular and molecular events, opening the field of host-virus interactions in pathological processes. *In vivo* investigations carried out with avian and murine retroviruses inoculated in their natural host (*i.e.*, chickens and mice) have largely contributed to decipher the initiation and development of numerous diseases. Concerning human retroviruses, experimental studies performed *in vitro* with human cells have clarified key events in cell-virus interactions. *In vivo* studies in small (rats, rabbits and mice) and large (monkeys) animals have led to an understanding of transmission, dissemination and persistence of infection.

Since the time of isolation and characterization of human retroviruses, the advent of transgenic and immunocompromised mice has provided investigators with new animal models to apprehend virus-induced diseases. More particularly, immunodeficient mouse strains developing a functional human hemato-lymphoid system (HHLS) after being transplanted with human hematopoietic stem cells (HSC) have been helpful for reaching significant achievements in studying HIV and HTLV-1 related diseases [5–7]. Such mouse models fulfill the conditions of reliable animal models ethically acceptable by society, easy to breed at a low cost and convenient to study the pathological processes linked to infection by lymphotropic viruses, such as HTLV-1 [8–11].

Infection by HTLV-1, a deltaretrovirus, is endemic in Japan, the Caribbean, Western Africa and South and Central America. It is estimated that 10 to 20 million individuals are infected worldwide. Most HTLV-1-infected individuals remain life-long asymptomatic carriers. However, in 3%–5% of cases, HTLV-1 is etiologically linked to a neoplastic syndrome, the adult T cell leukemia/lymphoma (ATLL) and to a spectrum of chronic inflammatory disorders, among which the most frequent is a chronic progressive encephalomyelopathy known as HTLV-1-associated myelopathy/tropical spastic paraparesis (HAM/TSP) [12–14].

## 2. The Leukemogenic Activity of HTLV-1

The main clinical feature of ATLL includes leukemic cells with multi-lobulated nuclei called "flower cells" which infiltrate various tissues (skin lesions are very common), abnormal high blood calcium level and opportunistic infections [14]. The CD3+, CD4+, CD8− and CD25+ phenotype of ATLL cells indicates that these cells derive from activated helper T cells. It was reported that in 10 of 17 ATLL cases, leukemic cells express forkhead box P3 (FoxP3), a marker of CD4+ and CD25+ regulatory T (Treg) cells that suppress the proliferation of bystander CD4+ T lymphocytes. Indeed, severe immunodeficiency and complicated opportunistic infections in ATLL patients may arise in part from the immunosuppressive properties of ATLL cells [15,16].

Epidemiological surveys have underlined that ATLL preferentially develops after transmission to neonates through maternal milk. After a prolonged asymptomatic period of 20–40 years, aneuploid leukemic cells emerge. ATLL has been classified into different subtypes: chronic, smoldering, acute and lymphoma. During the long chronic phase of infection, the virus is found integrated in the genome of T lymphocytes (more than 90% are CD4+ T cells). HTLV-1 expression remains undetectable, because of the development of a strong immune response, chiefly mediated by the anti-virus cytotoxic T-lymphocyte response (CTL) [17]. Several HTLV-1-positive CD4+ CD25+ T cell clones that progress from polyclonal to oligoclonal populations are observed. Finally, the outcome of several years of *in vivo* selection results in the dominance of one leukemic clone. At that stage, ATLL patients have a poor prognosis and a median survival time of less than one year. Anti-retroviral treatments, chemotherapies and stem-cell transplantations often fail to cure the disease [18]. Overall, preventing the infection of neonates by HTLV-1 infected mothers remains a crucial issue for the eradication of ATLL [19].

## 3. The Leukemogenic Potential of Tax and HBZ

The 5′ LTR of the HTLV-1 provirus has been shown to drive sense transcripts that encode structural and regulatory proteins and among the latter the Tax (transactivator of pX) protein [20]. Interestingly, the 3′ LTR of the HTLV-1 provirus drives antisense transcription involved in the translation of another regulatory protein HBZ (HTLV-1 basic leucine zipper factor) [21,22]. Cellular and molecular studies have emphasized that these two HTLV-1 regulatory proteins are exerting a critical role in HTLV-1-induced leukemogenesis (Figure 1).

The Tax protein is known to trans-activate the sense transcription from the 5′ LTR by interacting with members of the ATF/CREB (Activating Transcription Factor/Cyclic AMP Response Element Binding protein) family of transcription factors [23]. Tax is also defined as a modulator of cellular gene expression involved in the proliferation of T lymphocytes mainly via the activation of the NFκB and AP-1 pathways. This protein is able to bypass cell-cycle checkpoints, affects mechanisms involved in the DNA damage response and apoptosis pathways, and is associated with the accumulation of genetic and epigenetic alterations and RNA stability modifications [20].

**A**

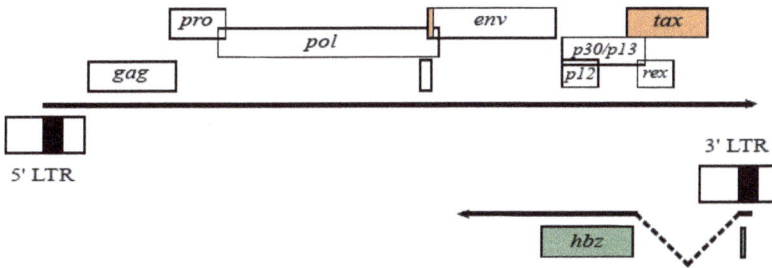

**B**

| Tax | HBZ |
|---|---|
| **Inducer of viral gene transcription** | *hbz* **RNA:**<br>◻ Proliferation of ATLL cells [29,30] |
| **Activation/Repression of cellular gene transcription** | **HBZ protein:** |
| ✓ T-cell proliferation and immune modulation [14,16]<br>✓ Alteration of cell cycle and DNA repair [20]<br>✓ Suppression of apoptosis and/or senescence; aneuploidy, chromosome instability [20]<br>✓ Abrogation of tumorigenesis barrier [23]<br>✓ Inducer of cell transformation [24] | ✓ Suppression of Tax-mediated viral gene transcription: HTLV-1 latency (Tax-) [31]<br>✓ Proliferation of HTLV-1 infected T-cells [32]<br>✓ Induction of hTERT transcription [33,34]<br>✓ Induction of Treg differentiation (TGF- β signaling) [35]<br>✓ Inhibition of the classic NFκB pathway [36]<br>✓ Inhibition of apoptosis [37] |
| **Essential for viral pathogenesis** | **Essential to induce persistent infection** |
| ✓ Chronic inflammation and NFκB-mediated tumors [25]<br>✓ LGL leukemia/lymphoma [26]<br>✓ Thymic lymphoma [27]<br>✓ Identification of cancer stem cells [28] | ✓ Tumor growth and organ infiltration [35,38]<br>✓ Inflammation and T-cell lymphoma [39,40]<br><br>✓ T-cell lymphoma in *hbz/tax* double transgenic [41]<br>✓ Increase of memory T-cells and FoxP3 Treg [41] |

**Figure 1.** Activities of HTLV-1 Tax and HBZ regulatory proteins *in vitro* and *in vivo*. (**A**) The scheme of HTLV-1 genome showing the sense and antisense genes; these genes are flanked by the long terminal repeats (LTR); (upper part) sense transcripts are initiated in the 5′ LTR containing the promoter region and terminate in the 3′ LTR; (lower part) antisense transcripts are initiated in the 3′ LTR. Coding exons for regulatory proteins are presented as orange box for Tax and green box for HBZ; (**B**) Major roles of Tax and HBZ regulatory proteins reported from experiments using cell culture (upper panel) or transgenic mouse models (lower panel).

A variety of transgenic mice have been generated to explore the activity of Tax in the initiation and development of HTLV-1-associated diseases [42] (Figure 2A). The first Tax transgenic mice, obtained in 1987 [43], with the *tax* gene being expressed under the control of the HTLV-1 LTR, resulted in the development of multicentric mesenchymal tumors with infiltration of granulocytes. This was the first demonstration that defines Tax as an oncoprotein *in vivo*. These data were later confirmed by the observation of an elevated expression of Tax in bone, associated with aberrant cell functions such as thymic atrophy [44], neurofibromatosis, muscle degeneration, lymphadenopathy, abnormal bone turnover [45] and mesenchymal tumors [25]. Other transgenic mice were generated in which *tax* was placed under the control of different promoters, either viral (Simian virus 40 and Mouse mammary tumor virus) or cellular (CD4, Ig, Granzyme B, Lck, TET, and CD3ε); for a review, see [46].

When *tax* was placed under the control of the Granzyme B (*GzmB*) promoter, which is expressed in mature T cells, transgenic mice exhibit large granular lymphocytic leukemia, associated with splenomegaly and lymphadenopathy, two main clinical features of ATLL [26]. Interestingly, by using a non-invasive imaging of Tax in *GzmB-tax*/LTR-luciferase transgenic mice, inflammation and the subsequent malignancy have been shown to be Tax-dependent through the deregulation of the NF-κB pathway [47]. The constitutive activation of that pathway is essential in the process of Tax-mediated oncogenesis underlining that it constitutes an ideal target for therapeutic treatment and that Tax transgenic mice represent good candidates for preclinical therapeutic *in vivo* trials.

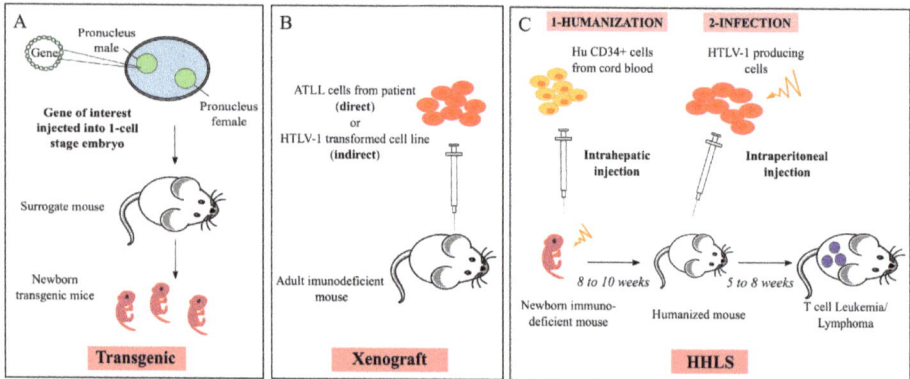

**Figure 2.** Schematic representation of the protocols used to generate HTLV-1 mouse models. (**A**) Generation of a transgenic mouse model. Briefly, the gene of interest is injected into the male pronucleus of a one-cell embryo. Micro-injected oocytes are introduced into a surrogate female and carried to term. The resulting offspring will then be screened by PCR and sequenced to find the transgenic ones; (**B**) Generation of direct and indirect xenograft mouse models: adult immunodeficient mice are injected either with cells isolated from ATLL patients (direct xenograft) or with HTLV-1 transformed cells (indirect xenograft); (**C**) Generation of Human Hemato-Lymphoid System (HHLS) mouse model: sublethally irradiated newborn immunodeficient mice are engrafted with hematopoietic stem cells. Those humanized mice are then infected with lethally irradiated HTLV-1 producing T cells (see Section 5.2.).

When *tax* was placed under the *Lck* proximal promoter that restricts its expression to developing thymocytes, T cell leukemia with characteristic "flower cells" and lymphoma with infiltrating malignant T lymphocytes highly expressing CD25 are observed [27]. These mice also display a constitutive activation of the NF-κB pathway and a marked hypercalcemia reminiscent to ATLL pathology. Collectively, these observations confirm that Tax expressed in immature thymocytes is sufficient to induce leukemogenesis in transgenic mice.

Yamazaki *et al.* [28] have used a Tax-transgenic mouse model that reproduces ATLL-like diseases. They have observed that the transfer of splenic lymphomatous cells to immunodeficient mice is followed by the regeneration of the original ATLL-like lymphoma. They then detected among lymphomatous cells the presence of a low number of chemotherapy-resistant stem cells. These cells that belong to a minor population of CD38− CD71− CD117+ hematopoietic progenitor cells were shown to be only responsible for the recapitulation of lymphoma in immunodeficient mice. This observation strongly suggests that ATLL leukemic clones exclusively originate in a minor population with stem cell-like properties.

The second regulatory protein HBZ, encoded by the antisense strand of the HTLV-1 provirus, has biologically important activities at both the RNA and protein levels [34]. *hbz* RNA promotes the proliferation of ATLL cells [29], whereas HBZ protein inhibits Tax-mediated viral transcription [21,31].

In addition, HBZ has been shown to modulate the AP-1 [48] and the classical NFκB signaling pathways [36] and to regulate the cell-mediated immune response to virus infection [49]. Nowadays, it is assumed that HBZ is playing an important role in the oncogenic process since it is able to drive infected cell proliferation [30,38], to increase hTERT transcription [33,34] and to inhibit apoptosis [37].

In HBZ transgenic mice, *hbz* RNA promotes CD4+ T cell proliferation [29]. In addition, HBZ protein was found to induce *foxp3* transcription, thus enhancing the number of CD4+FoxP3+ T cells. But a direct interaction between HBZ and FoxP3 proteins leads to an impairment of their regulatory function. Thus, the expression of HBZ in CD4+ T cells appears to be a key mechanism of HTLV-1-induced neoplastic and inflammatory diseases involving interferon-gamma (IFN-γ) [35,39,40]. Moreover, transgenic mice, in which *hbz* is expressed under the control of the CD4 promoter, have been used to test a new vaccine using a recombinant vaccinia virus expressing HBZ. That vaccine was able to induce a cytotoxic memory response against CD4+ T cells expressing HBZ [50]. Finally, double transgenic mice expressing both Tax and HBZ under the control of the CD4 promoter have increased memory T cells and FoxP3+ Treg cells leading to the development of T cell lymphoma and skin lesions [41]. This observation underlines that these two regulatory proteins exert a complementary effect on regulating signaling pathways.

The above observations indicate that the HTLV-1 provirus codes for two main regulatory proteins displaying an oncogenic potential. The question was raised to determine either if they act in a synergistic manner or if they are chronologically involved in the initiation, the maintenance and development of the leukemic process. The latter possibility should be considered since the expression of Tax is frequently disrupted in ATLL cells as indicated by the detection of Tax transcripts in only ~40% of ATLL cases [24]. Analyses of HTLV-1 proviruses and transcripts in ATLL cells revealed three ways in which cells can silence Tax expression: accumulation of nonsense mutations, insertions and deletions in *tax*, DNA methylation of the provirus that silences viral transcription and deletion of the proviral 5′ LTR. The last modification is especially prevalent in acute forms of ATLL. As Tax is the main immunogenic antigen, it is hypothesized that silencing of Tax allows infected cells to escape the CTL response against HTLV-1 [22]. In contrary to Tax, HBZ is expressed all along HTLV-1 infection and HBZ transcription is observed in all ATLL patients [21]. Accordingly, the leukemogenic process may be divided in two phases: the first one under the control of Tax that drives the proliferation of HTLV-1-positive CD4+ CD25+ T cell clones, the second one under the control of HBZ that mediates the proliferation and the maintenance of these clones.

The observations obtained with transgenic mice have provided valuable information about the involvement of these two regulatory proteins in HTLV-1-mediated leukemogenesis. However, these transgenic models do not allow exploring the natural history of HTLV-1 infection, and also the specific intervention of Tax and HBZ during the development of ATLL in humans.

## 4. Mouse Models

*From Immunodeficiency-*

The story of immunodeficient mice strains began fifty years ago with the report of BALB/c *nude* athymic mice that lack a fully developed T cell compartment (Figure 3) [51]. In CB17-SCID (severe combined immunodeficiency) mice, discovered in 1983, mature T and B cells do not develop. Indeed, these mice carry a spontaneous non-sense mutation in the gene coding the protein kinase DNA activated catalytic polypeptide (*Prkdc*), an enzyme necessary for the V(D)J recombination of the B and T cell receptors. However, innate immunity is still functional due to the presence of macrophages, antigen-presenting cells and natural killer (NK) cells [52]. Introducing the SCID mutation onto the non-obese diabetic (NOD) genetic background leads to NOD/SCID mice that display a severe innate immunodeficiency with neither complement system nor functional dendritic cells and macrophages. They provide a good *in vivo* environment for reconstitution with human HSC [53] (see part "to humanization"). In order to avoid thymic education of human HSC on mouse thymus in a MHC(H2)-restricted manner, a targeted mutation into the β*2-microglobulin* (β*2m*) gene was also introduced generating NOD/SCID β*2m^{null}* mice lacking the murine immune functions. Later on, a new strain (BALB/c.Cg-*Rag2^{null}*) of immunodeficient mice was created by deleting the recombinase-activating gene 2 (*Rag2*) in BALB/c mice [54].

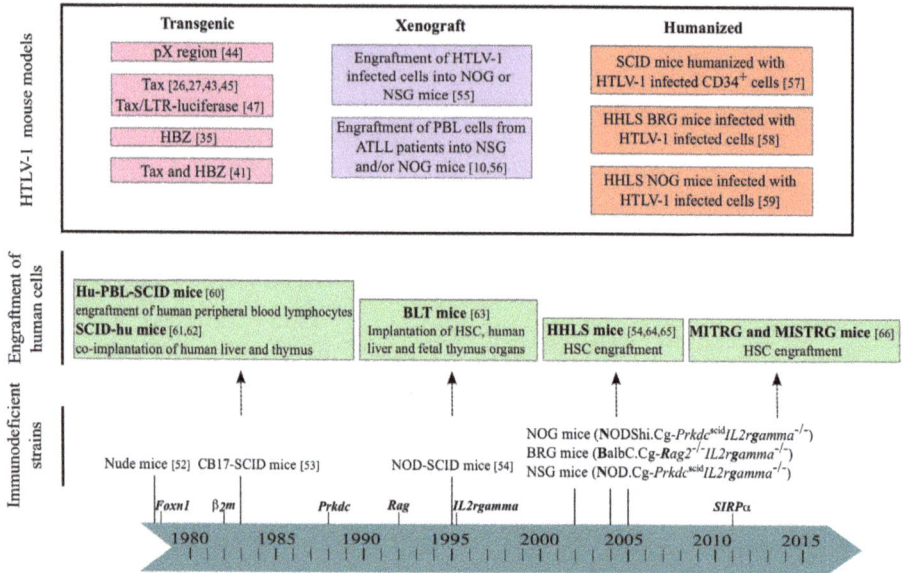

**Figure 3.** Mouse models in the study of HTLV-1 leukemogenesis. Transgenic immunocompetent mice were mainly used to investigate the role of HTLV-1 Tax and HBZ (see pink boxes). From the beginning of the 1980s, several strains of immunodeficient mice have been isolated and/or developed through the introduction of various gene mutations (in italics over the blue chronological scale). The tumorigenic potential of HTLV-1 infected T-cells or of ATLL cells has been studied by engrafting these cells in immunodeficient mice (xenograft, purple boxes). Likewise, the engraftment of immunodeficient mice with either human lymphocytes or stem cells or lymphoid tissues has led to the generation of humanized mice (green boxes), prone to investigate the role of HTLV-1 infection in leukemogenesis (orange boxes). Hu: humanized; HSC: hematopoietic stem cells; PBL: peripheral blood lymphocytes; HHLS: human hemato-lymphoid system; BLT: bone marrow-liver-thymus; MITRG and MISTRG: M-CSF, IL3, TPO, MG-CSF and/or SIRPα (signal regulatory protein α).

The next generation of immunocompromised mice was obtained by disruption of the IL-2 receptor common gamma chain (γ) gene [67]. These new strains of mice displayed a complete absence of murine T and B cells as well as NK cells. Currently, three major strains of immunodeficient mice are commonly used, NSG (NOD.Cg-*Prkdc^{SCID}*-γ*^{null}*) [64]; NOG (NODShi.Cg-*Prkdc^{SCID}*-γ*^{null}*) [65] and BRG (BALB/c.Cg-*Rag2^{null}* γ*^{null}*) [54]. Their advantages and limitations have been extensively reviewed earlier [5–7,68,69].

To improve the human innate immune cell development, MITRG mouse models were developed in which four genes encoding human cytokines (M-CSF, IL3, GM-CSF and TPO) were knocked into their respective mouse loci in *Rag2^{null}* γ*^{null}* mice. In MISTRG mice, an additional transgene encoding the human signal regulatory protein α (SIRPα) was introduced enabling mouse phagocytes to tolerate and not to phagocyte engrafted xenogeneic cells [66,70].

*-to Humanization*

The continuous improvements introduced in creating the immunocompromised mice to favor an efficient engraftment level of human tissue or cells have been exploited to generate humanized mouse models that carry a human functional immune system [10]. In this review, the term "humanized mice" is restricted to severely immunodeficient mice engrafted with human cells and/or tissues and developing a HHLS. These mice have been shown to be valuable to study human immune cell development under normal and pathological conditions.

Two humanized (hu) mouse models hu-PBL-SCID and SCID-hu Thy/Liv were described at the end of the 1980s. The first one is generated through the intraperitoneal inoculation of human peripheral blood lymphocytes (PBL) [60]. The second mouse model is developed after surgical implantation of fetal thymus/liver tissue under the renal capsule of SCID mice to form a conjoint thymus-like organ [61]. Such a model is cumbersome to generate and requires repeated biopsies of the organ. Furthermore, a wasting graft-versus-host disease develops within weeks after implantation of human cells, thus limiting experimentation to a few weeks.

Based on these models, another valuable humanized mouse model called "BLT" (Bone marrow, Liver, Thymus) has been described. NOD/SCID and NSG mice are first implanted with human fetal thymic and liver tissues and then with autologous human HSC. Several weeks later, they show long-term systemic repopulation with human T and B cells, monocytes, macrophages and dendritic cells (DC) [63]. T cells in these mice are educated in the human thymus generating human MHC class I- and II-restricted adaptive immune responses to Epstein-Barr Virus (EBV) infection and are activated by human DCs to mount a potent T cell immune response to superantigens. It represents a convenient model to study many aspects of T cell differentiation and function that could not be studied *in vitro*. But, technical, ethical and logistical reasons render this BLT model complex to carry out limiting its wide usage.

These difficulties sparked interest in the search of new protocols to generate convenient and efficient generations of humanized mice. Materials and methods used to reconstitute the HHLS in mice include many factors, such as choice of human tissue and/or cells, route of inoculation, age and gender of recipient mice and preconditioning regime (irradiation or busulfan) (Figure 2C). Immunodeficient newborn mice such as NSG and BRG mice transplanted with human purified CD34+ cells develop three to four months later a robust HHLS, through T cell thymopoiesis and B cell splenic and bone marrow lymphopoiesis. Interestingly, more T and B cells are found in NSG mice than in BRG ones, showing that NSG mice are more permissive to human cell engraftment than BRG mice. In fact, a reduced phagocytosis of human cells by mouse macrophages was observed in NSG mice. That property may be linked to the mouse SIRPα of the NOD genetic background that better recognizes the "don't eat me" signal of human CD47 than that of BALB/c background [54,71,72].

The MITRG/MISTRG immunodeficient mice are highly permissive for human cell engraftment and show an efficient development of human innate immune cells such as macrophages and NK cells. However, the increase in human myeloid cells correlated with the presence of human B and T cells at lower frequencies than in NSG mice. As B cells display an immature phenotype, humoral immune responses are low as in other strains of HHLS mice [66]. The development of human red blood cells (RBC) is inefficient, and as macrophages strongly phagocyte mouse RBC, anemia ultimately ensues two to three weeks after engraftment.

## 5. The Mouse Modeling of HTLV-1-Induced Leukemogenesis

The immortalization and transformation of HTLV-1 infected CD4+ T cells have been studied with great limitation in tissue culture and patients. Since ATLL develops through several oncogenic steps in a small percentage of HTLV-1-infected individuals, animal models of ATLL are urgently needed not only to understand the *in vivo* initiation and the progression of the leukemogenic process, but also to perform preclinical studies of potential therapeutic agents. Attempts to reach these objectives have been performed through the use of mouse xenograft models and of HHLS mice. In particular, human

T cells in HHLS mice display a phenotype of quiescent/activated and naive/memory cells and appear well suited for exploring HTLV-1 pathogenesis.

*5.1. Xenogeneic Transplantation Assays*

Upon the description of immunocompromised mice, their susceptibility to engraftment with either HTLV-1-infected cell lines or ATLL cells was evaluated (Figure 2B). These experiments confirmed that the engraftment efficacy directly correlated with the abrogation level of the murine immune responses and was dependent on a low NK cell activity, absence of complement activity and impaired macrophage and antigen presenting cell function [53]. Consequently, during the last twenty years, these mice were mainly used as xenogeneic engraftment models to apprehend critical aspects of the multistep development of ATLL [10,73,74]. More particularly, the following observations from three reports underline that the development of xenograft approaches in immunodeficient mice has largely contributed to understand kinetics, metastasis, disease progression as well as the origin of ATLL *in vivo*.

In immunodeficient mice inoculated with HTLV-1 infected MET-1 cells, T cell leukemia with tumors in organs such as liver and kidney and an increase of serum calcium level are observed similar to that in ATLL patients [55]. In these leukemic mice, the increase in serum calcium level correlated with expression of RANK-L (receptor activator of nuclear factor kappa-light-chain-enhancer of activated B cells ligand) and with secretion of parathyroid hormone-related protein and interleukin-6. As MET-1 cells expressed both the adhesion molecules CD11a (LFA-1$\alpha$) and CD49d (VLA-4$\alpha$) and produced several matrix metallo-proteinases, these observations underline the importance of these molecules in the spread of ATLL cells.

In the second study, primary ATLL cells from acute or smoldering ATLL patients were intravenously transplanted into neonatal NOD/SCID/$\beta 2m^{null}$ mice [75]. Acute-type ATLL cells were observed in the peripheral blood and in the lymph nodes of recipients. Engrafted ATLL cells were dually positive for human CD4 and CD25, and displayed patterns of HTLV-1 integration identical to those of donors by Southern blot analysis. These cells infiltrated into recipients' liver, and formed nodular lesions, recapitulating the clinical feature of each patient. In contrast, in smoldering-type ATLL cases, multiple clones of ATLL cells were efficiently engrafted in NOD/SCID/$\beta 2m^{null}$ mice. When these clones were retransplanted into secondary NOD/SCID/$\beta 2m^{null}$ recipients, single HTLV-1-infected clones became predominant, indicating the selection of clones with a dominant proliferative activity.

The third study has addressed the origin of ATLL cells. Nagai *et al.* [56] report that ATLL is sustained by a small population of transformed CD4+ CCR7+ CD45RA+ CD45RO− CD95+ T memory stem ($T_{SCM}$) cells, a unique population with stem cell-like properties, whereas the majority of ATLL cells are CD45RA− CD45RO+ conventional memory T cells. Indeed, in both HTLV-1 carriers and ATLL patients, HTLV-1 provirus was absent in naïve T cells, but was always detected in the three memory (stem, central and effector) subpopulations. *In vitro* culture assays performed with highly purified cells clearly demonstrate that the three memory subpopulations have equal susceptibility to HTLV-1 infection, since they express at least two cell surface receptors for HTLV-1, the heparan sulfate proteoglycans and the VEGF-165 receptor Neuropilin 1 [76]. But among the T memory cells, $T_{SCM}$ cells have a unique potential to self-renew while giving rise to T effector and central memory cells. Such an observation suggests that ATLL is hierarchically organized in the same manner as the normal memory T cell compartment. To further demonstrate the role played by $T_{SCM}$ in the initiation of ATLL, the authors proceeded to xenogeneic transplantation assays and inoculated the three subsets in adult irradiated NOG and NSG mice. They observed that a low number of $T_{SCM}$ cells efficiently repopulated identical ATLL clones and replenish downstream central and effector memory T cells, whereas these two other populations have no such capacities. Taken together, these findings reveal the phenotypic and functional heterogeneity of ATLL cells and identify that the $T_{SCM}$ population is the hierarchical apex of ATLL able to reconstitute identical ATLL clones. This study together with that of Yamazaki *et al.* [28] (see part 3) underline that like other cancers, ATLL may be sustained by a rare population with self-renewal capacity able to support accumulations of genetic abnormalities required for the development of this HTLV-1-induced disease.

Finally, xenogeneic transplantation assays have been performed to define specific therapeutic strategies against dysregulated pathways in HTLV-1-induced pathogenesis. Enhanced survival and reduction of tumor growth can be observed after treatment with inhibitors of NFκB-mediated pathway [77,78], of Bcl-2 family [79] or of histone deacetylase [10]. As HTLV-1 infection leads to genetic alterations, a drug inhibiting double strand break repair has been tested in a xenograft mouse model [80]. Oncolytic therapy using measles virus [81] and antibody therapy blocking CCR4 [82] or CD30 [83] also lead to increased survival in NOD/SCID and NOG mice inoculated with ATLL or HTLV-1 infected cells.

*5.2. HTLV-1 Infection of Humanized Mice*

Faithful recapitulation of ATLL in humanized mice has been challenging but required to further apprehend the natural history of HTLV-1 infection and to approach the importance of the immune response in the development and outcome of ATLL [84]. A first attempt to analyze the molecular and cellular events that control the HTLV-1 induced leukemogenesis was realized by inoculating CD34+ progenitor cells *ex vivo* infected with HTLV-1 in SCID mice engrafted with human fetal thymus and liver tissues [57]. An increased expression of the CD25 marker on thymocytes was observed together with a perturbation of the CD4+ and CD8+ thymocyte subset distribution indicating for the first time that hematopoietic progenitor cells and thymus may be targeted by HTLV-1 in humans. However, HTLV-1 infection of these SCID-hu mice failed to induce oncogenesis. In contrast, as reported by Banerjee *et al.* [85], NOD-SCID mice inoculated with CD34+ cells *ex vivo* infected with HTLV-1 have been shown to develop CD4+ T cell lymphoma. However, inoculation of *ex vivo* infected CD34+ cells might represent a bias since the presence of HTLV-1 infected cells among CD34+ cells in ATLL patients is still a matter of debate [56].

To come closer to the natural infection, we have investigated the *in vivo* effects of HTLV-1 infection in HHLS BRG mice [58]. Newborn mice were engrafted with human CD34+ cells and then infected with lethally irradiated HTLV-1-producing T cells at a time when the three main subpopulations of human thymocytes have been detected, *i.e.*, within a period of one to two months after engraftment (Figure 2C). As soon as three months after infection, significant alterations of human T cell development have been observed, the extent of which correlated with the proviral load. Human T cells from thymus and spleen were activated, as shown by the expression of the CD25 marker, that correlates with the presence of *tax* mRNA and with the increased expression of NFκB dependent genes such as *bfl-1*, an anti-apoptotic gene. Five months after HTLV-1 infection, hepato-splenomegaly, lymphadenopathy and T cell lymphoma/thymoma, in which Tax was detected, were observed in those mice. Thus, *in vivo* HTLV-1 infection of HHLS BRG mice perturbs human thymopoiesis at the level of immature cells, and propels T cell development towards the mature stages [86]. To note that these *in vivo* observations confirm results obtained *in vitro*, showing the ability of Tax to interfere with β-selection, an important checkpoint of early T cell differentiation in the thymus. These data suggest that the infection of immature target T cells in the thymus and the immunodeficient environment of these humanized mice favors the rapid development of a T cell malignancy. Interestingly, these observations suggesting that target cells of the leukemogenic activity of HTLV-1 are recruited among a stem cell population are in line with those showing the role played by $T_{SCM}$ in the initiation of ATLL [56].

Lastly, observations using mice that were generated through a different humanization protocol have been reported [59]. Indeed in that study, sub-lethally irradiated seven-week old NOG mice were submitted to an intra-bone marrow injection (IBMI) of human cord blood CD133+ cells. Three to eight months after engraftment, a stable B to T cell ratio was observed in the peripheral blood of these mice indicating the formation of a robust immune system. Four to five months after engraftment, these humanized NOG mice were infected by intra-peritoneal injection of lethally irradiated HTLV-1-producing T cells. Upon infection, the number of human CD4+ T cells in the periphery increased rapidly with the presence of abnormal T cells displaying lobulated nuclei resembling ATLL-specific "flower cells". Five months later, selective growth of a limited number of human CD25+ infected T cell clones was observed. Interestingly, HTLV-1-specific T cell mediated immune responses were induced in some infected mice, suggesting that an adequate thymic education has occurred in these IBMI-humanized NOG mice. Clearly, it is tempting to speculate that the NOG background may be at the origin of the development of adaptive immune responses. It is interesting to note that both Tax and HBZ are expressed in infected humanized mice. This is reminiscent to what is observed in the early phases of the infection process occurring in patients. However, this dual Tax/HBZ expression persists in humanized mice, probably because of the lack of an efficient immune response. Thus, induction of cellular and humoral immune responses against HTLV-1 in infected IBMI-huNOG mice might represent a valuable approach to investigate the natural history of HTLV-1 infection. It remains to be determined whether this immune response may lead to a down-regulation of Tax expression in these infected mice.

Collectively, studies performed to recapitulate HTLV-1 induced leukemogenesis in humanized mice are opening a new chapter in the *in vivo* understanding of pathological mechanisms mediated by the T cell lymphotropic virus. In addition, it is now evident that humanized mice represent a promising preclinical tool to study new therapeutic treatments since the nucleoside analogue reverse transcriptase inhibitor 3'-azido-3'deoxy-thymidine (AZT) was found to be effective at suppressing HIV replication in SCID-hu-mice [62]. Treatments to block the entry or the replication of HTLV-1 could be assessed in order to serve as a post-exposure way to prevent the persistent infection (Figure 4).

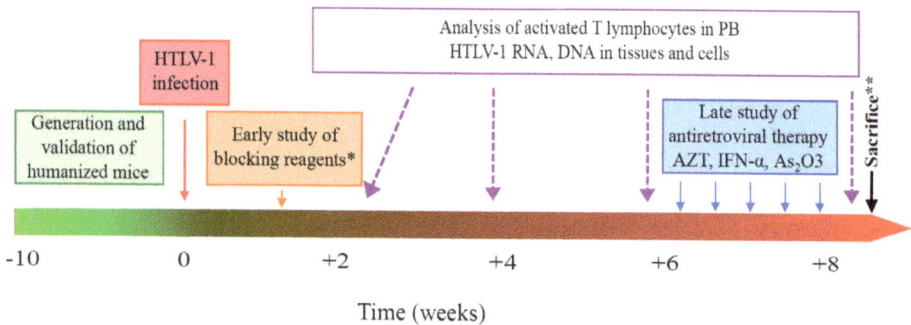

**Figure 4.** Humanized mice in the development of antiretroviral therapy. A schematic for modeling HTLV-1 infection and therapeutical approaches (orange and blue boxes) in humanized mice. * AZT, NFκB drugs and/or siRNA; ** analysis of activated T lymphocytes, of HTLV-1 DNA and RNA to evaluate the drug efficiency.

Currently, antiretroviral therapies including interferon α (IFN-α), zidovudine (AZT) and As$_2$O$_3$ have been tested as a first-line therapy for ATLL patients [87]. Furthermore, the anti-CCR4 monoclonal antibody mogamulizumab has been shown to have cytotoxic effects on ATLL cells and is now used in Japan to treat patients [19]. The demonstration of their ability to clear provirus and the understanding of the molecular and cellular mechanisms involved in humanized mice should accelerate clinical approaches for HTLV-1 eradication.

## 6. HHLS Mouse Models and HTLV-1 Pathogenesis: The Future Is Now

The advent of humanized mice to the HTLV-1 research field has offered a challenging opportunity to *in vivo* study ATLL development. Thus far, they have been helpful in elucidating the initial steps of the leukemogenic process induced by this human retrovirus. Concerning HAM/TSP and other immuno-inflammatory disorders associated with HTLV-1 infection, humanized mice have not yet been very useful mainly because of the lack of a strong immune response. One can speculate that enhancement of this immune response through new technologies (see paragraph *iii*) will definitively contribute to a real improvement in the understanding of these pathologies.

Consequently, advances have to be performed to further optimize this mouse model along these three possibilities:

(*i*)    To infect humanized mice with molecularly cloned HTLV-1 (unpublished data, Pérès *et al.*) opening a new way not only for understanding in detail the HTLV-1-pathogenesis, but also for delineating the importance of various viral genes on CD4+ T cell transformation and leukemogenesis. Inducible viral gene expression systems could also improve our knowledge [88].

(*ii*)   To mimic the way HTLV-1 is delivered (breast-feeding) and disseminated (through dendritic cells) in the body [89]. One can hypothesize that the gastrointestinal tract can serve as a secondary site of infection in which infected T cells present in the milk would be able to infect dendritic cells in the intestine. Clearly, new humanized mouse models engrafted with appropriate target tissues will be suitable for evaluation of HTLV-1 natural infection.

(*iii*)  To enhance the specific immune response, by using mouse strains transgenic for human HLAs. For example, in NSG-HLA-A2/HDD mice that possess the human HLA-A2 gene, T cell education is performed in a human HLA context. In these mice, a functional HLA-restricted cytotoxic response has been observed after EBV infection [90]. Likewise, in transgenic NOG/HLA-DR4 mice, T cell homeostasis was differentially regulated in HLA-matched humanized NOG mice compared with HLA-mismatched control mice. Furthermore, antibody class switching was induced after immunization of HLA-DR matched mice with exogenous antigens, underlining that this novel mouse strain will contribute to future studies of human humoral immune responses [91].

Thus, in the near future, it will be possible to infect humanized mice able to develop a fully functional human immune system after transgenic expression of human HLA molecules, cytokines and other species-specific factors and by targeting mouse genes to eliminate host MHC antigens and other genes to further reduce innate immunity [58]. Recently, new technologies for manipulation of the mouse genome have been described (CRISP/Cas9, clustered regularly interspaced short palindromic repeats) and provide exciting opportunities for rapidly generating new genetically modified mice in order to establish a robust small animal model to study the maintenance and development of ATLL [92].

In conclusion, together with observations obtained in immunodeficient mice through transplantation assays, studies performed with HTLV-1-infected mice have documented that the leukemogenic activity of HTLV-1 appears to be dependent on the infection of immature T cells with stem cell-like properties. Even when infected, these rare pre-leukemic cells can generate clonal populations of ATLL cells displaying phenotypic and functional heterogeneity. Therefore, reducing the number of these pre-leukemic cells in HTLV-1 carriers may represent a promising approach to prevent the development of ATLL. In that context, humanized mice would be very useful for testing the chimeric antigen receptor T cell therapy and to eliminate pre-leukemic cells, as recently demonstrated in refractive acute lymphoblastic leukemia [93].

**Acknowledgments:** This work was supported in part by INSERM and CNRS, by the European Union Project "The role of chronic infections in the development of cancer" (grant number: LSHC-CT-2005-018704) and by the Fondation de France, comité "Leucémie" (nuRAF09001CCA) to Madeleine Duc Dodon.

**Author Contributions:** Wrote the first draft of the manuscript: Eléonore Pérès and Eugénie Bagdassarian. Contributed to the writing of the manuscript: Eléonore Pérès, Eugénie Bagdassarian, Sébastien This, Julien Villaudy, Dominique Rigal, Louis Gazzolo, and Madeleine Duc Dodon. All authors reviewed and approved of the final manuscript.

**Conflicts of Interest:** The authors declare no conflict of interest.

## References

1. Baltimore, D. RNA-dependent DNA polymerase in virions of RNA tumour viruses. *Nature* **1970**, *226*, 1209–1211. [CrossRef] [PubMed]
2. Temin, H.M.; Mizutani, S. RNA-dependent DNA polymerase in virions of Rous sarcoma virus. *Nature* **1970**, *226*, 1211–1213. [CrossRef] [PubMed]
3. Poiesz, B.J.; Ruscetti, F.W.; Gazdar, A.F.; Bunn, P.A.; Minna, J.D.; Gallo, R.C. Detection and isolation of type C retrovirus particles from fresh and cultured lymphocytes of a patient with cutaneous T-cell lymphoma. *Proc. Natl. Acad. Sci. USA* **1980**, *77*, 7415–7419. [CrossRef] [PubMed]
4. Barre-Sinoussi, F.; Chermann, J.C.; Rey, F.; Nugeyre, M.T.; Chamaret, S.; Gruest, J.; Dauguet, C.; Axler-Blin, C.; Vezinet-Brun, F.; Rouzioux, C.; *et al.* Isolation of a T-lymphotropic retrovirus from a patient at risk for acquired immune deficiency syndrome (AIDS). *Science* **1983**, *220*, 868–871. [CrossRef] [PubMed]
5. Ito, R.; Takahashi, T.; Katano, I.; Ito, M. Current advances in humanized mouse models. *Cell. Mol. Immunol.* **2012**, *9*, 208–214. [CrossRef] [PubMed]
6. Rongvaux, A.; Takizawa, H.; Strowig, T.; Willinger, T.; Eynon, E.E.; Flavell, R.A.; Manz, M.G. Human hemato-lymphoid system mice: Current use and future potential for medicine. *Annu. Rev. Immunol.* **2013**, *31*, 635–674. [CrossRef] [PubMed]
7. Shultz, L.D.; Brehm, M.A.; Garcia-Martinez, J.V.; Greiner, D.L. Humanized mice for immune system investigation: Progress, promise and challenges. *Nat. Rev. Immunol.* **2012**, *12*, 786–798. [CrossRef] [PubMed]
8. Akkina, R. New generation humanized mice for virus research: Comparative aspects and future prospects. *Virology* **2013**, *435*, 14–28. [CrossRef] [PubMed]
9. Duc Dodon, M.; Villaudy, J.; Gazzolo, L.; Haines, R.; Lairmore, M. What we are learning on HTLV-1 pathogenesis from animal models. *Front. Microbiol.* **2012**, *3*. [CrossRef] [PubMed]
10. Zimmerman, B.; Niewiesk, S.; Lairmore, M.D. Mouse models of human T lymphotropic virus type-1-associated adult T-cell leukemia/lymphoma. *Vet. Pathol.* **2010**, *47*, 677–689. [CrossRef] [PubMed]
11. Panfil, A.R.; Al-Saleem, J.J.; Green, P.L. Animal models utilized in HTLV-1 research. *Virol. Res. Treat.* **2013**, *4*, 49–59.
12. Takatsuki, K. Discovery of adult T-cell leukemia. *Retrovirology* **2005**, *2*. [CrossRef] [PubMed]
13. Proietti, F.A.; Carneiro-Proietti, A.B.; Catalan-Soares, B.C.; Murphy, E.L. Global epidemiology of HTLV-I infection and associated diseases. *Oncogene* **2005**, *24*, 6058–6068. [CrossRef] [PubMed]
14. Yoshida, M. Discovery of HTLV-1, the first human retrovirus, its unique regulatory mechanisms, and insights into pathogenesis. *Oncogene* **2005**, *24*, 5931–5937. [CrossRef] [PubMed]
15. Kohno, T.; Yamada, Y.; Akamatsu, N.; Kamihira, S.; Imaizumi, Y.; Tomonaga, M.; Matsuyama, T. Possible origin of adult T-cell leukemia/lymphoma cells from human T lymphotropic virus type-1-infected regulatory T cells. *Cancer Sci.* **2005**, *96*, 527–533. [CrossRef] [PubMed]
16. Toulza, F.; Heaps, A.; Tanaka, Y.; Taylor, G.P.; Bangham, C.R. High frequency of CD4+FoxP3+ cells in HTLV-1 infection: Inverse correlation with HTLV-1-specific CTL response. *Blood* **2008**, *111*, 5047–5053. [CrossRef] [PubMed]
17. Rowan, A.G.; Bangham, C.R. Is there a role for HTLV-1-specific CTL in adult T-cell leukemia/lymphoma? *Leuk. Res. Treat.* **2012**. [CrossRef] [PubMed]
18. Uozumi, K. Treatment of adult T-cell leukemia. *J. Clin. Exp. Hematopathol.* **2010**, *50*, 9–25. [CrossRef]
19. Utsunomiya, A.; Choi, I.; Chihara, D.; Seto, M. Recent advances in the treatment of adult T-cell leukemia-lymphomas. *Cancer Sci.* **2015**, *106*, 344–351. [CrossRef] [PubMed]
20. Matsuoka, M.; Jeang, K.T. Human T-cell leukemia virus type 1 (HTLV-1) and leukemic transformation: Viral infectivity, Tax, HBZ and therapy. *Oncogene* **2011**, *30*, 1379–1389. [CrossRef] [PubMed]
21. Gaudray, G.; Gachon, F.; Basbous, J.; Biard-Piechaczyk, M.; Devaux, C.; Mesnard, J.M. The complementary strand of the human T-cell leukemia virus type 1 RNA genome encodes a bZIP transcription factor that down-regulates viral transcription. *J. Virol.* **2002**, *76*, 12813–12822. [CrossRef] [PubMed]
22. Barbeau, B.; Mesnard, J.M. Does chronic infection in retroviruses have a sense? *Trends Microbiol.* **2015**, *23*, 367–375. [CrossRef] [PubMed]
23. Lodewick, J.; Lamsoul, I.; Bex, F. Move or die: The fate of the Tax oncoprotein of HTLV-1. *Viruses* **2011**, *3*, 829–857. [CrossRef] [PubMed]

24. Matsuoka, M.; Jeang, K.T. Human T-cell leukaemia virus type 1 (HTLV-1) infectivity and cellular transformation. *Nat. Rev. Cancer* **2007**, *7*, 270–280. [CrossRef] [PubMed]

25. Coscoy, L.; Gonzalez-Dunia, D.; Tangy, F.; Syan, S.; Brahic, M.; Ozden, S. Molecular mechanism of tumorigenesis in mice transgenic for the human T cell leukemia virus Tax gene. *Virology* **1998**, *248*, 332–341. [CrossRef] [PubMed]

26. Grossman, W.J.; Kimata, J.T.; Wong, F.H.; Zutter, M.; Ley, T.J.; Ratner, L. Development of leukemia in mice transgenic for the tax gene of human T-cell leukemia virus type I. *Proc. Natl. Acad. Sci. USA* **1995**, *92*, 1057–1061. [CrossRef] [PubMed]

27. Hasegawa, H.; Sawa, H.; Lewis, M.J.; Orba, Y.; Sheehy, N.; Yamamoto, Y.; Ichinohe, T.; Katano, H.; Tsunetsugu-Yokota, Y.; Takahashi, H.; *et al.* Thymus-derived leukemia-lymphoma in mice transgenic for the Tax gene of human T-lymphotropic virus type I. *Nat. Med.* **2006**, *12*, 466–472. [CrossRef] [PubMed]

28. Yamazaki, J.; Mizukami, T.; Takizawa, K.; Kuramitsu, M.; Momose, H.; Masumi, A.; Ami, Y.; Hasegawa, H.; Hall, W.W.; Tsujimoto, H.; *et al.* Identification of cancer stem cells in a Tax-transgenic (Tax-Tg) mouse model of adult T-cell leukemia/lymphoma. *Blood* **2009**, *114*, 2709–2720. [CrossRef] [PubMed]

29. Satou, Y.; Yasunaga, J.; Yoshida, M.; Matsuoka, M. HTLV-I basic leucine zipper factor gene mRNA supports proliferation of adult T cell leukemia cells. *Proc. Natl. Acad. Sci. USA* **2006**, *103*, 720–725. [CrossRef] [PubMed]

30. Hagiya, K.; Yasunaga, J.; Satou, Y.; Ohshima, K.; Matsuoka, M. ATF3, an HTLV-1 bZip factor binding protein, promotes proliferation of adult T-cell leukemia cells. *Retrovirology* **2011**, *8*. [CrossRef] [PubMed]

31. Lemasson, I.; Lewis, M.R.; Polakowski, N.; Hivin, P.; Cavanagh, M.H.; Thebault, S.; Barbeau, B.; Nyborg, J.K.; Mesnard, J.M. Human T-cell leukemia virus type 1 (HTLV-1) bZIP protein interacts with the cellular transcription factor CREB to inhibit HTLV-1 transcription. *J. Virol.* **2007**, *81*, 1543–1553. [CrossRef] [PubMed]

32. Mitobe, Y.; Yasunaga, J.I.; Furuta, R.; Matsuoka, M. HTLV-1 bZIP factor RNA and protein impart distinct functions on T-cell proliferation and survival. *Cancer Res.* **2015**, *75*, 4143–4152. [CrossRef] [PubMed]

33. Borowiak, M.; Kuhlmann, A.S.; Girard, S.; Gazzolo, L.; Mesnard, J.M.; Jalinot, P.; Duc Dodon, M. HTLV-1 bZIP factor impedes the menin tumor suppressor and upregulates JunD-mediated transcription of the hTERT gene. *Carcinogenesis* **2013**, *34*, 2664–2672. [CrossRef] [PubMed]

34. Kuhlmann, A.S.; Villaudy, J.; Gazzolo, L.; Castellazzi, M.; Mesnard, J.M.; Duc Dodon, M. HTLV-1 HBZ cooperates with JunD to enhance transcription of the human telomerase reverse transcriptase gene (hTERT). *Retrovirology* **2007**, *4*. [CrossRef] [PubMed]

35. Satou, Y.; Yasunaga, J.; Zhao, T.; Yoshida, M.; Miyazato, P.; Takai, K.; Shimizu, K.; Ohshima, K.; Green, P.L.; Ohkura, N.; *et al.* HTLV-1 bZIP factor induces T-cell lymphoma and systemic inflammation *in vivo*. *PLoS Pathog.* **2011**, *7*, e1001274. [CrossRef] [PubMed]

36. Zhao, T.; Yasunaga, J.; Satou, Y.; Nakao, M.; Takahashi, M.; Fujii, M.; Matsuoka, M. Human T-cell leukemia virus type 1 bZIP factor selectively suppresses the classical pathway of NF-κB. *Blood* **2009**, *113*, 2755–2764. [CrossRef] [PubMed]

37. Tanaka-Nakanishi, A.; Yasunaga, J.; Takai, K.; Matsuoka, M. HTLV-1 bZIP factor suppresses apoptosis by attenuating the function of FoxO3a and altering its localization. *Cancer Res.* **2014**, *74*, 188–200. [CrossRef] [PubMed]

38. Arnold, J.; Zimmerman, B.; Li, M.; Lairmore, M.D.; Green, P.L. Human T-cell leukemia virus type-1 antisense-encoded gene, Hbz, promotes T-lymphocyte proliferation. *Blood* **2008**, *112*, 3788–3797. [CrossRef] [PubMed]

39. Yamamoto-Taguchi, N.; Satou, Y.; Miyazato, P.; Ohshima, K.; Nakagawa, M.; Katagiri, K.; Kinashi, T.; Matsuoka, M. HTLV-1 bZIP factor induces inflammation through labile Foxp3 expression. *PLoS Pathog.* **2013**, *9*, e1003630. [CrossRef] [PubMed]

40. Mitagami, Y.; Yasunaga, J.I.; Kinosada, H.; Ohshima, K.; Matsuoka, M. Interferon-gamma promotes inflammation and development of T-cell lymphoma in HTLV-1 bZIP factor transgenic mice. *PLoS Pathog.* **2015**, *11*, e1005120. [CrossRef] [PubMed]

41. Zhao, T.; Satou, Y.; Matsuoka, M. Development of T cell lymphoma in HTLV-1 bZIP factor and Tax double transgenic mice. *Arch. Virol.* **2014**, *159*, 1849–1856. [CrossRef] [PubMed]

42. Ohsugi, T. A transgenic mouse model of human T cell leukemia virus type 1-associated diseases. *Front. Microbiol.* **2013**, *4*. [CrossRef] [PubMed]

43. Nerenberg, M.; Hinrichs, S.H.; Reynolds, R.K.; Khoury, G.; Jay, G. The tat gene of human T-lymphotropic virus type 1 induces mesenchymal tumors in transgenic mice. *Science* **1987**, *237*, 1324–1329. [CrossRef] [PubMed]

44. Furuta, Y.; Aizawa, S.; Suda, Y.; Ikawa, Y.; Kishimoto, H.; Asano, Y.; Tada, T.; Hikikoshi, A.; Yoshida, M.; Seiki, M. Thymic atrophy characteristic in transgenic mice that harbor pX genes of human T-cell leukemia virus type I. *J. Virol.* **1989**, *63*, 3185–3189. [PubMed]

45. Ruddle, N.H.; Li, C.B.; Horne, W.C.; Santiago, P.; Troiano, N.; Jay, G.; Horowitz, M.; Baron, R. Mice transgenic for HTLV-I LTR-tax exhibit tax expression in bone, skeletal alterations, and high bone turnover. *Virology* **1993**, *197*, 196–204. [CrossRef] [PubMed]

46. Rauch, D.A.; Ratner, L. Targeting HTLV-1 activation of NFκB in mouse models and ATLL patients. *Viruses* **2011**, *3*, 886–900. [CrossRef] [PubMed]

47. Rauch, D.; Gross, S.; Harding, J.; Niewiesk, S.; Lairmore, M.; Piwnica-Worms, D.; Ratner, L. Imaging spontaneous tumorigenesis: Inflammation precedes development of peripheral NK tumors. *Blood* **2009**, *113*, 1493–1500. [CrossRef] [PubMed]

48. Thebault, S.; Basbous, J.; Hivin, P.; Devaux, C.; Mesnard, J.M. HBZ interacts with JunD and stimulates its transcriptional activity. *FEBS Lett.* **2004**, *562*, 165–170. [CrossRef]

49. Sugata, K.; Satou, Y.; Yasunaga, J.; Hara, H.; Ohshima, K.; Utsunomiya, A.; Mitsuyama, M.; Matsuoka, M. HTLV-1 bZIP factor impairs cell-mediated immunity by suppressing production of Th1 cytokines. *Blood* **2012**, *119*, 434–444. [CrossRef] [PubMed]

50. Sugata, K.; Yasunaga, J.I.; Mitobe, Y.; Miura, M.; Miyazato, P.; Kohara, M.; Matsuoka, M. Protective effect of cytotoxic T lymphocytes targeting HTLV-1 bZIP factor. *Blood* **2015**, *126*, 1095–1105. [CrossRef] [PubMed]

51. Flanagan, S.P. "Nude", a new hairless gene with pleiotropic effects in the mouse. *Genet. Res.* **1966**, *8*, 295–309. [CrossRef] [PubMed]

52. Bosma, G.C.; Custer, R.P.; Bosma, M.J. A severe combined immunodeficiency mutation in the mouse. *Nature* **1983**, *301*, 527–530. [CrossRef] [PubMed]

53. Shultz, L.D.; Schweitzer, P.A.; Christianson, S.W.; Gott, B.; Schweitzer, I.B.; Tennent, B.; McKenna, S.; Mobraaten, L.; Rajan, T.V.; Greiner, D.L.; *et al.* Multiple defects in innate and adaptive immunologic function in NOD/LtSz-scid mice. *J. Immunol.* **1995**, *154*, 180–191. [PubMed]

54. Traggiai, E.; Chicha, L.; Mazzucchelli, L.; Bronz, L.; Piffaretti, J.C.; Lanzavecchia, A.; Manz, M.G. Development of a human adaptive immune system in cord blood cell-transplanted mice. *Science* **2004**, *304*, 104–107. [CrossRef] [PubMed]

55. Parrula, C.; Zimmerman, B.; Nadella, P.; Shu, S.; Rosol, T.; Fernandez, S.; Lairmore, M.; Niewiesk, S. Expression of tumor invasion factors determines systemic engraftment and induction of humoral hypercalcemia in a mouse model of adult T-cell leukemia. *Vet. Pathol.* **2009**, *46*, 1003–1014. [CrossRef] [PubMed]

56. Nagai, Y.; Kawahara, M.; Hishizawa, M.; Shimazu, Y.; Sugino, N.; Fujii, S.; Kadowaki, N.; Takaori-Kondo, A. T memory stem cells are the hierarchical apex of adult T-cell leukemia. *Blood* **2015**, *125*, 3527–3535. [CrossRef] [PubMed]

57. Feuer, G.; Fraser, J.K.; Zack, J.A.; Lee, F.; Feuer, R.; Chen, I.S. Human T-cell leukemia virus infection of human hematopoietic progenitor cells: Maintenance of virus infection during differentiation *in vitro* and *in vivo*. *J. Virol.* **1996**, *70*, 4038–4044. [PubMed]

58. Villaudy, J.; Wencker, M.; Gadot, N.; Gillet, N.A.; Scoazec, J.Y.; Gazzolo, L.; Manz, M.G.; Bangham, C.R.; Duc Dodon, M. HTLV-1 propels thymic human T cell development in "human immune system" Rag2/gamma c/ mice. *PLoS Pathog.* **2011**, *7*, e1002231. [CrossRef] [PubMed]

59. Tezuka, K.; Xun, R.; Tei, M.; Ueno, T.; Tanaka, M.; Takenouchi, N.; Fujisawa, J. An animal model of adult T-cell leukemia: Humanized mice with HTLV-1-specific immunity. *Blood* **2014**, *123*, 346–355. [CrossRef] [PubMed]

60. Mosier, D.E.; Gulizia, R.J.; Baird, S.M.; Wilson, D.B. Transfer of a functional human immune system to mice with severe combined immunodeficiency. *Nature* **1988**, *335*, 256–259. [CrossRef] [PubMed]

61. Namikawa, R.; Weilbaecher, K.N.; Kaneshima, H.; Yee, E.J.; McCune, J.M. Long-term human hematopoiesis in the SCID-hu mouse. *J. Exp. Med.* **1990**, *172*, 1055–1063. [CrossRef] [PubMed]

62. McCune, J.M.; Namikawa, R.; Shih, C.C.; Rabin, L.; Kaneshima, H. Suppression of HIV infection in AZT-treated SCID-hu mice. *Science* **1990**, *247*, 564–566. [CrossRef] [PubMed]

63. Melkus, M.W.; Estes, J.D.; Padgett-Thomas, A.; Gatlin, J.; Denton, P.W.; Othieno, F.A.; Wege, A.K.; Haase, A.T.; Garcia, J.V. Humanized mice mount specific adaptive and innate immune responses to EBV and TSST-1. *Nat. Med.* **2006**, *12*, 1316–1322. [CrossRef] [PubMed]

64. Ishikawa, F.; Yasukawa, M.; Lyons, B.; Yoshida, S.; Miyamoto, T.; Yoshimoto, G.; Watanabe, T.; Akashi, K.; Shultz, L.D.; Harada, M. Development of functional human blood and immune systems in NOD/SCID/IL2 receptor γ chain*null* mice. *Blood* **2005**, *106*, 1565–1573. [CrossRef] [PubMed]

65. Ito, M.; Kobayashi, K.; Nakahata, T. NOD/Shi-scid IL2rγ*null* (NOG) mice more appropriate for humanized mouse models. *Curr. Top. Microbiol. Immunol.* **2008**, *324*, 53–76. [PubMed]

66. Rongvaux, A.; Willinger, T.; Martinek, J.; Strowig, T.; Gearty, S.V.; Teichmann, L.L.; Saito, Y.; Marches, F.; Halene, S.; Palucka, A.K.; *et al.* Development and function of human innate immune cells in a humanized mouse model. *Nat. Biotechnol.* **2014**, *32*, 364–372. [CrossRef] [PubMed]

67. Cao, X.; Shores, E.W.; Hu-Li, J.; Anver, M.R.; Kelsall, B.L.; Russell, S.M.; Drago, J.; Noguchi, M.; Grinberg, A.; Bloom, E.T.; *et al.* Defective lymphoid development in mice lacking expression of the common cytokine receptor gamma chain. *Immunity* **1995**, *2*, 223–238. [CrossRef]

68. Brehm, M.A.; Wiles, M.V.; Greiner, D.L.; Shultz, L.D. Generation of improved humanized mouse models for human infectious diseases. *J. Immunol. Methods* **2014**, *410*, 3–17. [CrossRef] [PubMed]

69. Cachat, A.; Villaudy, J.; Rigal, D.; Gazzolo, L.; Duc Dodon, M. Mice are not Men and yet . . . how humanized mice inform us about human infectious diseases. *Med. Sci.* **2012**, *28*, 63–68.

70. Legrand, N.; Huntington, N.D.; Nagasawa, M.; Bakker, A.Q.; Schotte, R.; Strick-Marchand, H.; de Geus, S.J.; Pouw, S.M.; Bohne, M.; Voordouw, A.; *et al.* Functional CD47/signal regulatory protein alpha (SIRPα) interaction is required for optimal human T-and natural killer-(NK) cell homeostasis *in vivo*. *Proc. Natl. Acad. Sci. USA* **2011**, *108*, 13224–13229. [CrossRef] [PubMed]

71. Brehm, M.A.; Cuthbert, A.; Yang, C.; Miller, D.M.; Dilorio, P.; Laning, J.; Burzenski, L.; Gott, B.; Foreman, O.; Kavirayani, A.; *et al.* Parameters for establishing humanized mouse models to study human immunity: Analysis of human hematopoietic stem cell engraftment in three immunodeficient strains of mice bearing the IL2rγ*null* mutation. *Clin. Immunol.* **2010**, *135*, 84–98. [CrossRef] [PubMed]

72. Takenaka, K.; Prasolava, T.K.; Wang, J.C.; Mortin-Toth, S.M.; Khalouei, S.; Gan, O.I.; Dick, J.E.; Danska, J.S. Polymorphism in Sirpa modulates engraftment of human hematopoietic stem cells. *Nat. Immunol.* **2007**, *8*, 1313–1323. [CrossRef] [PubMed]

73. Liu, Y.; Dole, K.; Stanley, J.R.; Richard, V.; Rosol, T.J.; Ratner, L.; Lairmore, M.; Feuer, G. Engraftment and tumorigenesis of HTLV-1 transformed T cell lines in SCID/bg and NOD/SCID mice. *Leuk. Res.* **2002**, *26*, 561–567. [CrossRef]

74. Takajo, I.; Umeki, K.; Morishita, K.; Yamamoto, I.; Kubuki, Y.; Hatakeyama, K.; Kataoka, H.; Okayama, A. Engraftment of peripheral blood mononuclear cells from human T-lymphotropic virus type 1 carriers in NOD/SCID/γ*null* (NOG) mice. *Int. J. Cancer* **2007**, *121*, 2205–2211. [CrossRef] [PubMed]

75. Kawano, N.; Ishikawa, F.; Shimoda, K.; Yasukawa, M.; Nagafuji, K.; Miyamoto, T.; Baba, E.; Tanaka, T.; Yamasaki, S.; Gondo, H.; *et al.* Efficient engraftment of primary adult T-cell leukemia cells in newborn NOD/SCID/β2-microglobulin*null* mice. *Leukemia* **2005**, *19*, 1384–1390. [CrossRef] [PubMed]

76. Jones, K.S.; Lambert, S.; Bouttier, M.; Benit, L.; Ruscetti, F.W.; Hermine, O.; Pique, C. Molecular aspects of HTLV-1 entry: Functional domains of the HTLV-1 surface subunit (SU) and their relationships to the entry receptors. *Viruses* **2011**, *3*, 794–810. [CrossRef] [PubMed]

77. Uota, S.; Zahidunnabi Dewan, M.; Saitoh, Y.; Muto, S.; Itai, A.; Utsunomiya, A.; Watanabe, T.; Yamamoto, N.; Yamaoka, S. An IκB kinase 2 inhibitor IMD-0354 suppresses the survival of adult T-cell leukemia cells. *Cancer Sci.* **2012**, *103*, 100–106. [CrossRef] [PubMed]

78. Satou, Y.; Nosaka, K.; Koya, Y.; Yasunaga, J.I.; Toyokuni, S.; Matsuoka, M. Proteasome inhibitor, bortezomib, potently inhibits the growth of adult T-cell leukemia cells both *in vivo* and *in vitro*. *Leukemia* **2004**, *18*, 1357–1363. [CrossRef] [PubMed]

79. Ishitsuka, K.; Kunami, N.; Katsuya, H.; Nogami, R.; Ishikawa, C.; Yotsumoto, F.; Tanji, H.; Mori, N.; Takeshita, M.; Miyamoto, S.; *et al.* Targeting Bcl-2 family proteins in adult T-cell leukemia/lymphoma: *In vitro* and *in vivo* effects of the novel Bcl-2 family inhibitor ABT-737. *Cancer Lett.* **2012**, *317*, 218–225. [CrossRef] [PubMed]

80. Hisatomi, T.; Sueoka-Aragane, N.; Sato, A.; Tomimasu, R.; Ide, M.; Kurimasa, A.; Okamoto, K.; Kimura, S.; Sueoka, E. NK314 potentiates antitumor activity with adult T-cell leukemia-lymphoma cells by inhibition of dual targets on topoisomerase IIα and DNA-dependent protein kinase. *Blood* **2011**, *117*, 3575–3584. [CrossRef] [PubMed]

81. Parrula, C.; Fernandez, S.A.; Zimmerman, B.; Lairmore, M.; Niewiesk, S. Measles virotherapy in a mouse model of adult T-cell leukaemia/lymphoma. *J. Gen. Virol.* **2011**, *92*, 1458–1466. [CrossRef] [PubMed]

82. Ito, A.; Ishida, T.; Utsunomiya, A.; Sato, F.; Mori, F.; Yano, H.; Inagaki, A.; Suzuki, S.; Takino, H.; Ri, M.; *et al.* Defucosylated anti-CCR4 monoclonal antibody exerts potent ADCC against primary ATLL cells mediated by autologous human immune cells in NOD/Shi-scid, IL-2R $\gamma^{null}$ mice *in vivo*. *J. Immunol.* **2009**, *183*, 4782–4791. [CrossRef] [PubMed]

83. Maeda, N.; Muta, H.; Oflazoglu, E.; Yoshikai, Y. Susceptibility of human T-cell leukemia virus type I-infected cells to humanized anti-CD30 monoclonal antibodies *in vitro* and *in vivo*. *Cancer Sci.* **2010**, *101*, 224–230. [CrossRef] [PubMed]

84. Bangham, C.R. CTL quality and the control of human retroviral infections. *Eur. J. Immunol.* **2009**, *39*, 1700–1712. [CrossRef] [PubMed]

85. Banerjee, P.; Crawford, L.; Samuelson, E.; Feuer, G. Hematopoietic stem cells and retroviral infection. *Retrovirology* **2010**, *7*. [CrossRef] [PubMed]

86. Wencker, M.; Gazzolo, L.; Duc Dodon, M. The leukemogenic activity of Tax HTLV-1 during αβ T cell development. *Front. Biosci.* **2009**, *1*, 194–204. [CrossRef]

87. El Hajj, H.; El-Sabban, M.; Hasegawa, H.; Zaatari, G.; Ablain, J.; Saab, S.T.; Janin, A.; Mahfouz, R.; Nasr, R.; Kfoury, Y.; *et al.* Therapy-induced selective loss of leukemia-initiating activity in murine adult T cell leukemia. *J. Exp. Med.* **2010**, *207*, 2785–2792. [CrossRef] [PubMed]

88. Centlivre, M.; Legrand, N.; Berkhout, B. A conditionally replicating human immunodeficiency virus in BRG-HIS mice. In *Humanized Mice for HIV Research*; Polakowski, N., Garcia, J.V., Koyanagi, Y., Manz, M.G., Tager, A.M., Eds.; Springer: New York, NY, USA, 2014; pp. 443–454.

89. Martin-Latil, S.; Gnadig, N.F.; Mallet, A.; Desdouits, M.; Guivel-Benhassine, F.; Jeannin, P.; Prevost, M.C.; Schwartz, O.; Gessain, A.; Ozden, S.; *et al.* Transcytosis of HTLV-1 across a tight human epithelial barrier and infection of subepithelial dendritic cells. *Blood* **2012**, *120*, 572–580. [CrossRef] [PubMed]

90. Shultz, L.D.; Saito, Y.; Najima, Y.; Tanaka, S.; Ochi, T.; Tomizawa, M.; Doi, T.; Sone, A.; Suzuki, N.; Fujiwara, H.; *et al.* Generation of functional human T-cell subsets with HLA-restricted immune responses in HLA class I expressing NOD/SCID/IL2r $\gamma^{null}$ humanized mice. *Proc. Natl. Acad. Sci. USA* **2010**, *107*, 13022–13027. [CrossRef] [PubMed]

91. Suzuki, M.; Takahashi, T.; Katano, I.; Ito, R.; Ito, M.; Harigae, H.; Ishii, N.; Sugamura, K. Induction of human humoral immune responses in a novel HLA-DR-expressing transgenic NOD/Shi-scid/gammacnull mouse. *Int. Immunol.* **2012**, *24*, 243–252. [CrossRef] [PubMed]

92. Yang, H.; Wang, H.; Shivalila, C.S.; Cheng, A.W.; Shi, L.; Jaenisch, R. One-step generation of mice carrying reporter and conditional alleles by CRISPR/Cas-mediated genome engineering. *Cell* **2013**, *154*, 1370–1379. [CrossRef] [PubMed]

93. Grupp, S.A. Advances in T-cell therapy for ALL. *Best Pract. Res. Clin. Haematol.* **2014**, *27*, 222–228. [CrossRef] [PubMed]

# viruses

*Review*

# Molecular Mechanisms of HTLV-1 Cell-to-Cell Transmission

Christine Gross and Andrea K. Thoma-Kress *

Institute of Clinical and Molecular Virology, Friedrich-Alexander-Universität Erlangen-Nürnberg (FAU), 91054 Erlangen, Germany; christine.gross@viro.med.uni-erlangen.de
* Correspondence: aakress@viro.med.uni-erlangen.de; Tel.: +49-9131-8526429

Academic Editor: Louis M. Mansky
Received: 18 December 2015; Accepted: 4 March 2016; Published: 9 March 2016

**Abstract:** The tumorvirus human T-cell lymphotropic virus type 1 (HTLV-1), a member of the delta-retrovirus family, is transmitted via cell-containing body fluids such as blood products, semen, and breast milk. *In vivo*, HTLV-1 preferentially infects CD4$^+$ T-cells, and to a lesser extent, CD8$^+$ T-cells, dendritic cells, and monocytes. Efficient infection of CD4$^+$ T-cells requires cell-cell contacts while cell-free virus transmission is inefficient. Two types of cell-cell contacts have been described to be critical for HTLV-1 transmission, tight junctions and cellular conduits. Further, two non-exclusive mechanisms of virus transmission at cell-cell contacts have been proposed: (1) polarized budding of HTLV-1 into synaptic clefts; and (2) cell surface transfer of viral biofilms at virological synapses. In contrast to CD4$^+$ T-cells, dendritic cells can be infected cell-free and, to a greater extent, via viral biofilms *in vitro*. Cell-to-cell transmission of HTLV-1 requires a coordinated action of steps in the virus infectious cycle with events in the cell-cell adhesion process; therefore, virus propagation from cell-to-cell depends on specific interactions between cellular and viral proteins. Here, we review the molecular mechanisms of HTLV-1 transmission with a focus on the HTLV-1-encoded proteins Tax and p8, their impact on host cell factors mediating cell-cell contacts, cytoskeletal remodeling, and thus, virus propagation.

**Keywords:** HTLV-1; Tax; p8; virus transmission; cell-to-cell transmission; cell-cell contacts; virological synapse; viral biofilm; cellular conduit

---

## 1. Introduction

Human T-cell lymphotropic virus type 1 (HTLV-1), a delta-retrovirus, is the causative agent of a severe and fatal lymphoproliferative disorder of CD4$^+$ T-cells, adult T-cell leukemia/lymphoma (ATL), and of a neurodegenerative, inflammatory disease, HTLV-1-associated myelopathy/tropical spastic paraparesis (HAM/TSP) [1–5]. Up to 5% of infected people develop one of the aforementioned diseases as a consequence of prolonged viral persistence after a clinical latency period that may last over decades [6–8]. Although the exact number of infected people is unknown [9], it is estimated that 5–10 million people worldwide are infected with HTLV-1 [10]. Endemic regions for HTLV-1 are Japan, Melanesia, South America, parts of sub-Saharan Africa, the Caribbean, central parts of Australia, and the Middle East [10,11]. In Europe, only Romania seems to be an endemic region [10,12,13].

Upon binding to its receptor, which is composed of the widely expressed glucose transporter 1 (Glut-1), neuropilin-1 (NRP-1, BDCA-4), and heparan sulfate proteoglycans (HSPG), HTLV-1 enters and infects its target cell [14–18]. After uncoating and reverse transcription, HTLV-1 integrates into the host cell genome and is predominantly maintained in its provirus form (9.1 kb), which is flanked by long terminal repeats (LTR) in both the 5' and 3' region carrying the viral promoter (reviewed by [19]). Next to genes common for retroviruses encoding structural proteins Gag, the enzymes protease, polymerase, integrase, and reverse transcriptase, HTLV-1 encodes regulatory (Tax, Rex) and accessory (p12/p8, p13, p30) proteins from the sense strand and the HTLV-1 basic leucine zipper (HBZ) from the antisense strand [7,19]. Tax and Rex are essential for viral replication. While Tax enhances viral mRNA synthesis by transactivating the HTLV-1 promoter located in the 5'-LTR, Rex controls the synthesis of the structural proteins on a post-transcriptional level [20]. The accessory proteins p12/p8, p13, and p30 are important for viral infectivity and persistence *in vivo*, but not for virus replication *in vitro* [21]. An important role in HTLV-1-induced cellular transformation has been attributed to the viral oncoprotein Tax, which is sufficient to immortalize T-cells *in vitro* [19]. During the last decade, important roles for promoting viral replication and cellular proliferation have been attributed to the HBZ protein and to HBZ RNA. It is thought that Tax is important for initiating immortalization of lymphocytes, while HBZ is essential for maintaining the immortalized phenotype [22].

HTLV-1 replicates either by infecting new target cells or by mitotic division and clonal proliferation of infected cells (for review see [23]). In this article, we review the molecular mechanisms of infectious HTLV-1 cell-to-cell transmission. We focus on the HTLV-1-encoded proteins Tax and p8, their impact on host factors mediating cell-cell contacts, cytoskeletal remodeling, and virus transmission.

## 2. Target Cells of HTLV-1 *in Vivo*

While HTLV-1 infects several cell types *in vitro* after binding of the viral envelope (Env) protein to the HTLV-1 receptor [14–18], CD4$^+$ T-cells are the main and preferential target for HTLV-1 infection *in vivo* [24]. Additionally, HTLV-1 proviral DNA can also be detected to a lesser extent in CD8$^+$ T-cells [25–27], dendritic cells (DC) [28], plasmacytoid dendritic cells (pDC) [29], and monocytes [26,30]. A recent study by Melamed *et al.* has shown that infected CD8$^+$ T-cells constitute about 5% of the total HTLV-1 proviral load found in peripheral blood mononuclear cells (PBMC) in a cohort of 12 HTLV-1-infected patients [27]. However, in clonally expanded populations of HTLV-1-infected cells, it seems unlikely that other cell types than CD4$^+$ and CD8$^+$ cells are present because almost all (99.7%) of the most highly abundant clones were CD4$^+$ or CD8$^+$ cells [27]. Another recent study reported the presence of HTLV-1 in classical, intermediate, and non-classical monocytes in PBMC of HTLV-1-infected individuals. HTLV-1 infection altered surface receptor expression, migratory function, and subset frequency of the monocytes [31]. The authors proposed the model that recruitment of classical monocytes to inflammation sites is increased in infected patients, which may result in virus acquisition and enhanced virus dissemination [30]. These *ex vivo* observations are in contrast to *in vitro* observations showing that monocytes are refractory to productive HTLV-1 infection, which initiates Caspase-3-dependent cell death [32]. Early work has also shown that HTLV-1-infected B-cell clones can be isolated from ATL patients and that B-cells are targets of HTLV-1 *in vitro* [31,33–36]. However, B-cells do not seem to constitute a major viral reservoir *in vivo*.

## 3. Routes of Viral Transmission *in Vivo*

In contrast to human immunodeficiency virus (HIV), cell-free infection of CD4$^+$ T-cells with HTLV-1 is very inefficient. Free virions can hardly be detected in the blood plasma of infected individuals and are poorly infectious for most cell types except DC [37–41]. Further, infected lymphocytes produce a limited amount of viral particles, amongst which 1 out of $10^5$ is infectious [38]. However, transmission is greatly improved upon establishment of cell-cell contacts [40]. Therefore, efficient virus transmission occurs via cell-containing body fluids such as blood, semen, and breast milk (for review, see [40]). In endemic regions, HTLV-1 is primarily transmitted from mother to child. Contrary to HIV, mother-to-child transmission of HTLV-1 predominantly occurs via breast-feeding, while transplacental transmission or transmission during delivery are rare [40,42,43]. The risk of viral transmission increases with longer breast-feeding periods and high maternal proviral load. Reduction in breast-feeding also reduces mother-to-child transmission [44]. Sexual transmission of HTLV-1 occurs more efficiently from men to women than *vice versa* [40], and transmission might be enhanced by other sexually transmitted diseases that cause ulcers and ruptures of the mucosa like syphilis or Herpes simplex type 2 [45]. Rarely, HTLV-1 can also be transmitted by organ transplantation and cause diseases in immunocompromised transplant recipients, like HTLV-1-associated lymphomas or HAM/TSP after kidney transplantation [46,47]. HTLV-1 is not only transmittable among humans, but also from non-human primates (NHP) to humans. Recent studies have reported that interspecies transmission of the simian counterpart STLV-1 through severe bites from NHP is an ongoing event in Central Africa [48,49].

It is still not settled whether cell-free or cell-associated HTLV-1 accounts for infectivity of the primary target cell *in vivo*. Moreover, the first host cell infected by HTLV-1 *in vivo* and the exact route of infection are currently unknown [50]. Since antigen-presenting cells such as DC (see Section 5) are naturally infected with HTLV-1, it is assumed that they could be involved in viral transmission to T-cells *in vivo*. To obtain insights into the first steps of HTLV-1 acquisition *in vivo*, e.g., during mother-to-child transmission by breastfeeding, Martin-Ladil *et al.* developed an *in vitro* model studying the transcytosis of HTLV-1 across a barrier of enterocytes [51]. Interestingly, the integrity of the epithelial barrier was maintained during co-culture with HTLV-1-infected lymphocytes, and enterocytes were not susceptible to HTLV-1 infection. However, free infectious HTLV-1 virions crossed the epithelial barrier via transcytosis and productively infected human DC located beneath the epithelial barrier [51]. Upon infection, DC could then pass the virus to T-cells. Surprisingly, DC are more susceptible to *in vitro* infection with viral biofilms than autologous CD4$^+$ T-cells, underlining their potential importance in virus dissemination [50].

The study of HTLV-1 infection *in vivo* has benefitted from small animal models (rabbits, rats, and mice) and from large animal models (macaques, sheep infected with the related bovine leukemia virus) [52,53]. Recently, HTLV-1-infected humanized mice that are reconstituted with a functional human immune system and that develop lymphomas have been described [54]. Humanized mice may provide the opportunity to visualize HTLV-1 transmission *in vivo* as it has been shown for transmission of the related retroviruses murine leukemia virus (MLV) and HIV [55]. Additionally, humanized mouse models have already been used to show the neutralizing function of anti-Env antibodies in preventing HTLV-1 transmission *in vivo* [56].

## 4. Molecular Mechanisms of HTLV-1 Cell-to-Cell Transmission between CD4$^+$ T-Cells

Cell-cell-mediated virus propagation requires coordination of steps of the virus infectious cycle with events in the cell-cell adhesion process. Therefore, the mechanism of cell-to-cell transmission depends on specific interactions between cellular and viral proteins [57]. Thus far, two types of cell-cell contacts have been described to be critical for HTLV-1 transmission, tight cell-cell contacts (see Section 4.1) and cellular conduits (see Section 4.2). For transmission at tight cell-cell contacts, two non-exclusive mechanisms of virus transmission at the virological synapse (VS) have been proposed: (1) polarized budding of HTLV-1 into synaptic clefts (see Section 4.1.1), and (2) cell surface transfer of viral biofilms (see Section 4.1.2). Thus far, the mechanism of HTLV-1 transmission via cellular conduits induced by the viral p8 protein is unclear (see Section 4.2) [58,59]. Independent of the route of HTLV-1 transmission, viral particles are transmitted in confined areas protected from the immune response of the host. Beyond, cytoskeletal remodeling and cell-cell contacts are a prerequisite for all routes of virus transmission as interference with both actin and tubulin polymerization strongly reduces HTLV-1 transmission [57,60].

*4.1. Transmission at Tight Cell-Cell Contacts*

4.1.1. Polarized Budding at the Virological Synapse (VS)

Imaging analysis revealed that HTLV-1 is transmitted from cell-to-cell at the so-called virological synapse (VS; Figure 1) [60]. The VS is defined as a "virus-induced, specialized area of cell-cell contact that promotes the directed transmission of the virus between cells" [61].

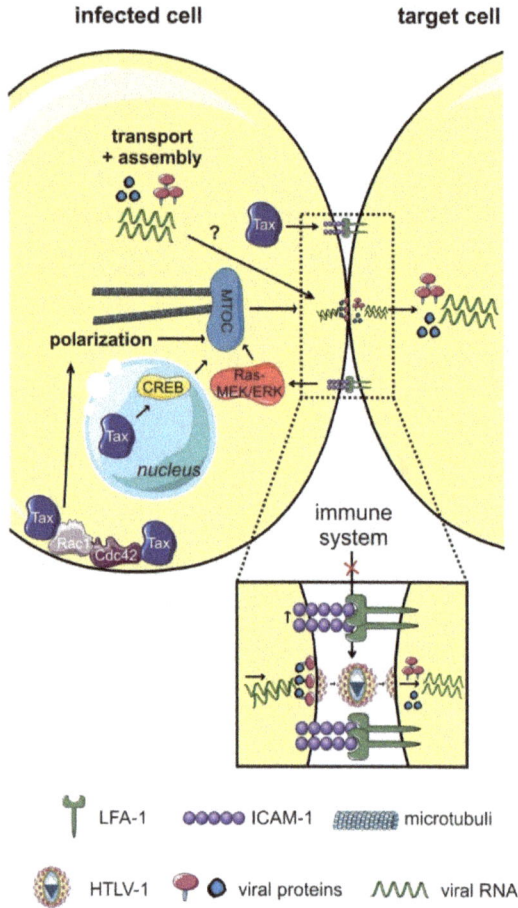

**Figure 1.** The virological synapse (VS). Interactions of intercellular adhesion molecule 1 (ICAM-1; on HTLV-1-infected T-cells) with lymphocyte function-associated antigen (LFA-1; on target cells), and signals induced by the viral Tax protein trigger polarization of the microtubule organizing center (MTOC) towards the cell-cell contact and formation of the VS at the cell-cell contact. Tax is not only located in the nucleus, but also at the MTOC and in the cell-cell contact region. Tax-induced CREB signaling (nuclear activity of Tax), the accumulation of Tax at the MTOC, and ICAM-1-induced Ras/MEK/ERK signaling are important for MTOC polarization. It is assumed that the VS allows for efficient polarized budding and virus transmission via a synaptic cleft, thus, avoiding recognition of HTLV-1 by the host immune system. Figure was realized thanks to *Servier Medical Art*.

Igakura *et al.* found that HTLV-1 Gag p19, Env, Gag p15 (nucleocapsid, important for incorporation of the viral RNA into the particle), and viral genomes accumulate at the interface between primary HTLV-1-infected and uninfected T-cells, followed by viral transfer to the uninfected cell [60]. This transfer was accompanied by polarization of the microtubule organizing center (MTOC) inside the infected cell towards the target cell. The cytoskeletal protein talin, which is important for cell adhesion, also accumulated at this specialized cell-cell contact, and inhibition of actin and tubulin polymerization diminished MTOC polarization [60]. The VS is distinct from the immunologic synapse (IS): contrary to the IS, where the cytoskeleton of the target cell polarizes towards the cell-cell contact, at the VS, the polarization of the cytoskeleton occurs inside the infected cell towards the target cell [61]. MTOC polarization and formation of the VS require at least two signals, one provided by the viral Tax protein, the other provided by the cell-cell contact as follows: (1) The presence of Tax located at the MTOC region and the ability of Tax located in the nucleus to stimulate CREB-dependent signaling pathways; and (2) cross-linking of intercellular adhesion molecule 1 (ICAM-1) at the cell-cell contact [62,63]. ICAM-1 binds to LFA-1 (lymphocyte function-associated antigen 1) on uninfected cells [63,64] at the site of the cell-cell contact, and this interaction could contribute to the preferred tropism of HTLV-1 for CD4$^+$ T-cells. Use of specific inhibitors revealed that the small GTPases Rac1 and Cdc42 are important for MTOC redistribution [63]. Electron tomography detected that cell membranes of infected and target cells are closely apposed at the VS, but interrupted by clefts. Gag-positive particles were detected inside the synaptic cleft, which resembled virions in size and morphology [65], suggesting that virions are transferred across this cleft to target cells. However, it is still questionable whether these particles were indeed infectious since no Env was detected at the surface of these particles [65].

Summed up, formation of the VS requires Tax to enhance expression of adhesion proteins (ICAM-1) in an HTLV-1-infected T-cell in contact with an uninfected T-cell [60]. After engagement of ICAM-1 on the infected T-cell and LFA-1 on uninfected T-cells, reorganization of the cytoskeleton in the infected cell occurs. Concomitant with polarization of the MTOC adjacent to the VS, viral proteins are concentrated in the center of the VS and surrounded by an outer ring of adhesion proteins [60]. Thereafter, it is assumed that viral particles are assembled and acquire the viral Env as they bud from the infected cell into the synaptic cleft. Upon induction and binding of the HTLV-1 receptor on the uninfected cell, viral particles cross the VS and enter the uninfected cell [61,66].

Interestingly, polarized assembly and transmission at the VS has also been described for other retroviruses like HIV and MLV [67–69]. Contrary to HTLV-1, both HIV and MLV can also spread cell-free. However, viral transmission under conditions of direct cell-cell contact is much more efficient [67,69]. Yet, the quantitative contribution of transmission via the VS for retroviral spread remains to be determined due to the lack of specific inhibitors of polarized budding processes. Taken together, transmission via the VS allows directed transmission of HTLV-1 to target cells whilst avoiding recognition by the host's immune response.

### 4.1.2. Transmission of Viral Biofilms at the VS

After infection of a host, microbes have evolved many issues to be protected from the host immune system. Bacteria developed an important way to hide from the immune system and to spread inside of the host by producing an extracellular biofilm, where bacteria are concentrated outside of infected cells. These distinct environments produced by the microbes themselves are rich in polysaccharides and carbohydrates [70]. Interestingly, biofilms have also been detected on cells infected with HTLV-1 and hence, were named "viral biofilms" (Figure 2) [71]. Biofilm-like, extracellular viral assemblies are composed of extracellular matrix (ECM) components and cellular lectins. In viral biofilms, virions are concentrated in a confined protective environment on the surface of infected cells and are transmitted to target cells at "virological synapses" [71]. HTLV-1 virions and clusters of viral proteins (Gag, Env) are accumulated in this specialized ECM on the surface of cells from infected patients and of chronically-infected cell lines. The biofilm is composed of carbohydrates, components of the ECM like collagen that form tight extracellular matrices, and the HSPG agrin [71]. Additionally, linker proteins (galectin-3, tetherin) [71], and O-glycosylated surface receptors (CD43, CD45) are part of the viral biofilm [72]. Tetherin, which was identified as an antiviral factor, prevents cell-free release of viruses from infected cells, maybe playing a role in the retention of HTLV-1 at the surface of infected cells [73].

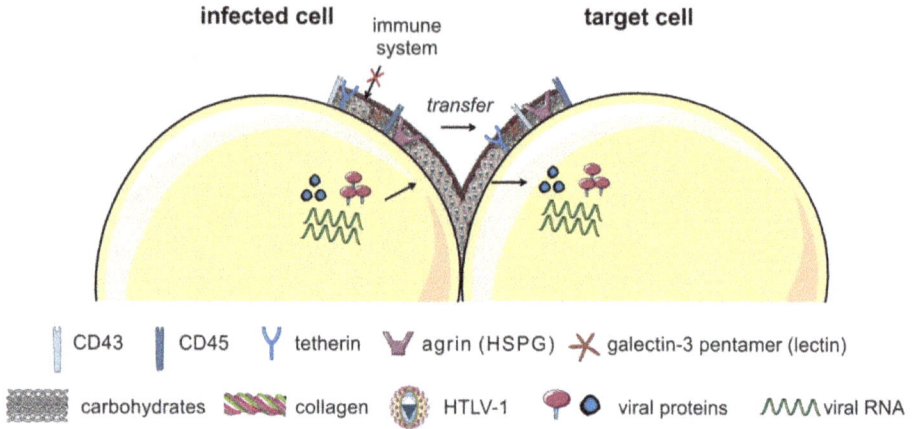

**Figure 2.** The viral biofilm. HTLV-1 virions are accumulated in a specialized extracellular matrix (ECM), the so-called viral biofilm, on the surface of infected cells. The viral biofilm is composed of carbohydrates, components of the ECM (collagen, agrin), linker proteins (galectin-3, tetherin), and O-glycosylated surface receptors (CD43, CD45). HTLV-1 particles are concentrated into large, highly infectious assemblies that cluster towards the cell-cell contact. HTLV-1 is transferred to target cells and guarded by the biofilm from immune recognition. Figure was realized thanks to *Servier Medical Art*.

HTLV-1 particles are assembled into large, highly infectious clusters and transferred to neighboring cells while being guarded by the biofilm from immune recognition [74]. Treatment with heparin or extensive pipetting removed the viral biofilm and strongly impaired the efficiency of HTLV-1 spreading to target cells by 80% [71], concluding that the viral biofilm is the major contributor of T-cell-associated infectivity. However, the involvement of polarized budding in biofilm formation is not excluded. Compared to the observations at the VS before [60], viral biofilms overlap cell-cell contacts and bridge the gap between both cell surfaces, rather than filling contact sites [71]. Thus, HTLV-1 transmission may not only occur across synaptic clefts, but also at the periphery of the cell contact [61]. The biofilm might also function as viral reservoir as viruses are highly concentrated within these biofilms in close proximity to their target cells. Additionally, cell-free preparations of viral biofilms infect monocyte-derived DC (MDDC) more efficiently than autologous CD4$^+$ T-lymphocytes *in vitro* [50]. The viral biofilm could also both provide a physical protection for the viral Env protein [50,75] and prevent recognition of Env by neutralizing host antibodies [76]. It is assumed, that after infection of new cells, viruses reprogram the protein expression of the host, amongst others, to form the viral biofilms [71,76]. Yet, the relative contribution of individual viral proteins to biofilm formation is not settled [74].

Both MLV and HIV also utilize virus-laden uropods for viral spreading at the VS [77,78]. Briefly, polarization of lymphocytes involves the formation of two distinct poles: (1) the leading edge, which attaches the cell to the substrate allowing directional movement of the cell; and, on the opposite side, (2) the uropod, which is mostly involved in cell-cell interactions [79]. The current model suggests that an infected cell will likely engage target cells to form virological synapses if uropods make the initial contact with the target cell [78]. Uropods contain adhesion molecules, Env-laden virions, and adhere to the receptor-expressing target cells, while the leading edge continues to drive cellular polarization of the migrating cells. Contrary, if the leading edge of a migrating lymphocyte makes the initial contact with a target cell, the leading edge will continue to migrate and bypass the target cells [77,78]. Since HTLV biofilms are found as one large or several smaller clusters of viruses bound to the uropod on isolated infected T-cells [71], the uropod might also participate in the formation of the VS during transmission of HTLV-1.

## 4.2. Transmission via Cellular Conduits

To allow for transmission of HTLV-1 over long distances, the transfer of virions via cellular conduits induced by the viral p8 protein has been proposed (Figure 3; for details on p8 see Section 6.5). Briefly, p8 is encoded by the open reading frame I of HTLV-1 located in the pX region as a cleavage product of the precursor protein p12 [80]. In co-culture assays with HTLV-1 reporter cells, Van Prooyen and colleagues found that overexpression of p8 rescues the infectivity of p12 knockout molecular clones, and enhances the infectivity of chronically-infected MT-2 cells [58]. Functionally, p8 increased T-cell conjugate formation, potentially through LFA-1 clustering on the surface of T-cells. Surprisingly, overexpression of p8 also enhanced the number and length of cellular conduits among T-cells [58]. Conduits are supposed to be formed by directed outgrowth of a filopodium-like protrusion towards a neighboring cell. In co-cultures between p8-expressing Jurkat T-cells and untransfected Jurkat T-cells, p8 was also detectable in untransfected cells, suggesting transfer of p8 via the conduits. The latter was corroborated by life-cell imaging, which detected fluorescently-labeled Gag and p8 in conduits between chronically-infected T-cells and uninfected target cells. However, it is not known, whether p8 and LFA-1 also cluster at the tip of the conduit, or only at the surface of the infected cell. Finally, transmission electron microscopy showed the presence of viral particles resembling HTLV-1 virions in shape and morphology either at the contact sites between two conduits, or between a conduit and a target T-cell [58]. The authors proposed the model that p8 enhances transmission of HTLV-1 by increasing cellular conduits and polysynapse formation (Figure 3) [58,81].

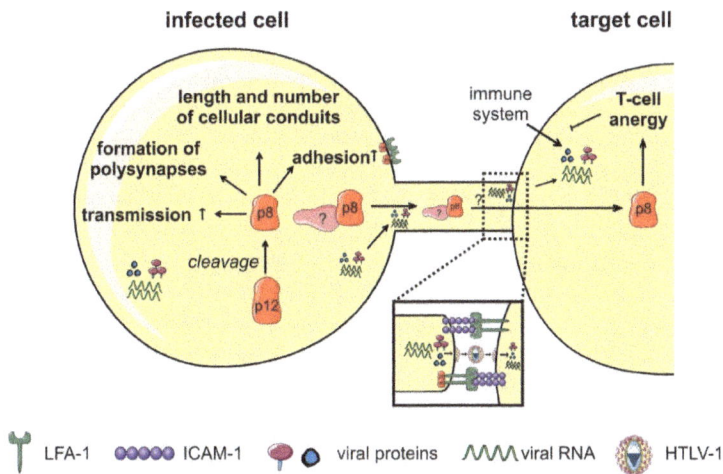

**Figure 3.** Cellular conduits. The viral accessory protein p12 is proteolytically cleaved into the p8 protein, which increases adhesion of T-cells through lymphocyte function-associated antigen-1 (LFA-1) clustering. Further, p8 induces polysynapse formation and enhances the number and length of cellular conduits between T-cells, thereby, enhancing HTLV-1-transmission. p8 is transferred to target cells through these conduits and it is hypothesized to induce T-cell anergy in the target cell. This might be a strategy for HTLV-1 to evade the host's immune surveillance during infection. Host cell proteins that interact with p8 to enhance conduit formation, p8 transfer, and HTLV-1 transmission are still unknown. Figure was realized thanks to *Servier Medical Art*.

In parallel, p8 is transferred to neighboring cells, invades target cells and is suggested to induce T-cell anergy by decreasing T-cell receptor (TCR) signaling in target cells, which could favor persistence of HTLV-1 in an immune competent host [58,81]. Taken together, p8-induced virus transmission seems to be a strategy of the virus to be transmitted via long distances. The presence of viral particles at the contact site between conduits and target cells leads to the assumption that HTLV-1 buds from the tip of the conduit towards the target cell via a "mini VS" [58,59]. However, it is not known whether p8 and LFA-1 also cluster at the tip of the conduit, or only at the surface of the infected cell. Formation of a VS between conduit and target cell suggests protected transfer of HTLV between cells and is in contrast to transmission of the related retroviruses HIV and MLV, where isolated viral particles were shown to surf on filopodial bridges before reaching the target cell [82,83]. For HTLV-1, surfing of isolated viral particles has not been observed yet [58]. The detailed molecular mechanism by which p8 promotes HTLV-1 transmission remains unknown. It is conceivable that cellular conduits account for HTLV-1 transmission, as suggested by the authors [58]. Nevertheless, it cannot be excluded that transfer occurs via virological synapses, polysynapses, syncytia, or viralbiofilms [59].

## 5. Cell-Free HTLV-1 Transmission to Dendritic Cells (DC)

Antigen-presenting DC and their precursor cells (monocytes) are found to be infected with HTLV-1 *in vivo* [28–30]. However, it is not clear whether DC play a role in establishing a chronic HTLV-1 infection. DC either capture virions and transfer them to target cells (*trans*-infection), or they are productively infected and infect other cells themselves (*cis*-infection) (Figure 4) [39,84]. The lectin DC-specific ICAM-3-grabbing nonintegrin (DC-SIGN) facilitates HTLV-1 binding and fusion of DC through an ICAM-dependent mechanism [84,85]. During HIV-transmission, most features previously associated with DC-SIGN-mediated trans-infection of DC are apparently fulfilled by CD169/Siglec-1 [86], whose role remains to be elucidated for HTLV-1.

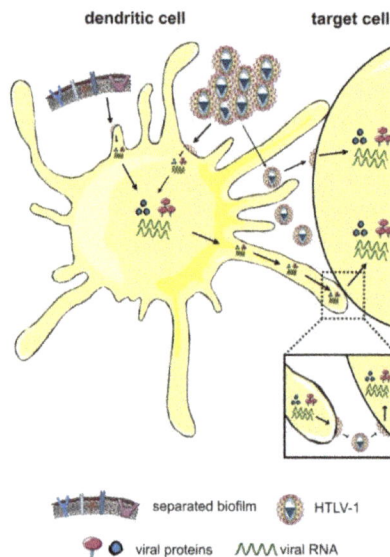

**Figure 4.** Transmission of HTLV-1 via dendritic cells (DC). DC either capture the virus and transmit it to target cells in the absence of infection (*trans*-infection), or they are productively-infected before viral transmission (*cis*-infection). Productive cell-free infection of DC is achieved *in vitro* by highly-concentrated preparations of cell-free HTLV-1 or by viral biofilms. Figure was realized thanks to *Servier Medical Art*.

*In vitro* studies have shown that DC can also be infected cell-free with highly-concentrated viral supernatants, and these infected DC mediate efficient cell-cell contact-dependent infection and transformation of CD4+ T-cells [84]. These findings and studies reporting the presence of viral genomes and proteins suggest a potential role of DC in transmission *in vivo* during initial acquisition of infection [61,84]. Interestingly, DC can be infected cell-free via transcytosis through an epithelial barrier [51]. The relevance of DC in viral transmission has been further strengthened by recent findings showing that MDDC are more susceptible to infection with viral biofilms than autologous CD4+ T-lymphocytes *in vitro*, which supports the model that infection of DC might be an important step during primary infection *in vivo* [50]. Searching the mechanism, Alais *et al.* found that MDDC express higher amounts of NRP-1 [50], which is part of the HTLV-1 entry receptor [16]. The study also revealed that infection of DC with virus-containing biofilm is much more efficient than infection with concentrated viral supernatants [50]. Thus far, it is not settled whether formation of the viral biofilm is restricted to lymphocytes, or whether it could also be formed upon DC-infection, or whether it could be transmitted via DC-mediated *trans*-infection to other cells. Moreover, infection of DC may also be required for the establishment and maintenance of HTLV-1 infection in primate species [87]. Since the maturation of DC is impaired in HTLV-1-infected patients [88,89], DC may not only contribute to viral dissemination, but also to immune dysregulation observed in HTLV-1-infected patients. Compared to HTLV-1, productive (*cis*) infection of DC with HIV is inefficient due to antiviral mechanisms like the presence of the restriction factor SAMHD1 in DC [86]. Infection of CD4+ T-cells occurs in *trans* by DC-captured HIV at the VS [67,90]. It is likely that also HTLV-1 is transmitted from DC to T-cells via polarized budding at the VS, but this has to be verified experimentally.

## 6. Viral Proteins Enhancing HTLV-1 Transmission

Amongst the HTLV-1-encoded proteins contributing to HTLV-1 transmission, we briefly sum up the roles of the structural proteins Env, Gag, and the regulatory protein Rex before we focus on Tax, which is important for formation of the VS (Figure 1) [60], and on p8, which enhances the number of cellular conduits between infected and uninfected T-cells (Figure 3) [58].

### 6.1. Env

Env plays a central role in HTLV-1 cell-to-cell transmission (for review, see [17,40,91–93] since Env is crucial for HTLV-1 infectivity. Briefly, Env encodes two different proteins, the transmembrane (TM) and the surface (SU) protein. The precursor protein of Env is highly glycosylated, proteolytically cleaved into SU and TM proteins, and afterwards transported to the cell membrane to initiate virus assembly and budding [17,92]. The SU subunit of Env binds to the host cell surface receptors Glut-1, NRP-1, and to HSPGs to trigger fusion of the membranes both of the virus and the host cell [40]. Env is also important for formation of the VS [66] and for transmission of HTLV-1 *in vitro* and *in vivo* [56,57].

### 6.2. Gag

The HTLV-1 group specific antigen (Gag, p55) is produced as a single precursor polyprotein. Upon posttranslational modification and myristoylation, the Gag polyprotein is targeted to the inner membrane of the cellular plasma membrane [91]. Subsequently, Gag is cleaved by viral proteases into its functional domains matrix (MA, p19), capsid (CA, p24), and nucleocapsid (NC, p15). Matrix is important for Gag targeting, membrane binding, and Env incorporation, while capsid interacts with itself to form the inner core of the virion. Nucleocapsid interacts with the genomic RNA inside the inner core of the virion. A proper spatial and temporal regulation of viral assembly and budding is crucial for HTLV-1 transmission [91,94].

### 6.3. Rex

Among the regulatory proteins, not only Tax, but also Rex is important for viral transmission. This is corroborated by at least two findings: (1) Use of a Rex-deficient HTLV-1 proviral clone showed that Rex is important for viral transmission *in vivo* [91]; (2) The chronically HTLV-1-infected T-cell line C8166-45, which is Rex-deficient, does not produce viral particles, and is not infectious [95]. Taken together, these results suggest that Rex's function to enhance trafficking of unspliced and single spliced RNA is important for ideal viral spread [91].

*6.4. Tax*

The regulatory protein Tax is essential for viral replication due to strong enhancement of viral mRNA synthesis by transactivating the HTLV-1 LTR (U3R) promoter. Further, Tax is a potent transactivator of cellular transcription and important for initiating oncogenic transformation. Tax shuttles between the nucleus and the cytoplasm and fulfills most of its functions by direct protein-protein interactions [6,19,96,97]. Thus far, not only a plethora of Tax interaction partners [98–100], but also of transcriptionally-induced Tax target genes has been identified [101–105]. The latter is attributed to Tax's function as activator of several signaling pathways including NF-κB, CREB, SRF, PI3K/AKT, and AP-1 [19,106].

Tax is important for HTLV-1 cell-to-cell transmission. First insights were obtained by fluorescent imaging analysis showing that Tax cooperates with ICAM-1 thereby inducing polarization of the MTOC at the VS (Figure 1) [63]. Use of Tax mutants revealed that Tax-induced CREB signaling is critical for MTOC polarization [62]. Interestingly, ICAM-1 is also induced by Tax on the surface of T-cells [107], thus, facilitating the formation of the VS and HTLV-1 transmission. Since engagement of ICAM-1 by interaction with its ligand LFA-1 on target T-cells is important for formation of the VS, Tax-induced ICAM-1 expression may also contribute to the T-cell tropism of HTLV-1 [61]. Use of chemical inhibitors revealed that activity of the small GTPases Cdc42 and Rac1 is critical for Tax-induced MTOC polarization [63]. Since Tax also complexes with these GTPases, Tax might connect Rho GTPases to their targets and affect cytoskeleton organization to favor HTLV-1 transmission [98,99].

Imaging-based methods were pioneering in defining the routes of viral transmission and identifying the localization of viral and cellular proteins involved in transmission. Later, Mazurov *et al.* developed an elegant single-cycle replication-dependent reporter system that allows quantitative evaluation of cell-to-cell transmission by measuring reporter gene expression in newly infected cells [57]. This system requires transient transfection of (1) plasmids carrying a replication-dependent reporter gene; and of (2) virus packaging plasmids. The packaging plasmids encode full-length HTLV-1, or they carry a deletion in the *env* gene and are pseudotyped with VSV-G (glycoprotein G of vesicular stomatitis virus). The reporter plasmids consist of a CMV-driven reporter gene in antisense orientation that is interrupted by a gamma-globin intron in sense orientation. After transcription, the intron is spliced, but the antisense orientation of the reporter gene precludes translation of the reporter mRNAs in transfected cells. These minus strand RNAs are packaged into virions. After infection of new cells, mRNAs are reversely transcribed and reporter gene activity is detectable [57]. Using this system, the authors found that both the cell type and the envelope type are critical for HTLV-1 cell-to-cell transmission: In co-cultures of transfected Jurkat T-cells with Raji/CD4+ B-cells, Tax enhanced transmission of HTLV-1 packaged with wildtype Env, but not with HTLV-1 packaged with VSV-G. [57]. On the contrary, the transmission of HTLV-1 reporter vectors in transfected 293T cells was not enhanced by Tax, suggesting that different host factors involved in transmission are induced by Tax in Jurkat T-cells than in 293T cells, possibly due to different signaling pathways being active in the respective cell type. Tax also enhanced cell-to-cell transmission of HIV reporter vectors, suggesting that Tax-induced changes in the infected donor cell are also beneficial for other retroviruses than HTLV-1 [57]. One obstacle when working with these reporter vectors was the lack of sufficient reporter signals in PBMC [57]. Recently, Mazurov and colleagues improved the reporter vectors by modifying the splice sites, and by enhancing packaging efficiency of spliced reporter vectors [108]. It will be interesting to see, which Tax-induced signaling pathways and host factors are required for viral transmission to PBMC.

With regard to pathways important for viral transmission, Tax transcriptionally alters the expression of cell adhesion and surface molecules [109], leading to cytoskeletal remodeling, and complexes with proteins involved in cytoskeleton structure and dynamics [99]. Table 1 lists host factors that are important for HTLV-1-transmission, amongst them are also interaction partners and transcriptional targets of Tax. Despite the knowledge of various Tax-targets involved in cell-cell interaction, adhesion and cytoskeletal organization, a comprehensive analysis evaluating the role of known and new Tax effectors on virus transmission is still lacking. Moreover, it is still not settled whether blocking Tax-induced pathways important for MTOC polarization also impairs cell-to-cell transmission of HTLV-1 reporter vectors.

**Table 1.** Host cell proteins important for HTLV-1 transmission.

| Host Cell Factor | Other Name; Protein Function | Function in Transmission | Modulation by Viral Protein | Reference |
|---|---|---|---|---|
| *Cell-Surface Associated Proteins* | | | | |
| Agrin | HSPG; cross-linker of cell surface receptors | biofilm formation | | [71] |
| CCL22 | chemokine ligand 22; binding to CCR4 | attraction of CCR4⁺ T-cells | induced by Tax | [110] |
| CCR4 | C-C chemokine receptor type 4 | on target cell; attracted by CCL22 (from infected cell) | | [110] |
| CD43 | leukosialin; sialophorin | adhesion; biofilm formation | | [72] |
| CD45 | protein-tyrosine phosphatase | adhesion; biofilm formation | | [72] |
| CD82 | Tetraspanin | inhibits syncytium formation | interacts with Gag and Env | [111,112] |
| Collagen | structural protein of ECM | biofilm formation | induced by Tax (collagen 1 alpha) | [71,113] |
| DC-SIGN | DC-specific ICAM-3-grabbing nonintegrin | syncytium formation (on target cell DC) | | [85] |
| GLUT-1 | glucose transporter 1 | virus entry | interacts with Env | [14] |
| Hsc70 | heat shock cognate protein 70 | syncytium formation (on target cell) | interacts with Env | [114] |
| HSPGs | heparan sulfate proteoglycans | virus entry | interact with Env | [16] |
| ICAM-1 | intercellular adhesion molecule 1; CD54 | VS formation; MTOC polarization; syncytium formation | induced by Tax | [60,62,107,115] |
| ICAM-3 | intercellular adhesion molecule 3 | syncytium formation | | [115] |
| Integrin β2/7 | CD18 | syncytium formation | | [115] |
| LFA-1 | lymphocyte function-associated antigen 1 | VS formation (target cell); adhesion (infected cell) | interacts with p8, p12 (infected cell) | [58,60,116] |
| NRP-1 | neuropilin-1 | virus entry | interacts with Env | [16] |
| SDC-1, SDC-2 | Syndecan-1/-2; transmembrane HSPGs | virus entry | | [117] |
| Talin | actin-anchor protein; clusters with LFA-1 | VS formation | | [60] |
| Tetherin | BST2: bone marrow stromal antigen 2; lipid raft associated protein | biofilm formation; virus attachment | | [71,73] |
| VCAM-1 | vascular cell adhesion molecule 1 | syncytium formation (on target cell) | induced by Tax (on infected cell) | [115,118,119] |
| *Cytoskeleton and Associated Factors* | | | | |
| Actin | structural protein | cytoskeleton remodeling; MTOC polarization; virus release | interacts with Tax | [57,63,98,99] |
| Cdc42 | cell division cycle 42; small GTPase | MTOC polarization | interacts with Tax | [63,98] |
| CRMP2 | collapsin response mediator protein 2 | migration, role in transmission unclear | induced by Tax | [120] |
| FSCN-1 | Fascin; actin-bundling protein | invasive migration; cytoskeleton remodeling; cell-to-cell transmission under investigation | induced by Tax | [121–123] |
| *Cytoskeleton and Associated Factors* | | | | |
| GEM | GTP-binding mitogen-induced T-cell protein | cytoskeleton remodeling; migration; conjugate formation | induced by Tax | [124] |
| Rac1 | Ras-related C3 botulinum toxin substrate 1; small GTPase | MTOC polarization | interacts with Tax | [63,98] |
| Tubulin | component of microtubule | cytoskeleton remodelling; MTOC polarization | | [57,63] |
| γ-Tubulin | component of centrosomes and spindle pole bodies | cytoskeleton remodelling; MTOC polarization | interacts with Tax | [60,63,99,125] |
| *Signaling Pathways and Associated Factors* | | | | |
| CREB | cAMP response element-binding protein | MTOC polarization | interacts with Tax | [62,126] |
| Jak/Stat | Janus kinase/signal transducer and activator of transcription | syncytium formation | | [127] |
| Ras-Raf-MEK-ERK | rat sarcoma/rat fibrosarcoma/mitogen-activated protein kinase/ERK kinase/extracellular-signal-regulated kinase | MTOC polarization | | [62] |
| *Other Proteins* | | | | |
| Dlg | disks large homolog | cell-to-cell fusion | interacts with Tax and Env | [128,129] |
| Galectin-3 | beta-galactoside-binding lectin, linker protein | biofilm formation | induced by Tax | [71,130] |

cAMP: cyclic adenosine monophosphate; CD: cluster of differentiation; DC: dendritic cell; Env: envelope protein of HTLV-1; ECM: extracellular matrix; GTP: guanosine-5′-triphosphate; MTOC: microtubule organizing center; VLP: virus-like particle; VS: virological synapse.

*6.5. p8*

The HTLV-1 p8 protein is a cleavage product of the viral accessory p12 protein encoded from the open reading frame I. The precursor protein p12 normally localizes to the endoplasmatic reticulum (ER) and to the golgi apparatus, and its functions have been reviewed earlier [21]. p12 is post-translationally modified by a two-step proteolytic cleavage: the first cleavage between amino acid (aa) 9/10 removes an ER-retention signal, which allows trafficking of the protein to the golgi. The second cleavage occurs between aa 29/30 resulting in the p8 protein [80]. p8 is a 70 aa comprising protein that localizes to the cytoplasm and is recruited to lipid rafts and the IS upon TCR ligation [131]. p8 enhances LFA-1-mediated cell adhesion on ICAM-1-coated plates [58]. Earlier work had attributed this function to p12-induced calcium-signaling and suggested that p12 could promote formation of the VS [116] until it became clear that p12 is processed to p8 [58,80]. It has been proposed that p8 enhances HTLV-1 transmission by increasing the number and length of cellular conduits among T-cells (see Figure 3 and Section 4.2). p8-enhanced polysynapse formation and virus transmission from HTLV-1-infected cells to uninfected T-cells [58] had previously been attributed to the precursor p12 [132,133]. Since p8 is also transferred to neighboring cells, invades target cells, and can induce T-cell anergy, it is proposed that p8 favors persistence of HTLV-1 in an immune competent host [58].

Both p8 and p12 form disulfide-linked dimers, and only the monomeric forms of p8 and p12 are palmitoylated at a conserved cysteine residue (C39). Albeit mutation of C39 to alanine abrogates dimerization and palmitoylation, these modifications are dispensable for p8 to increase adhesion and viral transmission [134]. *In vivo* studies in macaques support the notion that p8 and p12 are important for viral persistence and spread. Moreover, productive infection of monocytes depends on the expression of p8 and p12 proteins [87,135]. Cellular effectors and interaction partners of p8 other than LFA-1 that mediate conduit formation, p8-transfer, and viral transmission are still unknown. Interaction partners of p12 have been identified (reviewed by [21,91]), but none of them has been evaluated for a role in p8 transfer and viral transmission. Therefore, the composition of the host machinery that mediates transfer of p8 and HTLV-1 to the target cell remains to be determined.

## 7. Host Factors Involved in HTLV-1 Transmission

HTLV-1 has evolved strategies to manipulate the host cell for its transmission. Not only protein-protein interactions between viral and cellular proteins, but also specific transcriptional induction of host cell factors might facilitate viral transmission. Table 1 lists host proteins, that are involved in HTLV-1 transmission and, if indicated, their manipulation by HTLV-1-encoded proteins. For the sake of completeness, the table also lists proteins which are important for viral entry and syncytium formation.

*7.1. Cell Surface Receptors and Cell-Cell Contacts*

Since cell-cell contacts are a prerequisite for efficient HTLV-1 transmission, it is reasonable that cell surface receptors are critical for this step. Not only receptors on the target cells—like components of the HTLV-1 receptor (Glut-1, NRP-1, HSPGs, SDC-1/-2)—are important for viral transmission and tropism [18,117], but also secreted chemokines that could attract target cells. To attract CCR4[+]CD4[+] target T-cells, Tax expressing HTLV-1-infected T-cells produce large amounts of CCL22. Expression of CCL2 is stimulated by Tax and block of CCL22 using anti-CCL22 antibodies reduces viral transmission from HTLV-1-infected cells to CD4[+] T-cells [110].

Although a plethora of surface receptors is upregulated in HTLV-1-infected cells [109], only few of them play a role in virus transmission (Table 1). HTLV-1-induced syncytium formation is affected by Tax, and receptors like vascular cell adhesion molecule 1 (VCAM-1) or ICAM-1 have been shown to promote syncytium formation, and to be inducible by Tax [92,115,118,119,136]. For details about receptors being important for viral entry or syncytium formation, see [17,18,92].

The viral biofilm on the surface of infected cells contains clusters of virions in a cocoon-like structure, and its composition is shown in Figure 2. Thus far, it is not known in detail, whether individual viral proteins are important for biofilm formation. A study by Mazurov *et al.* indicates that large aggregates of HTLV-1 assemblies are more infectious than multiple clustered virions on the surface of infected cells [72]. Their data suggest that heavily O-glycosylated surface receptors CD43 and CD45 render cells less adhesive and prevent inappropriate cell-cell contacts and thus, favor the assembly of HTLV-1 particles into large, highly infectious structures on the surface of T-cells. The authors conclude that a balance between pro- and anti-adhesive molecules on the surface of the infected T-cell is important for the establishment of the VS and virus transmission [72].

### 7.2. Components and Regulators of the Cytoskeleton

Transmission of HTLV-1 and formation of the VS strongly depends on the functional integrity of the cytoskeleton [61]. Experiments using single-cycle replication dependent HTLV-1 reporter vectors confirmed these findings and showed that block of actin and tubulin polymerization strongly reduces HTLV-1 cell-to-cell transmission while transmission of HIV was only modestly impaired [57]. Beyond, Rho GTPases Rac1 and Cdc42, interaction partners of Tax, are involved in MTOC polarization at the VS [63,98]. However, a quantitative comparison of the contribution of individual cytoskeletal proteins and associated regulatory proteins on viral transmission has never been performed.

Host factors regulating cellular migration, invasion and conjugate formation could also be involved in HTLV-1 cell-to-cell transmission by favoring dissemination of infected cells *in vivo* (Figure 5). Among proteins enhancing cellular migration (Figure 5A), the Tax-induced small GTP-binding protein GEM plays an important role in HTLV-1 cell-to-cell transmission [124]. GEM is expressed in HTLV-1-infected T-cell lines and Tax regulates GEM transcription by recruiting CREB and CREB-binding protein (CBP) to the GEM-promoter. Interestingly, GEM is also important for conjugate formation between infected and uninfected T-cells (Figure 5B), which may explain its role in cell-to-cell transmission [124]. However, it is unknown whether GEM and other targets of Tax are required for formation of the VS. The semaphorin-signaling transducer collapsin response mediator protein 2 (CRMP2) has originally been identified in the nervous system where it mediates growth cone navigation induced by semaphorin 3A. Beyond, the phosphoprotein CRMP2 is also involved in cytoskeleton rearrangement controlling migration of human lymphocytes [137]. Activity of CRMP2 is modulated by Tax and correlates with migration of infected cells [120]. It is likely that CRMP2 plays a role in dissemination of infected cells *in vivo* and could thus enhance the probability to transmit viruses to uninfected cells. The actin-bundling protein Fascin is a tumor marker that is highly upregulated in many types of cancer and crucial for invasion and metastasis. We found that Fascin is also important for invasive migration of HTLV-1-infected cells [121]. Fascin is upregulated in chronically HTLV-1-infected T-cells and regulated by Tax through NF-κB signaling [121,123]. Interestingly, CRMP2 and Fascin function downstream of Rho kinases while GEM is an upstream negative regulator of ROCK-I Rho kinase [124]. Currently, we are investigating the role of Fascin in cell-to-cell transmission [122].

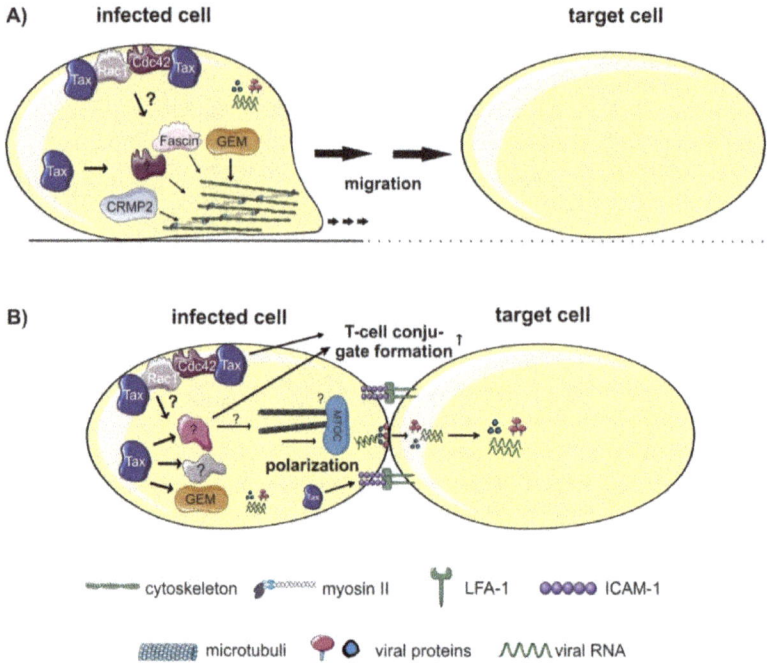

**Figure 5.** Host factors regulating cellular migration, invasion and conjugate formation. (**A**) Proteins enhancing cellular migration and/or invasion of HTLV-1-infected cells could favor dissemination of HTLV-1 to target cells. Expression of the Tax-induced small GTP-binding protein GEM enhances both migration of HTLV-1-infected cells and viral transmission. Activity of CRMP2, a phosphoprotein involved in cytoskeleton rearrangement, is modulated by Tax and correlates with migration of infected cells. The actin-bundling protein Fascin is induced by Tax and important for invasive migration of HTLV-1-infected cells. A role of CRMP2 and Fascin for viral transmission remains to be determined. Both Rac-1 and Cdc42 are interaction partners of Tax that are crucial for migration and for MTOC polarization. (**B**) T-cell conjugate formation, a prerequisite for cell-to-cell transmission depends on components of the cytoskeleton like the Tax-inducible GEM protein, and on Rac1 and Cdc42. Additionally, Tax regulates expression of surface receptors (see Table 1), which are important for cell-cell contact formation, and, potentially, for formation of the VS and HTLV-1 transmission. The influence of different host factors on polarized budding and formation of the VS remains to be determined. Figure was realized thanks to *Servier Medical Art*.

## 7.3. Signaling Pathways

Tax is a potent activator of different cellular signaling pathways [19] including CREB, PI3K/AKT, SRF, and NF-κB. However, only little is known about the relative contribution of these signaling pathways on Tax-induced formation of the VS. Using different Tax-mutants, Nejmeddine *et al.* found that CREB signals are important for triggering MTOC polarization, while Ras/MAPK/ERK signals mediate ICAM-1-induced MTOC polarization [62]. Interestingly, expression of the small GTP-binding protein GEM, which has been shown to induce conjugate formation between infected and uninfected T-cells, is also dependent on Tax-induced CREB signaling [124]. However, it remains to be determined whether GEM is involved in MTOC polarization. The contribution of different signaling pathways to formation of the viral biofilm or to p8-induced conduits is not known. It is also not settled whether Jak signaling contributes to p8-mediated virus transmission as has been shown for its precursor p12 [132].

Overall, the quantitative contribution of individual signaling pathways on different mechanisms of viral transmission remains an open question.

## 8. Conclusions

HTLV-1 has evolved several clever strategies to transmit via specialized routes from cell-to-cell, thus being protected from immune recognition. Significant progress has been made in elucidating molecular mechanisms of HTLV-1 cell-to-cell transmission. Nonetheless, the relative contribution of individual pathways on transmission *in vivo* remains to be determined.

**Acknowledgments:** Our work is supported by Deutsche Forschungsgemeinschaft (DFG; SFB796, C6), and we acknowledge support by DFG and Friedrich-Alexander-Universität Erlangen-Nürnberg (FAU) within the funding programme Open Access Publishing. This article exemplifies several findings of HTLV-1 cell-to-cell transmission; we apologize to investigators whose contributions were not included. We are grateful to the reviewers for valuable comments. All figures were designed using the medical image bank *Servier Medical Art*, which is available under the Creative Commons license CC-BY.

**Author Contributions:** Christine Gross and Andrea K. Thoma-Kress wrote the paper.

**Conflicts of Interest:** The authors declare no conflict of interest. The founding sponsors had no role in the writing of the manuscript.

## References

1. Poiesz, B.J.; Ruscetti, F.W.; Gazdar, A.F.; Bunn, P.A.; Minna, J.D.; Gallo, R.C. Detection and isolation of type C retrovirus particles from fresh and cultured lymphocytes of a patient with cutaneous T-cell lymphoma. *Proc. Natl. Acad. Sci. USA* **1980**, *77*, 7415–7419. [CrossRef] [PubMed]

2. Yoshida, M.; Seiki, M.; Yamaguchi, K.; Takatsuki, K. Monoclonal integration of human T-cell leukemia provirus in all primary tumors of adult T-cell leukemia suggests causative role of human T-cell leukemia virus in the disease. *Proc. Natl. Acad. Sci. USA* **1984**, *81*, 2534–2537. [CrossRef] [PubMed]

3. Yoshida, M.; Miyoshi, I.; Hinuma, Y. Isolation and characterization of retrovirus from cell lines of human adult T-cell leukemia and its implication in the disease. *Proc. Natl. Acad. Sci. USA* **1982**, *79*, 2031–2035. [CrossRef] [PubMed]

4. Osame, M.; Usuku, K.; Izumo, S.; Ijichi, N.; Amitani, H.; Igata, A.; Matsumoto, M.; Tara, M. HTLV-I associated myelopathy, a new clinical entity. *Lancet* **1986**, *1*, 1031–1032. [CrossRef]

5. Gessain, A.; Barin, F.; Vernant, J.C.; Gout, O.; Maurs, L.; Calender, A.; de The, G. Antibodies to human T-lymphotropic virus type-I in patients with tropical spastic paraparesis. *Lancet* **1985**, *2*, 407–410. [CrossRef]

6. Matsuoka, M.; Jeang, K.T. Human T-cell leukemia virus type 1 (HTLV-1) and leukemic transformation: Viral infectivity, Tax, HBZ and therapy. *Oncogene* **2011**, *30*, 1379–1389. [CrossRef] [PubMed]

7. Matsuoka, M.; Jeang, K.T. Human T-cell leukaemia virus type 1 (HTLV-1) infectivity and cellular transformation. *Nat. Rev. Cancer* **2007**, *7*, 270–280. [CrossRef] [PubMed]

8. Yasunaga, J.; Matsuoka, M. Molecular mechanisms of HTLV-1 infection and pathogenesis. *Int. J. Hematol.* **2011**, *94*, 435–442. [CrossRef] [PubMed]

9. Hlela, C.; Shepperd, S.; Khumalo, N.P.; Taylor, G.P. The prevalence of human T-cell lymphotropic virus type 1 in the general population is unknown. *AIDS Rev.* **2009**, *11*, 205–214. [PubMed]

10. Gessain, A.; Cassar, O. Epidemiological aspects and world distribution of HTLV-1 infection. *Front Microbiol.* **2012**, *3*. [CrossRef] [PubMed]

11. Proietti, F.A.; Carneiro-Proietti, A.B.; Catalan-Soares, B.C.; Murphy, E.L. Global epidemiology of HTLV-I infection and associated diseases. *Oncogene* **2005**, *24*, 6058–6068. [CrossRef] [PubMed]

12. Paun, L.; Ispas, O.; del, M.A.; Chieco-Bianchi, L. HTLV-I in Romania. *Eur. J. Haematol.* **1994**, *52*, 117–118. [CrossRef] [PubMed]

13. Veelken, H.; Kohler, G.; Schneider, J.; Dierbach, H.; Mertelsmann, R.; Schaefer, H.E.; Lubbert, M. HTLV-I-associated adult T cell leukemia/lymphoma in two patients from Bucharest, Romania. *Leukemia* **1996**, *10*, 1366–1369. [PubMed]

14. Manel, N.; Kim, F.J.; Kinet, S.; Taylor, N.; Sitbon, M.; Battini, J.L. The ubiquitous glucose transporter GLUT-1 is a receptor for HTLV. *Cell* **2003**, *115*, 449–459. [CrossRef]

15. Jones, K.S.; Petrow-Sadowski, C.; Bertolette, D.C.; Huang, Y.; Ruscetti, F.W. Heparan sulfate proteoglycans mediate attachment and entry of human T-cell leukemia virus type 1 virions into CD4⁺ T cells. *J. Virol.* **2005**, *79*, 12692–12702. [CrossRef] [PubMed]

16. Lambert, S.; Bouttier, M.; Vassy, R.; Seigneuret, M.; Petrow-Sadowski, C.; Janvier, S.; Heveker, N.; Ruscetti, F.W.; Perret, G.; Jones, K.S.; *et al.* HTLV-1 uses HSPG and neuropilin-1 for entry by molecular mimicry of VEGF165. *Blood* **2009**, *113*, 5176–5185. [CrossRef] [PubMed]

17. Jones, K.S.; Lambert, S.; Bouttier, M.; Benit, L.; Ruscetti, F.W.; Hermine, O.; Pique, C. Molecular aspects of HTLV-1 entry: Functional domains of the HTLV-1 surface subunit (SU) and their relationships to the entry receptors. *Viruses* **2011**, *3*, 794–810. [CrossRef] [PubMed]

18. Ghez, D.; Lepelletier, Y.; Jones, K.S.; Pique, C.; Hermine, O. Current concepts regarding the HTLV-1 receptor complex. *Retrovirology* **2010**, *7*. [CrossRef] [PubMed]

19. Currer, R.; van Duyne, R.; Jaworski, E.; Guendel, I.; Sampey, G.; Das, R.; Narayanan, A.; Kashanchi, F. HTLV tax: A fascinating multifunctional co-regulator of viral and cellular pathways. *Front Microbiol.* **2012**, *3*. [CrossRef] [PubMed]

20. Kashanchi, F.; Brady, J.N. Transcriptional and post-transcriptional gene regulation of HTLV-1. *Oncogene* **2005**, *24*, 5938–5951. [CrossRef] [PubMed]

21. Edwards, D.; Fenizia, C.; Gold, H.; de Castro-Amarante, M.F.; Buchmann, C.; Pise-Masison, C.A.; Franchini, G. Orf-I and orf-II-encoded proteins in HTLV-1 infection and persistence. *Viruses* **2011**, *3*, 861–885. [CrossRef] [PubMed]

22. Mesnard, J.M.; Barbeau, B.; Cesaire, R.; Peloponese, J.M. Roles of HTLV-1 basic Zip Factor (HBZ) in viral chronicity and leukemic transformation. Potential new therapeutic approaches to prevent and treat HTLV-1-related diseases. *Viruses* **2015**, *7*, 6490–6505. [CrossRef] [PubMed]

23. Carpentier, A.; Barez, P.Y.; Hamaidia, M.; Gazon, H.; de, B.A.; Perike, S.; Gillet, N.; Willems, L. Modes of human T cell leukemia virus type 1 transmission, replication and persistence. *Viruses* **2015**, *7*, 3603–3624. [CrossRef] [PubMed]

24. Richardson, J.H.; Edwards, A.J.; Cruickshank, J.K.; Rudge, P.; Dalgleish, A.G. *In vivo* cellular tropism of human T-cell leukemia virus type 1. *J. Virol.* **1990**, *64*, 5682–5687. [PubMed]

25. Nagai, M.; Brennan, M.B.; Sakai, J.A.; Mora, C.A.; Jacobson, S. CD8⁺ T cells are an *in vivo* reservoir for human T-cell lymphotropic virus type I. *Blood* **2001**, *98*, 1858–1861. [CrossRef] [PubMed]

26. Koyanagi, Y.; Itoyama, Y.; Nakamura, N.; Takamatsu, K.; Kira, J.; Iwamasa, T.; Goto, I.; Yamamoto, N. *In vivo* infection of human T-cell leukemia virus type I in non-T cells. *Virology* **1993**, *196*, 25–33. [CrossRef] [PubMed]

27. Melamed, A.; Laydon, D.J.; Al, K.H.; Rowan, A.G.; Taylor, G.P.; Bangham, C.R. HTLV-1 drives vigorous clonal expansion of infected CD8⁺ T cells in natural infection. *Retrovirology* **2015**, *12*. [CrossRef] [PubMed]

28. Macatonia, S.E.; Cruickshank, J.K.; Rudge, P.; Knight, S.C. Dendritic cells from patients with tropical spastic paraparesis are infected with HTLV-1 and stimulate autologous lymphocyte proliferation. *AIDS Res. Hum. Retrovir.* **1992**, *8*, 1699–1706. [CrossRef] [PubMed]

29. Hishizawa, M.; Imada, K.; Kitawaki, T.; Ueda, M.; Kadowaki, N.; Uchiyama, T. Depletion and impaired interferon-alpha-producing capacity of blood plasmacytoid dendritic cells in human T-cell leukaemia virus type I-infected individuals. *Br. J. Haematol.* **2004**, *125*, 568–575. [CrossRef] [PubMed]

30. De Castro-Amarante, M.F.; Pise-Masison, C.A.; McKinnon, K.; Washington, P.R.; Galli, V.; Omsland, M.; Andresen, V.; Massoud, R.; Brunetto, G.; Caruso, B.; *et al.* HTLV-1 infection of the three monocyte subsets contributes to viral burden in humans. *J. Virol.* **2015**. [CrossRef]

31. Longo, D.L.; Gelmann, E.P.; Cossman, J.; Young, R.A.; Gallo, R.C.; O'Brien, S.J.; Matis, L.A. Isolation of HTLV-transformed B-lymphocyte clone from a patient with HTLV-associated adult T-cell leukaemia. *Nature* **1984**, *310*, 505–506. [CrossRef] [PubMed]

32. Sze, A.; Belgnaoui, S.M.; Olagnier, D.; Lin, R.; Hiscott, J.; van Grevenynghe, J. Host restriction factor SAMHD1 limits human T cell leukemia virus type 1 infection of monocytes via STING-mediated apoptosis. *Cell Host Microbe* **2013**, *14*, 422–434. [CrossRef] [PubMed]

33. Mann, D.L.; Clark, J.; Clarke, M.; Reitz, M.; Popovic, M.; Franchini, G.; Trainor, C.D.; Strong, D.M.; Blattner, W.A.; Gallo, R.C. Identification of the human T cell lymphoma virus in B cell lines established from patients with adult T cell leukemia. *J. Clin. Investig.* **1984**, *74*, 56–62. [CrossRef] [PubMed]

34. Okada, M.; Koyanagi, Y.; Kobayashi, N.; Tanaka, Y.; Nakai, M.; Sano, K.; Takeuchi, K.; Hinuma, Y.; Hatanaka, M.; Yamamoto, N. *In vitro* infection of human B lymphocytes with adult T-cell leukemia virus. *Cancer Lett.* **1984**, *22*, 11–21. [CrossRef]

35. Ueda, S.; Maeda, Y.; Yamaguchi, T.; Hanamoto, H.; Hijikata, Y.; Tanaka, M.; Takai, S.; Hirase, C.; Morita, Y.; Kanamaru, A. Influence of Epstein-Barr virus infection in adult T-cell leukemia. *Hematology* **2008**, *13*, 154–162. [CrossRef] [PubMed]

36. Yamamoto, N.; Matsumoto, T.; Koyanagi, Y.; Tanaka, Y.; Hinuma, Y. Unique cell lines harbouring both Epstein-Barr virus and adult T-cell leukaemia virus, established from leukaemia patients. *Nature* **1982**, *299*, 367–369. [CrossRef] [PubMed]

37. Fan, N.; Gavalchin, J.; Paul, B.; Wells, K.H.; Lane, M.J.; Poiesz, B.J. Infection of peripheral blood mononuclear cells and cell lines by cell-free human T-cell lymphoma/leukemia virus type I. *J. Clin. Microbiol.* **1992**, *30*, 905–910. [PubMed]

38. Derse, D.; Hill, S.A.; Lloyd, P.A.; Chung, H.; Morse, B.A. Examining human T-lymphotropic virus type 1 infection and replication by cell-free infection with recombinant virus vectors. *J. Virol.* **2001**, *75*, 8461–8468. [CrossRef] [PubMed]

39. Jones, K.S.; Petrow-Sadowski, C.; Huang, Y.K.; Bertolette, D.C.; Ruscetti, F.W. Cell-free HTLV-1 infects dendritic cells leading to transmission and transformation of CD4$^+$ T cells. *Nat. Med.* **2008**, *14*, 429–436. [CrossRef] [PubMed]

40. Pique, C.; Jones, K.S. Pathways of cell-cell transmission of HTLV-1. *Front Microbiol.* **2012**, *3*. [CrossRef] [PubMed]

41. Demontis, M.A.; Sadiq, M.T.; Golz, S.; Taylor, G.P. HTLV-1 viral RNA is detected rarely in plasma of HTLV-1 infected subjects. *J. Med. Virol.* **2015**, *87*, 2130–2134. [CrossRef] [PubMed]

42. Carneiro-Proietti, A.B.; Amaranto-Damasio, M.S.; Leal-Horiguchi, C.F.; Bastos, R.H.; Seabra-Freitas, G.; Borowiak, D.R.; Ribeiro, M.A.; Proietti, F.A.; Ferreira, A.S.; Martins, M.L. Mother-to-child transmission of human T-cell lymphotropic viruses-1/2: What we know, and what are the gaps in understanding and preventing this route of infection. *J. Pediatric. Infect. Dis. Soc.* **2014**, *3*, S24–S29. [CrossRef] [PubMed]

43. Percher, F.; Jeannin, P.; Martin-Latil, S.; Gessain, A.; Afonso, P.V.; Vidy-Roche, A.; Ceccaldi, P.E. Mother-to-child transmission of HTLV-1 epidemiological aspects, mechanisms and determinants of mother-to-child transmission. *Viruses* **2016**, *2*, 40. [CrossRef] [PubMed]

44. Nerome, Y.; Kojyo, K.; Ninomiya, Y.; Ishikawa, T.; Ogiso, A.; Takei, S.; Kawano, Y.; Douchi, T.; Takezaki, T.; Owaki, T. Current human T-cell lymphotropic virus type 1 mother-to-child transmission prevention status in Kagoshima. *Pediatr. Int.* **2014**, *56*, 640–643. [CrossRef] [PubMed]

45. Paiva, A.; Casseb, J. Sexual transmission of human T-cell lymphotropic virus type 1. *Rev. Soc. Bras. Med. Trop.* **2014**, *47*, 265–274. [CrossRef] [PubMed]

46. Glowacka, I.; Korn, K.; Potthoff, S.A.; Lehmann, U.; Kreipe, H.H.; Ivens, K.; Barg-Hock, H.; Schulz, T.F.; Heim, A. Delayed seroconversion and rapid onset of lymphoproliferative disease after transmission of human T-cell lymphotropic virus type 1 from a multiorgan donor. *Clin. Infect. Dis.* **2013**, *57*, 1417–1424. [CrossRef] [PubMed]

47. Ramanan, P.; Deziel, P.J.; Norby, S.M.; Yao, J.D.; Garza, I.; Razonable, R.R. Donor-transmitted HTLV-1-associated myelopathy in a kidney transplant recipient—Case report and literature review. *Am. J. Transpl.* **2014**, *14*, 2417–2421. [CrossRef] [PubMed]

48. Kazanji, M.; Mouinga-Ondeme, A.; Lekana-Douki-Etenna, S.; Caron, M.; Makuwa, M.; Mahieux, R.; Gessain, A. Origin of HTLV-1 in hunters of nonhuman primates in Central Africa. *J. Infect. Dis.* **2015**, *211*, 361–365. [CrossRef] [PubMed]

49. Filippone, C.; Betsem, E.; Tortevoye, P.; Cassar, O.; Bassot, S.; Froment, A.; Fontanet, A.; Gessain, A. A severe bite from a nonhuman primate is a major risk factor for HTLV-1 infection in hunters from Central Africa. *Clin. Infect. Dis.* **2015**, *60*, 1667–1676. [CrossRef] [PubMed]

50. Alais, S.; Mahieux, R.; Dutartre, H. Viral source-independent high susceptibility of dendritic cells to human T-cell leukemia virus type 1 infection compared to that of T lymphocytes. *J. Virol.* **2015**, *89*, 10580–10590. [CrossRef] [PubMed]

51. Martin-Latil, S.; Gnadig, N.F.; Mallet, A.; Desdouits, M.; Guivel-Benhassine, F.; Jeannin, P.; Prevost, M.C.; Schwartz, O.; Gessain, A.; Ozden, S.; Ceccaldi, P.E. Transcytosis of HTLV-1 across a tight human epithelial barrier and infection of subepithelial dendritic cells. *Blood* **2012**, *120*, 572–580. [CrossRef] [PubMed]

52. Dodon, M.D.; Villaudy, J.; Gazzolo, L.; Haines, R.; Lairmore, M. What we are learning on HTLV-1 pathogenesis from animal models. *Front Microbiol.* **2012**, *3*. [CrossRef] [PubMed]
53. Barez, P.Y.; de, B.A.; Carpentier, A.; Gazon, H.; Gillet, N.; Gutierrez, G.; Hamaidia, M.; Jacques, J.R.; Perike, S.; Neelature, S.S.; *et al*. Recent Advances in BLV Research. *Viruses* **2015**, *7*, 6080–6088. [CrossRef] [PubMed]
54. Villaudy, J.; Wencker, M.; Gadot, N.; Gillet, N.A.; Scoazec, J.Y.; Gazzolo, L.; Manz, M.G.; Bangham, C.R.; Dodon, M.D. HTLV-1 propels thymic human T cell development in "human immune system" Rag2$^{-/-}$ gamma c$^{-/-}$ mice. *PLoS Pathog.* **2011**, *7*, e1002231. [CrossRef] [PubMed]
55. Sewald, X.; Ladinsky, M.S.; Uchil, P.D.; Beloor, J.; Pi, R.; Herrmann, C.; Motamedi, N.; Murooka, T.T.; Brehm, M.A.; Greiner, D.L.; *et al*. Retroviruses use CD169-mediated trans-infection of permissive lymphocytes to establish infection. *Science* **2015**, *350*, 563–567. [CrossRef] [PubMed]
56. Saito, M.; Tanaka, R.; Fujii, H.; Kodama, A.; Takahashi, Y.; Matsuzaki, T.; Takashima, H.; Tanaka, Y. The neutralizing function of the anti-HTLV-1 antibody is essential in preventing *in vivo* transmission of HTLV-1 to human T cells in NOD-SCID/gammacnull (NOG) mice. *Retrovirology* **2014**, *11*. [CrossRef]
57. Mazurov, D.; Ilinskaya, A.; Heidecker, G.; Lloyd, P.; Derse, D. Quantitative comparison of HTLV-1 and HIV-1 cell-to-cell infection with new replication dependent vectors. *PLoS Pathog.* **2010**, *6*, e1000788. [CrossRef] [PubMed]
58. Van Prooyen, N.; Gold, H.; Andresen, V.; Schwartz, O.; Jones, K.; Ruscetti, F.; Lockett, S.; Gudla, P.; Venzon, D.; Franchini, G. Human T-cell leukemia virus type 1 p8 protein increases cellular conduits and virus transmission. *Proc. Natl. Acad. Sci. USA* **2010**, *107*, 20738–20743. [CrossRef] [PubMed]
59. Malbec, M.; Roesch, F.; Schwartz, O. A new role for the HTLV-1 p8 protein: Increasing intercellular conduits and viral cell-to-cell transmission. *Viruses* **2011**, *3*, 254–259. [CrossRef] [PubMed]
60. Igakura, T.; Stinchcombe, J.C.; Goon, P.K.; Taylor, G.P.; Weber, J.N.; Griffiths, G.M.; Tanaka, Y.; Osame, M.; Bangham, C.R. Spread of HTLV-I between lymphocytes by virus-induced polarization of the cytoskeleton. *Science* **2003**, *299*, 1713–1716. [CrossRef] [PubMed]
61. Nejmeddine, M.; Bangham, C.R. The HTLV-1 virological synapse. *Viruses* **2010**, *2*, 1427–1447. [CrossRef] [PubMed]
62. Nejmeddine, M.; Negi, V.S.; Mukherjee, S.; Tanaka, Y.; Orth, K.; Taylor, G.P.; Bangham, C.R. HTLV-1-Tax and ICAM-1 act on T-cell signal pathways to polarize the microtubule-organizing center at the virological synapse. *Blood* **2009**, *114*, 1016–1025. [CrossRef] [PubMed]
63. Nejmeddine, M.; Barnard, A.L.; Tanaka, Y.; Taylor, G.P.; Bangham, C.R. Human T-lymphotropic virus, type 1, tax protein triggers microtubule reorientation in the virological synapse. *J. Biol. Chem.* **2005**, *280*, 29653–29660. [CrossRef] [PubMed]
64. Barnard, A.L.; Igakura, T.; Tanaka, Y.; Taylor, G.P.; Bangham, C.R. Engagement of specific T-cell surface molecules regulates cytoskeletal polarization in HTLV-1-infected lymphocytes. *Blood* **2005**, *106*, 988–995. [CrossRef] [PubMed]
65. Majorovits, E.; Nejmeddine, M.; Tanaka, Y.; Taylor, G.P.; Fuller, S.D.; Bangham, C.R. Human T-lymphotropic virus-1 visualized at the virological synapse by electron tomography. *PLoS ONE* **2008**, *3*, e2251. [CrossRef] [PubMed]
66. Derse, D.; Heidecker, G. Virology. Forced entry—Or does HTLV-I have the key? *Science* **2003**, *299*, 1670–1671. [CrossRef] [PubMed]
67. Jolly, C.; Sattentau, Q.J. Retroviral spread by induction of virological synapses. *Traffic* **2004**, *5*, 643–650. [CrossRef] [PubMed]
68. Jolly, C.; Kashefi, K.; Hollinshead, M.; Sattentau, Q.J. HIV-1 cell to cell transfer across an Env-induced, actin-dependent synapse. *J. Exp. Med.* **2004**, *199*, 283–293. [CrossRef] [PubMed]
69. Jin, J.; Sherer, N.M.; Heidecker, G.; Derse, D.; Mothes, W. Assembly of the murine leukemia virus is directed towards sites of cell-cell contact. *PLoS. Biol.* **2009**, *7*, e1000163. [CrossRef] [PubMed]
70. Stewart, P.S.; Franklin, M.J. Physiological heterogeneity in biofilms. *Nat. Rev. Microbiol.* **2008**, *6*, 199–210. [CrossRef] [PubMed]
71. Pais-Correia, A.M.; Sachse, M.; Guadagnini, S.; Robbiati, V.; Lasserre, R.; Gessain, A.; Gout, O.; Alcover, A.; Thoulouze, M.I. Biofilm-like extracellular viral assemblies mediate HTLV-1 cell-to-cell transmission at virological synapses. *Nat. Med.* **2010**, *16*, 83–89. [CrossRef] [PubMed]

72. Mazurov, D.; Ilinskaya, A.; Heidecker, G.; Filatov, A. Role of O-glycosylation and expression of CD43 and CD45 on the surfaces of effector T cells in human T cell leukemia virus type 1 cell-to-cell infection. *J. Virol.* **2012**, *86*, 2447–2458. [CrossRef] [PubMed]

73. Ilinskaya, A.; Derse, D.; Hill, S.; Princler, G.; Heidecker, G. Cell-cell transmission allows human T-lymphotropic virus 1 to circumvent tetherin restriction. *Virology* **2013**, *436*, 201–209. [CrossRef] [PubMed]

74. Jin, J.; Sherer, N.; Mothes, W. Surface transmission or polarized egress? Lessons learned from HTLV cell-to-cell transmission. *Viruses* **2010**, *2*, 601–605. [CrossRef] [PubMed]

75. Shinagawa, M.; Jinno-Oue, A.; Shimizu, N.; Roy, B.B.; Shimizu, A.; Hoque, S.A.; Hoshino, H. Human T-cell leukemia viruses are highly unstable over a wide range of temperatures. *J. Gen. Virol.* **2012**, *93*, 608–617. [CrossRef] [PubMed]

76. Thoulouze, M.I.; Alcover, A. Can viruses form biofilms? *Trends Microbiol.* **2011**, *19*, 257–262. [CrossRef] [PubMed]

77. Li, F.; Sewald, X.; Jin, J.; Sherer, N.M.; Mothes, W. Murine leukemia virus Gag localizes to the uropod of migrating primary lymphocytes. *J. Virol.* **2014**, *88*, 10541–10555. [CrossRef] [PubMed]

78. Llewellyn, G.N.; Hogue, I.B.; Grover, J.R.; Ono, A. Nucleocapsid promotes localization of HIV-1 gag to uropods that participate in virological synapses between T cells. *PLoS Pathog.* **2010**, *6*, e1001167. [CrossRef] [PubMed]

79. Fais, S.; Malorni, W. Leukocyte uropod formation and membrane/cytoskeleton linkage in immune interactions. *J. Leukoc. Biol.* **2003**, *73*, 556–563. [CrossRef] [PubMed]

80. Fukumoto, R.; Andresen, V.; Bialuk, I.; Cecchinato, V.; Walser, J.C.; Valeri, V.W.; Nauroth, J.M.; Gessain, A.; Nicot, C.; Franchini, G. *In vivo* genetic mutations define predominant functions of the human T-cell leukemia/lymphoma virus p12I protein. *Blood* **2009**, *113*, 3726–3734. [CrossRef] [PubMed]

81. Van Prooyen, N.; Andresen, V.; Gold, H.; Bialuk, I.; Pise-Masison, C.; Franchini, G. Hijacking the T-cell communication network by the human T-cell leukemia/lymphoma virus type 1 (HTLV-1) p12 and p8 proteins. *Mol. Aspects Med.* **2010**, *31*, 333–343. [CrossRef] [PubMed]

82. Sherer, N.M.; Lehmann, M.J.; Jimenez-Soto, L.F.; Horensavitz, C.; Pypaert, M.; Mothes, W. Retroviruses can establish filopodial bridges for efficient cell-to-cell transmission. *Nat. Cell Biol.* **2007**, *9*, 310–315. [CrossRef] [PubMed]

83. Sowinski, S.; Jolly, C.; Berninghausen, O.; Purbhoo, M.A.; Chauveau, A.; Kohler, K.; Oddos, S.; Eissmann, P.; Brodsky, F.M.; Hopkins, C.; *et al.* Membrane nanotubes physically connect T cells over long distances presenting a novel route for HIV-1 transmission. *Nat. Cell Biol.* **2008**, *10*, 211–219. [CrossRef] [PubMed]

84. Jain, P.; Manuel, S.L.; Khan, Z.K.; Ahuja, J.; Quann, K.; Wigdahl, B. DC-SIGN mediates cell-free infection and transmission of human T-cell lymphotropic virus type 1 by dendritic cells. *J. Virol.* **2009**, *83*, 10908–10921. [CrossRef] [PubMed]

85. Ceccaldi, P.E.; Delebecque, F.; Prevost, M.C.; Moris, A.; Abastado, J.P.; Gessain, A.; Schwartz, O.; Ozden, S. DC-SIGN facilitates fusion of dendritic cells with human T-cell leukemia virus type 1-infected cells. *J. Virol.* **2006**, *80*, 4771–4780. [CrossRef] [PubMed]

86. Gummuluru, S.; Pina Ramirez, N.G.; Akiyama, H. CD169-dependent cell-associated HIV-1 transmission: A driver of virus dissemination. *J. Infect. Dis.* **2014**, *210*, S641–S647. [CrossRef] [PubMed]

87. Valeri, V.W.; Hryniewicz, A.; Andresen, V.; Jones, K.; Fenizia, C.; Bialuk, I.; Chung, H.K.; Fukumoto, R.; Parks, R.W.; Ferrari, M.G.; *et al.* Requirement of the human T-cell leukemia virus p12 and p30 products for infectivity of human dendritic cells and macaques but not rabbits. *Blood* **2010**, *116*, 3809–3817. [CrossRef] [PubMed]

88. Nascimento, C.R.; Lima, M.A.; de Andrada Serpa, M.J.; Espindola, O.; Leite, A.C.; Echevarria-Lima, J. Monocytes from HTLV-1-infected patients are unable to fully mature into dendritic cells. *Blood* **2011**, *117*, 489–499. [CrossRef] [PubMed]

89. Inagaki, S.; Takahashi, M.; Fukunaga, Y.; Takahashi, H. HTLV-I-infected breast milk macrophages inhibit monocyte differentiation to dendritic cells. *Viral Immunol.* **2012**, *25*, 106–116. [CrossRef] [PubMed]

90. McDonald, D.; Wu, L.; Bohks, S.M.; KewalRamani, V.N.; Unutmaz, D.; Hope, T.J. Recruitment of HIV and its receptors to dendritic cell-T cell junctions. *Science* **2003**, *300*, 1295–1297. [CrossRef] [PubMed]

91. Lairmore, M.D.; Anupam, R.; Bowden, N.; Haines, R.; Haynes, R.A.; Ratner, L.; Green, P.L. Molecular determinants of human T-lymphotropic virus type 1 transmission and spread. *Viruses* **2011**, *3*, 1131–1165. [CrossRef] [PubMed]

92. Hoshino, H. Cellular factors involved in HTLV-1 entry and pathogenicit. *Front Microbiol.* **2012**, *3*. [CrossRef] [PubMed]

93. Maldonado, J.O.; Martin, J.L.; Mueller, J.D.; Zhang, W.; Mansky, L.M. New insights into retroviral Gag-Gag and Gag-membrane interactions. *Front Microbiol.* **2014**, *5*. [CrossRef] [PubMed]

94. Martin, J.L.; Maldonado, J.O.; Mueller, J.D.; Zhang, W.; Mansky, L.M. Molecular studies of HTLV-1 replication: An update. *Viruses* **2016**, *8*, 31. [CrossRef] [PubMed]

95. Pare, M.E.; Gauthier, S.; Landry, S.; Sun, J.; Legault, E.; Leclerc, D.; Tanaka, Y.; Marriott, S.J.; Tremblay, M.J.; Barbeau, B. A new sensitive and quantitative HTLV-I-mediated cell fusion assay in T cells. *Virology* **2005**, *338*, 309–322. [CrossRef] [PubMed]

96. Ciminale, V.; Rende, F.; Bertazzoni, U.; Romanelli, M.G. HTLV-1 and HTLV-2: Highly similar viruses with distinct oncogenic properties. *Front Microbiol.* **2014**, *5*. [CrossRef] [PubMed]

97. Grassmann, R.; Dengler, C.; Muller-Fleckenstein, I.; Fleckenstein, B.; McGuire, K.; Dokhelar, M.C.; Sodroski, J.G.; Haseltine, W.A. Transformation to continuous growth of primary human T lymphocytes by human T-cell leukemia virus type I X-region genes transduced by a *Herpesvirus saimiri* vector. *Proc. Natl. Acad. Sci. USA* **1989**, *86*, 3351–3355. [CrossRef] [PubMed]

98. Wu, K.; Bottazzi, M.E.; de la, F.C.; Deng, L.; Gitlin, S.D.; Maddukuri, A.; Dadgar, S.; Li, H.; Vertes, A.; Pumfery, A.; *et al.* Protein profile of tax-associated complexes. *J. Biol. Chem.* **2004**, *279*, 495–508. [CrossRef] [PubMed]

99. Boxus, M.; Twizere, J.C.; Legros, S.; Dewulf, J.F.; Kettmann, R.; Willems, L. The HTLV-1 Tax interactome. *Retrovirology* **2008**, *5*. [CrossRef] [PubMed]

100. Simonis, N.; Rual, J.F.; Lemmens, I.; Boxus, M.; Hirozane-Kishikawa, T.; Gatot, J.S.; Dricot, A.; Hao, T.; Vertommen, D.; Legros, S.; *et al.* Host-pathogen interactome mapping for HTLV-1 and -2 retroviruses. *Retrovirology* **2012**, *9*. [CrossRef] [PubMed]

101. de la Fuente, C.; Deng, L.; Santiago, F.; Arce, L.; Wang, L.; Kashanchi, F. Gene expression array of HTLV type 1-infected T cells: Up-regulation of transcription factors and cell cycle genes. *AIDS Res. Hum. Retrovir.* **2000**, *16*, 1695–1700. [CrossRef] [PubMed]

102. Pise-Masison, C.A.; Radonovich, M.; Mahieux, R.; Chatterjee, P.; Whiteford, C.; Duvall, J.; Guillerm, C.; Gessain, A.; Brady, J.N. Transcription profile of cells infected with human T-cell leukemia virus type I compared with activated lymphocytes. *Cancer Res.* **2002**, *62*, 3562–3571. [PubMed]

103. Pichler, K.; Kattan, T.; Gentzsch, J.; Kress, A.K.; Taylor, G.P.; Bangham, C.R.; Grassmann, R. Strong induction of 4-1BB, a growth and survival promoting costimulatory receptor, in HTLV-1-infected cultured and patients' T cells by the viral Tax oncoprotein. *Blood* **2008**, *111*, 4741–4751. [CrossRef] [PubMed]

104. Chevalier, S.A.; Durand, S.; Dasgupta, A.; Radonovich, M.; Cimarelli, A.; Brady, J.N.; Mahieux, R.; Pise-Masison, C.A. The transcription profile of Tax-3 is more similar to Tax-1 than Tax-2: Insights into HTLV-3 potential leukemogenic properties. *PLoS ONE* **2012**, *7*, e41003. [CrossRef] [PubMed]

105. Kress, A.K.; Schneider, G.; Pichler, K.; Kalmer, M.; Fleckenstein, B.; Grassmann, R. Elevated cyclic AMP levels in T lymphocytes transformed by human T-cell lymphotropic virus type 1. *J. Virol.* **2010**, *84*, 8732–8742. [CrossRef] [PubMed]

106. Grassmann, R.; Aboud, M.; Jeang, K.T. Molecular mechanisms of cellular transformation by HTLV-1 Tax. *Oncogene* **2005**, *24*, 5976–5985. [CrossRef] [PubMed]

107. Fukudome, K.; Furuse, M.; Fukuhara, N.; Orita, S.; Imai, T.; Takagi, S.; Nagira, M.; Hinuma, Y.; Yoshie, O. Strong induction of ICAM-1 in human T cells transformed by human T-cell-leukemia virus type 1 and depression of ICAM-1 or LFA-1 in adult T-cell-leukemia-derived cell lines. *Int. J. Cancer* **1992**, *52*, 418–427. [CrossRef] [PubMed]

108. Shunaeva, A.; Potashnikova, D.; Pichugin, A.; Mishina, A.; Filatov, A.; Nikolaitchik, O.; Hu, W.S.; Mazurov, D. Improvement of HIV-1 and human T cell lymphotropic virus type 1 replication-dependent vectors via optimization of reporter gene reconstitution and modification with intronic short hairpin RNA. *J. Virol.* **2015**, *89*, 10591–10601. [CrossRef] [PubMed]

109. Kress, A.K.; Grassmann, R.; Fleckenstein, B. Cell surface markers in HTLV-1 pathogenesis. *Viruses* **2011**, *3*, 1439–1459. [CrossRef] [PubMed]

110. Hieshima, K.; Nagakubo, D.; Nakayama, T.; Shirakawa, A.K.; Jin, Z.; Yoshie, O. Tax-inducible production of CC chemokine ligand 22 by human T cell leukemia virus type 1 (HTLV-1)-infected T cells promotes preferential transmission of HTLV-1 to CCR4-expressing CD4⁺ T cells. *J. Immunol.* **2008**, *180*, 931–939. [CrossRef] [PubMed]

111. Mazurov, D.; Heidecker, G.; Derse, D. The inner loop of tetraspanins CD82 and CD81 mediates interactions with human T cell lymphotrophic virus type 1 Gag protein. *J. Biol. Chem.* **2007**, *282*, 3896–3903. [CrossRef] [PubMed]

112. Pique, C.; Lagaudriere-Gesbert, C.; Delamarre, L.; Rosenberg, A.R.; Conjeaud, H.; Dokhelar, M.C. Interaction of CD82 tetraspanin proteins with HTLV-1 envelope glycoproteins inhibits cell-to-cell fusion and virus transmission. *Virology* **2000**, *276*, 455–465. [CrossRef] [PubMed]

113. Munoz, E.; Suri, D.; Amini, S.; Khalili, K.; Jimenez, S.A. Stimulation of alpha 1 (I) procollagen gene expression in NIH-3T3 cells by the human T cell leukemia virus type 1 (HTLV-1) Tax gene. *J. Clin. Investig.* **1995**, *96*, 2413–2420. [CrossRef] [PubMed]

114. Sagara, Y.; Ishida, C.; Inoue, Y.; Shiraki, H.; Maeda, Y. 71-Kilodalton heat shock cognate protein acts as a cellular receptor for syncytium formation induced by human T-cell lymphotropic virus type 1. *J. Virol.* **1998**, *72*, 535–541. [PubMed]

115. Daenke, S.; McCracken, S.A.; Booth, S. Human T-cell leukaemia/lymphoma virus type 1 syncytium formation is regulated in a cell-specific manner by ICAM-1, ICAM-3 and VCAM-1 and can be inhibited by antibodies to integrin beta2 or beta7. *J. Gen. Virol.* **1999**, *80*, 1429–1436. [CrossRef] [PubMed]

116. Kim, S.J.; Nair, A.M.; Fernandez, S.; Mathes, L.; Lairmore, M.D. Enhancement of LFA-1-mediated T cell adhesion by human T lymphotropic virus type 1 p12I1. *J. Immunol.* **2006**, *176*, 5463–5470. [CrossRef] [PubMed]

117. Tanaka, A.; Jinno-Oue, A.; Shimizu, N.; Hoque, A.; Mori, T.; Islam, S.; Nakatani, Y.; Shinagawa, M.; Hoshino, H. Entry of human T-cell leukemia virus type 1 is augmented by heparin sulfate proteoglycans bearing short heparin-like structures. *J. Virol.* **2012**, *86*, 2959–2969. [CrossRef] [PubMed]

118. Hildreth, J.E.; Subramanium, A.; Hampton, R.A. Human T-cell lymphotropic virus type 1 (HTLV-1)-induced syncytium formation mediated by vascular cell adhesion molecule-1: Evidence for involvement of cell adhesion molecules in HTLV-1 biology. *J. Virol.* **1997**, *71*, 1173–1180. [PubMed]

119. Valentin, H.; Lemasson, I.; Hamaia, S.; Casse, H.; Konig, S.; Devaux, C.; Gazzolo, L. Transcriptional activation of the vascular cell adhesion molecule-1 gene in T lymphocytes expressing human T-cell leukemia virus type 1 Tax protein. *J. Virol.* **1997**, *71*, 8522–8530. [PubMed]

120. Varrin-Doyer, M.; Nicolle, A.; Marignier, R.; Cavagna, S.; Benetollo, C.; Wattel, E.; Giraudon, P. Human T lymphotropic virus type 1 increases T lymphocyte migration by recruiting the cytoskeleton organizer CRMP2. *J. Immunol.* **2012**, *188*, 1222–1233. [CrossRef] [PubMed]

121. Kress, A.K.; Kalmer, M.; Rowan, A.G.; Grassmann, R.; Fleckenstein, B. The tumor marker Fascin is strongly induced by the Tax oncoprotein of HTLV-1 through NF-κB signals. *Blood* **2011**, *117*, 3609–3612. [CrossRef] [PubMed]

122. Gross, C.; Thoma-Kress, A.K.; Friedrich-Alexander-Universität Erlangen-Nürnberg, Erlangen, Germany. Unpublished work.

123. Mohr, C.F.; Gross, C.; Bros, M.; Reske-Kunz, A.B.; Biesinger, B.; Thoma-Kress, A.K. Regulation of the tumor marker Fascin by the viral oncoprotein Tax of human T-cell leukemia virus type 1 (HTLV-1) depends on promoter activation and on a promoter-independent mechanism. *Virology* **2015**, *485*, 481–491. [CrossRef] [PubMed]

124. Chevalier, S.A.; Turpin, J.; Cachat, A.; Afonso, P.V.; Gessain, A.; Brady, J.N.; Pise-Masison, C.A.; Mahieux, R. Gem-induced cytoskeleton remodeling increases cellular migration of HTLV-1-infected cells, formation of infected-to-target T-cell conjugates and viral transmission. *PLoS. Pathog.* **2014**, *10*, e1003917. [CrossRef] [PubMed]

125. Kfoury, Y.; Nasr, R.; Favre-Bonvin, A.; El-Sabban, M.; Renault, N.; Giron, M.L.; Setterblad, N.; Hajj, H.E.; Chiari, E.; Mikati, A.G.; *et al.* Ubiquitylated Tax targets and binds the IKK signalosome at the centrosome. *Oncogene* **2008**, *27*, 1665–1676. [CrossRef] [PubMed]

126. Zhao, L.J.; Giam, C.Z. Human T-cell lymphotropic virus type I (HTLV-I) transcriptional activator, Tax, enhances CREB binding to HTLV-I 21-base-pair repeats by protein-protein interaction. *Proc. Natl. Acad. Sci. USA* **1992**, *89*, 7070–7074. [CrossRef] [PubMed]

127. Cooper, S.A.; van der Loeff, M.S.; Taylor, G.P. The neurology of HTLV-1 infection. *Pract. Neurol.* **2009**, *9*, 16–26. [CrossRef] [PubMed]

128. Blot, V.; Delamarre, L.; Perugi, F.; Pham, D.; Benichou, S.; Benarous, R.; Hanada, T.; Chishti, A.H.; Dokhelar, M.C.; Pique, C. Human Dlg protein binds to the envelope glycoproteins of human T-cell leukemia virus type 1 and regulates envelope mediated cell-cell fusion in T lymphocytes. *J. Cell Sci.* **2004**, *117*, 3983–3993. [CrossRef] [PubMed]

129. Suzuki, T.; Ohsugi, Y.; Uchida-Toita, M.; Akiyama, T.; Yoshida, M. Tax oncoprotein of HTLV-1 binds to the human homologue of Drosophila discs large tumor suppressor protein, hDLG, and perturbs its function in cell growth control. *Oncogene* **1999**, *18*, 5967–5972. [CrossRef] [PubMed]

130. Hsu, D.K.; Hammes, S.R.; Kuwabara, I.; Greene, W.C.; Liu, F.T. Human T lymphotropic virus-I infection of human T lymphocytes induces expression of the beta-galactoside-binding lectin, galectin-3. *Am. J. Pathol.* **1996**, *148*, 1661–1670. [PubMed]

131. Fukumoto, R.; Dundr, M.; Nicot, C.; Adams, A.; Valeri, V.W.; Samelson, L.E.; Franchini, G. Inhibition of T-cell receptor signal transduction and viral expression by the linker for activation of T cells-interacting p12(I) protein of human T-cell leukemia/lymphoma virus type 1. *J. Virol.* **2007**, *81*, 9088–9099. [CrossRef] [PubMed]

132. Taylor, J.M.; Brown, M.; Nejmeddine, M.; Kim, K.J.; Ratner, L.; Lairmore, M.; Nicot, C. Novel role for interleukin-2 receptor-Jak signaling in retrovirus transmission. *J. Virol.* **2009**, *83*, 11467–11476. [CrossRef] [PubMed]

133. Albrecht, B.; Collins, N.D.; Burniston, M.T.; Nisbet, J.W.; Ratner, L.; Green, P.L.; Lairmore, M.D. Human T-lymphotropic virus type 1 open reading frame I p12(I) is required for efficient viral infectivity in primary lymphocytes. *J. Virol.* **2000**, *74*, 9828–9835. [CrossRef] [PubMed]

134. Edwards, D.; Fukumoto, R.; de Castro-Amarante, M.F.; Alcantara, L.C.; Galvao-Castro, B.; Washington, P.R.; Pise-Masison, C.; Franchini, G. Palmitoylation and p8-mediated human T-cell leukemia virus type 1 transmission. *J. Virol.* **2014**, *88*, 2319–2322. [CrossRef] [PubMed]

135. Pise-Masison, C.A.; de Castro-Amarante, M.F.; Enose-Akahata, Y.; Buchmann, R.C.; Fenizia, C.; Washington, P.R.; Edwards, D.; Fiocchi, M.; Alcantara, L.C., Jr.; Bialuk, I.; *et al.* Co-dependence of HTLV-1 p12 and p8 functions in virus persistence. *PLoS Pathog.* **2014**, *10*, e1004454. [CrossRef] [PubMed]

136. Tanaka, Y.; Fukudome, K.; Hayashi, M.; Takagi, S.; Yoshie, O. Induction of ICAM-1 and LFA-3 by Tax1 of human T-cell leukemia virus type 1 and mechanism of down-regulation of ICAM-1 or LFA-1 in adult-T-cell-leukemia cell lines. *Int. J. Cancer* **1995**, *60*, 554–561. [CrossRef] [PubMed]

137. Vincent, P.; Collette, Y.; Marignier, R.; Vuaillat, C.; Rogemond, V.; Davoust, N.; Malcus, C.; Cavagna, S.; Gessain, A.; Machuca-Gayet, I.; *et al.* A role for the neuronal protein collapsin response mediator protein 2 in T lymphocyte polarization and migration. *J. Immunol.* **2005**, *175*, 7650–7660. [CrossRef] [PubMed]

*viruses*

MDPI

*Review*

# The Role of HBZ in HTLV-1-Induced Oncogenesis

Tiejun Zhao [1,2]

[1]   College of Chemistry and Life Sciences, Zhejiang Normal University, 688 Yingbin Road, Jinhua 321004,
     China; tjzhao@zjnu.cn; Tel.: +86-579-8229-1410
[2]   Key Lab of Wildlife Biotechnology and Conservation and Utilization of Zhejiang Province,
     Zhejiang Normal University, 688 Yingbin Road, Jinhua 321004, China

Academic Editor: Louis M. Mansky
Received: 28 October 2015; Accepted: 28 January 2016; Published: 2 February 2016

**Abstract:** Human T-cell leukemia virus type 1 (HTLV-1) causes adult T-cell leukemia (ATL) and chronic inflammatory diseases. HTLV-1 bZIP factor (HBZ) is transcribed as an antisense transcript of the HTLV-1 provirus. Among the HTLV-1-encoded viral genes, HBZ is the only gene that is constitutively expressed in all ATL cases. Recent studies have demonstrated that HBZ plays an essential role in oncogenesis by regulating viral transcription and modulating multiple host factors, as well as cellular signaling pathways, that contribute to the development and continued growth of cancer. In this article, I summarize the current knowledge of the oncogenic function of HBZ in cell proliferation, apoptosis, T-cell differentiation, immune escape, and HTLV-1 pathogenesis.

**Keywords:** HBZ; HTLV-1; ATL

## 1. Introduction

Human T-cell leukemia virus type 1 (HTLV-1) is the first identified pathogenic retrovirus that is etiologically associated with two major diseases: adult T-cell leukemia (ATL) and a progressive myelopathy called HTLV-1-associated myelopathy/tropical spastic paraparesis (HAM/TSP) [1–4]. Four types of HTLVs, named HTLV-1, HTLV-2, HTLV-3, and HTLV-4, have been characterized [5]. However, their oncogenic properties are different. The HTLV-1 genome, in addition to the structural genes *gag*, *pol*, and *env*, carries a region at its 3' end, which is designated the pX region, and encodes several accessory genes, including *tax*, *rex*, *p12*, *p21*, *p30*, *p13*, and *HTLV-1bZIP factor* (*HBZ*) (Figure 1) [6,7]. In the 35 years since the discovery of HTLV-1, researchers have mainly focused on Tax, a *trans*-acting viral regulatory protein, due to its pleiotropic functions in viral replication and cellular transformation [8–10]. However, approximately 60% of ATL cases lack Tax expression because of genetic and epigenetic changes in the proviral genome of HTLV-1, suggesting that Tax is not essential for the maintenance of ATL [5,11]. In 2002, a new open reading frame (ORF) was identified on the minus strand of HTLV-1, corresponding to a region complementary to Tax, which encodes a novel basic leucine zipper factor named HBZ [6]. Studies have shown that HBZ is consistently expressed in all ATL cases [12]. In some of the ATL cases, HTLV-1 genes, except HBZ, are mutated by APOBEC3G [13]. Moreover, Tax is an immunodominant antigen of this virus and is the major target of cytotoxic T lymphocytes (CTLs) [14]. Thus, Tax expression is frequently silenced during the long clinically latent period of the virus [5,11]. In contrast, the immunogenicity of HBZ protein is low compared with that of Tax protein [15–17]. The constitutive expression of HBZ, by way of suppressing major HTLV-1 sense genes including Tax, could help the virus escape from the host's immune surveillance and thus promote spread of infection [5,18,19]. These observations suggest that the HBZ gene is essential for cellular transformation and leukemogenesis of HTLV-1. In this review, I will highlight recent advances in our understanding of how HBZ contributes to HTLV-1 oncogenesis.

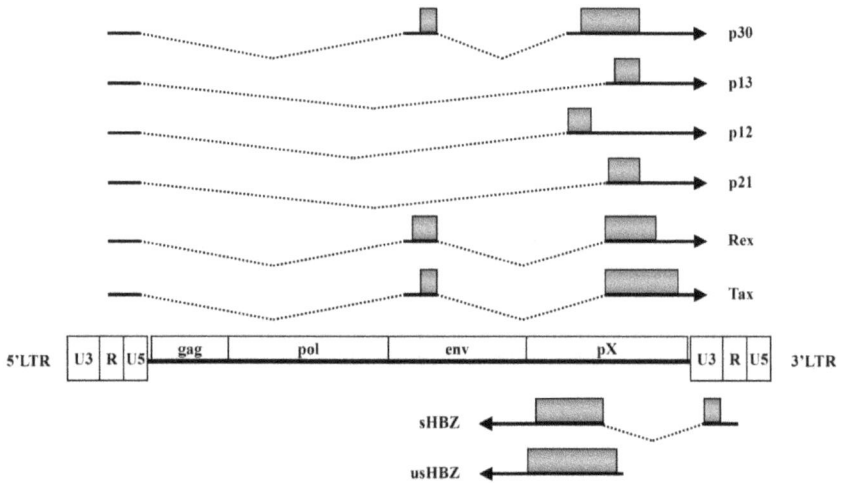

**Figure 1.** Regulatory and accessory genes encoded by HTLV-1. p12, p13, p30, Rex, p21, and Tax are transcribed from the 5' long terminal repeat (LTR). HBZ is located on the complementary proviral strand and transcribed from the 3' LTR. Spliced (s) and unspliced (us) HBZ are shown. Shaded boxes represent their coding regions.

## 2. Characteristics of HBZ

### 2.1. Expression of HBZ in ATL

In order to achieve successful transformation, HTLV-1 uses its genome very efficiently to encode multiple viral genes. Tax, an HTLV-1 plus strand-encoded viral gene, has been implicated in the leukemogenesis of HTLV-1 as it has growth-promoting activities and the ability to immortalize various types of cells *in vivo* [5,8,9,20]. However, Tax is the major target of CTLs [14]. Thus, ATL cells frequently lose Tax expression by several mechanisms. First, the 5' LTR of HTLV-1 provirus is reported to be deleted in 39% of ATL cases, resulting in the loss of Tax in ATL [21,22]. The second mechanism involves the nonsense mutation, deletion, and insertion of the *tax* gene in ATL cells [11,23]. The third mechanism includes DNA hypermethylation and histone modification of the 5' LTR of HTLV-1, which silences viral gene transcription [24,25]. Therefore, Tax may not be necessary for the development of ATL. On the contrary, the 3' LTR is conserved and unmethylated in all ATL cases [13,26]. Previous study demonstrated that the HBZ gene was expressed in all ATL cells and that HBZ gene knockdown inhibited the proliferation of HTLV-1-infected cells, indicating that HBZ may play a critical role in HTLV-1-mediated oncogenesis [12]. Sp1 binding sites in the 3' LTR of HTLV-1 have been demonstrated to be critical for HBZ promoter activity [26]. As Sp1 is a well-known regulator of housekeeping genes, the transcription of the HBZ gene may be relatively constant.

The expression of HBZ has been correlated with HTLV-1 proviral load [27,28]. Kinetic analyses of the HBZ transcript in HTLV-1-infected rabbits revealed that HBZ was detected at relatively low levels early after infection but gradually increased and stabilized, whereas other viral genes were maintained continuously at a low level [29]. Moreover, there is a correlation between the level of HBZ expression and the severity of HAM/TSP, indicating that high levels of HBZ may be associated with a greater risk of HAM/TSP [27].

A 5' RACE experiment identified two different HBZ transcripts in ATL; one is spliced (sHBZ) and the other is unspliced (usHBZ) (Figure 1) [12,30,31]. The s*HBZ* gene transcript level was approximately four times higher than the us*HBZ* gene transcript [28]. Consistently, the level of the usHBZ protein was much lower than that of sHBZ [28]. Furthermore, Western blot analyses could detect only the sHBZ protein in ATL cell lines [32,33].

## 2.2. Structure of HBZ

HBZ was first identified as a binding partner for the cAMP-response element binding protein-2 (CREB-2) by yeast two-hybrid screening [6]. Promoters for both the *sHBZ* and *usHBZ* genes are TATA-less [26]. The transcription factor Sp1 has been demonstrated to be important for TATA-less promoter activity [34,35]. Consistent with these observations, the sHBZ promoter is activated by the constitutively expressed Sp1 protein [26]. Moreover, Tax can activate the activity of sHBZ and usHBZ gene promoters through Tax-responsible elements (TREs) in the U3 region of HTLV-1 3' LTR [26,36]. However, Tax-mediated activation of antisense transcription is weaker than its activation of HTLV-1 sense transcription [26]. In addition, Tax-induced HBZ expression is influenced by the integration site in the host genome [36].

The sHBZ transcript is translated into a polypeptide of 206 amino acids, and the protein product of usHBZ is 209 amino acids long. Both HBZ isoforms contain three domains: activation domain (AD), central domain (CD), and basic leucine zipper domain (bZIP) [6,37]. The LXXLL-like motif, which is located at the N-terminal activation domain, is critical for HBZ-mediated activation of Smad3/TGF-$\beta$ pathway through binding to CBP/p300 [38]. There are three nuclear localization signals (NLSs) that are responsible for the nuclear localization of HBZ protein: two regions in the CD domain and a basic region in the bZIP domain [37]. A recent study demonstrated that HBZ contains a functional nuclear export signal (NES) sequence within its N-terminal region and disrupts the cellular autophagic response in the cytoplasm [39]. The sHBZ and unHBZ proteins differ in only seven amino acids at their N-terminal AD domains. The expression level of sHBZ is four times higher than that of usHBZ in ATL cells [28]. It is well known that different post-translational N-terminal modifications of proteins affect their half-lives. The half-life of sHBZ is much longer than that of usHBZ [26]. Thus, the N-terminal AD domain differences may be responsible for the distinct protein levels observed. In addition, Dissinger *et al.* [40] used an affinity-tagged protein and mass spectrometry method to identify seven modifications of HBZ protein. However, none of the identified post-translational modifications affected HBZ stability or its regulation of signaling pathways.

## 3. Oncogenic Properties of HBZ

### 3.1. Suppression of Viral Transcription

CREB-2 (ATF-4) was identified as a binding protein to HBZ [6]. Dimerization between HBZ and CREB-2 prevented CREB-2 from binding to a Tax-responsive element (TxRE) site in the HTLV-1 5' LTR, resulting in the suppression of Tax-mediated HTLV-1 5' LTR activation (Figure 2) [6]. Moreover, HBZ also represses CREB transcription from a cellular cyclic AMP-responsive element (CRE) in the cyclin D1 promoter, extending the inhibitory function of HBZ to CREB-dependent transcription of cellular genes [41]. The bZIP domain of HBZ contributes to this repression. In addition, Clerc *et al.* reported that HBZ interacts with p300/CBP and disrupts the interaction between Tax and p300/CBP, thereby inhibiting the Tax-dependent viral transcription [42]. Two LXXLL-like motifs located within the NH2-terminal region of HBZ mediate the interaction specifically through the KIX domain of the p300/CBP coactivator [42]. The two LXXLL motifs in the AD domain of HBZ promote binding to the mixed-lineage leukemia (MLL) surface of the KIX domain [43]. Formation of this interaction inhibits binding of MLL to the KIX domain while enhancing the binding of the transcription factor c-Myb to the opposite surface of KIX [43].

### 3.2. Promotion of T-Cell Proliferation

The development and continued growth of cancers involves altered rates of cell proliferation [44,45]. Multiple studies have demonstrated that HBZ plays critical roles in ATL leukemogenesis through pleiotropic actions, which include the promotion of cell proliferation [18,19]. Satou *et al.* reported that the HBZ gene enhances proliferation of T cells *in vitro* and *in vivo* [12]. Stable expression of HBZ gene increases Kit 225 cell proliferative capacity. Furthermore, repression of HBZ expression in ATL cell lines by shRNA inhibited the growth of ATL cells. Mutant analyses showed that HBZ promoted proliferation of T cells as a messenger RNA [12]. Growth-promoting activity was observed only in the cells expressing sHBZ and not in usHBZ-expressing cells [26]. Moreover, the percentage of CD4$^+$ T lymphocytes increased in splenocytes of HBZ transgenic (HBZ-Tg) mice, and HBZ-expressing T-cells proliferated more rapidly than those of non-transgenic mice [12,46]. In NOD/SCID$^{\gamma c-/-}$ (NOG) mice, HBZ-expressing SLB-1 cells engrafted to form solid tumor masses, but tumor formation was reduced significantly in animals challenged with HBZ-knockdown SLB-1 cells [32]. Thus, these data collectively indicate that HBZ expression enhances the proliferative capacity of HTLV-1-infected cells and plays a critical role in cell survival.

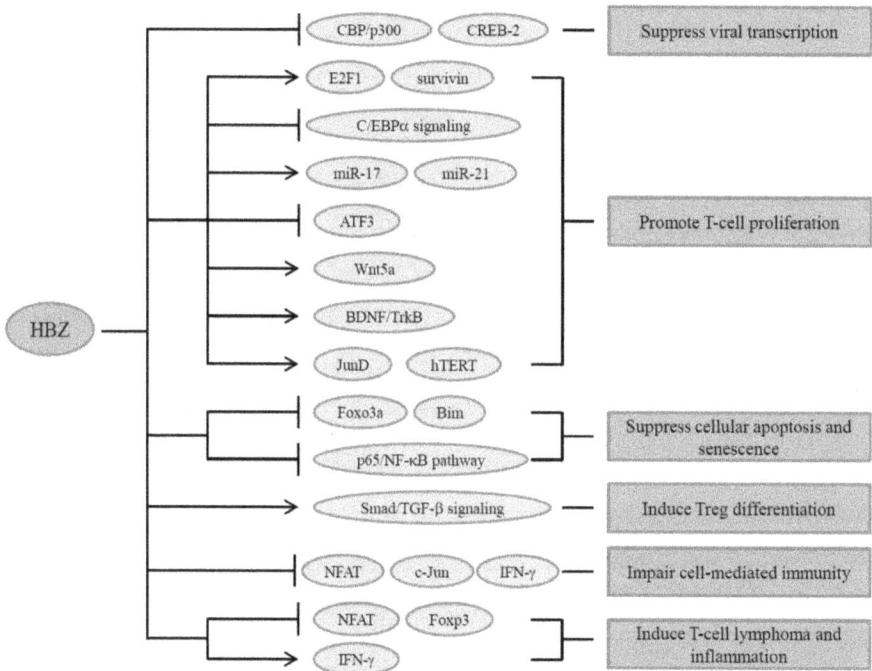

**Figure 2.** Oncogenic function of HBZ. HBZ fulfills its oncogenic functions mainly through regulating HTLV-1 5' LTR transcription and modulating a variety of cellular signaling pathways that are related to cell growth, apoptosis, immune escape, T-cell differentiation, and HTLV-1 pathogenesis. Detailed descriptions can be found in the text.

Recently, Mitobe *et al.* reported that *HBZ* RNA increased the number of CD4$^+$ T cells and attenuated cell death by activating the transcription of the *survivin* gene, which inhibits apoptosis (Figure 2). Moreover, the first 50 base pairs of the *HBZ* coding sequence are required for RNA-mediated cell proliferation [47].

Accumulating evidence suggests that the HBZ protein modulates cell growth through forming heterodimers with several host factors, such as C/EBPα and ATF3 (Figure 2) [48,49]. C/EBPα has emerged as an important negative regulator of cell proliferation in different cancers. In ATL cells, HBZ overcame the suppressive effect of C/EBPα on cell growth, leading to cell proliferation. A suggested underlying mechanism is that HBZ inhibits C/EBPα signaling activation by interacting with C/EBPα and diminishes its DNA binding capacity [48]. ATF3, an HBZ-binding protein, is constitutively expressed in ATL cell lines and fresh ATL cases. HBZ attenuates the negative effects of ATF3, allowing ATF3 to promote the proliferation of ATL cells by mechanisms that upregulate the expression of genes that are critical for mediating the cell cycle and cell death [49].

Increasing evidence has demonstrated that microRNAs (miRNA) play critical roles in the development of cancer [50]. In ATL, expression profiling of microRNA revealed that oncogenic miRNAs, including miR-17 and miR-21, are overexpressed in HTLV-1-infected T cells. These two miRNAs are post-transcriptionally upregulated by HBZ, and HBZ/miRNA-mediated downregulation of OBFC2A expression triggers both cell proliferation and genomic instability (Figure 2) [51].

Polakowski *et al.* reported that HBZ activates expression of neurotrophin BDNF. Moreover, HBZ promotes a BDNF/TrkB autocrine/paracrine signaling loop in HTLV-1-infected T cells, leading to the survival of these cells (Figure 2) [52].

In addition, HBZ suppresses the canonical Wnt pathway while enhancing the proliferation and migration of ATL cells by increasing expression of the noncanonical Wnt5a. These observations suggest that perturbation of the Wnt signaling pathways by HBZ is associated with the leukemogenesis of ATL (Figure 2) [53].

HBZ suppresses AP-1 signaling pathway, which is mediated by c-Jun and JunB [54–56]. However, HBZ can activate JunD-induced transcription by forming heterodimers with JunD, resulting in the activation of JunD-dependent cellular genes including human telomerase reverse transcriptase (hTERT) (Figure 2) [57–59]. The activation of telomerase by HBZ may contribute to the maintenance of leukemic cells.

### 3.3. Suppression of Cellular Apoptosis and Senescence

Defective apoptosis represents a major causative factor in the development and progression of cancer [44,45]. HTLV-1 suppresses apoptosis of infected cells by the interactions of viral proteins with host factors [60]. HBZ has been demonstrated to inhibit the transcription of a proapoptotic gene Bim, resulting in the decreased activation-induced cellular apoptosis. By interacting with FoxO3a, which is a transcriptional activator of the Bim gene, HBZ attenuates the DNA binding ability of FoxO3a and sequesters the inactive form of FoxO3a in the nucleus. Further study has identified that HBZ inhibited the expression of Bim by epigenetic alterations and histone modifications in the Bim promoter region. Thus, it may be advantageous for HBZ to suppress the apoptosis of ATL cells (Figure 2) [61].

Tax-induced nuclear factor-κB (NF-κB) activation plays a central role in HTLV-1-mediated transformation of human T cells [62,63]. However, hyper-activation of NF-κB by Tax triggers a defense mechanism that induces cellular senescence [64]. By contrast, HBZ delays or prevents the onset of Tax-induced cellular senescence by down-regulating NF-κB signaling (Figure 2). Zhao *et al.* reported that HBZ protein selectively inhibits Tax-mediated classical NF-κB activation by inhibiting p65 DNA binding capacity and by promoting expression of the PDLIM2 E3 ubiquitin ligase, which results in p65 degradation [65]. Inhibition of p65 acetylation by HBZ also contributes to the repression of the classic NF-κB pathway [66]. A recent study demonstrated that HBZ maintains viral latency by down-regulating Tax-mediated NF-κB activation and by inhibiting Rex-induced expression of viral proteins [67]. Taken together, these observations suggest that HBZ modulates Tax-mediated viral replication and NF-κB activation, thus allowing HTLV-1-infected cells to proliferate and persist.

In ATL, HBZ-mediated suppression of the classical NF-κB pathway decreases the expression of some genes associated with innate immunity and inflammatory responses [65]. NF-κB signaling is a well-established mediator of host immunity [68]. Therefore, HTLV-1 may facilitate escape from the host immune attack by suppressing the classical NF-κB pathway in such a manner.

Current data support the view that Tax may facilitate cell proliferation and survival in the early stage of HTLV-1 infection through activating NF-κB pathway. Nevertheless, Tax protein is the main target of the host's CTLs. Therefore, it is plausible that in the late stages of leukemogenesis, ATL cells that lack Tax expression are selected to emerge. In these cells, NF-κB-inducing kinase (NIK), a known activator of NF-κB, may replace Tax to maintain the constitutive activation of NF-κB, a hallmark of leukemic ATL cells [69]. At this stage, the mitogenic activity of HBZ may be required to maintain the proliferative nature of leukemic cells. This comprehensive scenario indicates that Tax and HBZ may cooperate for the long-term development and maintenance of leukemic cells in ATL.

*3.4. Induction of Regulatory T-Cell Differentiation*

Similar to regulatory T cells (Tregs), leukemic cells of ATL possess a CD4+CD25+ phenotype. The forkhead box P3 (FoxP3) is critical for the function of Tregs [5]. FoxP3 expression by HTLV-1 infected T cells is seen in two-thirds of ATL cases [70]. Previous reports indicate that the development and function of Tregs require the TGF-β signaling [71]. Notably, HTLV-1 infected T cells, unlike Tregs, are resistant to growth-inhibitory effect of TGF-β, thus the active TGF-β pathway does not impair the leukemic growth of ATL cells [72–75]. Recently, we observed that HBZ interacted with Smad2/3, key components of the TGF-β pathway, to form a ternary complex of HBZ/Smad3/p300 that enhanced TGF-β/Smad transcriptional responses in a p300-dependent manner. The enhancement of TGF-β signaling by HBZ results in the overexpression of Foxp3 in naïve T cells [38]. Increased CD4+Foxp3+ Treg cells were also observed in HBZ-transgenic mice (Figure 2). Furthermore, a luciferase assay validated that HBZ induces transcription of the *Foxp3* gene. Thus, HBZ-induced Foxp3 expression could be a mechanism for the increase of Foxp3+ Treg cells *in vitro* and *in vivo* [46].

Numerous viruses have developed strategies to modulate TGF-β signaling using viral proteins. Examples include hepatitis B virus pX; hepatitis C virus core protein, NS3 and NS5; Kaposi sarcoma-associated herpesvirus K-bZIP; and Epstein-Barr virus LMP1 [76–79]. Like HBZ, the HBV pX and severe acute respiratory syndrome N protein enhance the transcriptional responses of TGF-β. Curiously, these viruses seem to employ a common strategy to nullify the TGF-β signaling by having their viral proteins bind to Smad proteins [76–78].

*3.5. Impaired Cell-Mediated Immunity*

Impaired cell-mediated immunity has been demonstrated in HTLV-1 carriers and ATL patients, causing frequent opportunistic infections by various pathogens [80]. However, the mechanism by which HTLV-1 causes immune deficiency has not been well studied. Sugata *et al.* observed that HBZ transgenic mice were highly susceptible to intravaginal infection with herpes simplex virus type 2 (HSV-2) and displayed decreased immune responses to primary and secondary infection with *Listeria monocytogenes* (LM). The production of IFN-γ by CD4+ T cells was shown to be suppressed in HBZ-Tg mice [81]. Previous studies have reported that HBZ suppresses host cell signaling pathways that are critical for T-cell immune response, such as the NF-κB, AP-1, and NFAT (Figure 2) [46,54,56,65]. Indeed, HBZ suppresses IFN-γ transcription through interaction with NFAT and c-Jun. Thus, HBZ inhibits cell-mediated immunity *in vivo* by interfering with the host cell signaling pathway, suggesting important roles for HBZ in HTLV-1-induced immunodeficiency.

*3.6. Induction of T-Cell Lymphoma and Systemic Inflammation*

To study the function of HBZ *in vivo*, Satou *et al.* generated transgenic mice expressing HBZ under the control of the mouse CD4 promoter/enhancer, which induces HBZ gene expression specifically in CD4+ T cells [46]. Similar to HTLV-1 infected individuals, the majority of HBZ transgenic mice spontaneously developed chronic inflammation in the skin and lungs. Infiltration of CD3+CD4+ T cells into the dermis and epidermis in the lesions of HBZ-Tg mice was apparent. Moreover, one-third of HBZ-Tg mice developed T-cell lymphomas after a long latent period [46,82]. As observed in HTLV-1-infected individuals, more effector/memory and regulatory CD4+ T cells were detected in the HBZ transgenic mice. However, the function of CD4+Foxp3+ Treg cells in HBZ transgenic mice was impaired, whereas their proliferation increased. As a mechanism, HBZ impairs the suppressive function of Treg cells by binding to Foxp3 and NFAT (Figure 2) [46]. Further studies demonstrated that HBZ-mediated inflammation is closely linked to oncogenesis in CD4+ T cells and that IFN-γ is an accelerator of HBZ-induced inflammation [83]. Yamamoto-Taguchi *et al.* reported that iTreg cells increased in HBZ-Tg mice and that Treg cells of HBZ-Tg mice tend to lose Foxp3 expression, leading to increased IFN-γ-expressing proinflammatory cells [84]. Cell adhesion and migration are enhanced in the CD4$^+$ T cells of HBZ-Tg mice. Thus, HBZ seems to impair various functions of conventional and regulatory T cells and thereby critically contributes to the development of inflammation *in vivo*.

### 3.7. Differences between HBZ and APH-2

HTLV-1 and HTLV-2 are closely related human retroviruses that have been studied extensively [85,86]. HTLV-1 is associated with ATL and a variety of immune-mediated disorders including HAM/TSP. In contrast, HTLV-2 is much less pathogenic, with only a few cases of variant hairy cell leukemia and neurological disease reported. Similar to HBZ, HTLV-2 also generates an antisense transcript, termed APH-2 (antisense protein of HTLV-2) [87,88]. While most components of HTLV-1 and HTLV-2, including Tax-1 and Tax-2, show high degree of conservation, APH-2 exhibits less than 30% homology to HBZ. Although APH-2 does not harbor a bZIP domain, it can suppress Tax2-mediated viral transcription by interacting with CREB, similar to the effects displayed by HBZ [87]. APH-2 expression was found to correlate with the proviral load in HTLV-2-infected carriers; however, APH-2 did not appear to promote T cell proliferation and lymphocytosis. Arnold *et al.* reported that HBZ is not required for efficient infectivity or HTLV-1-mediated immortalization of primary human T lymphocytes *in vitro* [89]. However, HBZ enhanced infectivity and persistence of HTLV-1 in inoculated rabbits, indicating that HBZ is not required for cellular immortalization but enhances infectivity and persistence *in vivo* [32]. Unlike HBZ, APH-2 is dispensable for enhancing viral replication and persistent infection in the rabbit animal model [90]. In addition, APH-2 could contribute to the lower virulence of HTLV-2 [90]. Thus, the antisense transcripts of HTLV-1 and HTLV-2 exhibit different functions *in vivo*, and further studies are necessary to clarify their distinct pathobiologies.

### 4. Perspectives

It has been 35 years since the discovery of HTLV-1. Numerous studies have focused on elucidating the molecular mechanisms by which HTLV-1-encoded viral proteins induce viral replication, cellular transformation, and oncogenesis. Tax is thought to have an important role in the leukemogenesis of HTLV-1 because of its pleiotropic functions. However, Tax expression is frequently lost during the development of ATL suggesting that Tax may not be essential for this process. By contrast, HBZ is the only viral gene that is constitutively expressed in ATL cases and thus is a plausible player to critically control the ATL leukemogenesis. In addition, APOBEC3G frequently introduces mutations to HTLV-1 genes before proviral integration but HBZ seems to be spared from this alteration. As discussed in this review, current findings support the view that HBZ is indispensable for leukemogenesis by HTLV-1. Further studies are needed to elucidate the precise molecular mechanisms by which HBZ induces oncogenesis so that novel therapies targeting HBZ could be developed.

**Acknowledgments:** This work was supported by a grant from the National Natural Science Foundation of China to Tiejun Zhao (No. 31200128 and No. 31470262), a grant from the Science Technology Department of Zhejiang Province (China) to Tiejun Zhao (No. 2015C33149), and a grant from the Qianjiang Talent Program of Zhejiang Province (China) to Tiejun Zhao.

**Conflicts of Interest:** The author declares no conflict of interest.

### References

1. Uchiyama, T.; Yodoi, J.; Sagawa, K.; Takatsuki, K.; Uchino, H. Adult T-cell leukemia: Clinical and hematologic features of 16 cases. *Blood* **1977**, *50*, 481–492. [CrossRef]
2. Poiesz, B.J.; Ruscetti, F.W.; Gazdar, A.F.; Bunn, P.A.; Minna, J.D.; Gallo, R.C. Detection and isolation of type C retrovirus particles from fresh and cultured lymphocytes of a patient with cutaneous T-cell lymphoma. *Proc. Natl. Acad. Sci. USA* **1980**, *77*, 7415–7419. [CrossRef] [PubMed]

3.  Gessain, A.; Barin, F.; Vernant, J.C.; Gout, O.; Maurs, L.; Calender, A.; de The, G. Antibodies to human T-lymphotropic virus type-I in patients with tropical spastic paraparesis. *Lancet* **1985**, *2*, 407–410. [CrossRef]
4.  Osame, M.; Usuku, K.; Izumo, S.; Ijichi, N.; Amitani, H.; Igata, A.; Matsumoto, M.; Tara, M. HTLV-I associated myelopathy, a new clinical entity. *Lancet* **1986**, *1*, 1031–1032. [CrossRef]
5.  Matsuoka, M.; Jeang, K.T. Human T-cell leukaemia virus type 1 (HTLV-1) infectivity and cellular transformation. *Nat. Rev. Cancer* **2007**, *7*, 270–280. [CrossRef] [PubMed]
6.  Gaudray, G.; Gachon, F.; Basbous, J.; Biard-Piechaczyk, M.; Devaux, C.; Mesnard, J.M. The complementary strand of the human T-cell leukemia virus type 1 RNA genome encodes a bZIP transcription factor that down-regulates viral transcription. *J. Virol.* **2002**, *76*, 12813–12822. [CrossRef] [PubMed]
7.  Nicot, C.; Harrod, R.L.; Ciminale, V.; Franchini, G. Human T-cell leukemia/lymphoma virus type 1 nonstructural genes and their functions. *Oncogene* **2005**, *24*, 6026–6034. [CrossRef] [PubMed]
8.  Grassmann, R.; Aboud, M.; Jeang, K.T. Molecular mechanisms of cellular transformation by HTLV-1 Tax. *Oncogene* **2005**, *24*, 5976–5985. [CrossRef] [PubMed]
9.  Giam, C.Z.; Jeang, K.T. HTLV-1 Tax and adult T-cell leukemia. *Front. Biosci.* **2007**, *12*, 1496–1507. [CrossRef] [PubMed]
10. Yoshida, M. HTLV-1 Tax: Regulation of gene expression and disease. *Trends Microbiol.* **1993**, *1*, 131–135. [CrossRef]
11. Takeda, S.; Maeda, M.; Morikawa, S.; Taniguchi, Y.; Yasunaga, J.; Nosaka, K.; Tanaka, Y.; Matsuoka, M. Genetic and epigenetic inactivation of tax gene in adult T-cell leukemia cells. *Int. J. Cancer* **2004**, *109*, 559–567. [CrossRef]
12. Satou, Y.; Yasunaga, J.; Yoshida, M.; Matsuoka, M. HTLV-I basic leucine zipper factor gene mRNA supports proliferation of adult T cell leukemia cells. *Proc. Natl. Acad. Sci. USA* **2006**, *103*, 720–725. [CrossRef]
13. Fan, J.; Ma, G.; Nosaka, K.; Tanabe, J.; Satou, Y.; Koito, A.; Wain-Hobson, S.; Vartanian, J.P.; Matsuoka, M. APOBEC3G generates nonsense mutations in human T-cell leukemia virus type 1 proviral genomes *in vivo*. *J. Virol.* **2010**, *84*, 7278–7287. [CrossRef] [PubMed]
14. Kannagi, M.; Harada, S.; Maruyama, I.; Inoko, H.; Igarashi, H.; Kuwashima, G.; Sato, S.; Morita, M.; Kidokoro, M.; Sugimoto, M.; *et al*. Predominant recognition of human T cell leukemia virus type I (HTLV-I) pX gene products by human CD8+ cytotoxic T cells directed against HTLV-I-infected cells. *Int. Immunol.* **1991**, *3*, 761–767. [CrossRef]
15. Suemori, K.; Fujiwara, H.; Ochi, T.; Ogawa, T.; Matsuoka, M.; Matsumoto, T.; Mesnard, J.M.; Yasukawa, M. HBZ is an immunogenic protein, but not a target antigen for human T-cell leukemia virus type 1-specific cytotoxic T lymphocytes. *J. Gen. Virol.* **2009**, *90*, 1806–1811. [CrossRef] [PubMed]
16. Rowan, A.G.; Suemori, K.; Fujiwara, H.; Yasukawa, M.; Tanaka, Y.; Taylor, G.P.; Bangham, C.R. Cytotoxic T lymphocyte lysis of HTLV-1 infected cells is limited by weak HBZ protein expression, but non-specifically enhanced on induction of Tax expression. *Retrovirology* **2014**, *11*. [CrossRef]
17. Sugata, K.; Yasunaga, J.; Mitobe, Y.; Miura, M.; Miyazato, P.; Kohara, M.; Matsuoka, M. Protective effect of cytotoxic T lymphocytes targeting HTLV-1 bZIP factor. *Blood* **2015**, *126*, 1095–1105. [CrossRef] [PubMed]
18. Matsuoka, M.; Green, P.L. The HBZ gene, a key player in HTLV-1 pathogenesis. *Retrovirology* **2009**, *6*. [CrossRef] [PubMed]
19. Zhao, T.; Matsuoka, M. HBZ and its roles in HTLV-1 oncogenesis. *Front. Microbiol.* **2012**, *3*. [CrossRef] [PubMed]
20. Yamaoka, S.; Tobe, T.; Hatanaka, M. Tax protein of human T-cell leukemia virus type I is required for maintenance of the transformed phenotype. *Oncogene* **1992**, *7*, 433–437. [PubMed]
21. Tamiya, S.; Matsuoka, M.; Etoh, K.; Watanabe, T.; Kamihira, S.; Yamaguchi, K.; Takatsuki, K. Two types of defective human T-lymphotropic virus type I provirus in adult T-cell leukemia. *Blood* **1996**, *88*, 3065–3073. [PubMed]
22. Miyazaki, M.; Yasunaga, J.; Taniguchi, Y.; Tamiya, S.; Nakahata, T.; Matsuoka, M. Preferential selection of human T-cell leukemia virus type 1 provirus lacking the 5′ long terminal repeat during oncogenesis. *J. Virol.* **2007**, *81*, 5714–5723. [CrossRef] [PubMed]
23. Furukawa, Y.; Kubota, R.; Tara, M.; Izumo, S.; Osame, M. Existence of escape mutant in HTLV-I tax during the development of adult T-cell leukemia. *Blood* **2001**, *97*, 987–993. [CrossRef]

24. Koiwa, T.; Hamano-Usami, A.; Ishida, T.; Okayama, A.; Yamaguchi, K.; Kamihira, S.; Watanabe, T. 5'-long terminal repeat-selective CpG methylation of latent human T-cell leukemia virus type 1 provirus *in vitro* and *in vivo*. *J. Virol.* **2002**, *76*, 9389–9397. [CrossRef] [PubMed]

25. Taniguchi, Y.; Nosaka, K.; Yasunaga, J.; Maeda, M.; Mueller, N.; Okayama, A.; Matsuoka, M. Silencing of human T-cell leukemia virus type I gene transcription by epigenetic mechanisms. *Retrovirology* **2005**, *2*. [CrossRef] [PubMed]

26. Yoshida, M.; Satou, Y.; Yasunaga, J.; Fujisawa, J.; Matsuoka, M. Transcriptional control of spliced and unspliced human T-cell leukemia virus type 1 bZIP factor (HBZ) gene. *J. Virol.* **2008**, *82*, 9359–9368. [CrossRef] [PubMed]

27. Saito, M.; Matsuzaki, T.; Satou, Y.; Yasunaga, J.; Saito, K.; Arimura, K.; Matsuoka, M.; Ohara, Y. In vivo expression of the HBZ gene of HTLV-1 correlates with proviral load, inflammatory markers and disease severity in HTLV-1 associated myelopathy/tropical spastic paraparesis (HAM/TSP). *Retrovirology* **2009**, *6*. [CrossRef] [PubMed]

28. Usui, T.; Yanagihara, K.; Tsukasaki, K.; Murata, K.; Hasegawa, H.; Yamada, Y.; Kamihira, S. Characteristic expression of HTLV-1 basic zipper factor (HBZ) transcripts in HTLV-1 provirus-positive cells. *Retrovirology* **2008**, *5*. [CrossRef] [PubMed]

29. Li, M.; Kesic, M.; Yin, H.; Yu, L.; Green, P.L. Kinetic analysis of human T-cell leukemia virus type 1 gene expression in cell culture and infected animals. *J. Virol.* **2009**, *83*, 3788–3797. [CrossRef]

30. Cavanagh, M.H.; Landry, S.; Audet, B.; Arpin-Andre, C.; Hivin, P.; Pare, M.E.; Thete, J.; Wattel, E.; Marriott, S.J.; Mesnard, J.M.; Barbeau, B. HTLV-I antisense transcripts initiating in the 3'LTR are alternatively spliced and polyadenylated. *Retrovirology* **2006**, *3*. [CrossRef] [PubMed]

31. Murata, K.; Hayashibara, T.; Sugahara, K.; Uemura, A.; Yamaguchi, T.; Harasawa, H.; Hasegawa, H.; Tsuruda, K.; Okazaki, T.; Koji, T.; et al. A novel alternative splicing isoform of human T-cell leukemia virus type 1 bZIP factor (HBZ-SI) targets distinct subnuclear localization. *J. Virol.* **2006**, *80*, 2495–2505. [CrossRef] [PubMed]

32. Arnold, J.; Zimmerman, B.; Li, M.; Lairmore, M.D.; Green, P.L. Human T-cell leukemia virus type-1 antisense-encoded gene, Hbz, promotes T-lymphocyte proliferation. *Blood* **2008**, *112*, 3788–3797. [CrossRef] [PubMed]

33. Raval, G.U.; Bidoia, C.; Forlani, G.; Tosi, G.; Gessain, A.; Accolla, R.S. Localization, quantification and interaction with host factors of endogenous HTLV-1 HBZ protein in infected cells and ATL. *Retrovirology* **2015**, *12*. [CrossRef] [PubMed]

34. Boam, D.S.; Davidson, I.; Chambon, P. A TATA-less promoter containing binding sites for ubiquitous transcription factors mediates cell type-specific regulation of the gene for transcription enhancer factor-1 (TEF-1). *J. Biol. Chem.* **1995**, *270*, 19487–19494.

35. Liu, S.; Cowell, J.K. Cloning and characterization of the TATA-less promoter from the human GFI1 proto-oncogene. *Ann. Hum. Genet.* **2000**, *64*, 83–86. [CrossRef] [PubMed]

36. Landry, S.; Halin, M.; Vargas, A.; Lemasson, I.; Mesnard, J.M.; Barbeau, B. Upregulation of human T-cell leukemia virus type 1 antisense transcription by the viral Tax protein. *J. Virol.* **2009**, *83*, 2048–2054. [CrossRef] [PubMed]

37. Hivin, P.; Frederic, M.; Arpin-Andre, C.; Basbous, J.; Gay, B.; Thebault, S.; Mesnard, J.M. Nuclear localization of HTLV-I bZIP factor (HBZ) is mediated by three distinct motifs. *J. Cell Sci.* **2005**, *118*, 1355–1362. [CrossRef] [PubMed]

38. Zhao, T.; Satou, Y.; Sugata, K.; Miyazato, P.; Green, P.L.; Imamura, T.; Matsuoka, M. HTLV-1 bZIP factor enhances TGF-beta signaling through p300 coactivator. *Blood* **2011**, *118*, 1865–1876. [CrossRef] [PubMed]

39. Mukai, R.; Ohshima, T. HTLV-1 HBZ positively regulates the mTOR signaling pathway via inhibition of GADD34 activity in the cytoplasm. *Oncogene* **2014**, *33*, 2317–2328. [CrossRef] [PubMed]

40. Dissinger, N.; Shkriabai, N.; Hess, S.; Al-Saleem, J.; Kvaratskhelia, M.; Green, P.L. Identification and characterization of HTLV-1 HBZ post-translational modifications. *PLoS ONE* **2014**, *9*, e112762.

41. Ma, Y.; Zheng, S.; Wang, Y.; Zang, W.; Li, M.; Wang, N.; Li, P.; Jin, J.; Dong, Z.; Zhao, G. The HTLV-1 HBZ protein inhibits cyclin D1 expression through interacting with the cellular transcription factor CREB. *Mol. Biol. Rep.* **2013**, *40*, 5967–5975. [CrossRef] [PubMed]

42. Clerc, I.; Polakowski, N.; Andre-Arpin, C.; Cook, P.; Barbeau, B.; Mesnard, J.M.; Lemasson, I. An interaction between the human T cell leukemia virus type 1 basic leucine zipper factor (HBZ) and the KIX domain of p300/CBP contributes to the down-regulation of Tax-dependent viral transcription by HBZ. *J. Biol. Chem.* **2008**, *283*, 23903–23913. [CrossRef] [PubMed]

43. Cook, P.R.; Polakowski, N.; Lemasson, I. HTLV-1 HBZ protein deregulates interactions between cellular factors and the KIX domain of p300/CBP. *J. Mol. Biol.* **2011**, *409*, 384–398. [CrossRef] [PubMed]

44. Hanahan, D.; Weinberg, R.A. Hallmarks of cancer: The next generation. *Cell* **2011**, *144*, 646–674. [CrossRef] [PubMed]

45. Hanahan, D.; Weinberg, R.A. The hallmarks of cancer. *Cell* **2000**, *100*, 57–70. [CrossRef]

46. Satou, Y.; Yasunaga, J.; Zhao, T.; Yoshida, M.; Miyazato, P.; Takai, K.; Shimizu, K.; Ohshima, K.; Green, P.L.; Ohkura, N.; *et al.* HTLV-1 bZIP factor induces T-cell lymphoma and systemic inflammation *in vivo. PLoS Pathog.* **2011**, *7*, e1001274. [CrossRef] [PubMed]

47. Mitobe, Y.; Yasunaga, J.; Furuta, R.; Matsuoka, M. HTLV-1 bZIP factor RNA and protein impart distinct functions on T-cell proliferation and survival. *Cancer Res.* **2015**, *75*, 4143–4152. [CrossRef] [PubMed]

48. Zhao, T.; Coutts, A.; Xu, L.; Yu, J.; Ohshima, K.; Matsuoka, M. HTLV-1 bZIP factor supports proliferation of adult T cell leukemia cells through suppression of C/EBPα signaling. *Retrovirology* **2013**, *10*. [CrossRef] [PubMed]

49. Hagiya, K.; Yasunaga, J.; Satou, Y.; Ohshima, K.; Matsuoka, M. ATF3, an HTLV-1 bZip factor binding protein, promotes proliferation of adult T-cell leukemia cells. *Retrovirology* **2011**, *8*. [CrossRef] [PubMed]

50. Esquela-Kerscher, A.; Slack, F.J. Oncomirs—microRNAs with a role in cancer. *Nat. Rev. Cancer* **2006**, *6*, 259–269. [CrossRef] [PubMed]

51. Vernin, C.; Thenoz, M.; Pinatel, C.; Gessain, A.; Gout, O.; Delfau-Larue, M.H.; Nazaret, N.; Legras-Lachuer, C.; Wattel, E.; Mortreux, F. HTLV-1 bZIP factor HBZ promotes cell proliferation and genetic instability by activating OncomiRs. *Cancer Res.* **2014**, *74*, 6082–6093. [CrossRef] [PubMed]

52. Polakowski, N.; Terol, M.; Hoang, K.; Nash, I.; Laverdure, S.; Gazon, H.; Belrose, G.; Mesnard, J.M.; Cesaire, R.; Peloponese, J.M.; *et al.* HBZ stimulates brain-derived neurotrophic factor/TrkB autocrine/paracrine signaling to promote survival of human T-cell leukemia virus type 1-Infected T cells. *J. Virol.* **2014**, *88*, 13482–13494. [CrossRef] [PubMed]

53. Ma, G.; Yasunaga, J.; Fan, J.; Yanagawa, S.; Matsuoka, M. HTLV-1 bZIP factor dysregulates the Wnt pathways to support proliferation and migration of adult T-cell leukemia cells. *Oncogene* **2013**, *32*, 4222–4230. [CrossRef] [PubMed]

54. Basbous, J.; Arpin, C.; Gaudray, G.; Piechaczyk, M.; Devaux, C.; Mesnard, J.M. The HBZ factor of human T-cell leukemia virus type I dimerizes with transcription factors JunB and c-Jun and modulates their transcriptional activity. *J. Biol. Chem.* **2003**, *278*, 43620–43627. [CrossRef] [PubMed]

55. Hivin, P.; Basbous, J.; Raymond, F.; Henaff, D.; Arpin-Andre, C.; Robert-Hebmann, V.; Barbeau, B.; Mesnard, J.M. The HBZ-SP1 isoform of human T-cell leukemia virus type I represses JunB activity by sequestration into nuclear bodies. *Retrovirology* **2007**, *4*. [CrossRef] [PubMed]

56. Matsumoto, J.; Ohshima, T.; Isono, O.; Shimotohno, K. HTLV-1 HBZ suppresses AP-1 activity by impairing both the DNA-binding ability and the stability of c-Jun protein. *Oncogene* **2005**, *24*, 1001–1010. [CrossRef] [PubMed]

57. Thebault, S.; Basbous, J.; Hivin, P.; Devaux, C.; Mesnard, J.M. HBZ interacts with JunD and stimulates its transcriptional activity. *FEBS Lett.* **2004**, *562*, 165–170. [CrossRef]

58. Kuhlmann, A.S.; Villaudy, J.; Gazzolo, L.; Castellazzi, M.; Mesnard, J.M.; Duc Dodon, M. HTLV-1 HBZ cooperates with JunD to enhance transcription of the human telomerase reverse transcriptase gene (hTERT). *Retrovirology* **2007**, *4*. [CrossRef] [PubMed]

59. Borowiak, M.; Kuhlmann, A.S.; Girard, S.; Gazzolo, L.; Mesnard, J.M.; Jalinot, P.; Dodon, M.D. HTLV-1 bZIP factor impedes the menin tumor suppressor and upregulates JunD-mediated transcription of the hTERT gene. *Carcinogenesis* **2013**, *34*, 2664–2672. [CrossRef] [PubMed]

60. Saggioro, D.; Silic-Benussi, M.; Biasiotto, R.; D'Agostino, D.M.; Ciminale, V. Control of cell death pathways by HTLV-1 proteins. *Front. Biosci.* **2009**, *14*, 3338–3351. [CrossRef]

61. Tanaka-Nakanishi, A.; Yasunaga, J.; Takai, K.; Matsuoka, M. HTLV-1 bZIP factor suppresses apoptosis by attenuating the function of FoxO3a and altering its localization. *Cancer Res.* **2014**, *74*, 188–200. [CrossRef] [PubMed]

62. Yamaoka, S.; Inoue, H.; Sakurai, M.; Sugiyama, T.; Hazama, M.; Yamada, T.; Hatanaka, M. Constitutive activation of NF-κB is essential for transformation of rat fibroblasts by the human T-cell leukemia virus type I Tax protein. *EMBO J.* **1996**, *15*, 873–887. [PubMed]

63. Sun, S.C.; Yamaoka, S. Activation of NF-κB by HTLV-I and implications for cell transformation. *Oncogene* **2005**, *24*, 5952–5964. [CrossRef] [PubMed]

64. Zhi, H.; Yang, L.; Kuo, Y.L.; Ho, Y.K.; Shih, H.M.; Giam, C.Z. NF-κB hyper-activation by HTLV-1 tax induces cellular senescence, but can be alleviated by the viral anti-sense protein HBZ. *PLoS Pathog.* **2011**, *7*, e1002025. [CrossRef] [PubMed]

65. Zhao, T.; Yasunaga, J.; Satou, Y.; Nakao, M.; Takahashi, M.; Fujii, M.; Matsuoka, M. Human T-cell leukemia virus type 1 bZIP factor selectively suppresses the classical pathway of NF-κB. *Blood* **2009**, *113*, 2755–2764. [CrossRef] [PubMed]

66. Wurm, T.; Wright, D.G.; Polakowski, N.; Mesnard, J.M.; Lemasson, I. The HTLV-1-encoded protein HBZ directly inhibits the acetyl transferase activity of p300/CBP. *Nucleic Acids Res.* **2012**, *40*, 5910–5925. [CrossRef] [PubMed]

67. Philip, S.; Zahoor, M.A.; Zhi, H.; Ho, Y.K.; Giam, C.Z. Regulation of human T-lymphotropic virus type I latency and reactivation by HBZ and Rex. *PLoS Pathog.* **2014**, *10*, e1004040. [CrossRef] [PubMed]

68. Beinke, S.; Ley, S.C. Functions of NF-κB1 and NF-κB2 in immune cell biology. *Biochem. J.* **2004**, *382*, 393–409. [CrossRef]

69. Saitoh, Y.; Yamamoto, N.; Dewan, M.Z.; Sugimoto, H.; Martinez Bruyn, V.J.; Iwasaki, Y.; Matsubara, K.; Qi, X.; Saitoh, T.; Imoto, I.; *et al.* Overexpressed NF-κB-inducing kinase contributes to the tumorigenesis of adult T-cell leukemia and Hodgkin Reed-Sternberg cells. *Blood* **2008**, *111*, 5118–5129. [CrossRef]

70. Karube, K.; Ohshima, K.; Tsuchiya, T.; Yamaguchi, T.; Kawano, R.; Suzumiya, J.; Utsunomiya, A.; Harada, M.; Kikuchi, M. Expression of FoxP3, a key molecule in CD4CD25 regulatory T cells, in adult T-cell leukaemia/lymphoma cells. *Br. J. Haematol.* **2004**, *126*, 81–84. [CrossRef] [PubMed]

71. Bommireddy, R.; Doetschman, T. TGF-β, T-cell tolerance and anti-CD3 therapy. *Trends Mol. Med.* **2004**, *10*, 3–9. [CrossRef] [PubMed]

72. Niitsu, Y.; Urushizaki, Y.; Koshida, Y.; Terui, K.; Mahara, K.; Kohgo, Y.; Urushizaki, I. Expression of TGF-β gene in adult T cell leukemia. *Blood* **1988**, *71*, 263–266. [PubMed]

73. Arnulf, B.; Villemain, A.; Nicot, C.; Mordelet, E.; Charneau, P.; Kersual, J.; Zermati, Y.; Mauviel, A.; Bazarbachi, A.; Hermine, O. Human T-cell lymphotropic virus oncoprotein Tax represses TGF-beta 1 signaling in human T cells via c-Jun activation: A potential mechanism of HTLV-I leukemogenesis. *Blood* **2002**, *100*, 4129–4138. [CrossRef]

74. Lee, D.K.; Kim, B.C.; Brady, J.N.; Jeang, K.T.; Kim, S.J. Human T-cell lymphotropic virus type 1 tax inhibits transforming growth factor-beta signaling by blocking the association of Smad proteins with Smad-binding element. *J. Biol. Chem.* **2002**, *277*, 33766–33775. [CrossRef] [PubMed]

75. Mori, N.; Morishita, M.; Tsukazaki, T.; Giam, C.Z.; Kumatori, A.; Tanaka, Y.; Yamamoto, N. Human T-cell leukemia virus type I oncoprotein Tax represses Smad-dependent transforming growth factor β signaling through interaction with CREB-binding protein/p300. *Blood* **2001**, *97*, 2137–2144. [CrossRef] [PubMed]

76. Cheng, P.L.; Chang, M.H.; Chao, C.H.; Lee, Y.H. Hepatitis C viral proteins interact with Smad3 and differentially regulate TGF-β/Smad3-mediated transcriptional activation. *Oncogene* **2004**, *23*, 7821–7838. [CrossRef]

77. Lee, D.K.; Park, S.H.; Yi, Y.; Choi, S.G.; Lee, C.; Parks, W.T.; Cho, H.; de Caestecker, M.P.; Shaul, Y.; Roberts, A.B.; Kim, S.J. The hepatitis B virus encoded oncoprotein pX amplifies TGF-β family signaling through direct interaction with Smad4: Potential mechanism of hepatitis B virus-induced liver fibrosis. *Genes Dev.* **2001**, *15*, 455–466. [CrossRef] [PubMed]

78. Prokova, V.; Mosialos, G.; Kardassis, D. Inhibition of transforming growth factor beta signaling and Smad-dependent activation of transcription by the Latent Membrane Protein 1 of Epstein-Barr virus. *J. Biol. Chem.* **2002**, *277*, 9342–9350. [CrossRef] [PubMed]

79. Tomita, M.; Choe, J.; Tsukazaki, T.; Mori, N. The Kaposi's sarcoma-associated herpesvirus K-bZIP protein represses transforming growth factor β signaling through interaction with CREB-binding protein. *Oncogene* **2004**, *23*, 8272–8281. [CrossRef] [PubMed]

80. Takatsuki, K.; Matsuoka, M.; Yamaguchi, K. ATL and HTLV-1-related diseases. In *Adult T-Cell Leukemia*; Oxford University Press: New York, NY, USA, 1994; pp. 1–27.

81. Sugata, K.; Satou, Y.; Yasunaga, J.; Hara, H.; Ohshima, K.; Utsunomiya, A.; Mitsuyama, M.; Matsuoka, M. HTLV-1 bZIP factor impairs cell-mediated immunity by suppressing production of Th1 cytokines. *Blood* **2012**, *119*, 434–444. [CrossRef]

82. Zhao, T.; Satou, Y.; Matsuoka, M. Development of T cell lymphoma in HTLV-1 bZIP factor and Tax double transgenic mice. *Arch. Virol.* **2014**, *159*, 1849–1856. [CrossRef] [PubMed]

83. Mitagami, Y.; Yasunaga, J.; Kinosada, H.; Ohshima, K.; Matsuoka, M. Interferon-γ promotes inflammation and development of T-cell lymphoma in HTLV-1 bZIP factor transgenic mice. *PLoS Pathog.* **2015**, *11*, e1005120.

84. Yamamoto-Taguchi, N.; Satou, Y.; Miyazato, P.; Ohshima, K.; Nakagawa, M.; Katagiri, K.; Kinashi, T.; Matsuoka, M. HTLV-1 bZIP factor induces inflammation through labile Foxp3 expression. *PLoS Pathog.* **2013**, *9*, e1003630. [CrossRef] [PubMed]

85. Feuer, G.; Green, P.L. Comparative biology of human T-cell lymphotropic virus type 1 (HTLV-1) and HTLV-2. *Oncogene* **2005**, *24*, 5996–6004. [CrossRef] [PubMed]

86. Ciminale, V.; Rende, F.; Bertazzoni, U.; Romanelli, M.G. HTLV-1 and HTLV-2: Highly similar viruses with distinct oncogenic properties. *Front. Microbiol.* **2014**, *5*. [CrossRef] [PubMed]

87. Halin, M.; Douceron, E.; Clerc, I.; Journo, C.; Ko, N.L.; Landry, S.; Murphy, E.L.; Gessain, A.; Lemasson, I.; Mesnard, J.M.; *et al.* Human T-cell leukemia virus type 2 produces a spliced antisense transcript encoding a protein that lacks a classic bZIP domain but still inhibits Tax2-mediated transcription. *Blood* **2009**, *114*, 2427–3248. [CrossRef]

88. Barbeau, B.; Peloponese, J.M.; Mesnard, J.M. Functional comparison of antisense proteins of HTLV-1 and HTLV-2 in viral pathogenesis. *Front. Microbiol.* **2013**, *4*. [CrossRef] [PubMed]

89. Arnold, J.; Yamamoto, B.; Li, M.; Phipps, A.J.; Younis, I.; Lairmore, M.D.; Green, P.L. Enhancement of infectivity and persistence *in vivo* by HBZ, a natural antisense coded protein of HTLV-1. *Blood* **2006**, *107*, 3976–3982. [CrossRef] [PubMed]

90. Yin, H.; Kannian, P.; Dissinger, N.; Haines, R.; Niewiesk, S.; Green, P.L. Human T-cell leukemia virus type 2 antisense viral protein 2 is dispensable for *in vitro* immortalization but functions to repress early virus replication *in vivo*. *J. Virol.* **2012**, *86*, 8412–8421. [CrossRef] [PubMed]

*viruses*

MDPI

*Review*

# Roles of HTLV-1 basic Zip Factor (HBZ) in Viral Chronicity and Leukemic Transformation. Potential New Therapeutic Approaches to Prevent and Treat HTLV-1-Related Diseases

Jean-Michel Mesnard [1], Benoit Barbeau [2], Raymond Césaire [3] and Jean-Marie Péloponèse [1,*]

1  CPBS, CNRS FRE3689, Université Montpellier, 34293 Montpellier, France; jean-michel.mesnard@cpbs.cnrs.fr
2  Département des Sciences Biologiques, and Centre de Recherche BioMed Université du Québec à Montréal, Montréal, QC H2X 3X8, Canada; barbeau.benoit@uqam.ca
3  Laboratoire de Virologie-EA4537, Centre Hospitalier et Universitaire de Martinique, Fort de France, Martinique; Raymond.Cesaire@chu-fortdefrance.fr
*  Correspondence: jean-marie.peloponese@cpbs.cnrs.fr; Tel.: +33-434-359-440; Fax: +33-467-604-420

Academic Editor: Louis M. Mansky
Received: 15 October 2015; Accepted: 1 December 2015; Published: 9 December 2015

**Abstract:** More than thirty years have passed since human T-cell leukemia virus type 1 (HTLV-1) was described as the first retrovirus to be the causative agent of a human cancer, adult T-cell leukemia (ATL), but the precise mechanism behind HTLV-1 pathogenesis still remains elusive. For more than two decades, the transforming ability of HTLV-1 has been exclusively associated to the viral transactivator Tax. Thirteen year ago, we first reported that the minus strand of HTLV-1 encoded for a basic Zip factor factor (HBZ), and since then several teams have underscored the importance of this antisense viral protein for the maintenance of a chronic infection and the proliferation of infected cells. More recently, we as well as others have demonstrated that HBZ has the potential to transform cells both *in vitro* and *in vivo*. In this review, we focus on the latest progress in our understanding of HBZ functions in chronicity and cellular transformation. We will discuss the involvement of this paradigm shift of HTLV-1 research on new therapeutic approaches to treat HTLV-1-related human diseases.

**Keywords:** human T-cell leukemia virus type 1; adult T-cell leukemia; HTLV-1 bZip Factor; Valproate

## 1. Introduction

Thirty years ago, human T-cell leukemia virus type 1 (HTLV-1) was the first human retrovirus to be identified and is now known as the causative agent of a very aggressive form of leukemia termed adult T-cell leukemia (ATL). It was isolated in the early 1980s, first in the United States [1] and then in Japan [2,3]. Currently, HTLV-1 infects approximately 15 million individuals worldwide [4]. HTLV-1 is the etiological agent of both ATL and a slowly progressive neurologic disorder called HTLV-1-associated myelopathy/tropical spastic paraparesis (HAM/TSP). [5,6]. The role of HTLV-1 in HAM/TSP will not be discussed here. Below, we summarize and update insights relevant to human leukemogenesis induced by HTLV-1.

## 2. HTLV-1 Infectivity and Spread *in vivo*

Like any animal retrovirus, the HTLV-1 proviral genome encodes for the structural genes, *gag, pol* and *env*, and is bordered by two long terminal repeat sequences (LTR) [7]. The 5′ LTR serves as the main promoter for viral transcription. HTLV-1 is defined as a complex retrovirus because its genome also contains a region termed the pX region, which it located between the *env* gene and the 3′-LTR and contains genes encoding regulatory viral factors, Tax, Rex, $p12^I$, $p13^{II}$, $p30^{II}$ and $p21^I$. Furthermore, the minus strand of pX has been found to produce an antisense transcript, encoding HBZ [8–11] (Figure 1).

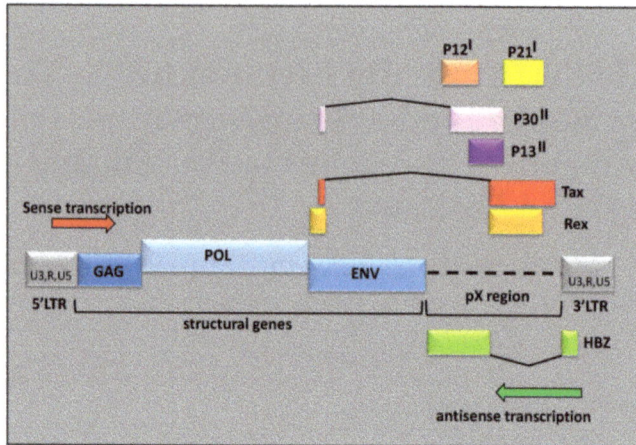

**Figure 1.** Structure of the HTLV-1 provirus: The human T-cell leukemia virus type 1 (HTLV-1) genome encodes for three structural proteins, Gag, Pol, and Env, and complex regulatory proteins such as Tax, which not only activates viral replication, but also induces the expression of several cellular genes. The *in vivo* expression of these viral proteins is suppressed by cytotoxic T lymphocyte (CTL) activity. HTLV-1 basic Zip factor (HBZ), produced by a minus-strand mRNA, likely plays a role in viral replication and T-cell proliferation as it is steadily expressed in most HTLV-1-infected cells and primary adult T-cell leukemia (ATL) cells, whereas Tax is not.

*In vitro*, HTLV-1 can infect a large variety of cells, including T- and B-cells, fibroblasts, macrophages, and dendritic cells [12]. These observations indicate that the receptor is common and expressed on a large number of cells. Studies show that glucose transporter 1 (Glut-1), heparan sulfate proteoglycans (HSPGs), and neuropilin-1 (NRP-1) are three proteins involved in the mechanism of HTLV-1 entry [13–15]. The current view on the entry event of HTLV-1 suggests that the virus first interacts with HSPG and then forms complexes with NRP-1 followed by an association with Glut-1 at the cell surface before final membrane fusion and entry into the cell. However, how these factors cooperate with each other requires further study. It is interesting to note that, despite the ubiquitous distribution of these membrane proteins, *in vivo*, the HTLV-1 provirus is mainly detected in CD4+ and in CD8+ T-cells [16]. There is nonetheless evidence that cellular receptors play an important role in determining the cellular tropism of HTLV-1 [17]. Thus, the differential outcome of HTLV-1 on CD4+ and CD8+ cell proliferation may be more important in dictating the apparent specificity for CD4+ cells than is receptor-binding and differences related to cellular entry [18].

*In vivo*, HTLV-1 is primarily transmitted by cell-to-cell contact, and not by cell-free virions [19]. Upon contact with an uninfected cell, HTLV-1-infected cells will transiently express high levels of Tax and intercellular adhesion molecule-1 (ICAM-1) to form a virological synapse [20] or a viral biofilm [21]. Enveloped viral particles can transfer through this synapse, thus propagating infection [22]. Recently, it has been reported that HTLV-1 cell-to-cell transmission is ten thousand times more efficient than cell-free infection, while, in comparison, similar experiments have shown that for human immunodeficiency virus type 1 (HIV-1) infection, this difference is only twofold [23]. However, the importance of *in vivo* cell-to-cell spread is tempered by findings that the administration of reverse transcriptase inhibitors (RTI) to HTLV-1-infected patients with HAM/TSP does not markedly influence the provirus load [24], and that RTI treatment immediately after HTLV-1 infection *in vivo* does not change subsequent proviral load. Thus, viral replication itself does not appear to be critical for the maintenance of persistent infection; rather, the proliferation of HTLV-1-infected cells seems to determine viral burden at the carrier state. In this regard, the viral strategy to increase the number of infected cells by promoting cellular proliferation is meaningful. Indeed, a long-standing observation is that HTLV-1 induces clonal proliferation of infected cells *in vivo* [18,25,26].

### 3. HTLV-1, Chronicity and Host Immune Response

In order to induce chronic infection, viruses need to establish an equilibrium between viral virulence and the host immunity [27]. Accordingly, human retroviruses, such as HTLV-1, have evolved several strategies to control the host immune system and temper viral replication, one of which is to directly deregulate the major histocompatibility complex (MHC) [27]. The function of MHC molecules is to bind peptide fragments derived from pathogens and display them on the cell surface for recognition by the appropriate T-cells. The consequences are often deleterious to the pathogen—virus-infected cells are killed and B-cells are activated to produce antibodies that eliminate or neutralize extracellular pathogens. Thus, there is a strong selective pressure in favor of any virus that has evolved mechanism allowing them to escape presentation of its antigens by MHC molecules. In its pX region, HTLV-1 encodes an accessory protein, p12 that interacts with MHC class I heavy chains, and leads to its degradation by the proteasome [28]. In HTLV-1-infected host, chronically activated cytotoxic T lymphocyte (CTL) response [29–31] and high titer of anti-HTLV-1 antibodies, mostly directed against the Tax protein [32–34], strongly support the idea that Tax is the main immunogenic target. Indeed, *in vivo* depletion of Tax-expressing CD4+ T-cells leads to moderate HTLV-1 replication [35]. CD8+ CTLs are in part responsible for this phenomenon because their depletion enhances Tax expression *in vivo* [35]. Furthermore, when a histone deacetylase inhibitor, valproate, was used to reactivate *tax* transcription in HTLV-infected host, their proviral load became reduced [36–38]. A similar observation of valproate-induced reduction of Simian T-Cell Leukemia Virus (STLV) proviral load has been also reported in a simian model [39]. Thus host's CTL response targets Tax-expressing cells, thereby reducing the number of infected cells *in vivo*. In fact, the HTLV-1 proviral load appears to be maintained, when an equilibrium is established with the immune response, allowing the maintenance and the proliferation of HTLV-1-infected cells [40]. While Tax is frequently targeted by CTL in HTLV-1 infection [32–34], the frequency of HBZ-specific CTL is low and could only be detected in 25%–40% of infected individuals [41,42]. However, in a systematic study, MacNamara *et al.* [42] showed that protective alleles A*0201 and C*0801 bound HBZ-derived peptides with significantly higher affinity in comparison to alleles which were associated with disease progression (B*5401). However, further analyses demonstrated that asymptomatic carriers ACs had human leukocyte antigen (HLA) alleles which bound HBZ peptides significantly more strongly than patients with HAM/TSP, and that this difference in binding was not simply attributable to A*0201, C*0801, and B*5401 [41,42].

In order to escape the host immune response, a proportion of cells that express Tax must subsequently shut down its expression. Recently, various molecular mechanisms accounting for suppression of Tax expression have been suggested, implicating viral—Rex [43], the pX protein p30 II [44] and HBZ [8]—and cellular proteins—histone deacetylases [45] and GLI-2/THP [46]. In each study, these data only indicate a partial rather than a complete shutdown of proviral transcription. Importantly, the extent of suppression of viral expression in natural HTLV-1 infection is not yet known. However, even partial suppression should provide significant survival advantage to an HTLV-1-infected cell since these cells might be less prone to elimination by the immune system, which would be particularly dependent on CTL activity. Furthermore, impairment of CTL surveillance may similarly allow HTLV-1-transformed leukemic cells to survive and proliferate [47,48].

### 4. Multifaceted Processes in the Transformation of Infected Cells

Over time, a subset (2%–6%) of HTLV-1-infected individuals will develop ATL [49]. One of the current models is that the Tax oncoprotein confers survival and proliferative properties to infected cells (Figure 1) [50–52]. Tax is post-translationally modified by phosphorylation, ubiquitination, and acetylation [53–59]. These post-translational modifications have been shown to be important for Tax function [54,58]. Expression of Tax alone has been postulated to be sufficient for immortalization, but not transformation, of human T-cells [60,61]. The *in vivo* transforming capacity of Tax has been extensively investigated using transgenic mouse models; results suggest that Tax expression can solely drive *in vivo* tumor formation [62–64]. However, frequent appearance of type 2 defective proviruses (*i.e.*, lacking the 5′ LTR and the Tax gene) in ATL cells lead into questioning these current models [65–67]. Investigation of the mechanisms underlying the generation of these defective proviruses by Miyazaki *et al.* [67] showed that 41% of type 2 defective proviruses lacking 5′ LTR were formed before proviral integration. Since Tax expression alone is not enough to transform primary human cells *in vitro* [60,61,68], it is likely that, similar to human papillomavirus E6 and E7, which cooperate for the development of tumors [69], tax functions cooperatively with other HTLV-1-encoded genes in HTLV-1 to induce human leukemogenesis [63]. Further studies are required to better clarify the roles of the Tax in ATL onset.

## 5. HTLV-1 Antisense HBZ Transcripts and Viral Pathogenesis

While HTLV-1 plus strand (sense) contains transcripts driven from the 5′ LTR, the 3′ LTR produces an antisense transcript, which encodes a protein called HTLV-1 basic Zip factor , HBZ (Figure 1) [7]. The promoter for the *hbz* gene is contained in the U5 sequence of the 3′ LTR [7]. Analyses showed that HTLV-1 LTR possess a bidirectional transcriptional activity. Interestingly, Sp1 sites within this region are critical for the control of bidirectional and HBZ transcription [70,71]. While tax transcripts are detected in few transformed ATL cells, *hbz* mRNA is present in all ATL cells [72–74].

Through its basic Zip (bZIP) domain, the HBZ protein has been described to interact with transcription factors the cAMP-response element binding proteins (CREB, CREB-2), the cAMP-responsive element modulator (CREM-Ia) the activating transcription factor ATF-1 [75], c-Jun [76,77], JunB [78], and JunD [79], and was originally reported to suppress Tax-mediated viral transcription [8]. Furthermore, HBZ selectively inhibits the classical nuclear factor-kappa B (NF-κB) pathway by inhibiting DNA binding of p65 and promoting its degradation [80]. On the other hand, Tax has been shown to rather activate both classical and alternative NF-κB pathways [81]. Since the two pathways differentially control the expression of genes with anti-apoptotic functions in lymphoma cell lines [82], preferential activation of the alternative pathway by Tax and HBZ might be implicated in the proliferation of ATL cells. Interestingly, a previous study had suggested that the *hbz* mRNA itself is also important for the induction of proliferation of HTLV-1-infected cells [9]. Moreover, it has been shown that HTLV-1 molecular clones harboring a mutation in the leucine zipper domain of HBZ exhibit reduced proviral load compared to wild type virus when inoculated into rabbits [83]. Furthermore, HBZ has been reported to increase the activity of human telomerase reverse transcriptase (hTERT) [84]. Collectively, all of the available data support that HBZ protein and RNA play important roles in promoting viral replication and cellular proliferation [85].

Genetic instability in infected cells might allow them to escape strong CTL response, and further protect them from clonal dominance [86]. On the other hand, HTLV-1-triggered alterations in cellular gene expression have been proposed to amount to a mutator phenotype that promotes leukemogenesis [87]. Thus far, Tax has been recognized as the main source of HTLV-1-associated genetic instability. In a recent study, Vernin *et al.* [88] showed that HBZ promotes onco-miR expression as well as DNA-strand breaks by downregulating the expression of OBFC2A protein via posttranscriptional activation of miR17 and miR21. OBFC2A intervenes with ATM signaling and subsequently activates DNA repair and cell-cycle checkpoints [88]. In their study, Vernin *et al.* further suggested that preleukemic phenotypes of HTLV-1-positive CD4+ T cells is portrayed by an oncogenic miRNA profile that is promoted by HBZ [88].

The transforming capacity of HBZ has been clearly demonstrated *in vitro* and *in vivo* using HBZ transgenic (Tg) mice [89,90]. Using the 5′ LTR deleted K30$_{4089}$ molecular clone [10], Gazon *et al.* have demonstrated that, in murine cells, HBZ expression on its own drives cellular proliferation and colony formation in soft agar [90]. Furthermore, *in vivo* studies showed that CD4+ Foxp3-positive T-cells from

HBZ-Tg mice had similar effector/memory [89] and regulatory phenotypes to infected CD4+ T cells from ATL patients or HTLV-1-infected carriers [91,92] In this study by Satou *et al.*, transgenic mice expressing Tax under the same promoter as the HBZ-Tg mice did not display any changes in their Treg phenotype [89]. These data suggest that HBZ, rather than Tax, is responsible for conferring a specific phenotype to HTLV-1-infected cells and ATL cells.

The interplay between Tax and HBZ in T-cell transformation might then be explained by the fact that the former would be needed to initiate transformation while the latter would be required to maintain the transformed phenotype of ATL cells at a time point when Tax expression is extinguished. However, when taking into consideration the accumulating evidence on HBZ functions, the notion that Tax only can initiate cellular transformation via its ability to induce genetic instability may require some revision. Indeed, it has been clearly established that HBZ is responsible for the specific phenotype, function and proliferation of HTLV-1-infected CD4+ T-cells and ATL cells, and that, in addition to Tax, HBZ plays an important role in the oncogenic activity of HTLV-1 (Figure 2). Furthermore, the long latent period observed by Satou *et al.* before the onset of T-cell lymphomas in HBZ-Tg mice suggests that additional epigenetic alterations in CD4+ T-cells are necessary for the development of T-cell lymphomas in HBZ-Tg mice as well as for ATL [89]. In conclusion, functions recently attributed to HBZ provide novel insights into the interaction between HTLV and its host and may be exploited to treat and prevent HTLV-1-induced diseases.

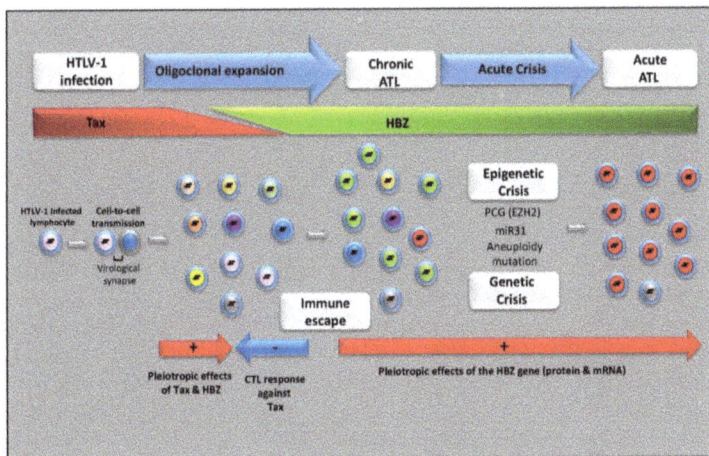

**Figure 2.** Model for ATL development. HTLV-1 is transmitted in a cell-to-cell fashion via a virological synapse. After infection, HTLV-1 promotes clonal proliferation of infected cells by pleiotropic actions of Tax and other viral proteins. Tax is considered crucial for the oligoclonal maintenance and expansion of HTLV-1-infected cells during the early phase but is only transiently expressed by HTLV-1-infected cells. Proliferation of HTLV-1-infected cells is controlled by cytotoxic T-cells *in vivo*. Thereafter, continuing expression of HBZ is followed by genetic/epigenetic loss of function of tumor suppressor genes and modulation of micro RNA levels. After a long latent period, ATL develops in about 5% of asymptomatic carriers. Diverse genetic abnormalities are acquired during the progression to ATL from an indolent to an aggressive disease form.

## 6. Therapeutic Approaches for ATL

ATL is an incurable and poorly treatable disease. Despite advances in both chemotherapy and supportive care, median survival time of patients remains less than one year [93]. As pointed out by Yamada and Tomonaga [94], an important amount of knowledge in molecular biology and oncogenesis of ATL has accumulated but has not yet been translated into improved prognosis of affected patients [94].

In fact, it has been reported that the prognosis of indolent subtypes, chronic and smoldering ATL, was 4.1 years, which is poorer than previously thought [95]. Therapeutic approaches using interferon-α combined with zidovudine have nonetheless been reported to be highly effective treatments for indolent ATL and, as they have been extensively reviewed, will not be further detailed here [96–98].

*6.1. Chemotherapy*

In Europe, US and Brazil, the recommended treatment of patients with acute or lymphoma-type ATL is based on the use of combined chemotherapy, as a first line therapy: cyclophosphamide, adriamycin, vincristine, and prednisolone (CHOP). 46.5% of patients treated with CHOP exhibit partial remission (PR), while only ~20% are achieving complete remission (CR) [99]. Intensification of CHOP with etoposide, vindesine, ranimustine, and mitoxantrone resulted in CR for 35.8% of ATL patients [100]. However, the median survival was only 8–8.5 months in these studies with predicted survivals of 13% after three years. In Japan, the first line of combined chemotherapy against ATL consists of VCAP-AMP-VECP) (i.e; vincristine, cyclophosphamide, doxorubicin, and prednisone (VCAP), doxorubicin, ranimustine, and prednisone (AMP), and vindesine, etoposide, carboplatin, and prednisone (VECP)) [101–103]. It has in fact been shown that the CR rate was better with VCAP-AMP-VECP (40%) than biweekly CHOP (~20%) and that a three-year survival rate of patients treated with VCAP-AMP-VECP therapy improved by 24% (*versus* 13% with CHOP treatment) [104]. The two main obstacles to combined chemotherapy are (1) the inherent drug resistance to chemotherapeutic agents observed in ATL cells [49] and (2) the profoundly weakened and immunodeficient state of ATL patients. Overall, ATL survival with various chemotherapy regimens is poor, with survival ranging between 5.5 and 13 months in several cohorts of patients, and more predominantly in patients with acute leukemia or lymphoma. At the moment, this approach does not represent a prospect for a cure.

*6.2. Molecular Targeted Therapy*

As an alternative to chemotherapy, a number of studies have addressed the potential use of nucleoside analogues for the treatment of ATL. The purine analog 2′-deoxycoformycin (DCF) that inhibits adenosine deaminase has been studied. In a phase II study using DCF, two CR (8%) and one PR (4%) cases were reported among 25 patients with ATL [105]. Unfortunately, these response rates are significantly lower than those with CHOP-based chemotherapy. Using DCF in conjunction with CHOP, 52% of ATL patients achieved CR, but the median survival of patients was only 7.4 months [106]. Interestingly, another study reported that a patient presenting resistant acute ATL had an improved and lasting partial response when treated with another adenosine analog, 2′-chlorodeoxyadenosine (cladribine) [107]. However the follow-up phase II study showed very limited benefit with this compound. It is nonetheless important to note that all patients under treatment showed resistant ATL prior to treatment in this study and therefore represented a poor prognosis group [108].

Parallel approaches, using analogs or inhibitors of topoisomerase, such as irinotecan hydrochloride (CPT-11) [109], a *bis* (2,6-dioxopiperazine) analog (MST-16) [110], Menogaril [111] or all-trans-retinoic acid (ATRA) [112,113], an analog of vitamin A have been similarly tested. The most promising results were obtained with the MST-16 treatment. In a cohort constituted of 21 acute-type ATL patients, treatment resulted in one CR and five PR. Among eight lymphoma-type ATL patients, two PR cases were identified, while out of two chronic-type ATL patients, one was diagnosed as being in CR while the other, as being in PR. Remissions were achieved within 23 days and lasted over two months (median, 68 days) [110]. These results do not clearly represent an improvement over conventional chemotherapy and further studies with MST-16 are needed.

Gene expression governed by epigenetic changes is crucial for the pathogenesis of cancer. Histone deacetylases are enzymes that are involved in the remodeling of chromatin and play a key role in the epigenetic regulation of gene expression. The use of histone deacetylase inhibitors (HDACi) to treat ATL has recently attracted attention. The HDACi LBH589 (panobinostat) exhibited significant anti-ATL activity by activating a novel RAIDD-caspase-2 pathway in mice and by modulating the expression of Tax and CCR4 [114]. However, a phase II study using panobinostat for cutaneous T-Cell Lymphoma (CTCL) and indolent ATL patients had to be terminated because of severe side effects and appearance of ulcers in patients with ATL [115].

Depsipeptide (HFR901228), another HDACi induces apoptosis in all tested HTLV-1-infected cell lines and in primary cells from patients with acute ATL, through a reduction of NF-κB and AP-1 transactivation activity, and downregulation of B-cell lymphoma-extra large (Bcl-xL) and cyclin D2 expression. Partial inhibition of tumor growth following transplantation of HTLV-1-infected cells, was seen in a severe combined immunodeficiency (SCID) mouse model [116]. Further studies are needed to evaluate the efficacy of HFR901228.

Sodium valproate (VPA) is another HDCAi under investigation, which is widely prescribed for the treatment of epilepsy, bipolar mood disorders, and migraine, and which shows HDACi activity among several other potential antitumor properties [117]. VPA is also being investigated for its inclusion in maintenance therapy after chemotherapy [118]. More importantly, dramatic clearance of both lymphoma and leukemic cells has been demonstrated in Bovine Leukemia Virus-induced B-cell malignancy in sheep upon treatment [37,119]. In a recent study, Belrose *et al.* [38] analyzed the impact of VPA treatment on the expression profile of Tax and HBZ in freshly cultured cells from HTLV-1-infected patients. It was then proposed that VPA relieved the epigenetic control over Tax expression, thereby exposing latently HTLV-1-infected cells to the immune response. Indeed, in the presence of VPA, Tax expression kinetics were profoundly modified, with Tax mRNA levels increasing constantly over time, suggesting dysregulation of the processes responsible for the control of its expression in lymphocytes from HAM/TSP patients, but not from asymptomatic carriers. One interesting finding from the Belrose *et al.* study was that VPA strongly impaired the expression of HBZ [38]. The authors suggested that the opposite effect of VPA on Tax and HBZ expression might be caused by the nature of HDAC complexes present on the 5'- and 3'-HTLV-1 promoters in relation to their selective down-modulating properties. Alternatively, it cannot be excluded that activation of sense transcription by Tax and VPA might have impaired antisense transcription, either by competition for transcription factors or interference with its initiation. Indeed, Cavanagh *et al.* [10] have showed that, by deleting the 5'-LTR, sense transcription has a negative impact on antisense transcription (*i.e.*, from the 3'-LTR) [10]. Interestingly, despite increased Tax expression, Belrose *et al.* did not observe the expected increase in proviral load (PVL) in VPA-treated samples from HAM/TSP. Instead, a significant decrease of the PVL in VPA-treated samples from acute ATL patients was observed, suggesting a decrease in the percentage of ATL cells [120]. Using VPA to augment the level of histone acetylation and increase HTLV-1 gene expression in cultured cells from HAM/TSP patients, Mosley *et al.* [121] demonstrated that, while the level of Tax expression doubled after overnight treatment, the rate of CD8+ T-cell-mediated lysis of Tax-expressing cells was reduced by 50% [121]. VPA thus appeared to inhibit CD8+ T-cell-mediated cell from killing itself. These observations indicate that HDCAis may reduce the efficiency of CTL surveillance of HTLV-1. Further studies are needed to evaluate the use of HDIs in nonmalignant cases of HTLV-1 infection. Taken together, these observations strongly suggest that HBZ is a very interesting therapeutic target and that a therapy using VPA as part of the management of patients with acute and lymphoma ATL should be considered for the prevention of progression of chronic and smoldering ATL. Should a protective effect be shown, the long-standing safety profile of this compound would justify a prospective study in which its efficacy in preventing ATL in patients considered to be at high risk of disease is evaluated [122].

*6.3. Immunotherapy*

Another alternative approach is to target ATL cells using specific markers on the surface of the malignant cells with monoclonal antibodies. One of the first potential tested target is the interleukin (IL)-2alpha receptor α chain, CD25. Indeed, ATL cells express high level of CD25 on their surface. In their recent phase I/II trial on 34 patients with ATL, Berkowitz *et al.* [123] reported that daclizumab, a humanized monoclonal antibody which blocks IL-2 binding by recognizing CD25 on ATL cells, was associated with effective clinical responses in patients with indolent disease, although no beneficial responses were observed in patients with acute or lymphomatous subtypes of ATL [123]. The finding that daclizumab has antitumor activity and demonstrates a potential in achieving long-term responses in patients with the indolent form of ATL, suggest that immunotherapy offers a therapeutic option to prevent indolent diseases to develop into aggressive ATL [123].

Another target for immuno-based therapy against ATL is the CC chemokine receptor 4 (CCR4). CCR4 is principally expressed on regulatory T-cells (Tregs) and helper T-cells (Th), where it functions in inducing homing of these leukocytes to sites of inflammation. Tregs play an essential role in maintaining immune balance; however, in malignancy, Tregs impair host antitumor immunity and provide a favorable environment for tumors to grow [49]. Furthermore, ATL cells express high levels of CCR4 on their surface [124]. Mogamulizumab (KW-0761) is the first approved glyco-engineered therapeutic monoclonal antibody to target CCR4. 30% of patients with acute forms of ATL (5 out of 15) treated with Mogamulizumab showed a positive response [125]. Several ongoing clinical trials in Japan are investigating if combining Mogamulizumab with a chemotherapy treatment could be beneficial [126–128].

*6.4. Stem Cell Transplantation*

As a treatment strategy for ATL, allogeneic hematopoietic stem cell transplantation (allo-HSCT) with reduced intensity conditioning regimens (RIC) was prospectively evaluated. Several teams have reported the safety and feasibility of allo-HSCT with RIC using peripheral blood stem cells from an HLA-matched sibling donor in patients with acute ATL, who achieved remission after chemotherapy [129–131]. These studies showed that the overall survival (OS) at three years after allo-HSCT with RIC treatment ranged from 33% to 49% [132]. Interestingly, a significant decrease of the proviral load was also reported in many of these patients. These findings suggest that cell-mediated immunity to HTLV-1 was augmented in these patients, which might account for the efficacy of this therapy. In fact, Graft-*versus*-host disease (GVHD) is a good prognostic factor for ATL patients [130], indicating that an immune attack by donor lymphocytes is critical for the efficacy of treatment. Kanda *et al.* [133] reported that grade I/II acute GVHD was associated with a longer OS. Beneficial effects of allo-HSCT on non-Japanese ATL patients were recently confirmed by a retrospective study from the European Group for Blood and Marrow Transplantation's Lymphoma Working Party [134].

## 7. Perspectives and Conclusions

Although new therapeutic options are emerging, treatment of ATL patients remains challenging. The initial pathogenic event for ATL is HTLV-1 genomic integration; however, additional genetic alterations have also been implicated in ATL pathogenesis. Umino *et al.* [135] reported on the importance of clonal heterogeneity of ATL cells involving different genomic alterations; they further demonstrated that these cells originated from a common cell. It was suggested that approximately 70% of ATL cases undergo clonal evolution, and that genetic instability may contribute to the accumulation of genomic alterations [135]. In fact, the existence of multiple clones with genomic instability is one factor that renders ATL cells resistant to conventional chemotherapy. Even if a proportion of cells are killed by chemotherapy, new resistant clones likely emerge. Therefore, allo-HSCT might be efficient in curing ATL patients by eliminating HTLV-1-integrated recipient ATL clones through strong immune response, and subsequent replacement of the hematopoietic system with donor cells.

Whole genome sequencing revealed that carriers have $10^3$ to $10^4$ distinct clones with different HTLV-1 integration sites, and that most clones harbored one copy of HTLV-1 proviral DNA [136]. This indicates that HTLV-1 carriers potentially have $10^3$ to $10^4$ malignant clones. If the number of infected cells increases, there is a greater possibility that malignant transformation might occur. In order to reduce the number of pre-malignant cells in HTLV-1 carriers and thus prevent the development of ATL, treatment with HDAC inhibitors seems to be a promising strategy. Indeed vorinostat (suberoylanilide hygroxamic acid: SAHA), panobinostat (LBH-589) and MS-275 have been demonstrated to impede the growth of HTLV-1-infected cell lines and freshly isolated infected cells [137]. Furthermore, as reported by Belrose *et al.* [38], the link between VPA-induced apoptosis of HTLV-1-infected cell lines, decrease of proviral load in freshly isolated infected cells and loss of HBZ expression indicates that HBZ is a promising therapeutic target. However, further studies are needed to clarify the effect of HDACi on HBZ, although the inclusion of HDACi in clinical trials for the treatment of ATL is expected. Nevertheless, to increase the likelihood of discovery of a cure for ATL, rigorous investigation remains necessary for optimizing therapeutic combinations, preventing ATL development in HTLV-1 carriers, and reducing the number of HTLV-1 carriers.

**Author Contributions:** Jean-Michel Mesnard and Jean-Marie Peloponese wrote the manuscript; Benoit Barbeau and Raymond Césaire critically revised the manuscript.

**Conflicts of Interest:** The authors declare no conflict of interest.

## References

1. Poiesz, B.J.; Ruscetti, F.W.; Gazdar, A.F.; Bunn, P.A.; Minna, J.D.; Gallo, R.C. Detection and isolation of type C retrovirus particles from fresh and cultured lymphocytes of a patient with cutaneous T-cell lymphoma. *Proc. Natl. Acad. Sci. USA* **1980**, *77*, 7415–7419. [CrossRef] [PubMed]

2. Popovic, M.; Reitz, M.S., Jr.; Sarngadharan, M.G.; Robert-Guroff, M.; Kalyanaraman, V.S.; Nakao, Y.; Miyoshi, I.; Minowada, J.; Yoshida, M.; Ito, Y.; *et al.* The virus of Japanese adult T-cell leukaemia is a member of the human T-cell leukaemia virus group. *Nature* **1982**, *300*, 63–66. [CrossRef] [PubMed]

3. Yoshida, M.; Miyoshi, I.; Hinuma, Y. Isolation and characterization of retrovirus from cell lines of human adult T-cell leukemia and its implication in the disease. *Proc. Natl. Acad. Sci. USA* **1982**, *79*, 2031–2035. [CrossRef] [PubMed]

4. Gessain, A.; Cassar, O. Epidemiological Aspects and World Distribution of HTLV-1 Infection. *Front. Microbiol.* **2012**, *3*, 388. [CrossRef] [PubMed]

5. Gessain, A.; Barin, F.; Vernant, J.C.; Gout, O.; Maurs, L.; Calender, A.; de Thé, G. Antibodies to human T-lymphotropic virus type-I in patients with tropical spastic paraparesis. *Lancet* **1985**, *2*, 407–410. [CrossRef]

6. Osame, M.; Igata, A. The history of discovery and clinico-epidemiology of HTLV-I-associated myelopathy(HAM). *Jpn. J. Med.* **1989**, *28*, 412–414. [CrossRef] [PubMed]

7. Barbeau, B.; Mesnard, J.M. Making Sense out of Antisense Transcription in Human T-Cell Lymphotropic Viruses (HTLVs). *Viruses* **2011**, *3*, 456–468. [CrossRef] [PubMed]

8. Gaudray, G.; Gachon, F.; Basbous, J.; Biard-Piechaczyk, M.; Devaux, C.; Mesnard, J.M. The complementary strand of the human T-cell leukemia virus type 1 RNA genome encodes a bZIP transcription factor that down-regulates viral transcription. *J. Virol.* **2002**, *76*, 12813–12822. [CrossRef] [PubMed]

9. Satou, Y.; Yasunaga, J.; Yoshida, M.; Matsuoka, M. HTLV-I basic leucine zipper factor gene mRNA supports proliferation of adult T cell leukemia cells. *Proc. Natl. Acad. Sci. USA* **2006**, *103*, 720–725. [CrossRef] [PubMed]

10. Cavanagh, M.H.; Landry, S.; Audet, B.; Arpin-Andre, C.; Hivin, P.; Pare, M.E.; Thete, J.; Wattel, E.; Marriott, S.J.; Mesnard, J.M.; *et al.* HTLV-I antisense transcripts initiating in the 3′LTR are alternatively spliced and polyadenylated. *Retrovirology* **2006**, *3*, 15. [CrossRef] [PubMed]

11. Matsuoka, M.; Green, P.L. The HBZ gene, a key player in HTLV-1 pathogenesis. *Retrovirology* **2009**, *6*, 71. [CrossRef] [PubMed]

12. Jones, K.S.; Petrow-Sadowski, C.; Huang, Y.K.; Bertolette, D.C.; Ruscetti, F.W. Cell-free HTLV-1 infects dendritic cells leading to transmission and transformation of CD4(+) T cells. *Nat. Med.* **2008**, *14*, 429–436. [CrossRef] [PubMed]

13. Manel, N.; Kim, F.J.; Kinet, S.; Taylor, N.; Sitbon, M.; Battini, J.L. The ubiquitous glucose transporter GLUT-1 is a receptor for HTLV. *Cell* **2003**, *115*, 449–459. [CrossRef]

14. Wielgosz, M.M.; Rauch, D.A.; Jones, K.S.; Ruscetti, F.W.; Ratner, L. Cholesterol dependence of HTLV-I infection. *AIDS Res. Hum. Retrovir.* **2005**, *21*, 43–50. [CrossRef] [PubMed]

15. Lambert, S.; Bouttier, M.; Vassy, R.; Seigneuret, M.; Petrow-Sadowski, C.; Janvier, S.; Heveker, N.; Ruscetti, F.W.; Perret, G.; Jones, K.S.; *et al.* HTLV-1 uses HSPG and neuropilin-1 for entry by molecular mimicry of VEGF165. *Blood* **2009**, *113*, 5176–5185. [CrossRef] [PubMed]

16. Yasunaga, J.; Matsuoka, M. Molecular mechanisms of HTLV-1 infection and pathogenesis. *Int. J. Hematol.* **2011**, *94*, 435–442. [CrossRef] [PubMed]

17. Jones, K.S.; Fugo, K.; Petrow-Sadowski, C.; Huang, Y.; Bertolette, D.C.; Lisinski, I.; Cushman, S.W.; Jacobson, S.; Ruscetti, F.W. Human T-cell leukemia virus type 1 (HTLV-1) and HTLV-2 use different receptor complexes to enter T cells. *J. Virol.* **2006**, *80*, 8291–8302. [CrossRef] [PubMed]

18. Zane, L.; Sibon, D.; Legras, C.; Lachuer, J.; Wierinckx, A.; Mehlen, P.; Delfau-Larue, M.H.; Gessain, A.; Gout, O.; Pinatel, C.; *et al.* Clonal expansion of HTLV-1 positive CD8+ cells relies on cIAP-2 but not on c-FLIP expression. *Virology* **2010**, *407*, 341–351. [CrossRef] [PubMed]

19. Igakura, T.; Stinchcombe, J.C.; Goon, P.K.; Taylor, G.P.; Weber, J.N.; Griffiths, G.M.; Tanaka, Y.; Osame, M.; Bangham, C.R. Spread of HTLV-I between lymphocytes by virus-induced polarization of the cytoskeleton. *Science* **2003**, *299*, 1713–1716. [CrossRef] [PubMed]

20. Nejmeddine, M.; Negi, V.S.; Mukherjee, S.; Tanaka, Y.; Orth, K.; Taylor, G.P.; Bangham, C.R. HTLV-1-Tax and ICAM-1 act on T-cell signal pathways to polarize the microtubule-organizing center at the virological synapse. *Blood* **2009**, *114*, 1016–1025. [CrossRef] [PubMed]

21. Pais-Correia, A.M.; Sachse, M.; Guadagnini, S.; Robbiati, V.; Lasserre, R.; Gessain, A.; Gout, O.; Alcover, A.; Thoulouze, M.I. Biofilm-like extracellular viral assemblies mediate HTLV-1 cell-to-cell transmission at virological synapses. *Nat. Med.* **2010**, *16*, 83–89. [CrossRef] [PubMed]

22. Majorovits, E.; Nejmeddine, M.; Tanaka, Y.; Taylor, G.P.; Fuller, S.D.; Bangham, C.R. Human T-lymphotropic virus-1 visualized at the virological synapse by electron tomography. *PLoS ONE* **2008**, *3*, e2251. [CrossRef] [PubMed]

23. Mazurov, D.; Ilinskaya, A.; Heidecker, G.; Lloyd, P.; Derse, D. Quantitative comparison of HTLV-1 and HIV-1 cell-to-cell infection with new replication dependent vectors. *PLoS Pathog.* **2010**, *6*, e1000788. [CrossRef] [PubMed]

24. Taylor, G.P.; Goon, P.; Furukawa, Y.; Green, H.; Barfield, A.; Mosley, A.; Nose, H.; Babiker, A.; Rudge, P.; Usuku, K.; *et al.* Zidovudine plus lamivudine in Human T-Lymphotropic Virus type-I-associated myelopathy: A randomised trial. *Retrovirology* **2006**, *3*, 63. [CrossRef] [PubMed]

25. Cavrois, M.; Leclercq, I.; Gout, O.; Gessain, A.; Wain-Hobson, S.; Wattel, E. Persistent oligoclonal expansion of human T-cell leukemia virus type 1-infected circulating cells in patients with Tropical spastic paraparesis/HTLV-1 associated myelopathy. *Oncogene* **1998**, *17*, 77–82. [CrossRef] [PubMed]

26. Etoh, K.; Tamiya, S.; Yamaguchi, K.; Okayama, A.; Tsubouchi, H.; Ideta, T.; Mueller, N.; Takatsuki, K.; Matsuoka, M. Persistent clonal proliferation of human T-lymphotropic virus type I-infected cells *in vivo*. *Cancer Res.* **1997**, *57*, 4862–4867. [PubMed]

27. Virgin, H.W.; Wherry, E.J.; Ahmed, R. Redefining chronic viral infection. *Cell* **2009**, *138*, 30–50. [CrossRef] [PubMed]

28. Johnson, J.M.; Nicot, C.; Fullen, J.; Ciminale, V.; Casareto, L.; Mulloy, J.C.; Jacobson, S.; Franchini, G. Free major histocompatibility complex class I heavy chain is preferentially targeted for degradation by human T-cell leukemia/lymphotropic virus type 1 p12(I) protein. *J. Virol.* **2001**, *75*, 6086–6094. [CrossRef] [PubMed]

29. Jacobson, S.; Shida, H.; McFarlin, D.E.; Fauci, A.S.; Koenig, S. Circulating CD8+ cytotoxic T lymphocytes specific for HTLV-I pX in patients with HTLV-I associated neurological disease. *Nature* **1990**, *348*, 245–248. [CrossRef] [PubMed]

30. Kannagi, M.; Harada, S.; Maruyama, I.; Inoko, H.; Igarashi, H.; Kuwashima, G.; Sato, S.; Morita, M.; Kidokoro, M.; Sugimoto, M.; *et al.* Predominant recognition of human T cell leukemia virus type I (HTLV-I) pX gene products by human CD8+ cytotoxic T cells directed against HTLV-I-infected cells. *Int. Immunol.* **1991**, *3*, 761–767. [CrossRef] [PubMed]

31. Goon, P.K.; Biancardi, A.; Fast, N.; Igakura, T.; Hanon, E.; Mosley, A.J.; Asquith, B.; Gould, K.G.; Marshall, S.; Taylor, G.P.; *et al.* Human T cell lymphotropic virus (HTLV) type-1-specific CD8+ T cells: Frequency and immunodominance hierarchy. *J. Infect. Dis.* **2004**, *189*, 2294–2298. [CrossRef] [PubMed]

32. Nagasato, K.; Nakamura, T.; Shirabe, S.; Shibayama, K.; Ohishi, K.; Ichinose, K.; Tsujihata, M.; Nagataki, S. Presence of serum anti-human T-lymphotropic virus type I (HTLV-I) IgM antibodies means persistent active replication of HTLV-I in HTLV-I-associated myelopathy. *J. Neurol. Sci.* **1991**, *103*, 203–208. [CrossRef]

33. Kira, J.; Nakamura, M.; Sawada, T.; Koyanagi, Y.; Ohori, N.; Itoyama, Y.; Yamamoto, N.; Sakaki, Y.; Goto, I. Antibody titers to HTLV-I-p40tax protein and gag-env hybrid protein in HTLV-I-associated myelopathy/tropical spastic paraparesis: Correlation with increased HTLV-I proviral DNA load. *J. Neurol. Sci.* **1992**, *107*, 98–104. [CrossRef]

34. Ishihara, S.; Okayama, A.; Stuver, S.; Horinouchi, H.; Shioiri, S.; Murai, K.; Kubota, T.; Yamashita, R.; Tachibana, N.; Tsubouchi, H.; *et al.* Association of HTLV-I antibody profile of asymptomatic carriers with proviral DNA levels of peripheral blood mononuclear cells. *J. Acquir. Immune Defic. Syndr.* **1994**, *7*, 199–203. [PubMed]

35. Hanon, E.; Hall, S.; Taylor, G.P.; Saito, M.; Davis, R.; Tanaka, Y.; Usuku, K.; Osame, M.; Weber, J.N.; Bangham, C.R. Abundant tax protein expression in CD4+ T cells infected with human T-cell lymphotropic virus type I (HTLV-I) is prevented by cytotoxic T lymphocytes. *Blood* **2000**, *95*, 1386–1392. [PubMed]

36. Lezin, A.; Gillet, N.; Olindo, S.; Signate, A.; Grandvaux, N.; Verlaeten, O.; Belrose, G.; de Carvalho Bittencourt, M.; Hiscott, J.; Asquith, B.; *et al.* Histone deacetylase mediated transcriptional activation reduces proviral loads in HTLV-1 associated myelopathy/tropical spastic paraparesis patients. *Blood* **2007**, *110*, 3722–3728. [CrossRef] [PubMed]

37. Lezin, A.; Olindo, S.; Belrose, G.; Signate, A.; Cesaire, R.; Smadja, D.; Macallan, D.; Asquith, B.; Bangham, C.; Bouzar, A.; *et al.* Gene activation therapy: From the BLV model to HAM/TSP patients. *Front. Biosci. (Schol. Ed.)* **2009**, *1*, 205–215. [CrossRef] [PubMed]

38. Belrose, G.; Gross, A.; Olindo, S.; Lezin, A.; Dueymes, M.; Komla-Soukha, I.; Smadja, D.; Tanaka, Y.; Willems, L.; Mesnard, J.M.; *et al.* Effects of valproate on Tax and HBZ expression in HTLV-1 and HAM/TSP T lymphocytes. *Blood* **2011**, *118*, 2483–2491. [CrossRef] [PubMed]

39. Afonso, P.V.; Mekaouche, M.; Mortreux, F.; Toulza, F.; Moriceau, A.; Wattel, E.; Gessain, A.; Bangham, C.R.; Dubreuil, G.; Plumelle, Y.; *et al.* Highly active antiretroviral treatment against STLV-1 infection combining reverse transcriptase and HDAC inhibitors. *Blood* **2010**, *116*, 3802–3808. [CrossRef] [PubMed]

40. Asquith, B.; Mosley, A.J.; Heaps, A.; Tanaka, Y.; Taylor, G.P.; McLean, A.R.; Bangham, C.R. Quantification of the virus-host interaction in human T lymphotropic virus I infection. *Retrovirology* **2005**, *2*, 75. [CrossRef] [PubMed]

41. Hilburn, S.; Rowan, A.; Demontis, M.A.; MacNamara, A.; Asquith, B.; Bangham, C.R.; Taylor, G.P. *In vivo* expression of human T-lymphotropic virus type 1 basic leucine-zipper protein generates specific CD8+ and CD4+ T-lymphocyte responses that correlate with clinical outcome. *J. Infect. Dis.* **2011**, *203*, 529–536. [CrossRef] [PubMed]

42. Macnamara, A.; Rowan, A.; Hilburn, S.; Kadolsky, U.; Fujiwara, H.; Suemori, K.; Yasukawa, M.; Taylor, G.; Bangham, C.R.; Asquith, B. HLA class I binding of HBZ determines outcome in HTLV-1 infection. *PLoS Pathog.* **2010**, *6*, e1001117. [CrossRef] [PubMed]

43. Hidaka, M.; Inoue, J.; Yoshida, M.; Seiki, M. Post-transcriptional regulator (rex) of HTLV-1 initiates expression of viral structural proteins but suppresses expression of regulatory proteins. *EMBO J.* **1988**, *7*, 519–523. [PubMed]

44. Nicot, C.; Dundr, M.; Johnson, J.M.; Fullen, J.R.; Alonzo, N.; Fukumoto, R.; Princler, G.L.; Derse, D.; Misteli, T.; Franchini, G. HTLV-1-encoded p30II is a post-transcriptional negative regulator of viral replication. *Nat. Med.* **2004**, *10*, 197–201. [CrossRef] [PubMed]

45. Lemasson, I.; Polakowski, N.J.; Laybourn, P.J.; Nyborg, J.K. Transcription regulatory complexes bind the human T-cell leukemia virus 5′ and 3′ long terminal repeats to control gene expression. *Mol. Cell. Biol.* **2004**, *24*, 6117–6126. [CrossRef] [PubMed]

46. Smith, M.J.; Gitlin, S.D.; Browning, C.M.; Lane, B.R.; Clark, N.M.; Shah, N.; Rainier, S.; Markovitz, D.M. GLI-2 modulates retroviral gene expression. *J. Virol.* **2001**, *75*, 2301–2313. [CrossRef] [PubMed]

47. Furukawa, Y.; Kubota, R.; Tara, M.; Izumo, S.; Osame, M. Existence of escape mutant in HTLV-I tax during the development of adult T-cell leukemia. *Blood* **2001**, *97*, 987–993. [CrossRef] [PubMed]

48. Nomura, M.; Ohashi, T.; Nishikawa, K.; Nishitsuji, H.; Kurihara, K.; Hasegawa, A.; Furuta, R.A.; Fujisawa, J.; Tanaka, Y.; Hanabuchi, S.; *et al.* Repression of tax expression is associated both with resistance of human T-cell leukemia virus type 1-infected T cells to killing by tax-specific cytotoxic T lymphocytes and with impaired tumorigenicity in a rat model. *J. Virol.* **2004**, *78*, 3827–3836. [CrossRef] [PubMed]

49. Matsuoka, M.; Jeang, K.T. Human T-cell leukaemia virus type 1 (HTLV-1) infectivity and cellular transformation. *Nat. Rev. Cancer* **2007**, *7*, 270–280. [CrossRef] [PubMed]

50. Grassmann, R.; Aboud, M.; Jeang, K.T. Molecular mechanisms of cellular transformation by HTLV-1 Tax. *Oncogene* **2005**, *24*, 5976–5985. [CrossRef] [PubMed]

51. Higuchi, M.; Fujii, M. Distinct functions of HTLV-1 Tax1 from HTLV-2 Tax2 contribute key roles to viral pathogenesis. *Retrovirology* **2009**, *6*, 117. [CrossRef] [PubMed]

52. Basbous, J.; Bazarbachi, A.; Granier, C.; Devaux, C.; Mesnard, J.M. The central region of human T-cell leukemia virus type 1 Tax protein contains distinct domains involved in subunit dimerization. *J. Virol.* **2003**, *77*, 13028–13035. [CrossRef] [PubMed]

53. Peloponese, J.M., Jr.; Iha, H.; Yedavalli, V.R.; Miyazato, A.; Li, Y.; Haller, K.; Benkirane, M.; Jeang, K.T. Ubiquitination of human T-cell leukemia virus type 1 tax modulates its activity. *J. Virol.* **2004**, *78*, 11686–11695. [CrossRef] [PubMed]

54. Peloponese, J.M., Jr.; Yasunaga, J.; Kinjo, T.; Watashi, K.; Jeang, K.T. Peptidylproline cis-trans-isomerase Pin1 interacts with human T-cell leukemia virus type 1 tax and modulates its activation of NF-kappaB. *J. Virol.* **2009**, *83*, 3238–3248. [CrossRef] [PubMed]

55. Durkin, S.S.; Ward, M.D.; Fryrear, K.A.; Semmes, O.J. Site-specific phosphorylation differentiates active from inactive forms of the human T-cell leukemia virus type 1 Tax oncoprotein. *J. Biol. Chem.* **2006**, *281*, 31705–31712. [CrossRef] [PubMed]

56. Chiari, E.; Lamsoul, I.; Lodewick, J.; Chopin, C.; Bex, F.; Pique, C. Stable ubiquitination of human T-cell leukemia virus type 1 tax is required for proteasome binding. *J. Virol.* **2004**, *78*, 11823–11832. [CrossRef] [PubMed]

57. Lodewick, J.; Lamsoul, I.; Polania, A.; Lebrun, S.; Burny, A.; Ratner, L.; Bex, F. Acetylation of the human T-cell leukemia virus type 1 Tax oncoprotein by p300 promotes activation of the NF-kappaB pathway. *Virology* **2009**, *386*, 68–78. [CrossRef] [PubMed]

58. Jeong, S.J.; Ryo, A.; Yamamoto, N. The prolyl isomerase Pin1 stabilizes the human T-cell leukemia virus type 1 (HTLV-1) Tax oncoprotein and promotes malignant transformation. *Biochem. Biophys. Res. Commun.* **2009**, *381*, 294–299. [CrossRef] [PubMed]

59. Marriott, S.J.; Semmes, O.J. Impact of HTLV-I Tax on cell cycle progression and the cellular DNA damage repair response. *Oncogene* **2005**, *24*, 5986–5995. [CrossRef] [PubMed]

60. Rosin, O.; Koch, C.; Schmitt, I.; Semmes, O.J.; Jeang, K.T.; Grassmann, R. A human T-cell leukemia virus Tax variant incapable of activating NF-kappaB retains its immortalizing potential for primary T-lymphocytes. *J. Biol. Chem.* **1998**, *273*, 6698–6703. [CrossRef] [PubMed]

61. Robek, M.D.; Ratner, L. Immortalization of CD4(+) and CD8(+) T lymphocytes by human T-cell leukemia virus type 1 Tax mutants expressed in a functional molecular clone. *J. Virol.* **1999**, *73*, 4856–4865. [PubMed]

62. Grossman, W.J.; Kimata, J.T.; Wong, F.H.; Zutter, M.; Ley, T.J.; Ratner, L. Development of leukemia in mice transgenic for the tax gene of human T-cell leukemia virus type I. *Proc. Natl. Acad. Sci. USA* **1995**, *92*, 1057–1061. [CrossRef] [PubMed]

63. Hasegawa, H.; Sawa, H.; Lewis, M.J.; Orba, Y.; Sheehy, N.; Yamamoto, Y.; Ichinohe, T.; Tsunetsugu-Yokota, Y.; Katano, H.; Takahashi, H.; *et al.* Thymus-derived leukemia-lymphoma in mice transgenic for the Tax gene of human T-lymphotropic virus type I. *Nat. Med.* **2006**, *12*, 466–472. [CrossRef] [PubMed]

64. Ohsugi, T.; Kumasaka, T.; Okada, S.; Urano, T. The Tax protein of HTLV-1 promotes oncogenesis in not only immature T cells but also mature T cells. *Nat. Med.* **2007**, *13*, 527–528. [CrossRef] [PubMed]

65. Matsuoka, M.; Tamiya, S.; Takemoto, S.; Yamaguchi, K.; Takatsuki, K. HTLV-I provirus in the clinical subtypes of ATL. *Leukemia* **1997**, *11* (Suppl. 3), 67–69. [PubMed]

66. Tamiya, S.; Matsuoka, M.; Etoh, K.; Watanabe, T.; Kamihira, S.; Yamaguchi, K.; Takatsuki, K. Two types of defective human T-lymphotropic virus type I provirus in adult T-cell leukemia. *Blood* **1996**, *88*, 3065–3073. [PubMed]

67. Miyazaki, M.; Yasunaga, J.; Taniguchi, Y.; Tamiya, S.; Nakahata, T.; Matsuoka, M. Preferential selection of human T-cell leukemia virus type 1 provirus lacking the 5' long terminal repeat during oncogenesis. *J. Virol.* **2007**, *81*, 5714–5723. [CrossRef] [PubMed]

68. Haller, K.; Kibler, K.V.; Kasai, T.; Chi, Y.H.; Peloponese, J.M.; Yedavalli, V.S.; Jeang, K.T. The N-terminus of rodent and human MAD1 confers species-specific stringency to spindle assembly checkpoint. *Oncogene* **2006**, *25*, 2137–2147. [CrossRef] [PubMed]

69. Duensing, S.; Munger, K. Mechanisms of genomic instability in human cancer: Insights from studies with human papillomavirus oncoproteins. *Int. J. Cancer* **2004**, *109*, 157–162. [CrossRef] [PubMed]

70. Yoshida, M.; Satou, Y.; Yasunaga, J.; Fujisawa, J.; Matsuoka, M. Transcriptional control of spliced and unspliced human T-cell leukemia virus type 1 bZIP factor (HBZ) gene. *J. Virol.* **2008**, *82*, 9359–9368. [CrossRef] [PubMed]

71. Arpin-Andre, C.; Laverdure, S.; Barbeau, B.; Gross, A.; Mesnard, J.M. Construction of a reporter vector for analysis of bidirectional transcriptional activity of retrovirus LTR. *Plasmid* **2014**, *74*, 45–51. [CrossRef] [PubMed]

72. Duc Dodon, M.; Mesnard, J.M.; Barbeau, B. [Adult T-cell leukemia induced by HTLV-1: Before and after HBZ]. *Med. Sci. (Paris)* **2010**, *26*, 391–396. [CrossRef] [PubMed]

73. Kataoka, K.; Nagata, Y.; Kitanaka, A.; Shiraishi, Y.; Shimamura, T.; Yasunaga, J.; Totoki, Y.; Chiba, K.; Sato-Otsubo, A.; Nagae, G.; et al. Integrated molecular analysis of adult T cell leukemia/lymphoma. *Nat. Genet.* **2015**, *47*, 1304–1315. [CrossRef] [PubMed]

74. Vicente, C.; Cools, J. The genomic landscape of adult T cell leukemia/lymphoma. *Nat. Genet.* **2015**, *47*, 1226–1227. [CrossRef] [PubMed]

75. Lemasson, I.; Lewis, M.R.; Polakowski, N.; Hivin, P.; Cavanagh, M.H.; Thebault, S.; Barbeau, B.; Nyborg, J.K.; Mesnard, J.M. Human T-cell leukemia virus type 1 (HTLV-1) bZIP protein interacts with the cellular transcription factor CREB to inhibit HTLV-1 transcription. *J. Virol.* **2007**, *81*, 1543–1553. [CrossRef] [PubMed]

76. Basbous, J.; Arpin, C.; Gaudray, G.; Piechaczyk, M.; Devaux, C.; Mesnard, J.M. The HBZ factor of human T-cell leukemia virus type I dimerizes with transcription factors JunB and c-Jun and modulates their transcriptional activity. *J. Biol. Chem.* **2003**, *278*, 43620–43627. [CrossRef] [PubMed]

77. Matsumoto, J.; Ohshima, T.; Isono, O.; Shimotohno, K. HTLV-1 HBZ suppresses AP-1 activity by impairing both the DNA-binding ability and the stability of c-Jun protein. *Oncogene* **2005**, *24*, 1001–1010. [CrossRef] [PubMed]

78. Hivin, P.; Basbous, J.; Raymond, F.; Henaff, D.; Arpin-Andre, C.; Robert-Hebmann, V.; Barbeau, B.; Mesnard, J.M. The HBZ-SP1 isoform of human T-cell leukemia virus type I represses JunB activity by sequestration into nuclear bodies. *Retrovirology* **2007**, *4*, 14. [CrossRef] [PubMed]

79. Thebault, S.; Basbous, J.; Hivin, P.; Devaux, C.; Mesnard, J.M. HBZ interacts with JunD and stimulates its transcriptional activity. *FEBS Lett.* **2004**, *562*, 165–170. [CrossRef]

80. Zhao, T.; Yasunaga, J.; Satou, Y.; Nakao, M.; Takahashi, M.; Fujii, M.; Matsuoka, M. Human T-cell leukemia virus type 1 bZIP factor selectively suppresses the classical pathway of NF-kappaB. *Blood* **2009**, *113*, 2755–2764. [CrossRef] [PubMed]

81. Peloponese, J.M.; Yeung, M.L.; Jeang, K.T. Modulation of nuclear factor-kappaB by human T cell leukemia virus type 1 Tax protein: Implications for oncogenesis and inflammation. *Immunol. Res.* **2006**, *34*, 1–12. [CrossRef]

82. Bernal-Mizrachi, L.; Lovly, C.M.; Ratner, L. The role of NF-{kappa}B-1 and NF-{kappa}B-2-mediated resistance to apoptosis in lymphomas. *Proc. Natl. Acad. Sci. USA* **2006**, *103*, 9220–9225. [CrossRef] [PubMed]

83. Arnold, J.; Zimmerman, B.; Li, M.; Lairmore, M.D.; Green, P.L. Human T-cell leukemia virus type-1 antisense-encoded gene, Hbz, promotes T-lymphocyte proliferation. *Blood* **2008**, *112*, 3788–3797. [CrossRef] [PubMed]

84. Kuhlmann, A.S.; Villaudy, J.; Gazzolo, L.; Castellazzi, M.; Mesnard, J.M.; Duc Dodon, M. HTLV-1 HBZ cooperates with JunD to enhance transcription of the human telomerase reverse transcriptase gene (hTERT). *Retrovirology* **2007**, *4*, 92. [CrossRef] [PubMed]

85. Barbeau, B.; Mesnard, J.M. Does chronic infection in retroviruses have a sense? *Trends Microbiol.* **2015**, *23*, 367–375. [CrossRef] [PubMed]

86. Mortreux, F.; Leclercq, I.; Gabet, A.S.; Leroy, A.; Westhof, E.; Gessain, A.; Wain-Hobson, S.; Wattel, E. Somatic mutation in human T-cell leukemia virus type 1 provirus and flanking cellular sequences during clonal expansion *in vivo*. *J. Natl. Cancer Inst.* **2001**, *93*, 367–377. [CrossRef] [PubMed]

87. Mortreux, F.; Gabet, A.S.; Wattel, E. Molecular and cellular aspects of HTLV-1 associated leukemogenesis *in vivo*. *Leukemia* **2003**, *17*, 26–38. [CrossRef] [PubMed]

88. Vernin, C.; Thenoz, M.; Pinatel, C.; Gessain, A.; Gout, O.; Delfau-Larue, M.H.; Nazaret, N.; Legras-Lachuer, C.; Wattel, E.; Mortreux, F. HTLV-1 bZIP factor HBZ promotes cell proliferation and genetic instability by activating OncomiRs. *Cancer Res.* **2014**, *74*, 6082–6093. [CrossRef] [PubMed]

89. Satou, Y.; Yasunaga, J.; Zhao, T.; Yoshida, M.; Miyazato, P.; Takai, K.; Shimizu, K.; Ohshima, K.; Green, P.L.; Ohkura, N.; *et al.* HTLV-1 bZIP factor induces T-cell lymphoma and systemic inflammation *in vivo*. *PLoS Pathog.* **2011**, *7*, e1001274. [CrossRef] [PubMed]

90. Gazon, H.; Lemasson, I.; Polakowski, N.; Cesaire, R.; Matsuoka, M.; Barbeau, B.; Mesnard, J.M.; Peloponese, J.M., Jr. Human T-cell leukemia virus type 1 (HTLV-1) bZIP factor requires cellular transcription factor JunD to upregulate HTLV-1 antisense transcription from the 3′ long terminal repeat. *J. Virol.* **2012**, *86*, 9070–9078. [CrossRef] [PubMed]

91. Richardson, J.H.; Edwards, A.J.; Cruickshank, J.K.; Rudge, P.; Dalgleish, A.G. *In vivo* cellular tropism of human T-cell leukemia virus type 1. *J. Virol.* **1990**, *64*, 5682–5687. [PubMed]

92. Karube, K.; Ohshima, K.; Tsuchiya, T.; Yamaguchi, T.; Kawano, R.; Suzumiya, J.; Utsunomiya, A.; Harada, M.; Kikuchi, M. Expression of FoxP3, a key molecule in CD4CD25 regulatory T cells, in adult T-cell leukaemia/lymphoma cells. *Br. J. Haematol.* **2004**, *126*, 81–84. [CrossRef] [PubMed]

93. Siegel, R.; Gartenhaus, R.; Kuzel, T. HTLV-I associated leukemia/lymphoma: Epidemiology, biology, and treatment. *Cancer Treat. Res.* **2001**, *104*, 75–88. [PubMed]

94. Yamada, Y.; Tomonaga, M. The current status of therapy for adult T-cell leukaemia-lymphoma in Japan. *Leuk Lymphoma* **2003**, *44*, 611–618. [CrossRef] [PubMed]

95. Takasaki, Y.; Iwanaga, M.; Imaizumi, Y.; Tawara, M.; Joh, T.; Kohno, T.; Yamada, Y.; Kamihira, S.; Ikeda, S.; Miyazaki, Y.; *et al.* Long-term study of indolent adult T-cell leukemia-lymphoma. *Blood* **2010**, *115*, 4337–4343. [CrossRef] [PubMed]

96. Bazarbachi, A.; Hermine, O. Treatment with a combination of zidovudine and alpha-interferon in naive and pretreated adult T-cell leukemia/lymphoma patients. *J. Acquir. Immune Defic. Syndr. Hum. Retrovirol.* **1996**, *13* (Suppl. 1), S186–S190. [CrossRef] [PubMed]

97. Nasr, R.; el Hajj, H.; Kfoury, Y.; de The, H.; Hermine, O.; Bazarbachi, A. Controversies in targeted therapy of adult T cell leukemia/lymphoma: ON target or OFF target effects? *Viruses* **2011**, *3*, 750–769. [CrossRef] [PubMed]

98. Bazarbachi, A.; Suarez, F.; Fields, P.; Hermine, O. How I treat adult T-cell leukemia/lymphoma. *Blood* **2011**, *118*, 1736–1745. [CrossRef] [PubMed]

99. Tsukasaki, K.; Ikeda, S.; Murata, K.; Maeda, T.; Atogami, S.; Sohda, H.; Momita, S.; Jubashi, T.; Yamada, Y.; Mine, M.; *et al.* Characteristics of chemotherapy-induced clinical remission in long survivors with aggressive adult T-cell leukemia/lymphoma. *Leuk Res.* **1993**, *17*, 157–166. [CrossRef]

100. Taguchi, H.; Kinoshita, K.I.; Takatsuki, K.; Tomonaga, M.; Araki, K.; Arima, N.; Ikeda, S.; Uozumi, K.; Kohno, H.; Kawano, F.; *et al.* An intensive chemotherapy of adult T-cell leukemia/lymphoma: CHOP followed by etoposide, vindesine, ranimustine, and mitoxantrone with granulocyte colony-stimulating factor support. *J. Acquir. Immune Defic. Syndr. Hum. Retrovirol.* **1996**, *12*, 182–186. [CrossRef] [PubMed]

101. Tobinai, K. Current management of adult T-cell leukemia/lymphoma. *Oncology (Williston Park)* **2009**, *23*, 1250–1256. [PubMed]

102. Makiyama, J.; Imaizumi, Y.; Tsushima, H.; Taniguchi, H.; Moriwaki, Y.; Sawayama, Y.; Imanishi, D.; Taguchi, J.; Hata, T.; Tsukasaki, K.; *et al.* Treatment outcome of elderly patients with aggressive adult T cell leukemia-lymphoma: Nagasaki University Hospital experience. *Int. J. Hematol.* **2014**, *100*, 464–472. [CrossRef] [PubMed]

103. Kawano, N.; Yoshida, S.; Kuriyama, T.; Tahara, Y.; Yamashita, K.; Nagahiro, Y.; Kawano, J.; Koketsu, H.; Toyofuku, A.; Manabe, T.; *et al.* Clinical Features and Treatment Outcomes of 81 Patients with Aggressive Type Adult T-cell Leukemia-lymphoma at a Single Institution over a 7-year Period (2006–2012). *Intern. Med.* **2015**, *54*, 1489–1498. [CrossRef] [PubMed]

104. Tsukasaki, K.; Utsunomiya, A.; Fukuda, H.; Shibata, T.; Fukushima, T.; Takatsuka, Y.; Ikeda, S.; Masuda, M.; Nagoshi, H.; Ueda, R.; *et al.* VCAP-AMP-VECP compared with biweekly CHOP for adult T-cell leukemia-lymphoma: Japan Clinical Oncology Group Study JCOG9801. *J. Clin. Oncol.* **2007**, *25*, 5458–5464. [CrossRef] [PubMed]

105. Mercieca, J.; Matutes, E.; Dearden, C.; MacLennan, K.; Catovsky, D. The role of pentostatin in the treatment of T-cell malignancies: Analysis of response rate in 145 patients according to disease subtype. *J. Clin. Oncol.* **1994**, *12*, 2588–2593. [PubMed]

106. Tsukasaki, K.; Tobinai, K.; Shimoyama, M.; Kozuru, M.; Uike, N.; Yamada, Y.; Tomonaga, M.; Araki, K.; Kasai, M.; Takatsuki, K.; *et al.* Deoxycoformycin-containing combination chemotherapy for adult T-cell leukemia-lymphoma: Japan Clinical Oncology Group Study (JCOG9109). *Int. J. Hematol.* **2003**, *77*, 164–170. [CrossRef] [PubMed]

107. Uike, N.; Choi, I.; Tokoro, A.; Goto, T.; Yufu, Y.; Kozuru, M.; Tobinai, K. Adult T-cell leukemia-lymphoma successfully treated with 2-chlorodeoxyadenosine. *Intern. Med.* **1998**, *37*, 411–413. [CrossRef] [PubMed]

108. Tobinai, K.; Uike, N.; Saburi, Y.; Chou, T.; Etoh, T.; Masuda, M.; Kawano, F.; Matsuoka, M.; Taguchi, H.; Makino, T.; *et al.* Phase II study of cladribine (2-chlorodeoxyadenosine) in relapsed or refractory adult T-cell leukemia-lymphoma. *Int. J. Hematol.* **2003**, *77*, 512–517. [CrossRef] [PubMed]

109. Makino, T.; Nakahara, K.; Takatsuka, Y.; Shimotakahara, S.; Utsunomiya, A.; Hanada, S.; Tokunaga, M.; Arima, T. [Successful treatment of chemotherapy-resistant adult T cell leukemia/lymphoma by irinotecan hydrochloride (CPT-11)]. *Rinsho Ketsueki* **1994**, *35*, 42–48. [PubMed]

110. Ohno, R.; Masaoka, T.; Shirakawa, S.; Sakamoto, S.; Hirano, M.; Hanada, S.; Yasunaga, K.; Yokomaku, S.; Mitomo, Y.; Nagai, K.; *et al.* Treatment of adult T-cell leukemia/lymphoma with MST-16, a new oral antitumor drug and a derivative of bis(2,6-dioxopiperazine). The MST-16 Study Group. *Cancer* **1993**, *71*, 2217–2221. [CrossRef]

111. Taguchi, T.; Ohta, K.; Hotta, T.; Shirakawa, S.; Masaoka, T.; Kimura, I. [Menogaril (TUT-7) late phase II study for malignant lymphoma, adult T-cell leukemia and lymphoma (ATLL)]. *Gan To Kagaku Ryoho* **1997**, *24*, 1263–1271. [PubMed]

112. Tsukasaki, K.; Tomonaga, M. ATRA, NF-kappaB and ATL. *Leuk Res.* **2001**, *25*, 407–408. [CrossRef]

113. Toshima, M.; Nagai, T.; Izumi, T.; Tarumoto, T.; Takatoku, M.; Imagawa, S.; Komatsu, N.; Ozawa, K. All-trans-retinoic acid treatment for chemotherapy-resistant acute adult T-cell leukemia. *Int. J. Hematol.* **2000**, *72*, 343–345. [PubMed]

114. Hasegawa, H.; Yamada, Y.; Tsukasaki, K.; Mori, N.; Tsuruda, K.; Sasaki, D.; Usui, T.; Osaka, A.; Atogami, S.; Ishikawa, C.; *et al.* LBH589, a deacetylase inhibitor, induces apoptosis in adult T-cell leukemia/lymphoma cells via activation of a novel RAIDD-caspase-2 pathway. *Leukemia* **2011**, *25*, 575–587. [CrossRef] [PubMed]

115. Tsukasaki, K.; Tobinai, K. Clinical Trials and Treatment of ATL. *Leuk Res. Treat.* **2012**, *2012*, 101754. [CrossRef] [PubMed]

116. Mori, N.; Matsuda, T.; Tadano, M.; Kinjo, T.; Yamada, Y.; Tsukasaki, K.; Ikeda, S.; Yamasaki, Y.; Tanaka, Y.; Ohta, T.; *et al.* Apoptosis induced by the histone deacetylase inhibitor FR901228 in human T-cell leukemia virus type 1-infected T-cell lines and primary adult T-cell leukemia cells. *J. Virol.* **2004**, *78*, 4582–4590. [CrossRef] [PubMed]

117. Blaheta, R.A.; Cinatl, J., Jr. Anti-tumor mechanisms of valproate: A novel role for an old drug. *Med. Res. Rev.* **2002**, *22*, 492–511. [CrossRef] [PubMed]

118. Bezecny, P. Histone deacetylase inhibitors in glioblastoma: Pre-clinical and clinical experience. *Med. Oncol.* **2014**, *31*, 985. [CrossRef] [PubMed]

119. Lagneaux, L.; Gillet, N.; Stamatopoulos, B.; Delforge, A.; Dejeneffe, M.; Massy, M.; Meuleman, N.; Kentos, A.; Martiat, P.; Willems, L.; *et al.* Valproic acid induces apoptosis in chronic lymphocytic leukemia cells through activation of the death receptor pathway and potentiates TRAIL response. *Exp. Hematol.* **2007**, *35*, 1527–1537. [CrossRef] [PubMed]

120. Belrose, G.; Gazon, H.; Meniane, J.-C.; Olindo, S.; Mesnard, J.-M.; Peloponese, J.-M.; Cesaire, R. Effects of valproate on Tax and HBZ expression in *ex vivo* cultured ATL cells. *Retrovirology* **2014**, *11* (Suppl. 1), P39. [CrossRef]

121. Mosley, A.J.; Meekings, K.N.; McCarthy, C.; Shepherd, D.; Cerundolo, V.; Mazitschek, R.; Tanaka, Y.; Taylor, G.P.; Bangham, C.R. Histone deacetylase inhibitors increase virus gene expression but decrease CD8+ cell antiviral function in HTLV-1 infection. *Blood* **2006**, *108*, 3801–3807. [CrossRef] [PubMed]

122. Olindo, S.; Belrose, G.; Gillet, N.; Rodriguez, S.; Boxus, M.; Verlaeten, O.; Asquith, B.; Bangham, C.; Signate, A.; Smadja, D.; *et al.* Safety of long-term treatment of HAM/TSP patients with valproic acid. *Blood* **2011**, *118*, 6306–6309. [CrossRef] [PubMed]

123. Berkowitz, J.L.; Janik, J.E.; Stewart, D.M.; Jaffe, E.S.; Stetler-Stevenson, M.; Shih, J.H.; Fleisher, T.A.; Turner, M.; Urquhart, N.E.; Wharfe, G.H.; *et al.* Safety, efficacy, and pharmacokinetics/pharmacodynamics of daclizumab (anti-CD25) in patients with adult T-cell leukemia/lymphoma. *Clin. Immunol.* **2014**, *155*, 176–187. [CrossRef] [PubMed]

124. Yoshie, O.; Fujisawa, R.; Nakayama, T.; Harasawa, H.; Tago, H.; Izawa, D.; Hieshima, K.; Tatsumi, Y.; Matsushima, K.; Hasegawa, H.; *et al.* Frequent expression of CCR4 in adult T-cell leukemia and human T-cell leukemia virus type 1-transformed T cells. *Blood* **2002**, *99*, 1505–1511. [CrossRef] [PubMed]

125. Yamamoto, K.; Utsunomiya, A.; Tobinai, K.; Tsukasaki, K.; Uike, N.; Uozumi, K.; Yamaguchi, K.; Yamada, Y.; Hanada, S.; Tamura, K.; *et al.* Phase I study of KW-0761, a defucosylated humanized anti-CCR4 antibody, in relapsed patients with adult T-cell leukemia-lymphoma and peripheral T-cell lymphoma. *J. Clin. Oncol.* **2010**, *28*, 1591–1598. [CrossRef] [PubMed]

126. Tsukasaki, K. [Mogamulizumab for the treatment of ATL and PTCL]. *Gan To Kagaku Ryoho* **2015**, *42*, 553–557. [PubMed]

127. Ishida, T.; Jo, T.; Takemoto, S.; Suzushima, H.; Uozumi, K.; Yamamoto, K.; Uike, N.; Saburi, Y.; Nosaka, K.; Utsunomiya, A.; *et al.* Dose-intensified chemotherapy alone or in combination with mogamulizumab in newly diagnosed aggressive adult T-cell leukaemia-lymphoma: A randomized phase II study. *Br. J. Haematol.* **2015**, *169*, 672–682. [CrossRef] [PubMed]

128. Kawashima, I.; Sueki, Y.; Yamamoto, T.; Nozaki, Y.; Nakajima, K.; Mitsumori, T.; Kirito, K. Adult T cell leukemia-lymphoma with allo-HSCT after treatment for pulmonary involvement with Mogamulizumab. *Rinsho Ketsueki* **2015**, *56*, 210–215. [PubMed]

129. Utsunomiya, A.; Miyazaki, Y.; Takatsuka, Y.; Hanada, S.; Uozumi, K.; Yashiki, S.; Tara, M.; Kawano, F.; Saburi, Y.; Kikuchi, H.; *et al.* Improved outcome of adult T cell leukemia/lymphoma with allogeneic hematopoietic stem cell transplantation. *Bone Marrow Transplant.* **2001**, *27*, 15–20. [CrossRef] [PubMed]

130. Okamura, J.; Uike, N.; Utsunomiya, A.; Tanosaki, R. Allogeneic stem cell transplantation for adult T-cell leukemia/lymphoma. *Int. J. Hematol.* **2007**, *86*, 118–125. [CrossRef] [PubMed]

131. Tanosaki, R.; Uike, N.; Utsunomiya, A.; Saburi, Y.; Masuda, M.; Tomonaga, M.; Eto, T.; Hidaka, M.; Harada, M.; Choi, I.; *et al.* Allogeneic hematopoietic stem cell transplantation using reduced-intensity conditioning for adult T cell leukemia/lymphoma: Impact of antithymocyte globulin on clinical outcome. *Biol. Blood Marrow Transplant.* **2008**, *14*, 702–708. [CrossRef] [PubMed]

132. Hishizawa, M.; Kanda, J.; Utsunomiya, A.; Taniguchi, S.; Eto, T.; Moriuchi, Y.; Tanosaki, R.; Kawano, F.; Miyazaki, Y.; Masuda, M.; *et al.* Transplantation of allogeneic hematopoietic stem cells for adult T-cell leukemia: A nationwide retrospective study. *Blood* **2010**, *116*, 1369–1376. [CrossRef] [PubMed]

133. Kanda, J.; Chiou, L.W.; Szabolcs, P.; Sempowski, G.D.; Rizzieri, D.A.; Long, G.D.; Sullivan, K.M.; Gasparetto, C.; Chute, J.P.; Morris, A.; *et al.* Immune recovery in adult patients after myeloablative dual umbilical cord blood, matched sibling, and matched unrelated donor hematopoietic cell transplantation. *Biol. Blood Marrow Transplant.* **2012**, *18*, 1664.e1–1676.e1. [CrossRef] [PubMed]

134. Bazarbachi, A.; Cwynarski, K.; Boumendil, A.; Finel, H.; Fields, P.; Raj, K.; Nagler, A.; Mohty, M.; Sureda, A.; Dreger, P.; *et al.* Outcome of patients with HTLV-1-associated adult T-cell leukemia/lymphoma after SCT: A retrospective study by the EBMT LWP. *Bone Marrow Transplant.* **2014**, *49*, 1266–1268. [CrossRef] [PubMed]

135. Umino, A.; Nakagawa, M.; Utsunomiya, A.; Tsukasaki, K.; Taira, N.; Katayama, N.; Seto, M. Clonal evolution of adult T-cell leukemia/lymphoma takes place in the lymph nodes. *Blood* **2011**, *117*, 5473–5478. [CrossRef] [PubMed]

136. Bangham, C.R.; Cook, L.B.; Melamed, A. HTLV-1 clonality in adult T-cell leukaemia and non-malignant HTLV-1 infection. *Semin. Cancer Biol.* **2014**, *26*, 89–98. [CrossRef] [PubMed]

137. Nishioka, C.; Ikezoe, T.; Yang, J.; Komatsu, N.; Bandobashi, K.; Taniguchi, A.; Kuwayama, Y.; Togitani, K.; Koeffler, H.P.; Taguchi, H. Histone deacetylase inhibitors induce growth arrest and apoptosis of HTLV-1-infected T-cells via blockade of signaling by nuclear factor kappaB. *Leuk. Res.* **2008**, *32*, 287–296. [CrossRef] [PubMed]

*viruses*

MDPI

*Review*

# Recent Advances in Therapeutic Approaches for Adult T-cell Leukemia/Lymphoma

**Koji Kato * and Koichi Akashi**

Department of Medicine and Biosystemic Science, Graduate School of Medical Science, Kyushu University, Fukuoka 812-8582, Japan; akashi@med.kyushu-u.ac.jp
* Correspondence: kojikato@intmed1.med.kyushu-u.ac.jp; Tel.: +81-92-642-5230

Academic Editor: Louis M. Mansky
Received: 30 September 2015; Accepted: 3 December 2015; Published: 14 December 2015

**Abstract:** Adult T-cell leukemia/lymphoma (ATLL) is a peripheral T-cell lymphoma caused by human T-cell leukemia/lymphoma virus type 1 (HTLV-1). ATLL occurs in approximately 3%–5% of HTLV-1 carriers during their lifetime and follows a heterogeneous clinical course. The Shimoyama classification has been frequently used for treatment decisions in ATLL patients, and antiviral therapy has been reportedly promising, particularly in patients with indolent type ATLL; however, the prognosis continues to be dismal for patients with aggressive-type ATLL. Recent efforts to improve treatment outcomes have been focused on the development of prognostic stratification and improved dosage, timing, and combination of therapeutic modalities, such as antiviral therapy, chemotherapy, allogeneic hematopoietic stem cell transplantation, and molecular targeted therapy.

**Keywords:** adult T-cell leukemia/lymphoma; allogeneic hematopoietic stem cell transplantation; graft-versus-host disease; mogamulizumab; chemotherapy; antiviral therapy

PACS: J0101

## 1. Introduction

Adult T-cell leukemia/lymphoma (ATLL) was first described in 1977 by Uchiyama *et al.* [1], as a distinct clinical entity frequently observed in southwestern Japan. The causative agent of ATLL is the retrovirus human T-cell leukemia virus type I (HTLV-1) [2], which also causes several immune-associated diseases, including HTLV-1-associated myelopathy/tropical spastic paraparesis (HAM/TSP) [3]. ATLL develops in approximately 3%–5% of HTLV-1 carriers and has a dismal prognosis. However, the clinical manifestations and the course of disease in ATLL patients vary to a great extent. Therefore, recent efforts to improve treatment outcomes in ATLL patients have been focused on the development of prognostic stratification and therapeutic modalities. In this review, recent advances in ATLL treatment including antiviral therapy, chemotherapy, allogeneic hematopoietic stem cell transplantation (allo-HSCT), and molecular targeted therapy are discussed.

## 2. Diagnosis and Prognostic Factors for ATLL

ATLL diagnosis is based on clinical features, serum anti-HTLV-1 antibody, and ATLL cell morphology. The clonality of ATLL as a mature T-cell malignancy is confirmed by identification of the monoclonal integration of HTLV-1 proviral DNA in malignant cells by Southern blot analysis. The quantification of HTLV-1 integration site clonality has been recently developed through deep sequence analysis [4]. A high proviral load in HTLV-1 carriers is suggested to be associated with the development of ATLL, although HTLV-1 proviral load is not used as a diagnostic criterion of ATLL. In 1991, the Japan Clinical Oncology Group (JCOG) proposed the Shimoyama classification that defines four clinical subtypes: acute, lymphoma, chronic, and smoldering (Table 1) [5]. The classification is based on the presence of organ involvement, leukemic manifestation, high lactate dehydrogenase (LDH) and hypercalcemia that altogether reflect the prognosis and natural history of the disease. Chronic-type ATLL can be further divided into favorable and unfavorable types based on LDH, blood urea nitrogen, and albumin concentration. Further, acute, lymphoma, and unfavorable chronic types are defined as aggressive-type ATLL, while favorable chronic and smoldering types are defined as indolent-type ATLL [6]. For the last two decades, this clinical classification has been widely used as a guide in ATLL treatment.

**Table 1.** Diagnostic criteria and classification (the Shimoyama classification).

|  | Smoldering | Chronic | Lymphoma | Acute |
|---|---|---|---|---|
| Anti-HTLV-I antibody | + | + | + | + |
| Lymphocyte ($\times 10^9$/L) | <4 | ≥4 [†] | <4 | * |
| Abnormal T lymphocytes | ≥5% | + [‡] | ≤1% | + [‡] |
| Flower cells with T-cell marker | Occasionally | Occasionally | No | + |
| LDH | ≤1.5 N | ≤2 N | * | * |
| Corrected Ca2+ (mEq/L) | <5.5 | <5.5 | * | * |
| Histology-proven lymphadenopathy | No | * | + | * |
| Tumor lesion |  |  |  |  |
| Skin | § | * | * | * |
| Lung | § | * | * | * |
| Lymph node | No | * | Yes | * |
| Liver | No | * | * | * |
| Spleen | No | * | * | * |
| Central nervous system | No | No | * | * |
| Bone | No | No | * | * |
| Ascites | No | No | * | * |
| Pleural effusion | No | No | * | * |
| Gastrointestinal tract | No | No | * | * |

HTLV-I, human T-lymphotropic virus type-I; LDH, lactate dehydrogenase; N, normal upper limit. * No essential qualification except terms required for other subtype(s); [†] Accompanied by T-lymphocytosis (3.5 × $10^9$/L or more); [‡] In case abnormal T-lymphocytes are less than 5% in peripheral blood, histology-proven tumor lesion is required; § No essential qualification if other terms are fulfilled, but histology-proven malignant lesion(s) is required in case abnormal T-lymphocytes are less than 5% in peripheral blood.

Although the prognosis of aggressive ATLL is dismal, there is marked diversity among patients. A prognostic index for acute- and lymphoma-type ATLL (ATL-PI) has been proposed based on a retrospective analysis of 807 newly diagnosed patients between January 2000 and May 2009 in Japan [7]. Ann Arbor stage (I–II *vs.* III–IV), Eastern Cooperative Oncology Group performance status (ECOG PS; 0–1 *vs.* 2–4), age, serum albumin, and soluble interleukin-2 receptor (sIL-2R) were statistically significant prognostic factors. A simplified ATL-PI was as follows: prognostic score; +2 (Ann Arbor stage = III or IV); +1 (ECOG PS > 1); +1 (age > 70); +1 (albumin < 35 g/L); and +1 (sIL2R > 20,000 U/mL). Scores from 0 to 2 were categorized as low risk, 3 to 4 as intermediate risk, and 5 to 6 as high risk. The median overall survival times (MST) were 16.2 months in low-risk patients, 7.0 months in intermediate-risk patients, and 4.6 months in high-risk patients. However, the Shimoyama classification and ATL-PI were established based on retrospectively collected data; thus, the patient characteristics, such as the type of treatment and prognostic factors, were not comparable between groups. The JCOG prognostic index (JCOG-PI) has recently been established based on data from 276 patients with aggressive ATLL in three prospective JCOG trials, which identified poor PS and hypercalcemia as significant prognostic factors [8]. In patients with corrected calcium of <2.75 mmol/L and a PS of 0 or 1 (moderate risk), the MST and five-year overall survival (OS) were 14 months and 18%, respectively; in patients with corrected calcium of ≥2.75 mmol/L and/or a PS of 2–4 (high-risk), the MST and five-year OS were eight months and 4%, respectively. The JCOG-PI may be useful in identifying aggressive ATLL patients with dismal prognosis. Assessment by both ATL-PI and JCOG-PI will certainly be useful in identifying patients with extremely poor prognosis among aggressive ATLL cases. In addition, several biomarkers, such as CC chemokine receptor 4 (CCR4), lung resistance-related protein, and p53 mutations, have been reported [9,10]; however, so far, prognostic models and biomarkers that are able to identify patients who may not need allogeneic hematopoietic stem cell transplantation (allo-HSCT) do not exist. Thus, further investigation is needed to establish robust prognostic models.

## 3. Treatment for ATLL

The treatment strategy for ATLL patients is based on the clinical subtype according to the Shimoyama classification [5,9–11]. The watchful waiting strategy or interferon-α (IFN-α)/zidovudine (AZT) are usually reserved for patients with indolent-type ATLL, whereas chemotherapy, allo-HSCT, and newer therapeutic agents are preferred for patients with aggressive-type ATLL. In Europe and the USA, antiviral therapy using IFN-α/AZT is the standard treatment for leukemic-type ATL. Importantly, a subset of patients with indolent type ATLL experience skin lesions that can be treated with either skin-directed therapy, such as topical steroids, ultraviolet light, and radiation, or systemic therapy, such as steroids, oral retinoids, or single agent chemotherapy. The current treatment strategies are summarized in Table 2.

**Table 2.** Treatment strategy for adult T-cell leukemia/lymphoma (ATLL).

| **1. Indolent-type ATLL: Smoldering- or favorable chronic-type** |
| --- |
| (1) Watchful waiting for asymptomatic patients |
| (2) Interferon-α (IFN-α)/zidovudine (AZT) or watchful waiting for symptomatic patients |
| (3) Skin lesion:<br>Local therapy; Topical steroids, Ultraviolet light, Radiation<br>Systemic therapy; Steroids, Oral retinoids, Single agent chemotherapy |
| **2. Aggressive-type ATLL: Unfavorable chronic-, lymphoma- or acute-type** |
| (1) Chemotherapy:<br>VCAP-AMP-VECP<br>CHOP or less-toxic regimen for elderly patients |
| (2) VCAP-AMP-VECP + mogamulizumab |
| (3) Allogeneic hematopoetic stem cell transplantation (allo-HSCT) |
| (4) IFN-α/AZT (except for lymphoma-type) |
| **3. Relapse or refractory ATLL** |
| (1) Mogamulizumab |
| (2) Allo-HSCT |
| (3) New agents under clinical trial:<br>Brentuximab vedotin, Bortezomib, Lenalidomide, Panobinostat, Forodesine<br>Pralatrexate, Denileukin diftitox |
| (4) Vaccine (autologous dendritic cells with tax-peptide) |

VCAP-AMP-VECP: vincristine, cyclophosphamide, doxorubicin, and prednisolone (VCAP); doxorubicin, ranimustine, and prednisolone (AMP); and vindesine, etoposide, carboplatin, and prednisolone (VECP). CHOP: doxorubicin, cyclophosphamide, vincristine and prednisone.

### 3.1. Interferon-α and Zidovudine

Combined IFN-α/AZT has been reported effective as an ATLL treatment [12]. An international consensus meeting recommended the IFN-α/AZT combination or watchful waiting in patients with indolent ATLL [13]. A meta-analysis of 254 ATLL patients, including 116 patients with acute type, 100 with lymphoma type, 18 with chronic type, and 11 with smoldering type, reported that patients with acute-, chronic-, and smoldering leukemic-type ATLL had better outcomes with first-line antiviral therapy alone, whereas chemotherapy was more effective in patients with lymphoma-type ATLL [14]. Specifically, the five-year OS of patients with chronic and smoldering indolent-type ATLL was 100% with antiviral therapy. Overall, the meta-analysis concluded that first-line antiviral therapy improved the survival of ATLL patients. JCOG has started a phase III study comparing IFN-$\alpha$/AZT with watchful waiting to determine any potential benefits from early intervention in patients with indolent-type ATLL.

### 3.2. Chemotherapy

To date, several chemotherapy regimens such as CHOP (cyclophosphamide, doxorubicin, vincristine, and prednisolone) and EPOCH (VP-16, prednisolone, vincristine, cyclophosphamide, doxorubicin) have been assessed in patients with aggressive ATLL [15]. Particularly, since 1978, the JCOG-Lymphoma Study Group (JCOG-LSG) has played a central role in the development of chemotherapy regimens for ATLL patients [16]. Most recently, JCOG-LSG conducted a randomized clinical trial (JCOG9801) to compare the modified LSG15 (mLSG15) regimen with a biweekly CHOP regimen (CHOP14) in untreated patients with aggressive-type ATLL [17]. The original LSG15, *i.e.*, VCAP-AMP-VECP, regimen sequentially consisted of vincristine, cyclophosphamide, doxorubicin, and prednisolone (VCAP); doxorubicin, ranimustine, and prednisolone (AMP); and vindesine, etoposide, carboplatin, and prednisolone (VECP). The mLSG15 regimen replaced one course of VCAP-AMP-VECP from the original LSG15 regimen with intrathecal administration of methotrexate and prednisone as prophylaxis against central nervous system relapse. The CHOP14 regimen consisted of doxorubicin, cyclophosphamide, vincristine and prednisone. The complete remission (CR) rate of 40% with mLSG15 was significantly better than the CR rate of 25% observed with CHOP14. The three-year OS rates were 24% and 13% for mLSG15 and CHOP14 regimens, respectively; however, the difference between the two groups was not statistically significant. In Japan, mLSG15, which has been recommended as the first-line treatment for aggressive-type ATLL at an international consensus meeting [13], is feasible as a standard regimen in patients with aggressive ATLL who are less than 56 years old. However, the mLSG15 regimen is not routinely available throughout the world due to restricted use of the medications such as ranimustine and vindesine. In addition, the 13-month MST with mLSG15 is unsatisfactory compared with other hematological malignancies. Currently, there are no salvage chemotherapy options established for patients with relapsed or refractory ATLL, and further investigation is needed.

### 3.3. Allogeneic Hematopoietic Stem Cell Transplantation

Allo-HSCT has become an important curative treatment modality in patients with aggressive-type ATLL during the last decade; however, intensified chemotherapy and autologous HSCT has not been successful [18]. Since Utsunomiya *et al.* [19] reported successful outcomes in 10 ATLL patients receiving allo-HSCT in 2001, the number of ATLL patients receiving allo-HSCT has been increasing. In earlier cases, patients received allo-HSCT almost always from human leukocyte antigen (HLA)-matched related donors (MRD) with full-intensity conditioning (FIC) [19–21]. Along with the well-developed Japanese marrow donor program/cord blood bank and improvements in supportive care, unrelated bone marrow (UBM) and umbilical cord blood (UCB) have been increasingly used as alternative donor sources [22–25]. In addition, with the introduction of reduced-intensity conditioning (RIC), the number of allo-HSCT has been steadily increasing [26]; so far, *i.e.*, by 2015, more than 1500 ATLL patients have received allo-HSCT in Japan. According to donor sources, the three-year OS rate was 41% in patients with MRD and 39% with UBM in a nationwide survey [27]. In contrast, the outcomes of allo-HSCT from UCB were unsatisfactory with the three-year OS rate of 17%, partially due to the overlap of the study period that was around 2005, with the developmental phase of allo-HSCT from UCB in adult patients. The recently updated data on ATLL patients showed that the three-year OS of allo-HSCT from UCB remained at 20.6% compared with the three-year OS rates of 34.4% and 37.1% with allo-HSCT from MRD and UBM, respectively [28]. It is certainly challenging to directly compare outcomes from different donor sources because the graft source selection is strongly influenced by donor availability. Nevertheless, the outcomes of CBT in ATLL patients continue to be unsatisfactory due to the high transplant-related mortality (TRM) of 46.1%. Novel interventions will be required, particularly during the early phase, to reduce TRM and control for graft-*versus*-host disease (GVHD) in patients receiving allo-HSCT from UCB [29,30]. Although most allo-HSCT outcomes have been reported in Japanese ATLL patients, the European Group for Blood and Marrow Transplantation's Lymphoma Working Party has recently shown similar results with allo-HSCT in ATLL patients in western countries [31]. However, most of these findings were based on retrospective analysis and patients with heterogeneous backgrounds, including chemosensitive and refractory diseases at transplantation. An ongoing prospective study is assessing the safety and efficacy of RIC followed by allo-HSCT in ATLL patients who achieved remission at transplantation and are stratified according to donor source. Okamura *et al.* [32,33] first reported that the five-year OS of allo-HSCT from MRD was 34% in this prospective study. Other prospective trials assessing RIC followed by allo-HSCT from UBM and UCB are also ongoing. In particular, the three-year OS of allo-HSCT from both MRD and UBM was approximately 30%, indicating that allo-HSCT is a curative treatment. However, survival rates of allo-HSCT have not dramatically improved during the last decade. The major risk factor affecting the survival of ATLL patients receiving allo-HSCT is disease status at transplantation. Based on the incidence rate of ATLL, about 80%–90% of ATLL patients are not able to receive allo-HSCT mostly due to disease resistance to chemotherapy. Therefore, further efforts are needed to increase the response rate prior to allo-HSCT.

*3.4. Novel Agents*

One recent promising therapeutic progress in ATLL is the introduction of mogamulizumab for the treatment of patients with relapsed or refractory ATLL. Mogamulizumab is an anti-CCR4 monoclonal antibody that markedly enhances antibody-dependent cellular cytotoxicity through high-affinity binding to effector cells. CCR4 is selectively expressed on regulatory T-cells and T-helper type 2 (Th2) cells and is expressed on the surface of most ATLL cells. In addition, CCR4 expression is highly associated with poorer prognosis. Based on the data of tolerability in a phase I study, [34] a phase II study for relapsed ATLL was subsequently conducted wherein 1.0 mg/kg of mogamulizumab as a single agent was intravenously administered once a week for eight weeks [35]. The overall response rate (ORR) was reported to be 50%. The median progression-free survival (PFS) was 5.2 months, and the OS was 13.7 months. Based on these results, mogamulizumab use was approved in Japan on March 2012, although mogamulizumab use is not available outside of Japan except in clinical trials. A second, randomized phase II clinical trial in newly diagnosed patients with aggressive ATLL has recently demonstrated that the CR rate was higher at 52% with mLSG15 in combination with mogamulizumab compared to 33% with mLSG15 alone, whereas there was no statistical difference in survival [36]. Since these responses were reported not to be long-lasting, allo-HSCT is still needed for a cure even with the introduction of mogamulizumab. Concurrently, mogamulizumab has been expected to serve as a bridge to transplantation to achieve better disease control during allo-HSCT and to improve survival. However, care should be taken with the use of mogamulizumab in patients with allo-HSCT as CCR4 is expressed not only on tumor cells but also in normal regulatory T-cells and Th2 cells. In a non-transplantation setting, severe skin reactions, such as Steven–Johnson syndrome have been reported [37]. Moreover, rare side effects after the administration of mogamulizumab, such as diffuse panbronchiolitis [38] and colitis [39], have recently been reported. In patients with allo-HSCT, the mogamulizumab treatment may accelerate GVHD by eradicating T-cells. Therefore, the safety and benefits of mogamulizumab both before and after allo-HSCT should be evaluated, and further clinical experience and accumulation of data is necessary [40,41]. To that end, a multicenter prospective observational study is now underway to evaluate the safety and efficacy of mogamulizumab use in ATLL patients with relapsed or refractory disease even after allo-HSCT. In addition to mogamulizumab, clinical trials are underway to determine the efficacy of other novel agents, including brentuximab vedotin, bortezomib, lenalidomide, panobinostat, forodesine, pralatrexate, and denileukin diftitox [9,42]. So far, it seems to be difficult to improve the dismal prognosis of ATLL

*Viruses* **2015**, *7*, 6604–6612

through these novel agent monotherapies. However, in future study, it is important to determine how new agents should be combined with conventional chemotherapy and allo-HSCT.

*3.5. Immunotherapy*

Allo-HSCT is a curative treatment approach in ATLL patients, partly through its graft-*versus*-ATLL (GvATLL) effect as described before. Grade I/II (mild to moderate) acute GVHD has been shown to be associated with improved survival rates [43,44]. The discontinuation of immunosuppressive agents or donor lymphocyte infusion was also effective in ATLL patients who relapsed even after allo-HSCT [45,46]. In addition, Tax- or HBZ-specific T-cells have been shown to play an important role in inducing a potent GvATLL effect [47,48]. The vaccine, which was developed by pulsing Tax peptide into autologous dendritic cells, was effective in ATLL patients [49]; a clinical trial is ongoing for the evaluation of this vaccine with mogamulizumab in Japan.

## 4. Conclusions

The prognosis of ATLL patients, despite an increase in the variety and potency of therapeutic options, has undeniably remained poor, and many obstacles still exist [50]. In future efforts, through studies that uncover the molecular mechanisms underlying ATLL, new treatment protocols integrating antiviral therapy, chemotherapy, allo-HSCT, and molecular targeted agents with optimized dosing and timing need to be developed.

**Acknowledgments:** This work was supported in part by the Japan Agency for Medical Research and Development Grant Number 15ck0106163h0001 (Koji Kato and Koichi Akashi).

**Author Contributions:** Contributions: Koji Kato and Koichi Akashi wrote the manuscript and developed the tables. The authors critically reviewed the manuscript and read and approved the final version of the manuscript.

**Conflicts of Interest:** The authors declare no competing financial interests.

## References

1. Uchiyama, T.; Yodoi, J.; Sagawa, K.; Takatsuki, K.; Uchino, H. Adult T-cell leukemia: Clinical and hematologic features of 16 cases. *Blood* **1977**, *50*, 481–492. [CrossRef]
2. Poiesz, B.J.; Ruscetti, F.W.; Gazdar, A.F.; Bunn, P.A.; Minna, J.D.; Gallo, R.C. Detection and isolation of type C retrovirus particles from fresh and cultured lymphocytes of a patient with cutaneous T-cell lymphoma. *Proc. Natl. Acad. Sci. USA* **1980**, *77*, 7415–7419. [CrossRef] [PubMed]
3. Osame, M.; Izumo, S.; Igata, A.; Matsumoto, M.; Matsumoto, T.; Sonoda, S.; Tara, M.; Shibata, Y. Blood transfusion and HTLV-I associated myelopathy. *Lancet* **1986**, *2*, 104–105. [CrossRef]
4. Laydon, D.J.; Melamed, A.; Sim, A.; Gillet, N.A.; Sim, K.; Darko, S.; Kroll, J.S.; Douek, D.C.; Price, D.A.; Bangham, C.R.; Asquith, B. Quantification of HTLV-1 clonality and TCR diversity. *PLoS Comput. Biol.* **2014**, *10*, e1003646. [CrossRef] [PubMed]
5. Shimoyama, M. Diagnostic criteria and classification of clinical subtypes of adult T-cell leukaemia-lymphoma. A report from the Lymphoma Study Group (1984–87). *Br. J. Haematol.* **1991**, *79*, 428–437. [CrossRef] [PubMed]
6. Takasaki, Y.; Iwanaga, M.; Imaizumi, Y.; Tawara, M.; Joh, T.; Kohno, T.; Yamada, Y.; Kamihira, S.; Ikeda, S.; Miyazaki, Y.; *et al.* Long-term study of indolent adult T-cell leukemia-lymphoma. *Blood* **2010**, *115*, 4337–4343. [CrossRef] [PubMed]
7. Katsuya, H.; Yamanaka, T.; Ishitsuka, K.; Utsunomiya, A.; Sasaki, H.; Hanada, S.; Eto, T.; Moriuchi, Y.; Saburi, Y.; Miyahara, M.; *et al.* Prognostic index for acute- and lymphoma-type adult T-cell leukemia/lymphoma. *J. Clin. Oncol.* **2012**, *30*, 1635–1640. [CrossRef] [PubMed]
8. Fukushima, T.; Nomura, S.; Shimoyama, M.; Shibata, T.; Imaizumi, Y.; Moriuchi, Y.; Tomoyose, T.; Uozumi, K.; Kobayashi, Y.; Fukushima, N.; *et al.* Japan Clinical Oncology Group (JCOG) prognostic index and characterization of long-term survivors of aggressive adult T-cell leukaemia-lymphoma (JCOG0902A). *Br. J. Haematol.* **2014**, *166*, 739–748. [CrossRef] [PubMed]
9. Ishitsuka, K.; Tamura, K. Human T-cell leukaemia virus type I and adult T-cell leukaemia-lymphoma. *Lancet Oncol.* **2014**, *15*, e517–e526. [CrossRef]

10. Utsunomiya, A.; Choi, I.; Chihara, D.; Seto, M. Recent advances in the treatment of adult T-cell leukemia-lymphomas. *Cancer Sci.* **2015**, *106*, 344–351. [CrossRef] [PubMed]
11. Bazarbachi, A.; Suarez, F.; Fields, P.; Hermine, O. How I treat adult T-cell leukemia/lymphoma. *Blood* **2011**, *118*, 1736–1745. [CrossRef] [PubMed]
12. Hermine, O.; Bouscary, D.; Gessain, A.; Turlure, P.; Leblond, V.; Franck, N.; Buzyn-Veil, A.; Rio, B.; Macintyre, E.; Dreyfus, F.; *et al.* Brief report: treatment of adult T-cell leukemia-lymphoma with zidovudine and interferon alfa. *N. Engl. J. Med.* **1995**, *332*, 1749–1751. [CrossRef] [PubMed]
13. Tsukasaki, K.; Hermine, O.; Bazarbachi, A.; Ratner, L.; Ramos, J.C.; Harrington, W., Jr.; O'Mahony, D.; Janik, J.E.; Bittencourt, A.L.; Taylor, G.P.; *et al.* Definition, prognostic factors, treatment, and response criteria of adult T-cell leukemia-lymphoma: a proposal from an international consensus meeting. *J. Clin. Oncol.* **2009**, *27*, 453–459. [CrossRef] [PubMed]
14. Bazarbachi, A.; Plumelle, Y.; Carlos Ramos, J.; Tortevoye, P.; Otrock, Z.; Taylor, G.; Gessain, A.; Harrington, W.; Panelatti, G.; Hermine, O. Meta-analysis on the use of zidovudine and interferon-$\alpha$ in adult T-cell leukemia/lymphoma showing improved survival in the leukemic subtypes. *J. Clin. Oncol.* **2010**, *28*, 4177–4183. [CrossRef] [PubMed]
15. Ratner, L.; Harrington, W.; Feng, X.; Grant, C.; Jacobson, S.; Noy, A.; Sparano, J.; Lee, J.; Ambinder, R.; Campbell, N.; *et al.* Human T cell leukemia virus reactivation with progression of adult T-cell leukemia-lymphoma. *PLoS ONE* **2009**, *4*, e4420. [CrossRef] [PubMed]
16. Yamada, Y.; Tomonaga, M.; Fukuda, H.; Hanada, S.; Utsunomiya, A.; Tara, M.; Sano, M.; Ikeda, S.; Takatsuki, K.; Kozuru, M.; *et al.* A new G-CSF-supported combination chemotherapy, LSG15, for adult T-cell leukaemia-lymphoma: Japan CLINICAL ONCOLOGY GROUP STUDY 9303. *Br. J. Haematol.* **2001**, *113*, 375–382. [CrossRef] [PubMed]
17. Tsukasaki, K.; Utsunomiya, A.; Fukuda, H.; Shibata, T.; Fukushima, T.; Takatsuka, Y.; Ikeda, S.; Masuda, M.; Nagoshi, H.; Ueda, R.; *et al.* VCAP-AMP-VECP compared with biweekly CHOP for adult T-cell leukemia-lymphoma: Japan Clinical Oncology Group Study JCOG9801. *J. Clin. Oncol.* **2007**, *25*, 5458–5464. [CrossRef] [PubMed]
18. Tsukasaki, K.; Maeda, T.; Arimura, K.; Taguchi, J.; Fukushima, T.; Miyazaki, Y.; Moriuchi, Y.; Kuriyama, K.; Yamada, Y.; Tomonaga, M. Poor outcome of autologous stem cell transplantation for adult T cell leukemia/lymphoma: A case report and review of the literature. *Bone Marrow Transpl.* **1999**, *23*, 87–89. [CrossRef] [PubMed]
19. Utsunomiya, A.; Miyazaki, Y.; Takatsuka, Y.; Hanada, S.; Uozumi, K.; Yashiki, S.; Tara, M.; Kawano, F.; Saburi, Y.; Kikuchi, H.; *et al.* Improved outcome of adult T cell leukemia/lymphoma with allogeneic hematopoietic stem cell transplantation. *Bone Marrow Transpl.* **2001**, *27*, 15–20. [CrossRef] [PubMed]
20. Kami, M.; Hamaki, T.; Miyakoshi, S.; Murashige, N.; Kanda, Y.; Tanosaki, R.; Takaue, Y.; Taniguchi, S.; Hirai, H.; Ozawa, K.; *et al.* Allogeneic haematopoietic stem cell transplantation for the treatment of adult T-cell leukaemia/lymphoma. *Br. J. Haematol.* **2003**, *120*, 304–309. [CrossRef] [PubMed]
21. Fukushima, T.; Miyazaki, Y.; Honda, S.; Kawano, F.; Moriuchi, Y.; Masuda, M.; Tanosaki, R.; Utsunomiya, A.; Uike, N.; Yoshida, S.; *et al.* Allogeneic hematopoietic stem cell transplantation provides sustained long-term survival for patients with adult T-cell leukemia/lymphoma. *Leukemia* **2005**, *19*, 829–834. [CrossRef] [PubMed]
22. Kato, K.; Kanda, Y.; Eto, T.; Muta, T.; Gondo, H.; Taniguchi, S.; Shibuya, T.; Utsunomiya, A.; Kawase, T.; Kato, S.; *et al.* Allogeneic bone marrow transplantation from unrelated human T-cell leukemia virus-I-negative donors for adult T-cell leukemia/lymphoma: Retrospective analysis of data from the Japan Marrow Donor Program. *Biol. Blood Marrow Transpl.* **2007**, *13*, 90–99. [CrossRef] [PubMed]
23. Nakase, K.; Hara, M.; Kozuka, T.; Tanimoto, K.; Nawa, Y. Bone marrow transplantation from unrelated donors for patients with adult T-cell leukaemia/lymphoma. *Bone Marrow Transpl.* **2006**, *37*, 41–44. [CrossRef] [PubMed]
24. Shiratori, S.; Yasumoto, A.; Tanaka, J.; Shigematsu, A.; Yamamoto, S.; Nishio, M.; Hashino, S.; Morita, R.; Takahata, M.; Onozawa, M.; *et al.* A retrospective analysis of allogeneic hematopoietic stem cell transplantation for adult T cell leukemia/lymphoma (ATL): Clinical impact of graft-versus-leukemia/lymphoma effect. *Biol. Blood Marrow Transpl.* **2008**, *14*, 817–823. [CrossRef] [PubMed]
25. Takizawa, J.; Aoki, S.; Kurasaki, T.; Higashimura, M.; Honma, K.; Kitajima, T.; Momoi, A.; Takahashi, H.; Nakamura, N.; Furukawa, T.; *et al.* Successful treatment of adult T-cell leukemia with unrelated cord blood transplantation. *Am. J. Hematol.* **2007**, *82*, 1113–1115. [CrossRef] [PubMed]

26. Ishida, T.; Hishizawa, M.; Kato, K.; Tanosaki, R.; Fukuda, T.; Taniguchi, S.; Eto, T.; Takatsuka, Y.; Miyazaki, Y.; Moriuchi, Y.; *et al.* Allogeneic hematopoietic stem cell transplantation for adult T-cell leukemia-lymphoma with special emphasis on preconditioning regimen: A nationwide retrospective study. *Blood* **2012**, *120*, 1734–1741. [CrossRef] [PubMed]

27. Hishizawa, M.; Kanda, J.; Utsunomiya, A.; Taniguchi, S.; Eto, T.; Moriuchi, Y.; Tanosaki, R.; Kawano, F.; Miyazaki, Y.; Masuda, M.; *et al.* Transplantation of allogeneic hematopoietic stem cells for adult T-cell leukemia: A nationwide retrospective study. *Blood* **2010**, *116*, 1369–1376. [CrossRef] [PubMed]

28. Kato, K.; Choi, I.; Wake, A.; Uike, N.; Taniguchi, S.; Moriuchi, Y.; Miyazaki, Y.; Nakamae, H.; Oku, E.; Murata, M.; *et al.* Treatment of patients with adult T cell leukemia/lymphoma with cord blood transplantation: A Japanese nationwide retrospective survey. *Biol. Blood Marrow Transpl.* **2014**, *20*, 1968–1974. [CrossRef] [PubMed]

29. Fukushima, T.; Itonaga, H.; Moriuchi, Y.; Yoshida, S.; Taguchi, J.; Imaizumi, Y.; Imanishi, D.; Tsushima, H.; Sawayama, Y.; Matsuo, E.; *et al.* Feasibility of cord blood transplantation in chemosensitive adult T-cell leukemia/lymphoma: A retrospective analysis of the Nagasaki Transplantation Network. *Int J. Hematol.* **2013**, *97*, 485–490. [CrossRef] [PubMed]

30. Nakamura, T.; Oku, E.; Nomura, K.; Morishige, S.; Takata, Y.; Seki, R.; Imamura, R.; Osaki, K.; Hashiguchi, M.; Yakushiji, K.; *et al.* Unrelated cord blood transplantation for patients with adult T-cell leukemia/lymphoma: experience at a single institute. *Int. J. Hematol.* **2012**, *96*, 657–663. [CrossRef] [PubMed]

31. Bazarbachi, A.; Cwynarski, K.; Boumendil, A.; Finel, H.; Fields, P.; Raj, K.; Nagler, A.; Mohty, M.; Sureda, A.; Dreger, P.; *et al.* Outcome of patients with HTLV-1-associated adult T-cell leukemia/lymphoma after SCT: A retrospective study by the EBMT LWP. *Bone Marrow Transpl.* **2014**, *49*, 1266–1268. [CrossRef] [PubMed]

32. Okamura, J.; Utsunomiya, A.; Tanosaki, R.; Uike, N.; Sonoda, S.; Kannagi, M.; Tomonaga, M.; Harada, M.; Kimura, N.; Masuda, M.; *et al.* Allogeneic stem-cell transplantation with reduced conditioning intensity as a novel immunotherapy and antiviral therapy for adult T-cell leukemia/lymphoma. *Blood* **2005**, *105*, 4143–4145. [CrossRef] [PubMed]

33. Choi, I.; Tanosaki, R.; Uike, N.; Utsunomiya, A.; Tomonaga, M.; Harada, M.; Yamanaka, T.; Kannagi, M.; Okamura, J. Long-term outcomes after hematopoietic SCT for adult T-cell leukemia/lymphoma: Results of prospective trials. *Bone Marrow Transpl.* **2011**, *46*, 116–118. [CrossRef] [PubMed]

34. Yamamoto, K.; Utsunomiya, A.; Tobinai, K.; Tsukasaki, K.; Uike, N.; Uozumi, K.; Yamaguchi, K.; Yamada, Y.; Hanada, S.; Tamura, K.; *et al.* Phase I study of KW-0761, a defucosylated humanized anti-CCR4 antibody, in relapsed patients with adult T-cell leukemia-lymphoma and peripheral T-cell lymphoma. *J. Clin. Oncol.* **2010**, *28*, 1591–1598. [CrossRef] [PubMed]

35. Ishida, T.; Joh, T.; Uike, N.; Yamamoto, K.; Utsunomiya, A.; Yoshida, S.; Saburi, Y.; Miyamoto, T.; Takemoto, S.; Suzushima, H.; *et al.* Defucosylated anti-CCR4 monoclonal antibody (KW-0761) for relapsed adult T-cell leukemia-lymphoma: A multicenter phase II study. *J. Clin. Oncol.* **2012**, *30*, 837–842. [CrossRef] [PubMed]

36. Ishida, T.; Jo, T.; Takemoto, S.; Suzushima, H.; Uozumi, K.; Yamamoto, K.; Uike, N.; Saburi, Y.; Nosaka, K.; Utsunomiya, A.; *et al.* Dose-intensified chemotherapy alone or in combination with mogamulizumab in newly diagnosed aggressive adult T-cell leukaemia-lymphoma: A randomized phase II study. *Br. J. Haematol.* **2015**, *169*, 672–682. [CrossRef] [PubMed]

37. Ishida, T.; Ito, A.; Sato, F.; Kusumoto, S.; Iida, S.; Inagaki, H.; Morita, A.; Akinaga, S.; Ueda, R. Stevens-Johnson Syndrome associated with mogamulizumab treatment of adult T-cell leukemia/lymphoma. *Cancer Sci.* **2013**, *104*, 647–650. [CrossRef] [PubMed]

38. Kato, K.; Miyamoto, T.; Numata, A.; Nakaike, T.; Oka, H.; Yurino, A.; Kuriyama, T.; Mori, Y.; Yamasaki, S.; Muta, T.; *et al.* Diffuse panbronchiolitis after humanized anti-CCR4 monoclonal antibody therapy for relapsed adult T-cell leukemia/lymphoma. *Int. J. Hematol.* **2013**, *97*, 430–432. [CrossRef] [PubMed]

39. Ishitsuka, K.; Murahashi, M.; Katsuya, H.; Mogi, A.; Masaki, M.; Kawai, C.; Goto, T.; Ishizu, M.; Ikari, Y.; Takamatsu, Y.; *et al.* Colitis mimicking graft-versus-host disease during treatment with the anti-CCR4 monoclonal antibody, mogamulizumab. *Int. J. Hematol.* **2015**, *102*, 493–497. [CrossRef] [PubMed]

40. Ito, Y.; Miyamoto, T.; Chong, Y.; Aoki, T.; Kato, K.; Akashi, K.; Kamimura, T. Successful treatment with anti-CC chemokine receptor 4 MoAb of relapsed adult T-cell leukemia/lymphoma after umbilical cord blood transplantation. *Bone Marrow Transpl.* **2013**, *48*, 998–999. [CrossRef] [PubMed]

41. Motohashi, K.; Suzuki, T.; Kishimoto, K.; Numata, A.; Nakajima, Y.; Tachibana, T.; Ohshima, R.; Kuwabara, H.; Tanaka, M.; Tomita, N.; *et al.* Successful treatment of a patient with adult T cell leukemia/lymphoma using anti-CC chemokine receptor 4 monoclonal antibody mogamulizumab followed by allogeneic hematopoietic stem cell transplantation. *Int. J. Hematol.* **2013**, *98*, 258–260. [CrossRef] [PubMed]

42. Ishitsuka, K.; Utsunomiya, A.; Katsuya, H.; Takeuchi, S.; Takatsuka, Y.; Hidaka, M.; Sakai, T.; Yoshimitsu, M.; Ishida, T.; Tamura, K. A phase II study of bortezomib in patients with relapsed or refractory aggressive adult T-cell leukemia/lymphoma. *Cancer Sci.* **2015**, *106*, 1219–1223. [CrossRef] [PubMed]

43. Yonekura, K.; Utsunomiya, A.; Takatsuka, Y.; Takeuchi, S.; Tashiro, Y.; Kanzaki, T.; Kanekura, T. Graft-versus-adult T-cell leukemia/lymphoma effect following allogeneic hematopoietic stem cell transplantation. *Bone Marrow Transpl.* **2008**, *41*, 1029–1035. [CrossRef] [PubMed]

44. Kanda, J.; Hishizawa, M.; Utsunomiya, A.; Taniguchi, S.; Eto, T.; Moriuchi, Y.; Tanosaki, R.; Kawano, F.; Miyazaki, Y.; Masuda, M.; *et al.* Impact of graft-versus-host disease on outcomes after allogeneic hematopoietic cell transplantation for adult T-cell leukemia: a retrospective cohort study. *Blood* **2012**, *119*, 2141–2148. [CrossRef] [PubMed]

45. Kamimura, T.; Miyamoto, T.; Kawano, N.; Numata, A.; Ito, Y.; Chong, Y.; Nagafuji, K.; Teshima, T.; Hayashi, S.; Akashi, K. Successful treatment by donor lymphocyte infusion of adult T-cell leukemia/lymphoma relapse following allogeneic hematopoietic stem cell transplantation. *Int. J. Hematol.* **2012**, *95*, 725–730. [CrossRef] [PubMed]

46. Itonaga, H.; Tsushima, H.; Taguchi, J.; Fukushima, T.; Taniguchi, H.; Sato, S.; Ando, K.; Sawayama, Y.; Matsuo, E.; Yamasaki, R.; *et al.* Treatment of relapsed adult T-cell leukemia/lymphoma after allogeneic hematopoietic stem cell transplantation: The Nagasaki Transplant Group experience. *Blood* **2013**, *121*, 219–225. [CrossRef] [PubMed]

47. Harashima, N.; Kurihara, K.; Utsunomiya, A.; Tanosaki, R.; Hanabuchi, S.; Masuda, M.; Ohashi, T.; Fukui, F.; Hasegawa, A.; Masuda, T.; *et al.* Graft-versus-Tax response in adult T-cell leukemia patients after hematopoietic stem cell transplantation. *Cancer Res.* **2004**, *64*, 391–399. [CrossRef] [PubMed]

48. Tamai, Y.; Hasegawa, A.; Takamori, A.; Sasada, A.; Tanosaki, R.; Choi, I.; Utsunomiya, A.; Maeda, Y.; Yamano, Y.; Eto, T.; *et al.* Potential contribution of a novel Tax epitope-specific CD4+ T cells to graft-versus-Tax effect in adult T cell leukemia patients after allogeneic hematopoietic stem cell transplantation. *J. Immunol.* **2013**, *190*, 4382–4392. [CrossRef] [PubMed]

49. Suehiro, Y.; Hasegawa, A.; Iino, T.; Sasada, A.; Watanabe, N.; Matsuoka, M.; Takamori, A.; Tanosaki, R.; Utsunomiya, A.; Choi, I.; *et al.* Clinical outcomes of a novel therapeutic vaccine with Tax peptide-pulsed dendritic cells for adult T cell leukaemia/lymphoma in a pilot study. *Br. J. Haematol.* **2015**, *169*, 356–367. [CrossRef] [PubMed]

50. Itonaga, H.; Taguchi, J.; Fukushima, T.; Tsushima, H.; Sato, S.; Ando, K.; Sawayama, Y.; Matsuo, E.; Yamasaki, R.; Onimaru, Y.; *et al.* Distinct clinical features of infectious complications in adult T cell leukemia/lymphoma patients after allogeneic hematopoietic stem cell transplantation: A retrospective analysis in the Nagasaki transplant group. *Biol. Blood Marrow Transpl.* **2013**, *19*, 607–615. [CrossRef] [PubMed]

*Review*

# Transcriptional and Epigenetic Regulatory Mechanisms Affecting HTLV-1 Provirus

Paola Miyazato, Misaki Matsuo, Hiroo Katsuya and Yorifumi Satou *

International Research Center for Medical Sciences, Center for AIDS Research, Priority Organization for Innovation and Excellence, Kumamoto University, Kumamoto 860-0811, Japan; pmiyazat@kumamoto-u.ac.jp (P.M.); 163r5152@st.kumamoto-u.ac.jp (M.M.); hkatsuya@kumamoto-u.ac.jp (H.K.)
* Correspondence: y-satou@kumamoto-u.ac.jp; Tel.: 81-96-373-6830

Academic Editor: Louis Mansky
Received: 3 May 2016; Accepted: 9 June 2016; Published: 16 June 2016

**Abstract:** Human T-cell leukemia virus type 1 (HTLV-1) is a retrovirus associated with human diseases, such as adult T-cell leukemia (ATL) and HTLV-1-associated myelopathy/Tropic spastic paraparesis (HAM/TSP). As a retrovirus, its life cycle includes a step where HTLV-1 is integrated into the host genomic DNA and forms proviral DNA. In the chronic phase of the infection, HTLV-1 is known to proliferate as a provirus via the mitotic division of the infected host cells. There are generally tens of thousands of infected clones within an infected individual. They exist not only in peripheral blood, but also in various lymphoid organs. Viral proteins encoded in HTLV-1 genome play a role in the proliferation and survival of the infected cells. As is the case with other chronic viral infections, HTLV-1 gene expression induces the activation of the host immunity against the virus. Thus, the transcription from HTLV-1 provirus needs to be controlled in order to evade the host immune surveillance. There should be a dynamic and complex regulation *in vivo*, where an equilibrium between viral antigen expression and host immune surveillance is achieved. The mechanisms regulating viral gene expression from the provirus are a key to understanding the persistent/latent infection with HTLV-1 and its pathogenesis. In this article, we would like to review our current understanding on this topic.

**Keywords:** HTLV-1 2; provirus 3; retroviral latency

## 1. Introduction

It has been estimated that Human T-cell leukemia virus type 1 (HTLV-1) has been infecting humans for several thousand years [1]. In ancient times, prior to the advent of blood transfusions or drug abuse, the virus spread either by vertical transmission, from mother to child via breast-feeding, or by horizontal transmission mainly from man to woman, through sexual intercourse. HTLV-1 generally induces *de novo* infection not via free viral particles but via cell-to-cell contact between infected and uninfected cells [2–4]. The presence of infected lymphocytes in breast milk or sperm is pivotal for *de novo* infection. In the case of vertical transmission, infected individuals acquire the virus during infancy and need to carry the virus for decades, before they are able to transfer the virus to their children. Also, infected women need to remain healthy for decades in order to become pregnant, in spite of the persistent HTLV-1 infection in their bodies. To achieve such a long-term persistent infection without having severe health problems, HTLV-1 seems to have developed a strategy to achieve an asymptomatic condition by minimizing the effect of viral infection on our vital systems. At the same time, as with other viruses that cause persistent infections, HTLV-1 needs to evade the host immune surveillance [5].

The main routes of *de novo* HTLV-1 infection are cell-to-cell transmission and/or extracellular biofilm-like-structure-mediated transmission [2,4]. In either case, the virus produces viral proteins that are necessary for reverse transcription and integration of the viral DNA into the host cellular DNA. HTLV-1 is required to keep a latent state in the chronic phase of infection, but it also needs to reactivate viral gene expression for *de novo* infection. This implies that the virus makes use of a reversible system to switch from the latent phase of infection to the active phase, where viruses are produced. For example, when the infected cells are transferred from mother to child, proviruses in the infected lymphocytes contained in the breast-milk would be transcriptionally reactivated, and producing infectious viral particles or inducing cell-to-cell transmission. Since an anti-virus immunity has not been established yet in the new host during the initial phase of infection, the virus would be able to spread via *de novo* infection (Figure 1).

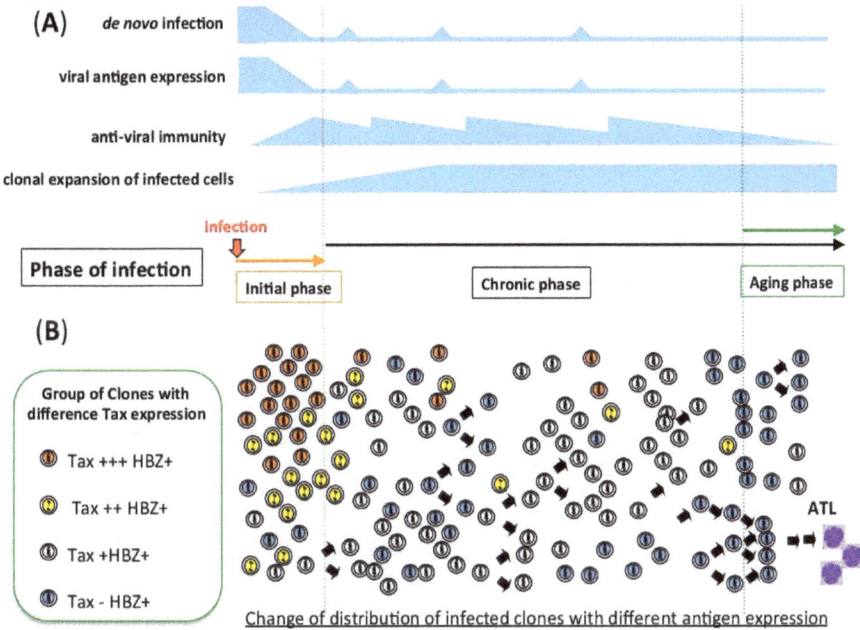

**Figure 1.** Schematic figure of Human T-cell leukemia virus type 1 (HTLV-1) infection from the initial to the chronic phase of infection. (**A**) During the initial phase of infection, before an anti-virus immunity has been established, *de novo* infection should be more dominant than the clonal expansion of infected cells. In the chronic phase of infection, antiviral immunity removes infected cells with high viral antigen expression. HTLV-1 increases the viral copy number by clonal expansion of the infected cells. There is sporadic viral antigen expression, which should maintain the activity of the anti-viral immunity. (**B**) Change in the distribution of infected clones with different antigen expression. In the initial phase, the proportion of infected clones with high Tax expression is high, because there is little anti-viral immunity. After the establishment of an anti-viral immunity, clones with high antigen expression are eliminated by the host immune system.

In a typical asymptomatic carrier, approximately 2% of peripheral mononuclear cells are infected with HTLV-1 [6], which is far more frequent when compared with the proviral load of HIV-1 in patients undergoing anti-retroviral therapy [7]. We can also observe cytotoxic T-lymphocytes (CTLs) specific for HTLV-1 antigens in asymptomatic carriers as well as in HTLV-1-associated myelopathy/Tropic spastic paraparesis (HAM/TSP) patients, suggesting that even in the absence of clinical symptoms there is a balance between the host immune surveillance and the persistent viral infection [8,9]. Therefore, to understand the regulatory mechanisms acting on HTLV-1 provirus integrated within the host genomic DNA is a key to elucidate the virological and pathophysiological aspect of HTLV-1 infection, including the mechanisms leading to transformation of the infected cells or the establishment of chronic inflammatory diseases.

## 2. Structure of Human T-cell Leukemia Virus Type 1 (HTLV-1)

HTLV-1 is a delta-type retrovirus [10,11]. Viral RNA is reverse-transcribed, integrated into the genomic DNA of the host cell, and thereafter remains as a provirus. The size of the proviral genome is approximately 9000 base pairs (bp) [12]. As is the case with other retroviruses, there are identical sequences, long terminal repeats (LTRs), at both ends of the provirus. The 5'-LTR is the promoter for the transcripts in the sense orientation, whereas the 3'-LTR is the promoter for antisense transcription. Most of the viral structural genes, such as *gag*, *pol*, and *env* are encoded in the 5' side of the provirus in the sense orientation, as is commonly observed in other retroviruses [13]. However, a unique characteristic of HTLV-1, also shared with the bovine leukemia virus (BLV), another delta-type retrovirus, is the presence of the pX region, which is located in the 3' side of the provirus. There are two regulatory proteins, Tax and Rex, encoded in the pX region. Tax, the most intensively characterized viral protein, is a strong transactivator of HTLV-1 5'-LTR. Rex is another positive regulator for the expression of viral antigens, which controls the nuclear export of viral mRNAs. There are several accessory proteins also encoded in the sense orientation in the pX region, including p13, p30, p12, p27, p21Rex and p8 [14–18]. In addition, HTLV-1 bZIP factor (HBZ) is encoded in the pX region in the anti-sense orientation [19].

HTLV-1 very efficiently utilizes its small genome via alternative splicing and bidirectional transcription. The regulatory and accessory viral proteins coordinately control viral antigen expression, contributing to achieve a persistent infection with HTLV-1.

## 3. Regulation of the 5'- and 3'-LTR Promoter Regions of HTLV-1 Provirus

There is sense- and antisense-transcription from HTLV-1 provirus, driven by sequences contained in the LTRs that serve as promoters. The sequence of the 5'- and 3'-LTRs is identical, so the directionality, sense or anti-sense orientation, confers the different promoter activity on the 5'- and 3'-LTRs. There is a DNA sequence found in the promoter region of some genes (TATA-box) in the sense orientation of the LTR, but not in the anti-sense orientation. Promoters containing a TATA-box structure generally exhibit high-plasticity in their promoter activity, whereas TATA-less promoters show low transcriptional plasticity [20,21]. In line with this notion, the plus strand of the 5'-LTR, a TATA-box-containing promoter, shows variable activity, whereas the minus strand of the 3'-LTR, a TATA-less promoter, shows a relatively stable promoter activity [22]. A recent study has shown that HTLV-1 LTR possesses a bidirectional transcriptional activity and that Tax could preferentially activate the sense transcription with no or limited effect on the antisense transcription in a reporter plasmid system [23]. It is well known that the sense transcription from the 5'-LTR is significantly induced in the presence of Tax. There are three copies of imperfect repeats of a 21 bp sequence called TRE (Tax-response element) that is responsive to the transactivation mediated by the viral protein Tax. Tax is a strong positive regulator of sense transcription from the 5'-LTR [12,24,25].

A recent interesting study has further extended our understanding on how both sense and antisense transcriptions are regulated within the provirus. They showed that sense transcription from the 5'-LTR did not interfere with antisense transcription from the 3'-LTR and *vice versa* [26]. They further showed that the cell cycle arrest induced by Tax expression might inhibit Tax-mediated activation of the sense transcription without affecting antisense transcription. As the authors pointed out, the mechanism may play a role in HTLV-1 latency. The 5'-LTR is regulated by cellular signaling pathways, such as the T-cell receptor (TCR)-mediated one, in addition to the viral regulatory/accessory proteins. TCR stimulation in combination with Tax strongly enhances HTLV-1 gene expression [27].

On the other hand, there are several negative regulatory systems acting on the 5'-LTR during transcription, translation, and even post-translation phases. HTLV-1 p30 has the potential to inhibit the interaction between Tax and p300, resulting in suppression of the 5'-LTR [28,29]. p30 additionally enhances the retention of mRNA in the nucleus and suppresses viral antigen expression [30]. HTLV-1 p13 is also known to exert an inhibitory effect on the physical interaction between Tax and p300 [15]. Furthermore, HBZ competes with Tax for cAMP response element binding (CREB) protein and p300 binding, so HBZ suppresses the 5'-LTR [19,31,32]. The minus strand of the 3'-LTR is the promoter of the spliced form of HBZ [33,34] and is controlled by Sp-1 and Jun-D [22,35]. The unspliced form of HBZ is transcribed from the promoter located within the pX region [22].

In summary, there is convergent transcription, in the sense- and antisense-orientations, and various viral transcripts with alternative splicing within HTLV-1 provirus. Growing evidence so far indicates that HTLV-1 maintains an equilibrium between viral antigen expression and the host immune surveillance by controlling both sense and antisense transcription, viral regulatory and accessory proteins' expression, and host cellular mechanisms, such as cellular-signaling pathways and cell cycling.

## 4. *In Vitro* and *in Vivo* Proviral Transcription Show Different Patterns

### 4.1. HTLV-1-Associated Cell Lines and adult T-cell Leukemia (ATL)-Derived Cell Lines

It has been reported that there is clearly a distinct level of proviral expression *in vitro* and *in vivo*. Even among cell lines infected with HTLV-1, there is a wide variation in transcription level. The distinct viral gene expression should be the consequence of different cellular transformation processes occurring during *in vitro* culture or in infected individuals. In vitro T-cell immortalization by HTLV-1 is induced by a high expression level of Tax protein. As shown in previous studies, high level of tax expression is sufficient to induce immortalization of T cells [36–38]. Because there is no immune surveillance in *in vitro* cell culture, viral antigen expression does not confer survival disadvantage on infected cells. In contrast, adult T-cell leukemia (ATL) cell lines derived from ATL patients are generated after a long latent period *in vivo* in the presence of anti-virus immune surveillance [3,39]. In the leukemogenesis of ATL *in vivo*, the process of transformation depends not only on HTLV-1 infection but also on several other factors, such as escape from the host immune surveillance [40,41], as well as genetic and epigenetic abnormalities of the host genome [42–50]. As a consequence of immune pressure, ATL cell lines generally have very low or no Tax expression [39,41].

### 4.2. Fresh Peripheral Blood Mononuclear Cells (PBMC) from Infected Individuals

Transcripts with sense orientation from the 5'-LTR are generally barely detectable, if not at all, in PBMCs freshly isolated from infected individuals [41,51]. In contrast, antisense transcripts from the 3'-LTR are constitutively detectable in fresh PBMCs [34,52]. Most viral antigens are encoded in the sense transcripts from the 5'-LTR, so infected cells with high expression of viral antigens would be eliminated by the host immune surveillance. This is quite consistent with the idea that selective pressure by the host immune system is evident in naturally infected individuals. There are several mechanisms related to suppression of the 5'-LTR, including deletion of the 5'-LTR and mutations in tax gene, a strong transactivator of the 5'-LTR [40,41]. In addition, epigenetic mechanisms acting on the 5'-LTR also contribute to the silencing effect, a point which we will discuss about in detail below. Antisense transcripts are known to encode the viral antigen HBZ [19]. A possible explanation why HBZ can be expressed *in vivo* is its low immunogenicity. A previous report demonstrated that the amino acid sequence of HBZ possesses substantially low immunogenicity [53]. Another important piece of evidence about proviral expression is that *ex vivo* culture induces expression of viral proteins [54–56]. Although we do not know the exact mechanism for the *in vivo* silencing, it is generally accepted that HTLV-1 expression is silenced but the virus keeps its capacity to re-start production of viral particles. Recent interesting studies have shown the kinetics of the sense and antisense proviral transcription during *ex vivo* culture, suggesting the existence of a time-dependent regulatory mechanism of HTLV-1 provirus, which should be important at the initial phase of infection [57–59].

*4.3. RNA-seq of Fresh ATL Cells*

Recently Kataoka *et al.* published the results of a comprehensive and genome-wide sequencing analysis of fresh ATL cells from several hundreds of ATL cases in Japan [47]. The RNA-seq analysis detected not only transcription from the human genome, but also from HTLV-1 provirus integrated into the host genome of ATL cells. In line with previous reports, sense transcripts from the 5′-LTR were frequently suppressed, whereas antisense transcripts were detectable in all the ATL cases they analyzed [34,52]. Minus strand transcripts encode HBZ protein and the transcript itself has a function [19,34,60]. Accumulating evidence has suggested that HBZ plays an important role in persistent infection and pathogenesis of HTLV-1 [61,62]. Interestingly, some antisense transcripts do not terminate at the polyadenylation site of HBZ, but there are read-through transcripts into the flanking host genomic regions [34,47,63]. Although we do not know the function of such read-through transcripts, the pattern of transcript and role of proviral transcription should be far more complicated than that of our current understanding.

## 5. Epigenetic Regulation of HTLV-1 Provirus

HTLV-1 provirus is integrated into the cellular DNA, chromatinized, and affected by genomic and epigenomic circumstances nearby the integration site [64–67]. An important characteristic of epigenetic regulation is that the epigenetic state is generally not static but variable depending on intra-cellular and/or extra-cellular conditions. For example, histone modifications are dynamically laid down and removed by chromatin-modifying enzymes [68,69]. As we discussed above, HTLV-1 has to be transcriptionally repressed but, at the same time, maintaining an ability to reactivate the proviral expression to achieve continuous infection within human beings. HTLV-1 should utilize a reversible epigenetic mechanism of the host cell to achieve a transiently suppressed condition of HTLV-1 provirus.

*5.1. DNA Methylation of HTLV-1 Provirus*

DNA methylation is a key mechanism to control gene expression of the host cells. Based on sequence analyses, HTLV-1 LTRs contain CpG islands, which are defined based on their high percentage of GC content (a GC percentage greater than 50%) and the length of the region (more than 200 bp) [70]. CpG islands are frequently present in gene promoter regions of the host cells and play a role in nucleosomal positioning and transcriptional regulation. DNA hypermethylation of CpG islands in promoter regions is associated with gene silencing. In the case of HTLV-1, DNA hypermethylation is what controls the activity of the 5′-LTR both in latently infected cell lines and ATL cells [54,71,72]. In contrast, the 3′-LTR is rarely methylated, suggesting that selective DNA methylation occurs in HTLV-1 provirus [41,71,73]. Since the sequence of the 5′- and the 3′-LTRs is identical, there should be some mechanism that induces selective DNA methylation of the 5′-LTR (or selective hypomethylation of 3′-LTR). That is a key question that remains to be answered about HTLV-1 infection.

*5.2. Histone Modifications in HTLV-1 Provirus*

The nucleosome is the fundamental unit of chromatin, and it is composed of an octamer of the four core histones (H3, H4, H2A, and H2B). Around 147 bp of DNA wrap around each histone octamer. The histone tails are modified by acetylation, methylation, phosphorylation, and ubiquitylation [68]. These modifications function as platforms for the recruitment of specific effector proteins, such as transcriptional factors, chromatin remodelers and the general transcription apparatus, including RNA Polymerase II (RNA Pol II) [74]. Thus, histone modifications are a critical determinant for gene transcription [69]. The viral protein Tax, together with the phosphorylated CREB, recruits both CBP and p300 to the 5′-LTR [75], resulting in histone acetylation and eviction of nucleosomes, thus contributing to a strong sense transcription from the 5′-LTR [76,77]. Tax also interacts with the chromatin remodeling complex, changes nucleosome positioning, and induces transcriptional activation from the 5′-LTR [78,79]. These findings are thought to be molecular mechanisms inducing strong sense-transcription in *in vitro* cell cultures, where there is abundant Tax expression. The Tax-mediated strong transcription from the 5′-LTR is the case *in vitro* but not *in vivo*, as discussed above.

In order to characterize the pattern of histone modifications *in vivo*, we have to analyze fresh ATL cells without doing any *ex vivo* culture. There are some reports on histone acetylation and methylation of ATL cells. The previous data suggested that histone acetylation is detectable both in 5′- and 3′-LTRs [80,81]. We recently demonstrated that H3K9ac is high in the 3′-LTR but very low in the 5′-LTR in the ATL-derived cell line ED [39,67]. Another histone modification, H3K4me3, a mark of active promoter regions, also shows a similar distribution pattern as H3K9ac. In a proportion of fresh ATL cells and ATL cell lines, H3K9ac and H3K4me3 are detectable at the 5′-LTR, although the distribution is limited only to the 5′-LTR. On the other hand, the active histone marks around the 3′LTR are not limited to the LTR region, but extend to the pX region, suggesting different histone modifications between 5′- and 3′-LTRs [67].

In general, there are two categories of promoters, active and poised promoters [82]. Active promoters exhibit active histone marks both on the promoter region and downstream of the transcriptional start site, suggesting the movement of RNA Pol II starts at the promoter and moves into the gene body [74]. In contrast, poised promoters show limited distribution of H3K9ac and H3K4me3 at promoter regions. Taken together these data suggest that RNA Pol II might be present at both 5′- and 3′-LTRs. But, while the 5′-LTR is a poised promoter, the 3′-LTR functions as an active one. This is consistent with the pattern of transcription where the 3′-LTR is constitutively active but the 5′-LTR is generally suppressed. Also, poised RNA Pol II is ready to start transcription. The possibility of a poised RNA Pol II at the 5′-LTR would also explain why fresh PBMCs, isolated from HTLV-1-infected individuals, start to express Tax after a very short time in *ex vivo* culture.

Most studies on histone modifications of HTLV-1 have, so far, been limited to the promoter regions, 5′- and 3′-LTRs. However transcription is a biological event with multiple steps, including transcriptional initiation, elongation, and termination [83]. To understand the whole picture of transcriptional regulation, we need to analyze all processes of the transcription. Recent technological advances, such as chromatin immunoprecipitation sequencing (ChIP-seq) analysis will be able to provide more evidence on the epigenetic regulation of HTLV-1.

*5.3. Insulator Region within HTLV-1 Provirus*

As mentioned above, there is a significant difference in the transcriptional activity and the epigenetic characteristics of the 5′ and the 3′ regions of HTLV-1 provirus. Since the size of the provirus is just about 9000 bp, there should be a positive regulatory mechanism to maintain such distinct transcriptional activity within the provirus [67]. We have recently reported that the insulator-binding protein CTCF (CCCTC-binding factor) directly binds to HTLV-1 provirus. Insulator regions are functional genomic regions that delimit an epigenetic border between transcriptionally active and inactive regions [84]. CTCF is also the most characterized insulator-binding protein, which is well conserved from flies to humans, and plays a fundamental role in the higher order chromatin structure of the genome [85,86]. Histone modifications, such as H3K4me3, H3K36me3, and H3K9ac, are significantly changed at the insulator region of HTLV-1. These data suggest that HTLV-1 utilizes this host insulator-binding molecule to maintain the appropriate pattern of proviral transcription to achieve persistent infection in infected individuals (Figure 2).

Another function of CTCF is chromatin loop formation by homodimerization [87]. CTCF, in concert with Cohesin, regulates the proximity of promoters and enhancers, and controls the transcriptional activity of genes [88–90]. This suggests the possibility that HTLV-1 provirus can form chromatin looping with distant host genomic sites through its CTCF-binding site. Therefore, CTCF-mediated chromatin looping with the host genome is another possible mechanism to regulate proviral expression. It is very interesting that some gamma herpes viruses, such as Epstein-Barr virus (EBV) and Kaposi's sarcoma-associated herpesvirus (KSHV), also use CTCF to switch the pattern of viral gene expression [91,92]. As is the case with HTLV-1, EBV is known as a virus with different *in vitro* and *in vivo* pattern of viral gene transcription [93]. A previous report demonstrated that CTCF plays a role in determining promoter usage at different latency states in EBV infection [92]. EBV is generally not integrated into the host genome, but HTLV-1 is integrated as a step of the viral life cycle of retroviruses. Thus integration of HTLV-1 generates ectopic CTCF-binding sites in the human genome. This could induce aberrant chromatin structure and gene transcription of the host genome [67,94].

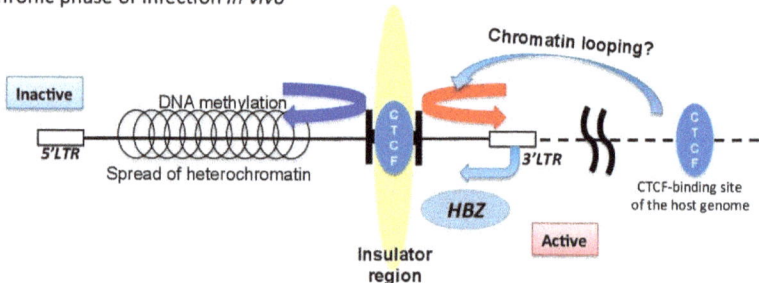

**Figure 2.** Schematic figure of mechanisms regulating HTLV-1 provirus, *in vitro* and *in vivo*. (**A**) Schematic figure of HTLV-1 provirus *in vitro* or during the initial phase of infection *in vivo*. Tax and viral structural proteins can be expressed, because there is little immune surveillance against the viral antigens. (**B**) Schematic figure of HTLV-1 provirus during the chronic phase of infection *in vivo*. HTLV-1 maintains a distinct transcription pattern between 5′ long terminal repeat (LTR) and 3′-LTR by recruiting the host insulator protein CTCF. The insulator region of HTLV-1 provirus is thought to prevent the spread of heterochromatin from 5′-LTR to 3′-LTR. This could also induce chromatin looping with the host's CTCF-binding sites.

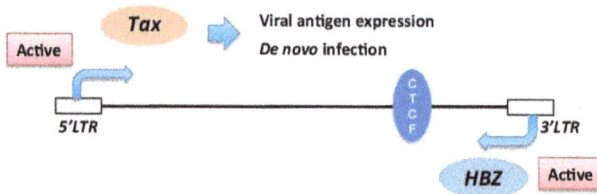

## 6. Integration Site and Its Role in Proviral Transcription

In principle, each infected clone has a different integration site (IS). IS data has been used to identify clonality of HTLV-1-infected cells. According to technological advances in DNA detection and sequencing, current analysis of clonality of infected cells is becoming far more accurate and quantitative than before [64,95–99]. Application of modified PCR and DNA sequencing enables us to determine the exact position of the viral integration site within the host genome. Novel DNA sequencing technologies have also enabled the characterization of the nature of HTLV-1 integration landscape with high resolution [66]. Now, we can characterize the distribution of integration sites and quantify the degree of clonal expansion of each individual infected cell with extremely high resolution by using next-generation sequencing technology. We know there are approximately ten of thousands of infected clones in a typical infected individual [66].

The distribution of HTLV-1 integration sites in the infected individuals is determined by two key factors, *i.e.*, initial integration-targeting and preferential survival of infected clones *in vivo*. There is clear difference in HTLV-1 IS between *in vitro*-infected samples and *in vivo*-infected clinical samples [64]. This idea is consistent with a recent report on the clonality analysis of cells infected with BLV, another member of the delta type retroviruses. The study showed significant depletion of BLV proviral clones located in transcriptionally active genomic sites during primary infection [100]. The data suggests that transcriptionally active regions are preferentially targeted by BLV integration, but the provirus integrated in transcriptionally active regions tends to express a viral antigen, which should result in removal by the host immune surveillance. During persistent infection *in vivo*, the infected clone that has an advantage in escaping from the host immune surveillance would be able to live for a long time. Although there is no clear hot spot of IS associated with development of ATL, there are some tendencies as listed below [64,65].

(i)     HTLV-1 tends to be integrated into open chromatin.

(ii)    Infected clones with ISs within a gene and with the same orientation as the host gene tend to expand.

(iii)   Presence of a transcription factor- or histone modifier-binding site, such as Brg1 and STAT-1, near the IS can affect the frequency of spontaneous Tax expression in *ex vivo* culture.

The data has also demonstrated that spontaneous Tax expression in *ex vivo* culture is inversely correlated with clonal abundance of infected clones, suggesting that Tax expression is not necessarily related with clonal expansion of infected cells *in vivo* [65]. Taken together, all these findings have demonstrated that the IS plays a role in proviral gene transcription.

## 7. Deletions and Mutations in HTLV-1 Proviral Genome

Compared with HIV-1 provirus, the HTLV-1 genome is highly conserved, which is evidence for the idea that HTLV-1 maintains its viral copy number via clonal expansion of infected cells rather than *de novo* infection with viral particles, in which error-prone reverse transcriptase generates genetic mutations in the viral genome [101]. Even so, evidence on deletions of HTLV-1 proviral genome has been provided [102–105] by several reports on defective HTLV-1 proviruses. There are two types, type 1 and type 2, of defective proviruses. Type 1 defective proviruses lack genomic segments within the provirus, whereas the type 2 lack the 5′-LTR [104]. Both types of defective proviruses should have less capacity to produce viral antigens than the complete ones, so the presence of the defective provirus is thought to be a consequence of selection by the host immune surveillance. Genetic mutations of the viral genome encoding viral genes were also reported previously. Furukawa *et al.* showed the presence of point mutations with a premature stop codon in *tax* gene in ATL patients [40]. Fan *et al.* performed a comprehensive analysis and found that there are mutations in various viral genes but not in the *HBZ* gene [106]. Such deletions and mutations of the provirus should significantly affect expression of viral genes.

## 8. Closing Remarks

Recent HTLV-1 research has made substantial progress in proviral regulation, but there are several issues that remain to be addressed. For example, the evidence we have so far is the result of analyzing a population of infected cells or ATL cells, so we cannot exclude the possibility that a part of the infected cells sporadically expresses plus-strand transcripts like Tax at the single-cell level. Also, it is still unclear how much the maintenance of ATL cells depends on HTLV-1 provirus. It is obvious that genetic and epigenetic alterations associated with oncogenesis in the host genome play a central role in the leukemogenesis of ATL [42–50]. Given that the incidence of T-cell malignancy in peripheral CD4 T cell is quite low in the absence of HTLV-1 infection, HTLV-1 does play a role in ATL generation in HTLV-1 infection. Intermittent expression of Tax possibly accelerates genomic and epigenomic abnormalities in the host cellular genome [25,107]. Continuous antisense transcription from the 3′-LTR supports the proliferation of ATL cells [34,52], which gives the host cells more of a chance to accumulate genomic and epigenomic alterations [3,13,61]. A wide variability of leukemogenesis among individual ATL cases is likely to exist. Further investigations are required to elucidate the role of proviral transcription in HTLV-1 pathogenesis.

**Acknowledgments:** This work was supported by grants from the JSPS KAKENHI Grant Number 26461428 (Y.S.), Takeda Science Foundation (Y.S.), the Japan Science of Technology Agency (Y.S.), and JSPS KAKENHI Grant Number 16K19580 (H.K.).

**Author Contributions:** P.M. and Y.S. wrote the paper; M.M. and H.K. contributed with helpful discussions.

**Conflicts of Interest:** The authors declare no conflict of interest.

## References

1. Van Dooren, S.; Salemi, M.; Vandamme, A.M. Dating the origin of the african human T-cell lymphotropic virus type-1 (HTLV-1) subtypes. *Mol. Biol. Evol.* **2001**, *18*, 661–671. [CrossRef] [PubMed]
2. Igakura, T.; Stinchcombe, J.C.; Goon, P.K.; Taylor, G.P.; Weber, J.N.; Griffiths, G.M.; Tanaka, Y.; Osame, M.; Bangham, C.R. Spread of HTLV-1 between lymphocytes by virus-induced polarization of the cytoskeleton. *Science* **2003**, *299*, 1713–1716. [CrossRef] [PubMed]
3. Matsuoka, M.; Jeang, K.T. Human T-cell leukaemia virus type 1 (HTLV-1) infectivity and cellular transformation. *Nat. Rev. Cancer* **2007**, *7*, 270–280. [CrossRef] [PubMed]
4. Pais-Correia, A.M.; Sachse, M.; Guadagnini, S.; Robbiati, V.; Lasserre, R.; Gessain, A.; Gout, O.; Alcover, A.; Thoulouze, M.I. Biofilm-like extracellular viral assemblies mediate HTLV-1 cell-to-cell transmission at virological synapses. *Nat. Med.* **2010**, *16*, 83–89. [CrossRef] [PubMed]
5. Bangham, C.R. CTL quality and the control of human retroviral infections. *Eur. J. Immunol.* **2009**, *39*, 1700–1712. [CrossRef] [PubMed]
6. Iwanaga, M.; Watanabe, T.; Utsunomiya, A.; Okayama, A.; Uchimaru, K.; Koh, K.R.; Ogata, M.; Kikuchi, H.; Sagara, Y.; Uozumi, K.; *et al.* Human T-cell leukemia virus type I (HTLV-1) proviral load and disease progression in asymptomatic HTLV-1 carriers: A nationwide prospective study in japan. *Blood* **2010**, *116*, 1211–1219. [CrossRef] [PubMed]
7. Ruelas, D.S.; Greene, W.C. An integrated overview of HIV-1 latency. *Cell* **2013**, *155*, 519–529. [CrossRef] [PubMed]
8. Jacobson, S.; Shida, H.; McFarlin, D.E.; Fauci, A.S.; Koenig, S. Circulating CD8+ cytotoxic T lymphocytes specific for HTLV-1 px in patients with HTLV-1 associated neurological disease. *Nature* **1990**, *348*, 245–248. [CrossRef] [PubMed]
9. Kannagi, M.; Harada, S.; Maruyama, I.; Inoko, H.; Igarashi, H.; Kuwashima, G.; Sato, S.; Morita, M.; Kidokoro, M.; Sugimoto, M.; *et al.* Predominant recognition of human T cell leukemia virus type i (HTLV-1) px gene products by human CD8+ cytotoxic T cells directed against HTLV-1-infected cells. *Int. Immunol.* **1991**, *3*, 761–767. [CrossRef] [PubMed]
10. Poiesz, B.J.; Ruscetti, F.W.; Gazdar, A.F.; Bunn, P.A.; Minna, J.D.; Gallo, R.C. Detection and isolation of type C retrovirus particles from fresh and cultured lymphocytes of a patient with cutaneous T-cell lymphoma. *Proc. Natl. Acad. Sci. USA* **1980**, *77*, 7415–7419. [CrossRef] [PubMed]
11. Gallo, R.C. The discovery of the first human retrovirus: HTLV-1 and HTLV-2. *Retrovirology* **2005**, *2*, 17. [CrossRef] [PubMed]
12. Seiki, M.; Hattori, S.; Hirayama, Y.; Yoshida, M. Human adult T-cell leukemia virus: Complete nucleotide sequence of the provirus genome integrated in leukemia cell DNA. *Proc. Natl. Acad. Sci. USA* **1983**, *80*, 3618–3622. [CrossRef] [PubMed]
13. Zhao, T. The role of HBZ in HTLV-1-induced oncogenesis. *Viruses* **2016**, *8*. [CrossRef] [PubMed]
14. Valeri, V.W.; Hryniewicz, A.; Andresen, V.; Jones, K.; Fenizia, C.; Bialuk, I.; Chung, H.K.; Fukumoto, R.; Parks, R.W.; Ferrari, M.G.; *et al.* Requirement of the human T-cell leukemia virus p12 and p30 products for infectivity of human dendritic cells and macaques but not rabbits. *Blood* **2010**, *116*, 3809–3817. [CrossRef] [PubMed]
15. Andresen, V.; Pise-Masison, C.A.; Sinha-Datta, U.; Bellon, M.; Valeri, V.; Washington Parks, R.; Cecchinato, V.; Fukumoto, R.; Nicot, C.; Franchini, G. Suppression of HTLV-1 replication by tax-mediated rerouting of the p13 viral protein to nuclear speckles. *Blood* **2011**, *118*, 1549–1559. [CrossRef] [PubMed]

16. Zazopoulos, E.; Sodroski, J.G.; Haseltine, W.A. P21rex protein of HTLV-1. *J. Acquir. Immune Defic. Syndr.* **1990**, *3*, 1135–1139. [PubMed]

17. Silic-Benussi, M.; Biasiotto, R.; Andresen, V.; Franchini, G.; D'Agostino, D.M.; Ciminale, V. HTLV-1 p13, a small protein with a busy agenda. *Mol. Asp. Med.* **2010**, *31*, 350–358. [CrossRef] [PubMed]

18. Silic-Benussi, M.; Cavallari, I.; Vajente, N.; Vidali, S.; Chieco-Bianchi, L.; Di Lisa, F.; Saggioro, D.; D'Agostino, D.M.; Ciminale, V. Redox regulation of T-cell turnover by the p13 protein of human T-cell leukemia virus type 1: Distinct effects in primary *versus* transformed cells. *Blood* **2010**, *116*, 54–62. [CrossRef] [PubMed]

19. Gaudray, G.; Gachon, F.; Basbous, J.; Biard-Piechaczyk, M.; Devaux, C.; Mesnard, J.M. The complementary strand of the human T-cell leukemia virus type 1 RNA genome encodes a bZIP transcription factor that down-regulates viral transcription. *J. Virol.* **2002**, *76*, 12813–12822. [CrossRef] [PubMed]

20. Tirosh, I.; Weinberger, A.; Carmi, M.; Barkai, N. A genetic signature of interspecies variations in gene expression. *Nat. Genet.* **2006**, *38*, 830–834. [CrossRef] [PubMed]

21. Landry, C.R.; Lemos, B.; Rifkin, S.A.; Dickinson, W.J.; Hartl, D.L. Genetic properties influencing the evolvability of gene expression. *Science* **2007**, *317*, 118–121. [CrossRef] [PubMed]

22. Yoshida, M.; Satou, Y.; Yasunaga, J.; Fujisawa, J.; Matsuoka, M. Transcriptional control of spliced and unspliced human T-cell leukemia virus type 1 bZIP factor (HBZ) gene. *J. Virol.* **2008**, *82*, 9359–9368. [CrossRef] [PubMed]

23. Arpin-Andre, C.; Laverdure, S.; Barbeau, B.; Gross, A.; Mesnard, J.M. Construction of a reporter vector for analysis of bidirectional transcriptional activity of retrovirus LTR. *Plasmid* **2014**, *74*, 45–51. [CrossRef] [PubMed]

24. Fujisawa, J.; Seiki, M.; Kiyokawa, T.; Yoshida, M. Functional activation of the long terminal repeat of human T-cell leukemia virus type 1 by a trans-acting factor. *Proc. Natl. Acad. Sci. USA* **1985**, *82*, 2277–2281. [CrossRef] [PubMed]

25. Yoshida, M. Multiple viral strategies of HTLV-1 for dysregulation of cell growth control. *Annu. Rev. Immunol.* **2001**, *19*, 475–496. [CrossRef] [PubMed]

26. Laverdure, S.; Polakowski, N.; Hoang, K.; Lemasson, I. Permissive sense and antisense transcription from the 5′ and 3′ long terminal repeats of human T-cell leukemia virus type 1. *J. Virol.* **2016**, *90*, 3600–3610. [CrossRef] [PubMed]

27. Lin, H.C.; Hickey, M.; Hsu, L.; Medina, D.; Rabson, A.B. Activation of human T cell leukemia virus type 1 LTR promoter and cellular promoter elements by T cell receptor signaling and HTLV-1 tax expression. *Virology* **2005**, *339*, 1–11. [CrossRef] [PubMed]

28. Zhang, W.; Nisbet, J.W.; Bartoe, J.T.; Ding, W.; Lairmore, M.D. Human t-lymphotropic virus type 1 p30(ii) functions as a transcription factor and differentially modulates CREB-responsive promoters. *J. Virol.* **2000**, *74*, 11270–11277. [CrossRef] [PubMed]

29. Zhang, W.; Nisbet, J.W.; Albrecht, B.; Ding, W.; Kashanchi, F.; Bartoe, J.T.; Lairmore, M.D. Human t-lymphotropic virus type 1 p30(II) regulates gene transcription by binding CREB binding protein/p300. *J. Virol.* **2001**, *75*, 9885–9895. [CrossRef] [PubMed]

30. Nicot, C.; Dundr, M.; Johnson, J.M.; Fullen, J.R.; Alonzo, N.; Fukumoto, R.; Princler, G.L.; Derse, D.; Misteli, T.; Franchini, G. HTLV-1-encoded p30ii is a post-transcriptional negative regulator of viral replication. *Nat. Med.* **2004**, *10*, 197–201. [CrossRef] [PubMed]

31. Lemasson, I.; Lewis, M.R.; Polakowski, N.; Hivin, P.; Cavanagh, M.H.; Thebault, S.; Barbeau, B.; Nyborg, J.K.; Mesnard, J.M. Human T-cell leukemia virus type 1 (HTLV-1) bZIP protein interacts with the cellular transcription factor CREB to inhibit HTLV-1 transcription. *J. Virol.* **2007**, *81*, 1543–1553. [CrossRef] [PubMed]

32. Clerc, I.; Polakowski, N.; Andre-Arpin, C.; Cook, P.; Barbeau, B.; Mesnard, J.M.; Lemasson, I. An interaction between the human T cell leukemia virus type 1 basic leucine zipper factor (HBZ) and the KIX domain of p300/CBP contributes to the down-regulation of tax-dependent viral transcription by HBZ. *J. Biol. Chem.* **2008**, *283*, 23903–23913. [CrossRef] [PubMed]

33. Murata, K.; Hayashibara, T.; Sugahara, K.; Uemura, A.; Yamaguchi, T.; Harasawa, H.; Hasegawa, H.; Tsuruda, K.; Okazaki, T.; Koji, T.; *et al.* A novel alteRNAtive splicing isoform of human T-cell leukemia virus type 1 bZIP factor (HBZ-SI) targets distinct subnuclear localization. *J. Virol.* **2006**, *80*, 2495–2505. [CrossRef] [PubMed]

34. Satou, Y.; Yasunaga, J.; Yoshida, M.; Matsuoka, M. HTLV-1 basic leucine zipper factor gene mRNA supports proliferation of adult T cell leukemia cells. *Proc. Natl. Acad. Sci. USA* **2006**, *103*, 720–725. [CrossRef] [PubMed]

35. Gazon, H.; Lemasson, I.; Polakowski, N.; Cesaire, R.; Matsuoka, M.; Barbeau, B.; Mesnard, J.M.; Peloponese, J.M., Jr. Human T-cell leukemia virus type 1 (HTLV-1) bZIP factor requires cellular transcription factor jund to upregulate HTLV-1 antisense transcription from the 3′ long terminal repeat. *J. Virol.* **2012**, *86*, 9070–9078. [CrossRef] [PubMed]

36. Grassmann, R.; Berchtold, S.; Radant, I.; Alt, M.; Fleckenstein, B.; Sodroski, J.G.; Haseltine, W.A.; Ramstedt, U. Role of human T-cell leukemia virus type 1 x region proteins in immortalization of primary human lymphocytes in culture. *J. Virol.* **1992**, *66*, 4570–4575. [PubMed]

37. Akagi, T.; Shimotohno, K. Proliferative response of tax1-transduced primary human T cells to anti-cd3 antibody stimulation by an interleukin-2-independent pathway. *J. Virol.* **1993**, *67*, 1211–1217. [PubMed]

38. Akagi, T.; Ono, H.; Shimotohno, K. Characterization of T cells immortalized by tax1 of human T-cell leukemia virus type 1. *Blood* **1995**, *86*, 4243–4249. [PubMed]

39. Maeda, M.; Shimizu, A.; Ikuta, K.; Okamoto, H.; Kashihara, M.; Uchiyama, T.; Honjo, T.; Yodoi, J. Origin of human t-lymphotrophic virus i-positive T cell lines in adult T cell leukemia. Analysis of T cell receptor gene rearrangement. *J. Exp. Med.* **1985**, *162*, 2169–2174. [CrossRef] [PubMed]

40. Furukawa, Y.; Kubota, R.; Tara, M.; Izumo, S.; Osame, M. Existence of escape mutant in HTLV-1 tax during the development of adult T-cell leukemia. *Blood* **2001**, *97*, 987–993. [CrossRef] [PubMed]

41. Takeda, S.; Maeda, M.; Morikawa, S.; Taniguchi, Y.; Yasunaga, J.; Nosaka, K.; Tanaka, Y.; Matsuoka, M. Genetic and epigenetic inactivation of *tax* gene in adult T-cell leukemia cells. *Int. J. Cancer* **2004**, *109*, 559–567. [CrossRef] [PubMed]

42. Nosaka, K.; Maeda, M.; Tamiya, S.; Sakai, T.; Mitsuya, H.; Matsuoka, M. Increasing methylation of the cdkn2a gene is associated with the progression of adult T-cell leukemia. *Cancer Res.* **2000**, *60*, 1043–1048. [PubMed]

43. Yasunaga, J.; Taniguchi, Y.; Nosaka, K.; Yoshida, M.; Satou, Y.; Sakai, T.; Mitsuya, H.; Matsuoka, M. Identification of aberrantly methylated genes in association with adult T-cell leukemia. *Cancer Res.* **2004**, *64*, 6002–6009. [CrossRef] [PubMed]

44. Yoshida, M.; Nosaka, K.; Yasunaga, J.; Nishikata, I.; Morishita, K.; Matsuoka, M. Aberrant expression of the *MEL1S* gene identified in association with hypomethylation in adult T-cell leukemia cells. *Blood* **2004**, *103*, 2753–2760. [CrossRef] [PubMed]

45. Hidaka, T.; Nakahata, S.; Hatakeyama, K.; Hamasaki, M.; Yamashita, K.; Kohno, T.; Arai, Y.; Taki, T.; Nishida, K.; Okayama, A.; *et al.* Down-regulation of TCF8 is involved in the leukemogenesis of adult T-cell leukemia/lymphoma. *Blood* **2008**, *112*, 383–393. [CrossRef] [PubMed]

46. Yamagishi, M.; Nakano, K.; Miyake, A.; Yamochi, T.; Kagami, Y.; Tsutsumi, A.; Matsuda, Y.; Sato-Otsubo, A.; Muto, S.; Utsunomiya, A.; *et al.* Polycomb-mediated loss of mir-31 activates nik-dependent nf-kappab pathway in adult T cell leukemia and other cancers. *Cancer Cell* **2012**, *21*, 121–135. [CrossRef] [PubMed]

47. Kataoka, K.; Nagata, Y.; Kitanaka, A.; Shiraishi, Y.; Shimamura, T.; Yasunaga, J.; Totoki, Y.; Chiba, K.; Sato-Otsubo, A.; Nagae, G.; *et al.* Integrated molecular analysis of adult T cell leukemia/lymphoma. *Nat. Genet.* **2015**, *47*, 1304–1315. [CrossRef] [PubMed]

48. Fujikawa, D.; Nakagawa, S.; Hori, M.; Kurokawa, N.; Soejima, A.; Nakano, K.; Yamochi, T.; Nakashima, M.; Kobayashi, S.; Tanaka, Y.; *et al.* Polycomb-dependent epigenetic landscape in adult T-cell leukemia. *Blood* **2016**, *127*, 1790–1802. [CrossRef] [PubMed]

49. Nagata, Y.; Kontani, K.; Enami, T.; Kataoka, K.; Ishii, R.; Totoki, Y.; Kataoka, T.R.; Hirata, M.; Aoki, K.; Nakano, K.; *et al.* Variegated RHOA mutations in adult T-cell leukemia/lymphoma. *Blood* **2016**, *127*, 596–604. [CrossRef] [PubMed]

50. Nakagawa, M.; Schmitz, R.; Xiao, W.; Goldman, C.K.; Xu, W.; Yang, Y.; Yu, X.; Waldmann, T.A.; Staudt, L.M. Gain-of-function ccr4 mutations in adult T cell leukemia/lymphoma. *J. Exp. Med.* **2014**, *211*, 2497–2505. [CrossRef] [PubMed]

51. Belrose, G.; Gross, A.; Olindo, S.; Lezin, A.; Dueymes, M.; Komla-Soukha, I.; Smadja, D.; Tanaka, Y.; Willems, L.; Mesnard, J.M.; *et al.* Effects of valproate on tax and HBZ expression in HTLV-1 and ham/tsp T lymphocytes. *Blood* **2011**, *118*, 2483–2491. [CrossRef] [PubMed]

52. Usui, T.; Yanagihara, K.; Tsukasaki, K.; Murata, K.; Hasegawa, H.; Yamada, Y.; Kamihira, S. Characteristic expression of HTLV-1 basic zipper factor (HBZ) transcripts in HTLV-1 provirus-positive cells. *Retrovirology* **2008**, *5*, 34. [CrossRef] [PubMed]

53. Macnamara, A.; Rowan, A.; Hilburn, S.; Kadolsky, U.; Fujiwara, H.; Suemori, K.; Yasukawa, M.; Taylor, G.; Bangham, C.R.; Asquith, B. Hla class i binding of HBZ determines outcome in HTLV-1 infection. *PLoS Pathog.* **2010**, *6*. [CrossRef] [PubMed]

54. Clarke, M.F.; Trainor, C.D.; Mann, D.L.; Gallo, R.C.; Reitz, M.S. Methylation of human T-cell leukemia virus proviral DNA and viral RNA expression in short- and long-term cultures of infected cells. *Virology* **1984**, *135*, 97–104. [CrossRef]

55. Hanon, E.; Hall, S.; Taylor, G.P.; Saito, M.; Davis, R.; Tanaka, Y.; Usuku, K.; Osame, M.; Weber, J.N.; Bangham, C.R. Abundant tax protein expression in CD4+ T cells infected with human T-cell lymphotropic virus type i (HTLV-1) is prevented by cytotoxic T lymphocytes. *Blood* **2000**, *95*, 1386–1392. [PubMed]

56. Satou, Y.; Utsunomiya, A.; Tanabe, J.; Nakagawa, M.; Nosaka, K.; Matsuoka, M. HTLV-1 modulates the frequency and phenotype of foxp3+CD4+ T cells in virus-infected individuals. *Retrovirology* **2012**, *9*, 46. [CrossRef] [PubMed]

57. Rende, F.; Cavallari, I.; Corradin, A.; Silic-Benussi, M.; Toulza, F.; Toffolo, G.M.; Tanaka, Y.; Jacobson, S.; Taylor, G.P.; D'Agostino, D.M.; *et al.* Kinetics and intracellular compartmentalization of HTLV-1 gene expression: Nuclear retention of HBZ mRNAs. *Blood* **2011**, *117*, 4855–4859. [CrossRef] [PubMed]

58. Cavallari, I.; Rende, F.; Ciminale, V. Quantitative analysis of human t-lymphotropic virus type 1 (HTLV-1) gene expression using nucleo-cytoplasmic fractionation and splice junction-specific real-time rt-pcr (qrt-pcr). *Methods Mol. Biol.* **2014**, *1087*, 325–337. [PubMed]

59. Cavallari, I.; Rende, F.; Bona, M.K.; Sztuba-Solinska, J.; Silic-Benussi, M.; Tognon, M.; LeGrice, S.F.; Franchini, G.; D'Agostino, D.M.; Ciminale, V. Expression of alternatively spliced human T-cell leukemia virus type 1 mRNAs is influenced by mitosis and by a novel cis-acting regulatory sequence. *J. Virol.* **2016**, *90*, 1486–1498. [CrossRef] [PubMed]

60. Mitobe, Y.; Yasunaga, J.; Furuta, R.; Matsuoka, M. HTLV-1 bZIP factor RNA and protein impart distinct functions on T-cell proliferation and survival. *Cancer Res.* **2015**, *75*, 4143–4152. [CrossRef] [PubMed]

61. Satou, Y.; Matsuoka, M. Implication of the HTLV-1 bZIP factor gene in the leukemogenesis of adult T-cell leukemia. *Int. J. Hematol.* **2007**, *86*, 107–112. [CrossRef] [PubMed]

62. Ma, G.; Yasunaga, J.; Matsuoka, M. Multifaceted functions and roles of HBZ in HTLV-1 pathogenesis. *Retrovirology* **2016**, *13*, 16. [CrossRef] [PubMed]

63. Cavanagh, M.H.; Landry, S.; Audet, B.; Arpin-Andre, C.; Hivin, P.; Pare, M.E.; Thete, J.; Wattel, E.; Marriott, S.J.; Mesnard, J.M.; *et al.* HTLV-1 antisense transcripts initiating in the 3'LTR are alternatively spliced and polyadenylated. *Retrovirology* **2006**, *3*, 15. [CrossRef] [PubMed]

64. Gillet, N.A.; Malani, N.; Melamed, A.; Gormley, N.; Carter, R.; Bentley, D.; Berry, C.; Bushman, F.D.; Taylor, G.P.; Bangham, C.R. The host genomic environment of the provirus determines the abundance of HTLV-1-infected T-cell clones. *Blood* **2011**, *117*, 3113–3122. [CrossRef] [PubMed]

65. Melamed, A.; Laydon, D.J.; Gillet, N.A.; Tanaka, Y.; Taylor, G.P.; Bangham, C.R. Genome-wide determinants of proviral targeting, clonal abundance and expression in natural HTLV-1 infection. *PLoS Pathog.* **2013**, *9*, e1003271. [CrossRef] [PubMed]

66. Bangham, C.R.; Cook, L.B.; Melamed, A. HTLV-1 clonality in adult T-cell leukaemia and non-malignant HTLV-1 infection. *Semin. Cancer Biol.* **2014**, *26*, 89–98. [CrossRef] [PubMed]

67. Satou, Y.; Miyazato, P.; Ishihara, K.; Yaguchi, H.; Melamed, A.; Miura, M.; Fukuda, A.; Nosaka, K.; Watanabe, T.; Rowan, A.G.; *et al.* The retrovirus HTLV-1 inserts an ectopic CTCF-binding site into the human genome. *Proc. Natl. Acad. Sci. USA* **2016**, *113*, 3054–3059. [CrossRef] [PubMed]

68. Kouzarides, T. Chromatin modifications and their function. *Cell* **2007**, *128*, 693–705. [CrossRef] [PubMed]

69. Li, B.; Carey, M.; Workman, J.L. The role of chromatin during transcription. *Cell* **2007**, *128*, 707–719. [CrossRef] [PubMed]

70. Deaton, A.M.; Bird, A. CpG islands and the regulation of transcription. *Genes Dev.* **2011**, *25*, 1010–1022. [CrossRef] [PubMed]

71. Kitamura, T.; Takano, M.; Hoshino, H.; Shimotohno, K.; Shimoyama, M.; Miwa, M.; Takaku, F.; Sugimura, T. Methylation pattern of human T-cell leukemia virus *in vivo* and *in vitro*: Px and LTR regions are hypomethylated *in vivo*. *Int. J. Cancer* **1985**, *35*, 629–635. [CrossRef] [PubMed]

72. Saggioro, D.; Panozzo, M.; Chieco-Bianchi, L. Human t-lymphotropic virus type i transcriptional regulation by methylation. *Cancer Res.* **1990**, *50*, 4968–4973. [PubMed]

73. Koiwa, T.; Hamano-Usami, A.; Ishida, T.; Okayama, A.; Yamaguchi, K.; Kamihira, S.; Watanabe, T. 5'-long terminal repeat-selective CpG methylation of latent human T-cell leukemia virus type 1 provirus *in vitro* and *in vivo*. *J. Virol.* **2002**, *76*, 9389–9397. [CrossRef] [PubMed]

74. Zhou, Q.; Li, T.; Price, D.H. RNA polymerase ii elongation control. *Annu. Rev. Biochem.* **2012**, *81*, 119–143. [CrossRef] [PubMed]

75. Geiger, T.R.; Sharma, N.; Kim, Y.M.; Nyborg, J.K. The human T-cell leukemia virus type 1 Tax protein confers CBP/p300 recruitment and transcriptional activation properties to phosphorylated CREB. *Mol. Cell. Biol.* **2008**, *28*, 1383–1392. [CrossRef] [PubMed]

76. Lemasson, I.; Polakowski, N.J.; Laybourn, P.J.; Nyborg, J.K. Tax-dependent displacement of nucleosomes during transcriptional activation of human T-cell leukemia virus type 1. *J. Biol. Chem.* **2006**, *281*, 13075–13082. [CrossRef] [PubMed]

77. Sharma, N.; Nyborg, J.K. The coactivators CBP/p300 and the histone chaperone nap1 promote transcription-independent nucleosome eviction at the HTLV-1 promoter. *Proc. Natl. Acad. Sci. USA* **2008**, *105*, 7959–7963. [CrossRef] [PubMed]

78. Easley, R.; Carpio, L.; Guendel, I.; Klase, Z.; Choi, S.; Kehn-Hall, K.; Brady, J.N.; Kashanchi, F. Human t-lymphotropic virus type 1 transcription and chromatin-remodeling complexes. *J. Virol.* **2010**, *84*, 4755–4768. [CrossRef] [PubMed]

79. Lemasson, I.; Polakowski, N.J.; Laybourn, P.J.; Nyborg, J.K. Transcription factor binding and histone modifications on the integrated proviral promoter in human T-cell leukemia virus-i-infected T-cells. *J. Biol. Chem.* **2002**, *277*, 49459–49465. [CrossRef] [PubMed]

80. Lemasson, I.; Polakowski, N.J.; Laybourn, P.J.; Nyborg, J.K. Transcription regulatory complexes bind the human T-cell leukemia virus 5' and 3' long terminal repeats to control gene expression. *Mol. Cell. Biol.* **2004**, *24*, 6117–6126. [CrossRef] [PubMed]

81. Taniguchi, Y.; Nosaka, K.; Yasunaga, J.; Maeda, M.; Mueller, N.; Okayama, A.; Matsuoka, M. Silencing of human T-cell leukemia virus type i gene transcription by epigenetic mechanisms. *Retrovirology* **2005**, *2*, 64. [CrossRef] [PubMed]

82. Barski, A.; Cuddapah, S.; Cui, K.; Roh, T.Y.; Schones, D.E.; Wang, Z.; Wei, G.; Chepelev, I.; Zhao, K. High-resolution profiling of histone methylations in the human genome. *Cell* **2007**, *129*, 823–837. [CrossRef] [PubMed]

83. Moore, M.J.; Proudfoot, N.J. Pre-mRNA processing reaches back to transcription and ahead to translation. *Cell* **2009**, *136*, 688–700. [CrossRef] [PubMed]

84. Bell, A.C.; Felsenfeld, G. Methylation of a CTCF-dependent boundary controls imprinted expression of the igf2 gene. *Nature* **2000**, *405*, 482–485. [PubMed]

85. Hark, A.T.; Schoenherr, C.J.; Katz, D.J.; Ingram, R.S.; Levorse, J.M.; Tilghman, S.M. CTCF mediates methylation-sensitive enhancer-blocking activity at the H19/IGF2 locus. *Nature* **2000**, *405*, 486–489. [PubMed]

86. Ong, C.T.; Corces, V.G. CTCF: An architectural protein bridging genome topology and function. *Nat. Rev. Genet.* **2014**, *15*, 234–246. [CrossRef] [PubMed]

87. de Wit, E.; Vos, E.S.; Holwerda, S.J.; Valdes-Quezada, C.; Verstegen, M.J.; Teunissen, H.; Splinter, E.; Wijchers, P.J.; Krijger, P.H.; de Laat, W. CTCF binding polarity determines chromatin looping. *Mol. Cell* **2015**, *60*, 676–684. [CrossRef] [PubMed]

88. Wendt, K.S.; Yoshida, K.; Itoh, T.; Bando, M.; Koch, B.; Schirghuber, E.; Tsutsumi, S.; Nagae, G.; Ishihara, K.; Mishiro, T.; *et al.* Cohesin mediates transcriptional insulation by ccctc-binding factor. *Nature* **2008**, *451*, 796–801. [CrossRef]

89. Rao, S.S.; Huntley, M.H.; Durand, N.C.; Stamenova, E.K.; Bochkov, I.D.; Robinson, J.T.; Sanborn, A.L.; Machol, I.; Omer, A.D.; Lander, E.S.; *et al.* A 3D map of the human genome at kilobase resolution reveals principles of chromatin looping. *Cell* **2014**, *159*, 1665–1680. [CrossRef] [PubMed]

90. Tang, Z.; Luo, O.J.; Li, X.; Zheng, M.; Zhu, J.J.; Szalaj, P.; Trzaskoma, P.; Magalska, A.; Wlodarczyk, J.; Ruszczycki, B.; *et al.* CTCF-mediated human 3D genome architecture reveals chromatin topology for transcription. *Cell* **2015**, *163*, 1611–1627. [CrossRef] [PubMed]

91. Kang, H.; Cho, H.; Sung, G.H.; Lieberman, P.M. CTCF regulates kaposi's sarcoma-associated herpesvirus latency transcription by nucleosome displacement and RNA polymerase programming. *J. Virol.* **2013**, *87*, 1789–1799. [CrossRef] [PubMed]

92. Tempera, I.; Lieberman, P.M. Epigenetic regulation of ebv persistence and oncogenesis. *Semin. Cancer Biol.* **2014**, *26*, 22–29. [CrossRef] [PubMed]

93. Lieberman, P.M. Chromatin structure of epstein-barr virus latent episomes. *Curr. Top. Microbiol. Immunol.* **2015**, *390*, 71–102. [PubMed]

94. Kim, T.H.; Abdullaev, Z.K.; Smith, A.D.; Ching, K.A.; Loukinov, D.I.; Green, R.D.; Zhang, M.Q.; Lobanenkov, V.V.; Ren, B. Analysis of the vertebrate insulator protein CTCF-binding sites in the human genome. *Cell* **2007**, *128*, 1231–1245. [CrossRef] [PubMed]

95. Yoshida, M.; Seiki, M.; Yamaguchi, K.; Takatsuki, K. Monoclonal integration of human T-cell leukemia provirus in all primary tumors of adult T-cell leukemia suggests causative role of human T-cell leukemia virus in the disease. *Proc. Natl. Acad. Sci. USA* **1984**, *81*, 2534–2537. [CrossRef] [PubMed]

96. Etoh, K.; Tamiya, S.; Yamaguchi, K.; Okayama, A.; Tsubouchi, H.; Ideta, T.; Mueller, N.; Takatsuki, K.; Matsuoka, M. Persistent clonal proliferation of human t-lymphotropic virus type 1-infected cells *in vivo*. *Cancer Res.* **1997**, *57*, 4862–4867. [PubMed]

97. Doi, K.; Wu, X.; Taniguchi, Y.; Yasunaga, J.; Satou, Y.; Okayama, A.; Nosaka, K.; Matsuoka, M. Preferential selection of human T-cell leukemia virus type 1 provirus integration sites in leukemic *versus* carrier states. *Blood* **2005**, *106*, 1048–1053. [CrossRef] [PubMed]

98. Meekings, K.N.; Leipzig, J.; Bushman, F.D.; Taylor, G.P.; Bangham, C.R. HTLV-1 integration into transcriptionally active genomic regions is associated with proviral expression and with HAM/TSP. *PLoS Pathog.* **2008**, *4*, e1000027. [CrossRef] [PubMed]

99. Firouzi, S.; Lopez, Y.; Suzuki, Y.; Nakai, K.; Sugano, S.; Yamochi, T.; Watanabe, T. Development and validation of a new high-throughput method to investigate the clonality of HTLV-1-infected cells based on provirus integration sites. *Genome Med.* **2014**, *6*, 46. [CrossRef] [PubMed]

100. Gillet, N.A.; Gutierrez, G.; Rodriguez, S.M.; de Brogniez, A.; Renotte, N.; Alvarez, I.; Trono, K.; Willems, L. Massive depletion of bovine leukemia virus proviral clones located in genomic transcriptionally active sites during primary infection. *PLoS Pathog.* **2013**, *9*, e1003687. [CrossRef] [PubMed]

101. Daenke, S.; Nightingale, S.; Cruickshank, J.K.; Bangham, C.R. Sequence variants of human T-cell lymphotropic virus type i from patients with tropical spastic paraparesis and adult T-cell leukemia do not distinguish neurological from leukemic isolates. *J. Virol.* **1990**, *64*, 1278–1282. [PubMed]

102. Konishi, H.; Kobayashi, N.; Hatanaka, M. Defective human T-cell leukemia virus in adult T-cell leukemia patients. *Mol. Biol. Med.* **1984**, *2*, 273–283. [PubMed]

103. Manzari, V.; Wong-Staal, F.; Franchini, G.; Colombini, S.; Gelmann, E.P.; Oroszlan, S.; Staal, S.; Gallo, R.C. Human T-cell leukemia-lymphoma virus (HTLV): Cloning of an integrated defective provirus and flanking cellular sequences. *Proc. Natl. Acad. Sci. USA* **1983**, *80*, 1574–1578. [CrossRef] [PubMed]

104. Tamiya, S.; Matsuoka, M.; Etoh, K.; Watanabe, T.; Kamihira, S.; Yamaguchi, K.; Takatsuki, K. Two types of defective human t-lymphotropic virus type i provirus in adult T-cell leukemia. *Blood* **1996**, *88*, 3065–3073. [PubMed]

105. Miyazaki, M.; Yasunaga, J.; Taniguchi, Y.; Tamiya, S.; Nakahata, T.; Matsuoka, M. Preferential selection of human T-cell leukemia virus type 1 provirus lacking the 5' long terminal repeat during oncogenesis. *J. Virol.* **2007**, *81*, 5714–5723. [CrossRef] [PubMed]

106. Fan, J.; Ma, G.; Nosaka, K.; Tanabe, J.; Satou, Y.; Koito, A.; Wain-Hobson, S.; Vartanian, J.P.; Matsuoka, M. Apobec3g generates nonsense mutations in human T-cell leukemia virus type 1 proviral genomes *in vivo*. *J. Virol.* **2010**, *84*, 7278–7287. [CrossRef] [PubMed]

107. Grassmann, R.; Aboud, M.; Jeang, K.T. Molecular mechanisms of cellular transformation by HTLV-1 tax. *Oncogene* **2005**, *24*, 5976–5985. [CrossRef] [PubMed]

*Review*

# The Emerging Role of miRNAs in HTLV-1 Infection and ATLL Pathogenesis

Ramona Moles and Christophe Nicot *

Department of Pathology and Laboratory Medicine, Center for Viral Oncology, University of Kansas Medical Center, 3901 Rainbow Boulevard, Kansas City, KS 66160, USA; rmoles@kumc.edu

* Author to whom correspondence should be addressed; cnicot@kumc.edu; Tel.: +1-913-588-6724; Fax: +1-913-945-6836.

Academic Editor: Louis M. Mansky

Received: 11 June 2015; Accepted: 7 July 2015; Published: 20 July 2015

**Abstract:** Human T-cell leukemia virus (HTLV)-1 is a human retrovirus and the etiological agent of adult T-cell leukemia/lymphoma (ATLL), a fatal malignancy of CD4/CD25+ T lymphocytes. In recent years, cellular as well as virus-encoded microRNA (miRNA) have been shown to deregulate signaling pathways to favor virus life cycle. HTLV-1 does not encode miRNA, but several studies have demonstrated that cellular miRNA expression is affected in infected cells. Distinct mechanisms such as transcriptional, epigenetic or interference with miRNA processing machinery have been involved. This article reviews the current knowledge of the role of cellular microRNAs in virus infection, replication, immune escape and pathogenesis of HTLV-1.

**Keywords:** human; HTLV-I infections; T-lymphotrophic virus 1; leukemia-lymphoma; adult T-cell; microRNAs; virus replication; cell line; cell transformation; gene expression regulation

## 1. Introduction

The transmission of the human T-cell leukemia virus (HTLV-1) retrovirus requires close contact with infected T cells, and occurs from mother to child, predominantly through breastfeeding as well as through sexual contact and blood transfusion [1,2]. The HTLV-1 infection is also associated with other diseases, such as: a chronic and progressive neurologic disorder named HTLV-1-associated myelopathy/tropical spastic paraparesis (HAM/TSP), polymyositis, infective dermatitis, HTLV-1-associated arthropathy, and HTLV-1-associated uveitis [2,3]. According to the Shimoyama classification, adult T-cell leukemia/lymphoma (ATLL) can be distinguished into four subtypes: smoldering, chronic and acute leukemic forms and ATLL lymphoma [4]. The overall survival of ATLL with different regimens of chemotherapy is poor, ranging between 5.5 and 13 months in patients presenting acute leukemia or lymphoma [5]. HTLV-1 mediates T lymphocyte transformation using a multistep process in which the virus promotes genomic instability, accumulation of genetic defects, and chronic proliferation of infected cells [6]. The genome of HTLV-1 encodes common retrovirus structural and enzymatic proteins, Gag, Pro, Pol, and Env, and additional accessory and regulatory proteins such as Tax, Rex, P30, p12, p13, and HTLV-1 basic leucine zipper factor protein (HBZ). Tax and HBZ regulatory proteins have been reported to play a central role in regulation of viral and cellular genes that lead to proliferation of infected cells [7]. Tax is a transcriptional trans-activator that promotes the expression of genes linked to the 5′ long terminal repeat promoter (LTR) element of the HTLV-1 genome [8]. Tax induces genomic instability [9] and promotes cell-cycle progression, survival and growth of HTLV-1-positive T cells [10]. HBZ is involved in the proliferation of infected cells *in vitro* and *in vivo* and plays an essential role in oncogenesis mediated by HTLV-1 in late stages of the disease when Tax is not expressed [11]. Consistently, HBZ was found to be expressed in ATLL cells through the whole period of ATLL development, suggesting that it might be involved in maintenance of HTLV-1-transformed cells [12]. Rex is a post-transcriptional regulator of viral expression, which activates viral replication in the early phase of HTLV-1 infection by promoting the nuclear export of HTLV-1 mRNA [13]. Several studies have shown altered expression of microRNAs (miRNAs) in HTLV-1/ATLL cell lines and primary peripheral blood mononuclear cells (PBMCs) from ATLL patients, suggesting that miRNA deregulation is involved in HTLV-1 infection and adult T-cell leukemia/lymphoma pathogenesis. MicroRNAs play an essential role in a wide range of biological processes, including development, differentiation, cell cycle, apoptosis and oncogenesis [14–16].

## 2. MiRNA Biogenesis

MicroRNAs (miRNAs) are small, non-coding RNA molecules that transcriptionally regulate gene expression. The first miRNA identified in animals is *Lin-4*, discovered in 1993 by Ambros and colleagues. *Lin-4* was identified as heterochronic genes in *Caenorhabditis elegans* involved in cell fate [17,18]. Subsequent studies have shown the involvement of miRNAs in different biological processes, including tumorigenesis by targeting oncogenes or tumor suppressor genes [16]. MiRNA sequences are localized in different genomic contexts. Some miRNAs are encoded by exon; however, the majority are encoded by the intronic region of non-coding and coding transcripts [19]. MiRNAs are transcribed by the RNA polymerase II or III into the nucleus as primary miRNAs (pri-miRNAs). Pri-miRNAs are normally over 1 kilobase and contain a local steam-loop structure in which mature miRNA sequences are included. The nuclear RNase III Drosha recognized and processed pri-miRNAs into a hairpin-shaped RNA of nearly 65 nucleotides in length, named precursor miRNAs (pre-miRNAs). After transport to the cytoplasm by the RanGTP-dependent dsRNA-binding protein Exportin 5, pre-miRNAs are processed by the cytoplasmic RNase III Dicer, liberating a mature 20–24 nucleotide long duplex. Argonaute family proteins, AGO, and Trans-Activation Responsive RNA-Binding Protein (TARBP2), together with the duplex form a complex named RNA-Induced Silencing Complex (RISC) [19,20]. One strand of the duplex, called guide strand, is incorporated into the RISC complex while the other strand, named passenger strand, is targeted for degradation [21]. Apart from the canonical miRNA biogenesis described above, different alternative mechanisms, which bypass Drosha processing, were described [22]. MiRNAs can be generated through non-canonical pathways, wherein the precursor miRNAs are cleavaged by Dicer. Mirtrons represent an example of miRNA processed by a non-canonical pathway. They are generated from intron lariats serving as pri-miRNAs, which is processed by Spliceosome that function as Drosha, to release pre-miRNAs [22,23]. MiRNAs bind complementary sequences usually localized at 3′UTR of messenger RNA and guide RISC to target mRNA. MiRNAs used different mechanisms to regulate post-transcriptional gene expression: inhibition of translation and/or messenger RNA degradation. The repression of many miRNA targets is frequently associated with their destabilization. Degradation of target mRNA is characterized by gradual shortening of the mRNA poly-Adenine tail, which is catalyzed by the exosome or exonuclease XRN1. MiRNAs might also induce gene silencing by interfering with protein translation [24]. Several pieces of evidence show that miRNA silencing is observed with either no change in the mRNA level or with a significantly smaller decrease of mRNA compared to the protein level [25,26]. Deregulated MiRNAs in HTLV-1 context will be discussing in the next section of the review.

## 3. MiRNA Profile in HTLV-1-Transformed Cell Lines and ATLL Patients

Four studies have characterized miRNA expression profiles in HTLV-1/ATLL cell lines and ATLL patients. Pichler [27] and colleagues chose the phenotype of regulatory T cells (Treg) as a starting point to study miRNA expression in HTLV-1-transformed cells. The authors have selected and analyzed the expression of a set of miRNAs characteristic of murine Treg and downregulated in different tumors. The analysis identified five deregulated miRNAs: miR-21, miR-24, miR-146a, and miR-155 were found upregulated, whereas miR-223 was downregulated. Bellon [28] and colleagues analyzed miRNA profiles from ATLL patients compared to HTLV-1-negative donors by using microarray. The results were confirmed by Real Time (RT)-PCR of mature miRNAs in uncultured ATLL cells and

HTLV-1-transformed cell lines. Microarray analysis and RT-PCR demonstrated downregulation of miR-181a, miR-132 and miR-125a and upregulation of miR-155 and miR-142-3p. This study identifies two miRNAs differently expressed *in vitro* and *in vivo*. MiR-150 and miR-223 were both upregulated in uncultured ATLL cells and downregulated in HTLV-1-transformed cell lines. Yeung [29] and colleagues examined miRNA profiles in several ATLL-derived cell lines and primary peripheral blood mononuclear cells (PBMCs) from acute ATLL patients using miRNA microarray. Several HTLV-1/ATLL cell lines and four ATLL patients were studied. Thirteen miRNAs were found to be upregulated and thirty downregulated among the different cell lines. In parallel, 22 upregulated and 22 downregulated miRNAs were identified in acute ATLL patients. Among those, miR-9, miR-17-3p, miR-20b, miR-93, miR-130b and miR-18a were found to be induced; in contrast, miR-1, miR-144, miR-126, miR-130a, miR-199a, miR-338, miR-432, miR-335 and miR-337 were found to be downregulated. Yamagishi and colleagues [30] studied the miRNA expression signature in primary ATL cells by using microarray analysis compared to CD4+ T cells from healthy donors. The results show that 59 of the miRNAs tested were found with a decrease in ATL primary cells. Among them, miR-31 was the one most profoundly repressed.

## 4. HTLV-1 Interferes with Cellular miRNA Machinery

The dysregulation of miRNA pathways has been reported across several viruses, including HIV, Ebola, Epstein–Barr, Influenza, HBV, HCV, Adenovirus, and HTLV-1 [31–38]. Drosha was reported to be downregulated in HTLV-1-infected cell lines, HTLV-1-transfected cells, and infected primary cells [38]. Van Duyne [38] and colleagues proposed that HTLV-1 deregulates the cellular RNAi pathway, including miRNAs, by suppressing the function and degrading Drosha (Figure 1).

The authors have demonstrated a direct interaction between the Tax oncoprotein and Drosha, which is responsible for its downregulation. The *N*-terminal region of Tax presents two putative motifs, the Zinc finger motif and leucine-zipper-like region, which interact with Drosha. The Tax *N*-terminal region is reported to interact with the proteasome complex. Van Duyne and colleagues demonstrated that the binding between Tax and Drosha leads to its degradation mediated by proteasome complex. In addition, Drosha increases HTLV-1 replication and is not efficient in processing miRNAs when Tax is expressed, suggesting that the dysregulation of miRNA machinery might be involved in the rate of HTLV-1 infection [38]. The HTLV-1 regulatory protein, Rex, is reported to directly interact with Dicer. Abe [39] and colleagues have demonstrated that Rex suppresses the ribonuclease-directed processing activity of Dicer, protecting against the cleavage Rex-mRNA (Figure 1). Inhibition of Dicer activation might represent an additional mechanism used by HTLV-1 to deregulate cellular miRNA expression.

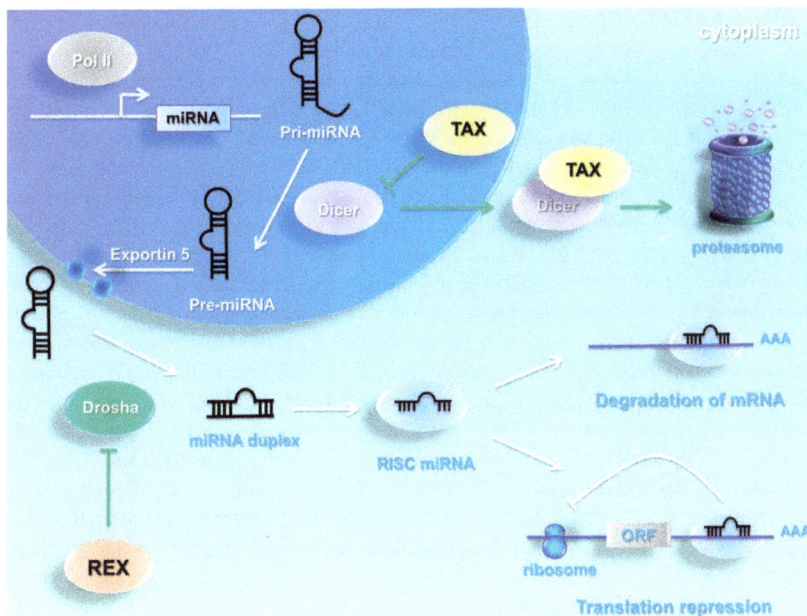

**Figure 1.** Human T-cell leukemia virus HTLV-1 interferes with cellular miRNA machinery. MiRNAs are transcribed by the RNA polymerase II or III into the nucleus as primary miRNAs (pri-miRNAs) from coding or non-coding part of genes. The nuclear RNase III Drosha recognized and processed pri-miRNAs into a hairpin-shaped RNA, named precursor miRNAs. Pre-miRNAs are transported to the cytoplasm by Exportin 5, and processed by the cytoplasmic RNase III Dicer in the mature miRNA duplex. The duplex forms a complex named RNA-Induced Silencing Complex (RISC). MiRNAs bind complementary sequences usually localized at 3′UTR of messenger RNA and this binding results in the inhibition of translation and/or messenger RNA degradation. HTLV-1 deregulates the cellular miRNA pathway by suppressing the function of Drosha and Dicer. Tax directly interacts with Drosha and the binding leads to Drosha degradation mediated by proteasome complex. The regulatory protein, Rex, is reported to directly interact with Dicer. Rex suppresses the ribonuclease-directed processing activity of Dicer, protecting against the cleavage Rex-mRNA.

## 5. MiRNAs Target the HTLV-1 Genome

The cellular environment has an essential role in virus infection and replication. Many cellular genes prevent replication and virus dissemination by acting as innate immunity factors. However, viruses have evolved strategies to avoid activation of an antiviral state: virus-derived miRNAs can enhance viral gene expression, replication, and infectivity [40] or suppress the IFN response [41]. The genome of the Epstein–Barr virus (EBV), Kaposi sarcoma-associated herpesvirus (KSHV), human cytomegalovirus (hCMV) and bovine leukemia virus (BLV) encodes for virus-derived miRNAs [42–45]. BLV shares many characteristics in disease pathogenesis with HTLV-1 and is associated with the development of B-cell tumors. Kincaid and colleagues [45] show that BLV is capable of producing miRNAs *in vitro*. A subsequent study demonstrated that BLV encodes a conserved cluster of miRNAs located in a specific BLV proviral region, which is essential for *in vivo* infectivity [46]. BLV has different common features in genomic organization with HTLV-1, however HTLV-1-encoded miRNAs have not been reported. Cellular miRNAs can promote virus replication or negatively regulate virus expression and infectivity [47]. MiR-28, miR-125b, together with miR-150, miR-223 and miR-382 target 3′ ends of HIV-1 messenger, promoting viral latency [47]. Bai and colleagues identified a binding site in the

HTLV-1 genome for miR-28-3p and demonstrated a mechanism used by a cellular miRNA to prevent HTLV-1 gene expression and viral transmission (Figure 2) [48].

MiR-28-3p was found to target a sequence localized within the viral gag/pol HTLV-1 mRNA and reduced viral replication and gene expression. MiR-28-3p-expressing cells are characterized by reduced levels of HTLV-1 gag p19 and p24 products and they are resistant to infection. MiR-28-3p expression leads to abortive infection by inhibiting HTLV-1 reverse transcription and preventing the formation of the pre-integration complex. MiR-28-3p suppresses HTLV-1 expression and infection; this is consistent with the high levels of miR-28-3p reported in resting T cells and their inability to be infected by HTLV-1 without prior activation. Bai and colleagues [48] demonstrated a natural feedback loop that regulated miR-28-3p expression in response to virus infection. It is well established that *de novo* infection in T cells activates the interferon anti-viral response. MiR-28-3p expression was found to be induced after IFN-α or -γ stimulation, suggesting that miR-28-3p might contribute to restricting virus expansion to neighboring cells by reducing local inflammation and the initial establishment of latent infection. The miR-28-3p site is highly conserved in HTLV-1 subtypes B and C, at nearly 90%. However, the subtype 1A, Japanese ATK-1, presents a natural polymorphism (T to C substitution) within the miR-28-3p target site. The mutation is silent and more resistant to miR-28-3p inhibition of viral replication. Bai and colleagues [48] proposed a model where the modulation of miR-28-3p expression affected HTLV-1 virus spreading. Virus particles can transiently activate resting T cells by reducing miR-28-3p expression and favoring infection. Because IFN response is a potent inducer of miR-28-3p expression, the initial antiviral response might backfire, helping to protect newly infected cells from being eliminated by the immune system.

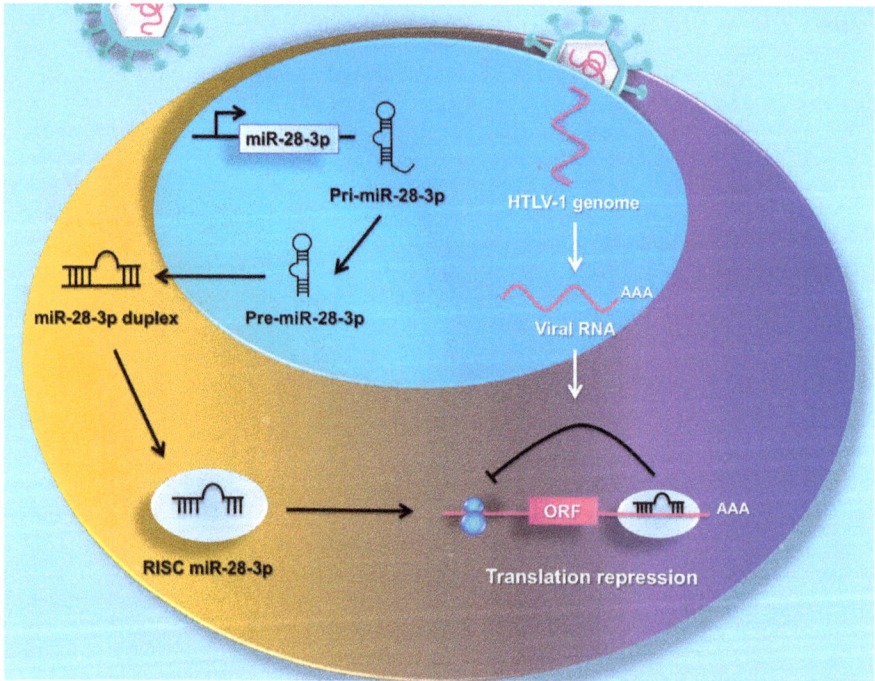

**Figure 2.** MiR-28-3p targets the HTLV-1 genome. The figure illustrates a natural feedback loop that regulated cellular miRNA expression in response to virus infection. MiR-28-3p suppresses HTLV-1 expression by targeting a sequence localized within the viral gag/pol HTLV-1 sequence. MiR-28-3p expression leads to abortive infection by inhibiting HTLV-1 reverse transcription and preventing the formation of the pre-integration complex.

## 6. MiRNAs Promote Cell Proliferation

### 6.1. MiR-146a

MiR-146a has a central role in the regulation of immune response and its expression is induced by NF-κB signaling. MiR-146a is deregulated in different cancers. A high level of expression was reported in papillary thyroid carcinoma, anaplastic thyroid cancer, breast cancer, glioblastoma and cervical cancer [49–53]. In contrast, low-expressing levels were described in pancreatic carcinoma, gastric cancer, prostate cancer, acute myeloid leukemia (AML), myeloblastic syndrome and chronic myeloid leukemia (CML) [54–59]. MiR-146a was found to be upregulated in HTLV-1-transformed cell lines [27]. Ectopic expression of Tax in HTLV-1-negative T cells, Jurkat, induced miR-146a expression. Promotor analysis showed a 15-fold activation of miR-146a by Tax [60], suggesting that HTLV-1 infection might be involved in the regulation of miR-146a expression. Pichler [27] and colleagues used a mutated form of Tax and dominant active NF-κB inhibitor to show that miR-146a transactivation is mediated by NF-κB (Figure 3).

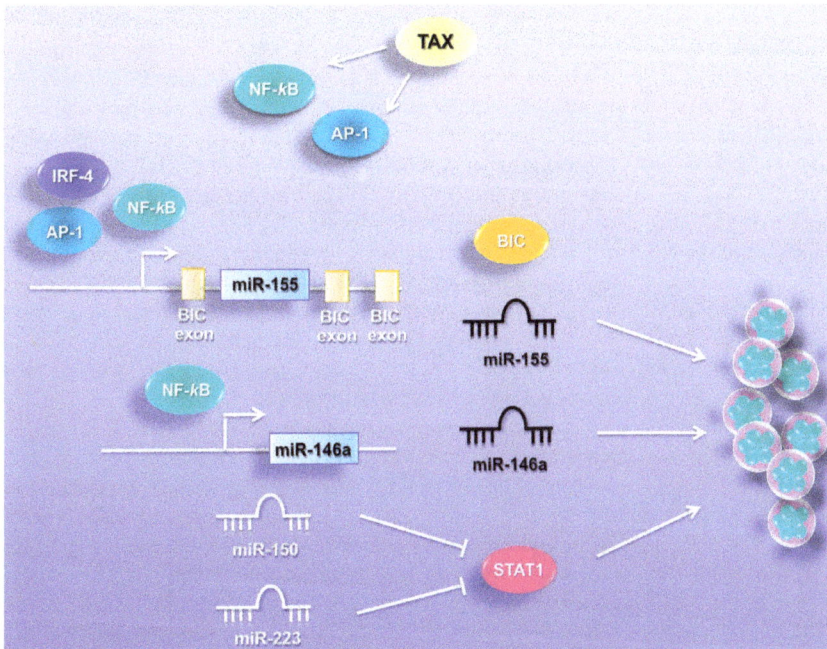

**Figure 3.** MiRNAs promote cell proliferation. MiR-155 and miR-146a were found elevated in HTLV-1-infected cells *in vitro*. Tax induces the transcription factors NF-κB and AP-1, which promote miR-155 expression by binding the miRNA promoter. This binding resulted in an increased expression of the B-cell integration cluster (BIC) gene whose transcript is processed into miR-155. The interferon regulatory factor-4, IRF4, which is induced in HTLV-1-infected cells, promotes BIC/miR-155 expression. NF-κB also mediates miR-146a transactivation; both miRNAs enhance cellular growth in HTLV-1-infected cells. MiR-150 and miR-223 are differentially regulated in ATLL samples and in HTLV-1-transformed cells. MiR-150 and miR-223 were found upregulated in acute ATLL patients and downregulated in HTLV-1-transformed cell lines. MiR-150 and miR-223 target the STAT1 3′UTR. Inhibition of STAT1 expression, through miR-150, miR-223 reduced proliferation of HTLV-1-transformed and ATLL-derived cell lines. MiR-150 and miR-223, by decreasing STAT1 expression and dampening STAT1-dependent signaling in human T cells, regulated proliferation in an HTLV-1 context.

Tomita *et al.* [60] also reported an NF-κB binding site on the miR-146a gene. In addition, it has been described as having a suppressive effect of miR-146a on NF-κB signaling [51]. This might represent a negative feedback loop, which seems to be ineffective in HTLV-1-infected cells. MiR-146a has also been shown to induce proliferation in several human cancers, including cervical cancer [53], breast cancer cells [61], gastric cancer cells [62] and mesenchymal stem cells (MSCs) [63]. Consistent with this report, treatment with an anti-miR-146a inhibitor suppressed the proliferation of HTLV-1-transformed cell lines but not uninfected T-cell lines. In addition, overexpression of miR-146a increased the growth of HTLV-1-transformed cell lines [27]. Because overexpression of miR-146a has also been described in an EBV context, Tomita [60] and colleagues suggested that miR-146a up-regulation might represent a common mechanism in the pathogenesis of persistent viruses. Wang *et al.* [64] identified 622 putative target genes of miR-146a that are predicted by using different prediction programs. Gene ontology analysis shows that these genes are involved in the inhibition of cell growth and promotion of apoptosis, and this partially explains the role of miR-146a in the proliferation of HTLV-1-transformed cells.

## 6.2. MiR-155

MiR-155 has been implicated in normal hematopoiesis [65], immune response [66], and in the carcinogenesis of different human tumors [67,68]. Mouse studies have reported that transgenic overexpression of miR-155 results in the increased frequency of tumor formation [69]. Overexpression of miR-155 was found in breast cancer [70], pancreatic cancer [71], lung cancer [72], B-cell lymphoma [67], MALT lymphoma [73] and acute myeloid leukemia (AML) [74]. MiR-155 was found elevated in HTLV-1-infected cells *in vitro* and *in vivo* [28,75], suggesting that this miRNA might play an important role in the biology and pathogenesis of HTLV-1. Babar [76] and colleagues used an inducible knock-in mouse model to show that miR-155 induction in the lymphoid tissue led to disseminated lymphoma. In contrast, reduction of miR-155 resulted in the decrease of tumor size. In humans, Calin [77] and colleagues identified a miRNA signature associated with progression and prognosis in chronic lymphocytic leukemia (CLL) and showed an association between miR-155 upregulation and poor prognosis. Several lymphoma-associated viruses, including the Epstein-Barr virus, Kaposi sarcoma-associated herpesvirus and Marek's disease virus, are characterized by overexpression of miR-155 [73,78], suggesting that HTLV-1 infection might be responsible for the induction of miR-155 in infected T cells. MiR-155 upregulation has been reported in HTLV-1 cell lines and ATLL patients [28,75]. Tomita [75] and colleagues demonstrated that transcription factors NF-κB and AP-1 induced miR-155 expression by binding the miRNA promoter in an HTLV-1 context (Figure 3). This binding resulted in an increased expression of the B-cell integration cluster (*BIC*) gene whose transcript is processed into miR-155 (Figure 3). Tomita and colleagues demonstrated that miR-155 overexpression enhanced the growth in HTLV-1-transformed cells. Consistently, treatment with anti-miR-155 reduced the proliferation of these cells and had no effect on HTLV-1-negative T cells. Wang [79] and colleagues demonstrated that interferon regulatory factor-4, IRF4, which is reported to be oncogenic [80], induces BIC/miR-155 expression in HTLV-1-transformed cells (Figure 3). In normal lymphocytes, IRF4 is involved in cellular proliferation and differentiation [80]. In mature human CD4+ T cells, IRF4 is essential for cytokine production and survival [81,82]. Several studies show that IRF4 is overexpressed in HTLV-1-transformed and primary ATLL/L cells and associated with poor prognosis [81–83], suggesting that IRF4 might be involved in HTLV-1 pathogenesis. Wang and colleagues show that depletion of IRF4 drastically reduced cell proliferation of HTLV-1-transformed cell lines, suggesting that the IRF4/miR-155 pathway might play a central role in the malignant proliferation of HTLV-1-infected cells [80]. In addition, miR-155 is reported to target Tumor Protein 53-Induced Nuclear Protein 1 (TP53INP1) in liver cancer stem cells [84], which promotes cell cycle arrest and apoptosis, suggesting a possible mechanism that could enhance cellular proliferation in an HTLV-1 context.

### 6.3. MiR-150 and MiR-223

MiR-150 and miR-223 were reported to be differentially regulated in HTLV-1-transformed cells and in ATLL samples. MiR-150 and miR-223 were found upregulated in acute ATLL patients and downregulated in HTLV-1-transformed cell lines, suggesting that different selective pressure *in vitro* and *in vivo* might regulate the expression of those miRNAs. MiR-150 can have either oncogenic or tumor suppressor activity in different human tumors. It is overexpressed in chronic lymphocytic leukemia (CLL) [85,86] and downregulated in chronic myeloid leukemia (CML) [87,88], acute lymphoblastic leukemia (ALL) [89] and mantle cell lymphoma (MCL) [90]. Additional studies show that miR-150 promotes the proliferation and migration in lung cancer by targeting SRC kinase signaling inhibitor 1 (SRCIN1) and SRC activity [91]. In contrast, miR-150 expression was reported to inhibit cell migration and invasion in breast cancer [92,93]. C-MYB, NOTCH3, CBL, EGR2, AKT2 and DKC1 are established targets of miR-150 [94–98]. MiR-223 was reported to be differentially regulated in human cancers; it is downregulated in hepatocellular carcinoma, B-cell chronic lymphocytic leukemia (B-CLL), acute myeloid leukemia (AML), gastric MALT lymphoma and recurrent ovarian cancer [99–103]. In contrast, miR-223 is upregulated in T-cell acute lymphocytic leukemia (T-ALL), EBV-positive diffuse large B-cell lymphoma, and metastatic gastric cancer [104–108]. FBXW7/Cdc4, RhoB, STMN1, E2F1, STAT3, C/EBPβ, FOXO1 and NFI-A are validated targets of miR-223 [106–111]. It has previously been shown that E2F1 represses the miR-223 promoter [110–112]. Interestingly, viral HBZ mRNA increases the expression and transcriptional activity of E2F1. HBZ expression is consistently increased in ATLL cells *in vivo* [11]. These observations can partially explain the differential regulation of miR-223 *in vitro* and *in vivo*. MiR-150 and miR-223 target the STAT1 3′UTR in an HTLV-1 context (Figure 3). STAT1 plays an essential role in immune modulatory functions, anti-viral responses, apoptosis and anti-proliferative responses [113]. In addition, several studies have shown that STAT1 can also act as a potent tumor promoter for leukemia development [114] and that many T-ALL leukemic cells are dependent upon the TYK2-STAT1-BCL2 pathway for continued survival [115]. Inverse correlation between STAT1 expression and miR-150 and miR-223 was identified in HTLV-1-transformed and IL-2-independent ATLL-derived cells [116]. IL-2-dependent ATLL cells display a high level of miR-150 expression, but low miR-223, suggesting that miR-150 might be regulated through the IL-2 signaling pathway. Absence of IL-2 signaling results in miR-150 downregulation in IL-2-dependent ATLL cells. In contrast, IL-2 stimulation in IL-2-independent ATLL-derived cells leads to miR-150 induction. This evidence suggests that miR-150 is regulated by the IL-2 signaling pathway. It was reported that ATLL tumor cells *in vivo* produce IL-2 or IL-15 and express IL-2 receptor alpha chain, CD25. These observations partially explain the higher levels of miR-150 in ATLL patients compared with HTLV-1 cell lines. Despite the miR-150 and miR-223 overexpression in freshly isolated ATLL samples, STAT1 was found to be induced in a majority of ATLL samples, suggesting that miR-150 and miR-223 cannot efficiently suppress STAT1 expression in ATLL patient cells. STAT1 has been reported to have tumor promoting activities. Inhibition of STAT1 expression, through miR-150, miR-223 or directly by shRNA targeting, reduced proliferation of HTLV-1-transformed and ATLL-derived cell lines. MiR-150 and miR-223, by decreasing STAT1 expression and dampening STAT1-dependent signaling in human T cells, regulated proliferation in an HTLV-1 context.

### 7. MiRNAs Induce Resistance to Apoptosis

### 7.1. MiR-31

Yamagishi [30] and colleagues identified miR-31 as one of the most profoundly repressed miRNAs in primary ATLL cells. MiR-31 is reported as a tumor suppressor and correlates inversely with metastasis in breast cancer [117]. MiR-31 *in vivo* targets several genes, such as Fzd3, ITGA5, MMP16, RDX, RhoA, WAVE3 and integrin α5 subunit, that contribute to cell migration and metastatic invasion [117,118]. The Polycomb protein complex has been reported to be a strong suppressor of miR-31 in breast cancer and adult T-cell leukemia [30,117]. Polycomb group proteins are overexpressed in ATLL cells [119] and have an important role in cellular development and regeneration by controlling histone methylation, especially at histone H3 Lys27 (H3K27), which induces chromatin compaction. The Polycomb family is associated with cancer phenotypes and malignancy in breast cancer, prostate cancer, bladder tumors, and other neoplasms [120,121]. MiR-31 negatively regulates NF-κB-inducing kinase (NIK) expression and activity in adult T-cell leukemia and other cancers [30]. NIK has an important role in tumor progression and the aggressive phenotypes of various cancers. It is well established that the NIK level directly regulates NF-κB activity in various cell types [122]. Constitutive activation of the nuclear factor NF-κB is observed in the ATLL cell lines and primary isolated tumor cells from ATLL patients [123]. NF-κB activation contributes to cell propagation and anti-apoptotic responses in ATLL [124]. An inverse correlation has been reported between the expression level of miR-31 and NIK in ATLL patients. Rescue of miR-31 represses NF-κB expression and leads to increased proliferation and apoptosis resistance. The inhibition of NF-kB promotes tumor cell death in HTLV-1-transformed cells and primary ATLL cells. The model proposed by Yamagishi and colleagues show that the Polycomb group regulates miR-31 expression and leads to NF-κB activation via NIK-miR-31 regulation and apoptosis resistance in HTLV-1 context (Figure 4). The downregulation of miR-31 might play an important role in ATLL pathogenesis.

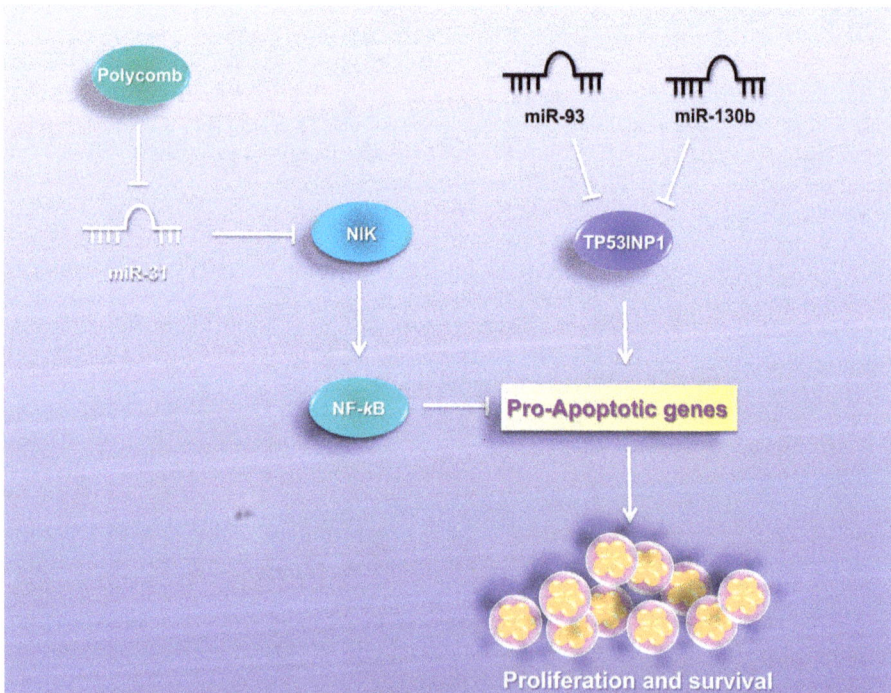

**Figure 4.** MiRNAs induce resistance to apoptosis. MiR-31 is one of the most profoundly repressed miRNAs in primary ATLL cells. The Polycomb protein complex is overexpressed in ATLL cells and suppresses miR-31 expression. MiR-31 negatively regulates NF-κB-inducing kinase (NIK) and leads to apoptosis resistance. MiR-130b and miR-93 are upregulated in HTLV-1 cell lines and ATLL patients and both target Tumor protein p53-inducible nuclear protein (TP53INP1). TP53INP1 is a tumor suppressor gene that has anti-proliferative and pro-apoptotic activities via both p53-dependent and p53-independent means. TP53INP1 has in its 3′ UTR two binding sites for miR-93 and two sites for miR-130b.

*Viruses* **2015**, *7*, 4047–4074

*7.2. MiR-130b and MiR-93*

Microarray analyses demonstrated that miR-130b and miR-93 were consistently upregulated in HTLV-1 cell lines and ATLL patients and both target Tumor protein p53-inducible nuclear protein, TP53INP1 [29]. MiR-130b was found to be deregulated in several human cancers. Overexpression of miR-130b has been reported in colorectal cancer, gastric cancer, bladder cancer, cutaneous malignant melanoma, and head and neck squamous cell carcinoma [125–128]. In contrast, miR-130b is downregulated in papillary thyroid carcinoma, ovarian cancer and endometrial cancer [129–132]. Identified targets of miR-130b are STAT3, PTEN and TGF-b1 [133–135]. MiR-93 belongs to the miR-106b-25 cluster, which also includes miR-106b and miR-25 [136]. The miR-106b-25 cluster is overexpressed in neuroblastoma, multiple myeloma, and lung, prostate and gastric tumors [136–138]. Reported targets of miR-93 are PTEN, VEGF, ITGB8, DAB2 and LATS2 [139–143]. TP53INP1 is a tumor suppressor gene that has anti-proliferative and pro-apoptotic activities via both p53-dependent [144] and p53-independent means [145]. TP53INP1 has in its 3' UTR two binding sites for miR-93 and two sites for miR-130b. Yeung [29] and colleagues have shown that transfection of antagomirs against miR-93 and miR-130b into an HTLV-1-transformed cell line increased the expression of TP53INP1 and decreased cellular viability by promoting apoptosis (Figure 4). These results show that TP53INP1 has anti-proliferative properties and can be regulated by miR-130b and miR-93. Transfection of miR-93 or miR-130b in HTLV-1-negative T-cell lines reduced TP53INP1 expression and increased cellular proliferation. It has been reported that loss of TP53INP1 correlates with the development of cancers [146,147] and its induction promotes G1 cell cycle arrest and apoptosis [144,145,148]. This evidence suggests that up-regulation of miR-130b and miR-93 reduces TP53INP1 levels in ATLL cells and promotes cellular proliferation. TP53INP1 is also able to reduce cell migration in pancreatic cancer cells [149] and this might be significant because it is well established that HTLV-1 infection promotes T-lymphocyte migration [150].

## 8. MiRNAs Promote Chromatin Remodeling

The Tax protein promotes HTLV-1 gene expression by its interaction with the long terminal repeat (LTR) or U3 region of the viral promoter [151,152]. To activate the transcription, Tax recruits the p300/CREB-binding protein (p300/CBP) and p300/CBP-associated factor (P/CAF), which bind two different regions of Tax, resulting in histone acetylation and chromatin remodeling (Figure 5) [153–158]. Rahman [157] and colleagues identified the chromatin remodeling factors, p300 and p/CAF, as a target of miR-149 and miR-873. MiR-149 has been reported to have a role as an oncogene and tumor suppressor in different human cancers [158,159]. Downregulation of miR-149 has been described in prostatic cancer, astrocytomas and renal carcinoma [160–162]. In contrast, miR-873 was found to be suppressed in colorectal cancer, glioblastoma and breast cancer [163–165]. Recent evidence has established the role of miR-873 in cell proliferation, tumor growth and tamoxifen resistance in breast cancer [165]. MiR-149 and miR-873 were found to be profoundly downregulated in HTLV-1-transformed cell lines, MT-2, compared to an uninfected control, Jurkat [157]. To verify that miR-149 and miR-873 could target p/CAF and p300, the authors over-expressed these miRNAs in HTLV-1-transformed cells and observed a significant reduction in the expression of chromatin-remodeling enzymes. In addition, the cell culture supernatant was analyzed for viral protein p19 before and after transfection. The results show a decrease in the levels of viral progeny production in cells transfected with miR-149 and miR-873, suggesting that these miRNAs, by targeting chromatin remodeling factors p/CAF and p300, might play a role in HTLV-1 infection and pathogenesis (Figure 5).

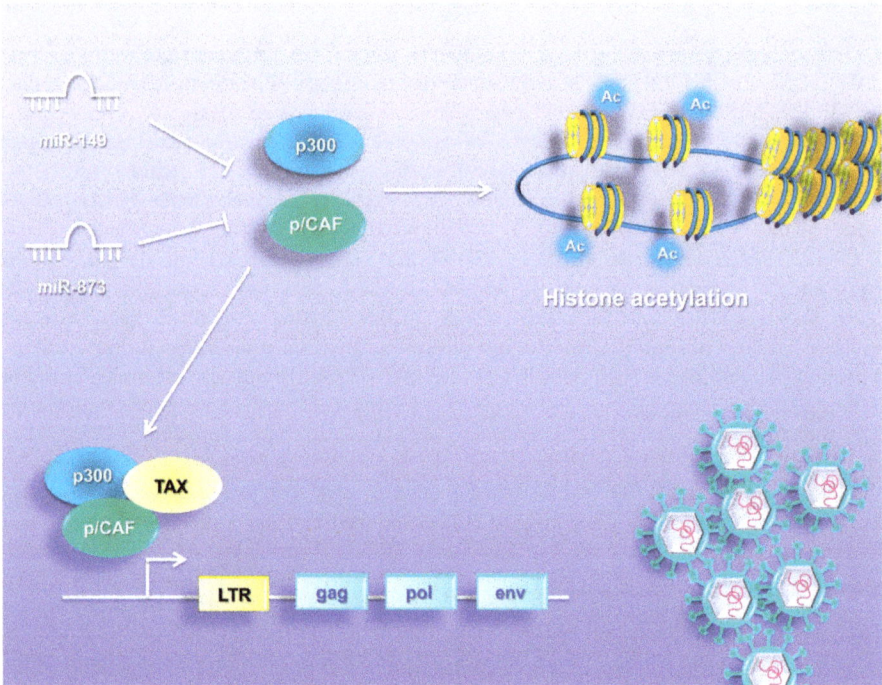

**Figure 5.** MiR-149 and miR-873 promote chromatin remodeling. The Tax protein promotes HTLV-1 gene expression by its interaction with the long terminal repeat (LTR) or U3 region of the viral promoter. To activate the transcription, Tax recruits the p300/CREB-binding protein (p300/CBP) and p300/CBP-associated factor (P/CAF), which bind two different regions of Tax, resulting in histone acetylation and chromatin remodeling. MiR-149 and miR-873 are downregulated in HTLV-1-transformed cell lines and target the chromatin remodeling factors p300 and p/CAF.

## 9. MiRNAs Induce Genetic Instability

MiRNA expression analysis in CD4+ lymphocytes, derived from HAM/TSP patients, has identified a high expression level of miR-17 and miR-21 [166]. Spry 1, Spry 2, PTEN, TPM1 and Pdcd4 have been reported to be miR-21 targets, suggesting its central role in cell proliferation, apoptosis, and invasion [167–171]. MiR-17, instead, is the main effector of the miR-17-92 cluster component, which has been identified as a member of the miRNA signature in solid tumors [172]. MiR-17 regulates E2F1 and c-Myc, p21, PTEN and BIM expression [173–176], suggesting its potential functions in cell migration, invasion and proliferation. Vernin [166] and colleagues identified OBFC2A as a potential target of miR-17 and miR-21 in an HTLV-1 context. OBFC2A encodes for hSSB2, which is involved in the ATM signaling pathway, the activation of the cell cycle checkpoint and promotes DNA repair. The down-regulation of OBFC2A and a positive correlation between miR-17, miR-21 and HBZ expression has been reported in HTLV-1-infected cells [166]. Vernin and colleagues suggested that HBZ inactivates OBFC2A via miR-17 and miR-21, promoting genetic instability and cell proliferation (Figure 6).

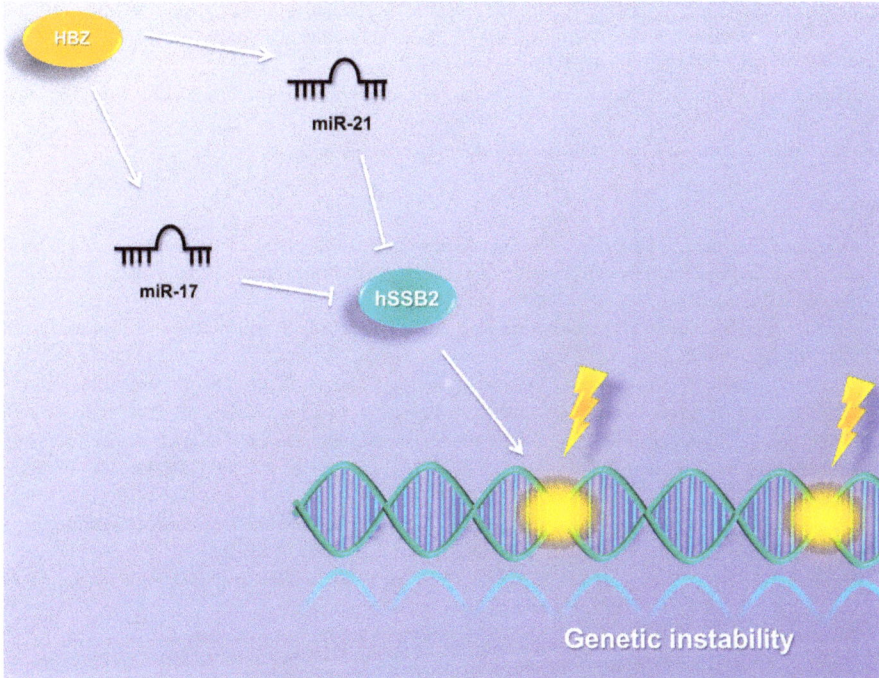

**Figure 6.** MiRNAs induce genetic instability. MiR-17 and miR-21 are upregulated in an HTLV-1 context. HBZ inactivates OBFC2A via miR-17 and miR-21, promoting genetic instability and cell proliferation. OBFC2A encodes for hSSB2, which is involved in the ATM signaling pathway, the activation of the cell cycle checkpoint and promotes DNA repair.

The authors have shown that ectopic expression of HBZ does not decrease cellular growth in DNA-damaged cells. HBZ-expressing cells continued to proliferate when treated with a DNA-damaging agent, neocarzinostatin. This phenotype can be reversed by ectopic expression of OBFC2A, which leads to a decrease of proliferation rates and restores the DNA damage response. This evidence suggested a potential role of miR-17 and miR-21 in genetic instability and cell proliferation in HTLV-1-infected cells.

## 10. Conclusions and Prospective

The role of miRNAs in HTLV-1 infection and ATLL pathogenesis is beginning to emerge. Available evidence shows a complex interplay between cellular miRNA machinery and virus infection. HTLV-1 inhibits proteins involved in biogenesis and maturation of cellular miRNAs, resulting in a perturbation of the expression profile of host miRNAs. In this review, we focused on miRNAs, which are involved in virus production, establishment of latency, tumor cell transformation and proliferation. A potential role of MiRNA modulation could represent a therapeutic approach for ATLL patients. The combination delivery of miRNAs with chemotherapy drugs might provide a promising strategy to overcome chemo-resistance. Different studies have shown that co-delivery of miRNA and chemotherapeutic agents are effective to inhibit tumor growth by targeting genes, which are involved in tumor cell proliferation and/or survival [177–179]. In addition, in combination with antitumor drugs, miRNAs might have an important role by targeting genes involved in drug resistance, thus overcoming the chemo-resistance in ATLL patients.

**Acknowledgments:** Author would like to thank Brandi Miller for editorial assistance. This work was supported by grant 106258 to Christophe Nicot. The content is solely the responsibility of the authors and does not necessarily represent the official views of the National Institutes of Health.

**Author Contributions:** Ramona Moles created the figures and wrote the manuscript. Christophe Nicot wrote the manuscript.

**Conflicts of Interest:** The authors declare no competing financial interests.

## References

1. Yoshida, M.; Miyoshi, I.; Hinuma, Y. Isolation and characterization of retrovirus from cell lines of human adult T-cell leukemia and its implication in the disease. *Proc. Natl. Acad. Sci. USA* **1982**, *79*, 2031–2035. [CrossRef] [PubMed]

2. Nicot, C. Current views in HTLV-I-associated adult T-cell leukemia/lymphoma. *Am. J. Hematol.* **2005**, *78*, 232–239. [CrossRef] [PubMed]

3. Gessain, A.; Barin, F.; Vernant, J.C.; Gout, O.; Maurs, L.; Calender, A.; de The, G. Antibodies to human T-lymphotropic virus type-I in patients with tropical spastic paraparesis. *Lancet* **1985**, *2*, 407–410. [CrossRef]

4. Shimoyama, M. Diagnostic criteria and classification of clinical subtypes of adult T-cell leukaemia-lymphoma. A report from the lymphoma study group (1984–1987). *Br. J. Haematol.* **1991**, *79*, 428–437. [CrossRef] [PubMed]

5. Bazarbachi, A.; Ghez, D.; Lepelletier, Y.; Nasr, R.; de The, H.; el-Sabban, M.E.; Hermine, O. New therapeutic approaches for adult T-cell leukaemia. *Lancet Oncol.* **2004**, *5*, 664–672. [CrossRef]

6. Franchini, G.; Nicot, C.; Johnson, J.M. Seizing of T cells by human T-cell leukemia/lymphoma virus type 1. *Adv. Cancer Res.* **2003**, *89*, 69–132. [PubMed]

7. Matsuoka, M.; Jeang, K.T. Human T-cell leukemia virus type 1 (HTLV-1) and leukemic transformation: Viral infectivity, tax, HBZ and therapy. *Oncogene* **2011**, *30*, 1379–1389. [CrossRef] [PubMed]

8. Kashanchi, F.; Brady, J.N. Transcriptional and post-transcriptional gene regulation of HTLV-1. *Oncogene* **2005**, *24*, 5938–5951. [CrossRef] [PubMed]

9. Giam, C.Z.; Jeang, K.T. HTLV-1 tax and adult T-cell leukemia. *Front. Biosci.* **2007**, *12*, 1496–1507. [CrossRef] [PubMed]

10. Boxus, M.; Twizere, J.C.; Legros, S.; Dewulf, J.F.; Kettmann, R.; Willems, L. The HTLV-1 tax interactome. *Retrovirology* **2008**, *5*, e76. [CrossRef] [PubMed]

11. Satou, Y.; Yasunaga, J.; Yoshida, M.; Matsuoka, M. HTLV-I basic leucine zipper factor gene mRNA supports proliferation of adult T cell leukemia cells. *Proc. Natl. Acad. Sci. USA* **2006**, *103*, 720–725. [CrossRef] [PubMed]

12. Zhao, T.; Matsuoka, M. HBZ and its roles in HTLV-1 oncogenesis. *Front. Microbial.* **2012**, *3*, e247. [CrossRef] [PubMed]

13. Nakano, K.; Watanabe, T. HTLV-1 rex: The courier of viral messages making use of the host vehicle. *Front. Microbiol.* **2012**, *3*, e330. [CrossRef] [PubMed]

14. Ivey, K.N.; Srivastava, D. Micrornas as regulators of differentiation and cell fate decisions. *Cell Stem Cell* **2010**, *7*, 36–41. [CrossRef] [PubMed]

15. O'Connell, R.M.; Rao, D.S.; Chaudhuri, A.A.; Baltimore, D. Physiological and pathological roles for micrornas in the immune system. *Nat. Rev. Immunol.* **2010**, *10*, 111–122. [CrossRef] [PubMed]

16. Bouyssou, J.M.; Manier, S.; Huynh, D.; Issa, S.; Roccaro, A.M.; Ghobrial, I.M. Regulation of micrornas in cancer metastasis. *Biochim. Biophys. Acta* **2014**, *1845*, 255–265. [CrossRef] [PubMed]

17. Lee, R.C.; Feinbaum, R.L.; Ambros, V. The c. Elegans heterochronic gene lin-4 encodes small RNAs with antisense complementarity to lin-14. *Cell* **1993**, *75*, 843–854. [CrossRef]

18. Wightman, B.; Ha, I.; Ruvkun, G. Posttranscriptional regulation of the heterochronic gene lin-14 by lin-4 mediates temporal pattern formation in C. Elegans.. *Cell* **1993**, *75*, 855–862. [CrossRef]

19. Ha, M.; Kim, V.N. Regulation of microRNA biogenesis. *Nat. Rev. Mol. Cell Biol.* **2014**, *15*, 509–524. [CrossRef] [PubMed]

20. Lee, Y.; Jeon, K.; Lee, J.T.; Kim, S.; Kim, V.N. Microrna maturation: Stepwise processing and subcellular localization. *EMBO J.* **2002**, *21*, 4663–4670. [CrossRef] [PubMed]

21. Holley, C.L.; Topkara, V.K. An introduction to small non-coding RNAs: MiRNA and snoRNA. *Cardiovasc. Drugs Ther.* **2011**, *25*, 151–159. [CrossRef] [PubMed]

22. Havens, M.A.; Reich, A.A.; Duelli, D.M.; Hastings, M.L. Biogenesis of mammalian micrornas by a non-canonical processing pathway. *Nucleic Acids Res.* **2012**, *40*, 4626–4640. [CrossRef] [PubMed]

23. Xia, J.; Zhang, W. Noncanonical micrornas and endogenous siRNAs in lytic infection of murine gammaherpesvirus. *PLoS ONE* **2012**, *7*, e47863. [CrossRef] [PubMed]

24. Valencia-Sanchez, M.A.; Liu, J.; Hannon, G.J.; Parker, R. Control of translation and mRNA degradation by mirnas and sirnas. *Genes Dev.* **2006**, *20*, 515–524. [CrossRef] [PubMed]

25. Brennecke, J.; Hipfner, D.R.; Stark, A.; Russell, R.B.; Cohen, S.M. Bantam encodes a developmentally regulated microRNA that controls cell proliferation and regulates the proapoptotic gene hid in drosophila. *Cell* **2003**, *113*, 25–36. [CrossRef]

26. Cimmino, A.; Calin, G.A.; Fabbri, M.; Iorio, M.V.; Ferracin, M.; Shimizu, M.; Wojcik, S.E.; Aqeilan, R.I.; Zupo, S.; Dono, M.; *et al.* Mir-15 and mir-16 induce apoptosis by targeting bcl2. *Proc. Natl. Acad. Sci. USA* **2005**, *102*, 13944–13949. [CrossRef] [PubMed]

27. Pichler, K.; Schneider, G.; Grassmann, R. Microrna mir-146a and further oncogenesis-related cellular micrornas are dysregulated in HTLV-1-transformed T lymphocytes. *Retrovirology* **2008**, *5*, e100. [CrossRef] [PubMed]

28. Bellon, M.; Lepelletier, Y.; Hermine, O.; Nicot, C. Deregulation of microrna involved in hematopoiesis and the immune response in HTLV-I adult T-cell leukemia. *Blood* **2009**, *113*, 4914–4917. [CrossRef] [PubMed]

29. Yeung, M.L.; Yasunaga, J.; Bennasser, Y.; Dusetti, N.; Harris, D.; Ahmad, N.; Matsuoka, M.; Jeang, K.T. Roles for micrornas, miR-93 and miR-130b, and tumor protein 53-induced nuclear protein 1 tumor suppressor in cell growth dysregulation by human T-cell lymphotrophic virus 1. *Cancer Res.* **2008**, *68*, 8976–8985. [CrossRef] [PubMed]

30. Yamagishi, M.; Nakano, K.; Miyake, A.; Yamochi, T.; Kagami, Y.; Tsutsumi, A.; Matsuda, Y.; Sato-Otsubo, A.; Muto, S.; Utsunomiya, A.; *et al.* Polycomb-mediated loss of miR-31 activates nik-dependent NF- κb pathway in adult T cell leukemia and other cancers. *Cancer Cell* **2012**, *21*, 121–135. [CrossRef] [PubMed]

31. Bennasser, Y.; Yeung, M.L.; Jeang, K.T. HIV-1 tar RNA subverts RNA interference in transfected cells through sequestration of tar RNA-binding protein, TRBP. *J. Biol. Chem.* **2006**, *281*, 27674–27678. [CrossRef] [PubMed]

32. Haasnoot, J.; de Vries, W.; Geutjes, E.J.; Prins, M.; de Haan, P.; Berkhout, B. The ebola virus vp35 protein is a suppressor of RNA silencing. *PLoS Pathog.* **2007**, *3*, e86. [CrossRef] [PubMed]

33. Godshalk, S.E.; Bhaduri-McIntosh, S.; Slack, F.J. Epstein-barr virus-mediated dysregulation of human microRNA expression. *Cell Cycle* **2008**, *7*, 3595–3600. [CrossRef] [PubMed]

34. De Vries, W.; Haasnoot, J.; Fouchier, R.; de Haan, P.; Berkhout, B. Differential RNA silencing suppression activity of NS 1 proteins from different influenza a virus strains. *J. Gen. Virol.* **2009**, *90*, 1916–1922. [CrossRef] [PubMed]

35. Xie, K.L.; Zhang, Y.G.; Liu, J.; Zeng, Y.; Wu, H. Micrornas associated with HBV infection and HBV-related HCC. *Theranostics* **2014**, *4*, 1176–1192. [CrossRef] [PubMed]

36. Wang, Y.; Kato, N.; Jazag, A.; Dharel, N.; Otsuka, M.; Taniguchi, H.; Kawabe, T.; Omata, M. Hepatitis C virus core protein is a potent inhibitor of RNA silencing-based antiviral response. *Gastroenterology* **2006**, *130*, 883–892. [CrossRef] [PubMed]

37. Lu, S.; Cullen, B.R. Adenovirus va1 noncoding RNA can inhibit small interfering RNA and microRNA biogenesis. *J. Virol.* **2004**, *78*, 12868–12876. [CrossRef] [PubMed]

38. Van Duyne, R.; Guendel, I.; Klase, Z.; Narayanan, A.; Coley, W.; Jaworski, E.; Roman, J.; Popratiloff, A.; Mahieux, R.; Kehn-Hall, K.; *et al.* Localization and sub-cellular shuttling of HTLV-1 tax with the mirna machinery. *PLoS ONE* **2012**, *7*, e40662. [CrossRef] [PubMed]

39. Abe, M.; Suzuki, H.; Nishitsuji, H.; Shida, H.; Takaku, H. Interaction of human T-cell lymphotropic virus type I rex protein with dicer suppresses RNAi silencing. *FEBS Lett.* **2010**, *584*, 4313–4318. [CrossRef] [PubMed]

40. Gottwein, E. Roles of micrornas in the life cycles of mammalian viruses. *Curr. Top. Microbiol. Immunol.* **2013**, *371*, 201–227. [PubMed]

41. Sedger, L.M. MicroRNA control of interferons and interferon induced anti-viral activity. *Mol. Immunol.* **2013**, *56*, 781–793. [CrossRef] [PubMed]

42. Lei, X.; Bai, Z.; Ye, F.; Huang, Y.; Gao, S.J. MicroRNAs control herpesviral dormancy. *Cell Cycle* **2010**, *9*, 1225–1226. [CrossRef] [PubMed]

43. Swaminathan, S. Noncoding RNAs produced by oncogenic human herpesviruses. *J. Cell. Physiol.* **2008**, *216*, 321–326. [CrossRef] [PubMed]

44. Sun, L.; Li, Q. The miRNAs of herpes simplex virus (HSV). *Virol. Sin.* **2012**, *27*, 333–338. [CrossRef] [PubMed]

45. Kincaid, R.P.; Burke, J.M.; Sullivan, C.S. RNA virus microrna that mimics a B-cell oncomir. *Proc. Natl. Acad. Sci. USA* **2012**, *109*, 3077–3082. [CrossRef] [PubMed]

46. Rosewick, N.; Momont, M.; Durkin, K.; Takeda, H.; Caiment, F.; Cleuter, Y.; Vernin, C.; Mortreux, F.; Wattel, E.; Burny, A.; *et al.* Deep sequencing reveals abundant noncanonical retroviral micrornas in B-cell leukemia/lymphoma. *Proc. Natl. Acad. Sci. USA* **2013**, *110*, 2306–2311. [CrossRef] [PubMed]

47. Huang, J.; Wang, F.; Argyris, E.; Chen, K.; Liang, Z.; Tian, H.; Huang, W.; Squires, K.; Verlinghieri, G.; Zhang, H. Cellular micrornas contribute to HIV-1 latency in resting primary CD4+ T lymphocytes. *Nat. Med.* **2007**, *13*, 1241–1247. [CrossRef] [PubMed]

48. Bai, X.T.; Nicot, C. MiR-28-3p is a cellular restriction factor that inhibits human T cell leukemia virus, type 1 (HTLV-1) replication and virus infection. *J. Biol. Chem.* **2015**, *290*, 5381–5390. [CrossRef] [PubMed]

49. He, H.; Jazdzewski, K.; Li, W.; Liyanarachchi, S.; Nagy, R.; Volinia, S.; Calin, G.A.; Liu, C.G.; Franssila, K.; Suster, S.; *et al.* The role of microrna genes in papillary thyroid carcinoma. *Proc. Natl. Acad. Sci. USA* **2005**, *102*, 19075–19080. [CrossRef] [PubMed]

50. Pacifico, F.; Crescenzi, E.; Mellone, S.; Iannetti, A.; Porrino, N.; Liguoro, D.; Moscato, F.; Grieco, M.; Formisano, S.; Leonardi, A. Nuclear factor-κb contributes to anaplastic thyroid carcinomas through up-regulation of mir-146a. *J. Clin. Endocrinol. Metab.* **2010**, *95*, 1421–1430. [CrossRef] [PubMed]

51. Bhaumik, D.; Scott, G.K.; Schokrpur, S.; Patil, C.K.; Campisi, J.; Benz, C.C. Expression of microRNA-146 suppresses NF-κb activity with reduction of metastatic potential in breast cancer cells. *Oncogene* **2008**, *27*, 5643–5647. [CrossRef] [PubMed]

52. Lavon, I.; Zrihan, D.; Granit, A.; Einstein, O.; Fainstein, N.; Cohen, M.A.; Cohen, M.A.; Zelikovitch, B.; Shoshan, Y.; Spektor, S.; *et al.* Gliomas display a microrna expression profile reminiscent of neural precursor cells. *Neuro Oncol.* **2010**, *12*, 422–433. [PubMed]

53. Wang, X.; Tang, S.; Le, S.Y.; Lu, R.; Rader, J.S.; Meyers, C.; Zheng, Z.M. Aberrant expression of oncogenic and tumor-suppressive micrornas in cervical cancer is required for cancer cell growth. *PLoS ONE* **2008**, *3*, e2557. [CrossRef] [PubMed]

54. Li, L.; Chen, X.P.; Li, Y.J. Microrna-146a and human disease. *Scand. J. Immunol.* **2010**, *71*, 227–231. [CrossRef] [PubMed]

55. Kogo, R.; Mimori, K.; Tanaka, F.; Komune, S.; Mori, M. Clinical significance of miR-146a in gastric cancer cases. *Clin. Cancer Res.* **2011**, *17*, 4277–4284. [CrossRef] [PubMed]

56. Lin, S.L.; Chiang, A.; Chang, D.; Ying, S.Y. Loss of miR-146a function in hormone-refractory prostate cancer. *RNA* **2008**, *14*, 417–424. [CrossRef] [PubMed]

57. Garzon, R.; Volinia, S.; Liu, C.G.; Fernandez-Cymering, C.; Palumbo, T.; Pichiorri, F.; Fabbri, M.; Coombes, K.; Alder, H.; Nakamura, T.; *et al.* Microrna signatures associated with cytogenetics and prognosis in acute myeloid leukemia. *Blood* **2008**, *111*, 3183–3189. [CrossRef] [PubMed]

58. Starczynowski, D.T.; Kuchenbauer, F.; Argiropoulos, B.; Sung, S.; Morin, R.; Muranyi, A.; Hirst, M.; Hogge, D.; Marra, M.; Wells, R.A.; *et al.* Identification of miR-145 and miR-146a as mediators of the 5q- syndrome phenotype. *Nat. Med.* **2010**, *16*, 49–58. [CrossRef] [PubMed]

59. Visone, R.; Rassenti, L.Z.; Veronese, A.; Taccioli, C.; Costinean, S.; Aguda, B.D.; Volinia, S.; Ferracin, M.; Palatini, J.; Balatti, V.; *et al.* Karyotype-specific microRNA signature in chronic lymphocytic leukemia. *Blood* **2009**, *114*, 3872–3879. [CrossRef] [PubMed]

60. Tomita, M.; Tanaka, Y.; Mori, N. Microrna mir-146a is induced by HTLV-1 tax and increases the growth of HTLV-1-infected T-cells. *Int. J. Cancer* **2012**, *130*, 2300–2309. [CrossRef] [PubMed]

61. Sandhu, R.; Rein, J.; D'Arcy, M.; Herschkowitz, J.I.; Hoadley, K.A.; Troester, M.A. Overexpression of miR-146a in basal-like breast cancer cells confers enhanced tumorigenic potential in association with altered p53 status. *Carcinogenesis* **2014**, *35*, 2567–2575. [CrossRef] [PubMed]

62. Xiao, B.; Zhu, E.D.; Li, N.; Lu, D.S.; Li, W.; Li, B.S.; Zhao, Y.L.; Mao, X.H.; Guo, G.; Yu, P.W.; *et al.* Increased miR-146a in gastric cancer directly targets SMAD4 and is involved in modulating cell proliferation and apoptosis. *Oncol. Rep.* **2012**, *27*, 559–566. [PubMed]

63.  Hsieh, J.Y.; Huang, T.S.; Cheng, S.M.; Lin, W.S.; Tsai, T.N.; Lee, O.K.; Wang, H.W. MiR-146a-5p circuitry uncouples cell proliferation and migration, but not differentiation, in human mesenchymal stem cells. *Nucleic Acids Res.* **2013**, *41*, 9753–9763. [CrossRef] [PubMed]
64.  Wang, Y.; Li, Z.; He, C.; Wang, D.; Yuan, X.; Chen, J.; Jin, J. Micrornas expression signatures are associated with lineage and survival in acute leukemias. *Blood Cells Mol. Dis.* **2010**, *44*, 191–197. [CrossRef] [PubMed]
65.  Vasilatou, D.; Papageorgiou, S.; Pappa, V.; Papageorgiou, E.; Dervenoulas, J. The role of micro RNAs in normal and malignant hematopoiesis. *Eur. J. Haematol.* **2010**, *84*, 1–16. [CrossRef] [PubMed]
66.  Baltimore, D.; Boldin, M.P.; O'Connell, R.M.; Rao, D.S.; Taganov, K.D. Micrornas: New regulators of immune cell development and function. *Nat. Immunol.* **2008**, *9*, 839–845. [CrossRef] [PubMed]
67.  Kluiver, J.; Poppema, S.; de Jong, D.; Blokzijl, T.; Harms, G.; Jacobs, S.; Kroesen, B.J.; van den Berg, A. Bic and mir-155 are highly expressed in hodgkin, primary mediastinal and diffuse large B cell lymphomas. *J. Pathol.* **2005**, *207*, 243–249. [CrossRef] [PubMed]
68.  Tili, E.; Croce, C.M.; Michaille, J.J. Mir-155: On the crosstalk between inflammation and cancer. *Int. Rev. Immunol.* **2009**, *28*, 264–284. [CrossRef] [PubMed]
69.  Costinean, S.; Zanesi, N.; Pekarsky, Y.; Tili, E.; Volinia, S.; Heerema, N.; Croce, C.M. Pre-B cell proliferation and lymphoblastic leukemia/high-grade lymphoma in e(mu)-miR-155 transgenic mice. *Proc. Natl. Acad. Sci. USA* **2006**, *103*, 7024–7029. [CrossRef] [PubMed]
70.  Chen, Z.; Ma, T.; Huang, C.; Hu, T.; Li, J. The pivotal role of microRNA-155 in the control of cancer. *J. Cell. Physiol.* **2014**, *229*, 545–550. [CrossRef] [PubMed]
71.  Habbe, N.; Koorstra, J.B.; Mendell, J.T.; Offerhaus, G.J.; Ryu, J.K.; Feldmann, G.; Mullendore, M.E.; Goggins, M.G.; Hong, S.M.; Maitra, A. Microrna miR-155 is a biomarker of early pancreatic neoplasia. *Cancer Biol. Ther.* **2009**, *8*, 340–346. [CrossRef] [PubMed]
72.  Zang, Y.S.; Zhong, Y.F.; Fang, Z.; Li, B.; An, J. MiR-155 inhibits the sensitivity of lung cancer cells to cisplatin via negative regulation of APAF-1 expression. *Cancer Gene Ther.* **2012**, *19*, 773–778. [CrossRef] [PubMed]
73.  Saito, Y.; Suzuki, H.; Tsugawa, H.; Imaeda, H.; Matsuzaki, J.; Hirata, K.; Hosoe, N.; Nakamura, M.; Mukai, M.; Saito, H.; *et al.* Overexpression of mir-142–5p and mir-155 in gastric mucosa-associated lymphoid tissue (malt) lymphoma resistant to helicobacter pylori eradication. *PLoS ONE* **2012**, *7*, e47396. [CrossRef] [PubMed]
74.  Garzon, R.; Garofalo, M.; Martelli, M.P.; Briesewitz, R.; Wang, L.; Fernandez-Cymering, C.; Volinia, S.; Liu, C.G.; Schnittger, S.; Haferlach, T.; *et al.* Distinctive microrna signature of acute myeloid leukemia bearing cytoplasmic mutated nucleophosmin. *Proc. Natl. Acad. Sci. USA* **2008**, *105*, 3945–3950. [CrossRef] [PubMed]
75.  Tomita, M. Important roles of cellular microrna mir-155 in leukemogenesis by human T-cell leukemia virus type 1 infection. *ISRN Microbiol.* **2012**, *2012*, e978607. [CrossRef] [PubMed]
76.  Babar, I.A.; Cheng, C.J.; Booth, C.J.; Liang, X.; Weidhaas, J.B.; Saltzman, W.M.; Slack, F.J. Nanoparticle-based therapy in an *in vivo* micro RNA-155 (miR-155)-dependent mouse model of lymphoma. *Proc. Natl. Acad. Sci. USA* **2012**, *109*, E1695–E1704. [CrossRef] [PubMed]
77.  Calin, G.A.; Ferracin, M.; Cimmino, A.; di Leva, G.; Shimizu, M.; Wojcik, S.E.; Iorio, M.V.; Visone, R.; Sever, N.I.; Fabbri, M.; *et al.* A micro RNA signature associated with prognosis and progression in chronic lymphocytic leukemia. *N. Engl. J. Med.* **2005**, *353*, 1793–1801. [CrossRef] [PubMed]
78.  Lawrie, C.H. Micrornas and lymphomagenesis: A functional review. *Br. J. Haematol.* **2013**, *160*, 571–581. [CrossRef] [PubMed]
79.  Wang, L.; Toomey, N.L.; Diaz, L.A.; Walker, G.; Ramos, J.C.; Barber, G.N.; Ning, S. Oncogenic IRFS provide a survival advantage for epstein-barr virus- or human T-cell leukemia virus type 1-transformed cells through induction of Bic expression. *J. Virol.* **2011**, *85*, 8328–8337. [CrossRef] [PubMed]
80.  Shaffer, A.L.; Emre, N.C.; Romesser, P.B.; Staudt, L.M. Irf4: Immunity. Malignancy! Therapy? *Clin. Cancer Res.* **2009**, *15*, 2954–2961. [CrossRef] [PubMed]
81.  Ramos, J.C.; Ruiz, P.; Ratner, L.; Reis, I.M.; Brites, C.; Pedroso, C.; Byrne, G.E.; Toomey, N.L.; Andela, V.; Harhaj, E.W.; *et al.* IRF -4 and c-Rel expression in antiviral-resistant adult T-cell leukemia/lymphoma. *Blood* **2007**, *109*, 3060–3068. [CrossRef] [PubMed]
82.  Sharma, S.; Mamane, Y.; Grandvaux, N.; Bartlett, J.; Petropoulos, L.; Lin, R.; Hiscott, J. Activation and regulation of interferon regulatory factor 4 in HTLV type 1-infected T lymphocytes. *AIDS Res. Hum. Retroviruses* **2000**, *16*, 1613–1622. [CrossRef] [PubMed]

83. Suzuki, S.; Zhou, Y.; Refaat, A.; Takasaki, I.; Koizumi, K.; Yamaoka, S.; Tabuchi, Y.; Saiki, I.; Sakurai, H. Human T cell lymphotropic virus 1 manipulates interferon regulatory signals by controlling the TAK1-IRF3 and IRF4 pathways. *J. Biol. Chem.* **2010**, *285*, 4441–4446. [CrossRef] [PubMed]

84. Liu, F.; Kong, X.; Lv, L.; Gao, J. MiR-155 targets tp53inp1 to regulate liver cancer stem cell acquisition and self-renewal. *FEBS Lett.* **2015**, *589*, 500–506. [CrossRef] [PubMed]

85. Papakonstantinou, N.; Ntoufa, S.; Chartomatsidou, E.; Papadopoulos, G.; Hatzigeorgiou, A.; Anagnostopoulos, A.; Chlichlia, K.; Ghia, P.; Muzio, M.; Belessi, C.; *et al.* Differential microrna profiles and their functional implications in different immunogenetic subsets of chronic lymphocytic leukemia. *Mol. Med.* **2013**, *19*, 115–123. [CrossRef] [PubMed]

86. Mraz, M.; Chen, L.; Rassenti, L.Z.; Ghia, E.M.; Li, H.; Jepsen, K.; Smith, E.N.; Messer, K.; Frazer, K.A.; Kipps, T.J. MiR-150 influences B-cell receptor signaling in chronic lymphocytic leukemia by regulating expression of GAB1 and FOXP1. *Blood* **2014**, *124*, 84–95. [CrossRef] [PubMed]

87. Morris, V.A.; Zhang, A.; Yang, T.; Stirewalt, D.L.; Ramamurthy, R.; Meshinchi, S.; Oehler, V.G. MicroRNA-150 expression induces myeloid differentiation of human acute leukemia cells and normal hematopoietic progenitors. *PLoS ONE* **2013**, *8*, e75815. [CrossRef] [PubMed]

88. Machova Polakova, K.; Lopotova, T.; Klamova, H.; Burda, P.; Trneny, M.; Stopka, T.; Moravcova, J. Expression patterns of microRNAs associated with cml phases and their disease related targets. *Mol. Cancer* **2011**, *10*, e41. [CrossRef] [PubMed]

89. Xu, L.; Liang, Y.N.; Luo, X.Q.; Liu, X.D.; Guo, H.X. Association of mirnas expression profiles with prognosis and relapse in childhood acute lymphoblastic leukemia. *Zhonghua Xue Ye Xue Za Zhi* **2011**, *32*, 178–181. [PubMed]

90. Zhao, J.J.; Lin, J.; Lwin, T.; Yang, H.; Guo, J.; Kong, W.; Dessureault, S.; Moscinski, L.C.; Rezania, D.; Dalton, W.S.; *et al.* Microrna expression profile and identification of miR-29 as a prognostic marker and pathogenetic factor by targeting CDK6 in mantle cell lymphoma. *Blood* **2010**, *115*, 2630–2639. [CrossRef] [PubMed]

91. Cao, M.; Hou, D.; Liang, H.; Gong, F.; Wang, Y.; Yan, X.; Jiang, X.; Wang, C.; Zhang, J.; Zen, K.; *et al.* MiR-150 promotes the proliferation and migration of lung cancer cells by targeting SRC kinase signalling inhibitor 1. *Eur. J. Cancer* **2014**, *50*, 1013–1024. [CrossRef] [PubMed]

92. Avery-Kiejda, K.A.; Braye, S.G.; Mathe, A.; Forbes, J.F.; Scott, R.J. Decreased expression of key tumour suppressor micrornas is associated with lymph node metastases in triple negative breast cancer. *BMC Cancer* **2014**, *14*, e51. [CrossRef] [PubMed]

93. Huang, S.; Chen, Y.; Wu, W.; Ouyang, N.; Chen, J.; Li, H.; Liu, X.; Su, F.; Lin, L.; Yao, Y. MiR-150 promotes human breast cancer growth and malignant behavior by targeting the pro-apoptotic purinergic p2x7 receptor. *PLoS ONE* **2013**, *8*, e80707. [CrossRef] [PubMed]

94. Xiao, C.; Calado, D.P.; Galler, G.; Thai, T.H.; Patterson, H.C.; Wang, J.; Rajewsky, N.; Bender, T.P.; Rajewsky, K. MiR-150 controls B cell differentiation by targeting the transcription factor c-myb. *Cell* **2007**, *131*, 146–159. [CrossRef] [PubMed]

95. Ghisi, M.; Corradin, A.; Basso, K.; Frasson, C.; Serafin, V.; Mukherjee, S.; Mussolin, L.; Ruggero, K.; Bonanno, L.; Guffanti, A.; *et al.* Modulation of microRNA expression in human T-cell development: Targeting of notch3 by miR-150. *Blood* **2011**, *117*, 7053–7062. [CrossRef] [PubMed]

96. Bousquet, M.; Zhuang, G.; Meng, C.; Ying, W.; Cheruku, P.S.; Shie, A.T.; Wang, S.; Ge, G.; Wong, P.; Wang, G.; *et al.* MiR-150 blocks MLL-AF9-associated leukemia through oncogene repression. *Mol. Cancer Res.* **2013**, *11*, 912–922. [CrossRef] [PubMed]

97. Wu, Q.; Jin, H.; Yang, Z.; Luo, G.; Lu, Y.; Li, K.; Ren, G.; Su, T.; Pan, Y.; Feng, B.; *et al.* MiR-150 promotes gastric cancer proliferation by negatively regulating the pro-apoptotic gene EGR2. *Biochem. Biophys. Res. Commun.* **2010**, *392*, 340–345. [CrossRef] [PubMed]

98. Watanabe, A.; Tagawa, H.; Yamashita, J.; Teshima, K.; Nara, M.; Iwamoto, K.; Kume, M.; Kameoka, Y.; Takahashi, N.; Nakagawa, T.; *et al.* The role of microRNA-150 as a tumor suppressor in malignant lymphoma. *Leukemia* **2011**, *25*, 1324–1334. [CrossRef] [PubMed]

99. Gessain, A.; Cassar, O. Epidemiological aspects and world distribution of HTLV-1 infection. *Front. Microbial.* **2012**, *3*, e388. [CrossRef] [PubMed]

100. Stamatopoulos, B.; Meuleman, N.; Haibe-Kains, B.; Saussoy, P.; van den Neste, E.; Michaux, L.; Heimann, P.; Martiat, P.; Bron, D.; Lagneaux, L. MicroRNA-29c and microRNA-223 down-regulation has *in vivo* significance in chronic lymphocytic leukemia and improves disease risk stratification. *Blood* **2009**, *113*, 5237–5245. [CrossRef] [PubMed]

101. Mi, S.; Lu, J.; Sun, M.; Li, Z.; Zhang, H.; Neilly, M.B.; Wang, Y.; Qian, Z.; Jin, J.; Zhang, Y.; et al. Microrna expression signatures accurately discriminate acute lymphoblastic leukemia from acute myeloid leukemia. *Proc. Natl. Acad. Sci. USA* **2007**, *104*, 19971–19976. [CrossRef] [PubMed]

102. Liu, T.Y.; Chen, S.U.; Kuo, S.H.; Cheng, A.L.; Lin, C.W. E2a-positive gastric malt lymphoma has weaker plasmacytoid infiltrates and stronger expression of the memory B-cell-associated miR-223: Possible correlation with stage and treatment response. *Mod. Pathol.* **2010**, *23*, 1507–1517. [CrossRef] [PubMed]

103. Laios, A.; O'Toole, S.; Flavin, R.; Martin, C.; Kelly, L.; Ring, M.; Finn, S.P.; Barrett, C.; Loda, M.; Gleeson, N.; et al. Potential role of mir-9 and mir-223 in recurrent ovarian cancer. *Mol. Cancer* **2008**, *7*, e35. [CrossRef] [PubMed]

104. Kumar, V.; Palermo, R.; Talora, C.; Campese, A.F.; Checquolo, S.; Bellavia, D.; Tottone, L.; Testa, G.; Miele, E.; Indraccolo, S.; et al. Notch and NF-κb signaling pathways regulate mir-223/fbxw7 axis in T-cell acute lymphoblastic leukemia. *Leukemia* **2014**, *28*, 2324–2335. [CrossRef] [PubMed]

105. Lee, J.E.; Hong, E.J.; Nam, H.Y.; Kim, J.W.; Han, B.G.; Jeon, J.P. Microrna signatures associated with immortalization of EBV-transformed lymphoblastoid cell lines and their clinical traits. *Cell Prolif.* **2011**, *44*, 59–66. [CrossRef] [PubMed]

106. Li, J.; Guo, Y.; Liang, X.; Sun, M.; Wang, G.; De, W.; Wu, W. Microrna-223 functions as an oncogene in human gastric cancer by targeting FBXW7/HCDC4. *J. Cancer Res. Clin. Oncol.* **2012**, *138*, 763–774. [CrossRef] [PubMed]

107. Li, X.; Zhang, Y.; Zhang, H.; Liu, X.; Gong, T.; Li, M.; Sun, L.; Ji, G.; Shi, Y.; Han, Z.; et al. MirRNA-223 promotes gastric cancer invasion and metastasis by targeting tumor suppressor EPB41l3. *Mol. Cancer Res.* **2011**, *9*, 824–833. [CrossRef] [PubMed]

108. Sun, G.; Li, H.; Rossi, J.J. Sequence context outside the target region influences the effectiveness of mir-223 target sites in the rhob 3'utr. *Nucleic Acids Res.* **2010**, *38*, 239–252. [CrossRef] [PubMed]

109. Wong, Q.W.; Lung, R.W.; Law, P.T.; Lai, P.B.; Chan, K.Y.; To, K.F.; Wong, N. Microrna-223 is commonly repressed in hepatocellular carcinoma and potentiates expression of stathmin1. *Gastroenterology* **2008**, *135*, 257–269. [CrossRef] [PubMed]

110. Pulikkan, J.A.; Dengler, V.; Peramangalam, P.S.; Peer Zada, A.A.; Muller-Tidow, C.; Bohlander, S.K.; Tenen, D.G.; Behre, G. Cell-cycle regulator E2F1 and microRNA-223 comprise an autoregulatory negative feedback loop in acute myeloid leukemia. *Blood* **2010**, *115*, 1768–1778. [CrossRef] [PubMed]

111. Haneklaus, M.; Gerlic, M.; O'Neill, L.A.; Masters, S.L. Mir-223: Infection, inflammation and cancer. *J. Intern. Med.* **2013**, *274*, 215–226. [CrossRef] [PubMed]

112. McGirt, L.Y.; Adams, C.M.; Baerenwald, D.A.; Zwerner, J.P.; Zic, J.A.; Eischen, C.M. Mir-223 regulates cell growth and targets proto-oncogenes in mycosis fungoides/cutaneous t-cell lymphoma. *J. Investig. Dermatol.* **2014**, *134*, 1101–1107. [CrossRef] [PubMed]

113. Decker, T.; Stockinger, S.; Karaghiosoff, M.; Muller, M.; Kovarik, P. Ifns and stats in innate immunity to microorganisms. *J. Clin. Invest.* **2002**, *109*, 1271–1277. [CrossRef] [PubMed]

114. Kovacic, B.; Stoiber, D.; Moriggl, R.; Weisz, E.; Ott, R.G.; Kreibich, R.; Levy, D.E.; Beug, H.; Freissmuth, M.; Sexl, V. Stat1 acts as a tumor promoter for leukemia development. *Cancer Cell* **2006**, *10*, 77–87. [CrossRef] [PubMed]

115. Sanda, T.; Tyner, J.W.; Gutierrez, A.; Ngo, V.N.; Glover, J.; Chang, B.H.; Yost, A.; Ma, W.; Fleischman, A.G.; Zhou, W.; et al. TYK2-STAT1-BCL2 pathway dependence in T-cell acute lymphoblastic leukemia. *Cancer Discov.* **2013**, *3*, 564–577. [CrossRef] [PubMed]

116. Moles, R.; Bellon, M.; Nicot, C. Stat1: A novel target of mir-150 and mir-223 is involved in the proliferation of HTLV-I-transformed and ATL cells. *Neoplasia* **2015**, *17*, 449–462. [CrossRef] [PubMed]

117. Valastyan, S.; Reinhardt, F.; Benaich, N.; Calogrias, D.; Szasz, A.M.; Wang, Z.C.; Brock, J.E.; Richardson, A.L.; Weinberg, R.A. A pleiotropically acting microrna, miR-31, inhibits breast cancer metastasis. *Cell* **2009**, *137*, 1032–1046. [CrossRef] [PubMed]

118. Augoff, K.; Das, M.; Bialkowska, K.; McCue, B.; Plow, E.F.; Sossey-Alaoui, K. MiR-31 is a broad regulator of β 1-integrin expression and function in cancer cells. *Mol. Cancer Res.* **2011**, *9*, 1500–1508. [CrossRef] [PubMed]

119. Fujikawa, D.; Yamagishi, M.; Kurokawa, N.; Soejima, A.; Ishida, T.; Tanaka, Y.; Nakano, K.; Watanabe, T. HTLV-1 Tax disrupts the host epigenome by interacting with a Polycomb group protein EZH2. *Retrovirology* **2014**, *11*, e144. [CrossRef]

120. Sparmann, A.; van Lohuizen, M. Polycomb silencers control cell fate, development and cancer. *Nat. Rev. Cancer* **2006**, *6*, 846–856. [CrossRef] [PubMed]

121. Richly, H.; Aloia, L.; di Croce, L. Roles of the polycomb group proteins in stem cells and cancer. *Cell Death Dis.* **2011**, *2*, e204. [CrossRef] [PubMed]

122. Thu, Y.M.; Richmond, A. Nf-κb inducing kinase: A key regulator in the immune system and in cancer. *Cytokine Growth Factor Rev.* **2010**, *21*, 213–226. [CrossRef] [PubMed]

123. Mori, N.; Fujii, M.; Ikeda, S.; Yamada, Y.; Tomonaga, M.; Ballard, D.W.; Yamamoto, N. Constitutive activation of NF-κb in primary adult T-cell leukemia cells. *Blood* **1999**, *93*, 2360–2368. [PubMed]

124. Prasad, S.; Ravindran, J.; Aggarwal, B.B. NF-κb and cancer: How intimate is this relationship. *Mol. Cell. Biochem.* **2010**, *336*, 25–37. [CrossRef] [PubMed]

125. Colangelo, T.; Fucci, A.; Votino, C.; Sabatino, L.; Pancione, M.; Laudanna, C.; Binaschi, M.; Bigioni, M.; Maggi, C.A.; Parente, D.; *et al.* MicroRNA-130b promotes tumor development and is associated with poor prognosis in colorectal cancer. *Neoplasia* **2013**, *15*, 1086–1099. [CrossRef] [PubMed]

126. Lai, K.W.; Koh, K.X.; Loh, M.; Tada, K.; Subramaniam, M.M.; Lim, X.Y.; Vaithilingam, A.; Salto-Tellez, M.; Iacopetta, B.; Ito, Y.; *et al.* MicroRNA-130b regulates the tumour suppressor RUNX3 in gastric cancer. *Eur. J. Cancer* **2010**, *46*, 1456–1463. [CrossRef] [PubMed]

127. Scheffer, A.R.; Holdenrieder, S.; Kristiansen, G.; von Ruecker, A.; Muller, S.C.; Ellinger, J. Circulating micrornas in serum: Novel biomarkers for patients with bladder cancer? *World J. Urol.* **2014**, *32*, 353–358. [CrossRef] [PubMed]

128. Sand, M.; Skrygan, M.; Georgas, D.; Sand, D.; Gambichler, T.; Altmeyer, P.; Bechara, F.G. The miRNA machinery in primary cutaneous malignant melanoma, cutaneous malignant melanoma metastases and benign melanocytic nevi. *Cell Tissue Res.* **2012**, *350*, 119–126. [CrossRef] [PubMed]

129. Chen, Z.; Jin, Y.; Yu, D.; Wang, A.; Mahjabeen, I.; Wang, C.; Liu, X.; Zhou, X. Down-regulation of the microrna-99 family members in head and neck squamous cell carcinoma. *Oral Oncol.* **2012**, *48*, 686–691. [CrossRef] [PubMed]

130. Yip, L.; Kelly, L.; Shuai, Y.; Armstrong, M.J.; Nikiforov, Y.E.; Carty, S.E.; Nikiforova, M.N. Microrna signature distinguishes the degree of aggressiveness of papillary thyroid carcinoma. *Ann. Surg. Oncol.* **2011**, *18*, 2035–2041. [CrossRef] [PubMed]

131. Yang, C.; Cai, J.; Wang, Q.; Tang, H.; Cao, J.; Wu, L.; Wang, Z. Epigenetic silencing of miR-130b in ovarian cancer promotes the development of multidrug resistance by targeting colony-stimulating factor 1. *Gynecol. Oncol.* **2012**, *124*, 325–334. [CrossRef] [PubMed]

132. Dong, P.; Karaayvaz, M.; Jia, N.; Kaneuchi, M.; Hamada, J.; Watari, H.; Sudo, S.; Ju, J.; Sakuragi, N. Mutant p53 gain-of-function induces epithelial-mesenchymal transition through modulation of the miR-130b-ZEB1 axis. *Oncogene* **2013**, *32*, 3286–3295. [CrossRef] [PubMed]

133. Zhao, G.; Zhang, J.G.; Shi, Y.; Qin, Q.; Liu, Y.; Wang, B.; Tian, K.; Deng, S.C.; Li, X.; Zhu, S.; *et al.* MiR-130b is a prognostic marker and inhibits cell proliferation and invasion in pancreatic cancer through targeting STAT3. *PLoS ONE* **2013**, *8*, e73803. [CrossRef] [PubMed]

134. Yu, T.; Cao, R.; Li, S.; Fu, M.; Ren, L.; Chen, W.; Zhu, H.; Zhan, Q.; Shi, R. MiR-130b plays an oncogenic role by repressing pten expression in esophageal squamous cell carcinoma cells. *BMC Cancer* **2015**, *15*, e29. [CrossRef] [PubMed]

135. Castro, N.E.; Kato, M.; Park, J.T.; Natarajan, R. Transforming growth factor beta1 (TGF-β1) enhances expression of profibrotic genes through a novel signaling cascade and micrornas in renal mesangial cells. *J. Biol. Chem.* **2014**, *289*, 29001–29013. [CrossRef] [PubMed]

136. Petrocca, F.; Vecchione, A.; Croce, C.M. Emerging role of miR-106b-25/mir-17–92 clusters in the control of transforming growth factor beta signaling. *Cancer Res.* **2008**, *68*, 8191–8194. [CrossRef] [PubMed]

137. Mendell, J.T. Miriad roles for the miR-17–92 cluster in development and disease. *Cell* **2008**, *133*, 217–222. [CrossRef] [PubMed]

138. Hayashita, Y.; Osada, H.; Tatematsu, Y.; Yamada, H.; Yanagisawa, K.; Tomida, S.; Yatabe, Y.; Kawahara, K.; Sekido, Y.; Takahashi, T. A polycistronic microrna cluster, miR-17–92, is overexpressed in human lung cancers and enhances cell proliferation. *Cancer Res.* **2005**, *65*, 9628–9632. [CrossRef] [PubMed]

139. Fu, X.; Tian, J.; Zhang, L.; Chen, Y.; Hao, Q. Involvement of microrna-93, a new regulator of PTEN/AKT signaling pathway, in regulation of chemotherapeutic drug cisplatin chemosensitivity in ovarian cancer cells. *FEBS Lett.* **2012**, *586*, 1279–1286. [CrossRef] [PubMed]

140. Long, J.; Wang, Y.; Wang, W.; Chang, B.H.; Danesh, F.R. Identification of microrna-93 as a novel regulator of vascular endothelial growth factor in hyperglycemic conditions. *J. Biol. Chem.* **2010**, *285*, 23457–23465. [CrossRef] [PubMed]

141. Fang, L.; Deng, Z.; Shatseva, T.; Yang, J.; Peng, C.; Du, W.W.; Yee, A.J.; Ang, L.C.; He, C.; Shan, S.W.; *et al.* Microrna miR-93 promotes tumor growth and angiogenesis by targeting integrin-beta8. *Oncogene* **2011**, *30*, 806–821. [CrossRef] [PubMed]

142. Du, L.; Zhao, Z.; Ma, X.; Hsiao, T.H.; Chen, Y.; Young, E.; Suraokar, M.; Wistuba, I.; Minna, J.D.; Pertsemlidis, A. Mir-93-directed downregulation of dab2 defines a novel oncogenic pathway in lung cancer. *Oncogene* **2014**, *33*, 4307–4315. [CrossRef] [PubMed]

143. Fang, L.; Du, W.W.; Yang, W.; Rutnam, Z.J.; Peng, C.; Li, H.; O'Malley, Y.Q.; Askeland, R.W.; Sugg, S.; Liu, M.; *et al.* MiR-93 enhances angiogenesis and metastasis by targeting LATS2. *Cell Cycle* **2012**, *11*, 4352–4365. [CrossRef] [PubMed]

144. Okamura, S.; Arakawa, H.; Tanaka, T.; Nakanishi, H.; Ng, C.C.; Taya, Y.; Monden, M.; Nakamura, Y. p53DINP1, a p53-inducible gene, regulates p53-dependent apoptosis. *Mol. Cell* **2001**, *8*, 85–94. [CrossRef]

145. Tomasini, R.; Seux, M.; Nowak, J.; Bontemps, C.; Carrier, A.; Dagorn, J.C.; Pebusque, M.J.; Iovanna, J.L.; Dusetti, N.J. TP53INP1 is a novel p73 target gene that induces cell cycle arrest and cell death by modulating p73 transcriptional activity. *Oncogene* **2005**, *24*, 8093–8104. [CrossRef] [PubMed]

146. Gironella, M.; Seux, M.; Xie, M.J.; Cano, C.; Tomasini, R.; Gommeaux, J.; Garcia, S.; Nowak, J.; Yeung, M.L.; Jeang, K.T.; *et al.* Tumor protein 53-induced nuclear protein 1 expression is repressed by miR-155, and its restoration inhibits pancreatic tumor development. *Proc. Natl. Acad. Sci. USA* **2007**, *104*, 16170–16175. [CrossRef] [PubMed]

147. Jiang, F.; Liu, T.; He, Y.; Yan, Q.; Chen, X.; Wang, H.; Wan, X. MiR-125b promotes proliferation and migration of type II endometrial carcinoma cells through targeting TP53INP1 tumor suppressor *in vitro* and *in vivo*. *BMC Cancer* **2011**, *11*, e425. [CrossRef] [PubMed]

148. Hershko, T.; Chaussepied, M.; Oren, M.; Ginsberg, D. Novel link between e2f and p53: Proapoptotic cofactors of p53 are transcriptionally upregulated by e2f. *Cell Death Differ.* **2005**, *12*, 377–383. [CrossRef] [PubMed]

149. Seux, M.; Peuget, S.; Montero, M.P.; Siret, C.; Rigot, V.; Clerc, P.; Gigoux, V.; Pellegrino, E.; Pouyet, L.; N'Guessan, P.; *et al.* TP53INP1 decreases pancreatic cancer cell migration by regulating sparc expression. *Oncogene* **2011**, *30*, 3049–3061. [CrossRef] [PubMed]

150. Varrin-Doyer, M.; Nicolle, A.; Marignier, R.; Cavagna, S.; Benetollo, C.; Wattel, E.; Giraudon, P. Human T lymphotropic virus type 1 increases T lymphocyte migration by recruiting the cytoskeleton organizer CRMP2. *J. Immunol.* **2012**, *188*, 1222–1233. [CrossRef] [PubMed]

151. Beimling, P.; Moelling, K. Direct interaction of CREB protein with 21 bp tax-response elements of HTLV-Iltr. *Oncogene* **1992**, *7*, 257–262. [PubMed]

152. Zhao, L.J.; Giam, C.Z. Human T-cell lymphotropic virus type I (HTLV-I) transcriptional activator, tax, enhances CREB binding to HTLV-I 21-base-pair repeats by protein-protein interaction. *Proc. Natl. Acad. Sci. USA* **1992**, *89*, 7070–7074. [CrossRef] [PubMed]

153. Adya, N.; Giam, C.Z. Distinct regions in human T-cell lymphotropic virus type I tax mediate interactions with activator protein CREB and basal transcription factors. *J. Virol.* **1995**, *69*, 1834–1841. [PubMed]

154. Harrod, R.; Tang, Y.; Nicot, C.; Lu, H.S.; Vassilev, A.; Nakatani, Y.; Giam, C.Z. An exposed kid-like domain in human T-cell lymphotropic virus type 1 tax is responsible for the recruitment of coactivators CBP/P300. *Mol. Cell. Biol.* **1998**, *18*, 5052–5061. [PubMed]

155. Harrod, R.; Kuo, Y.L.; Tang, Y.; Yao, Y.; Vassilev, A.; Nakatani, Y.; Giam, C.Z. P300 and P300/CAMP-responsive element-binding protein associated factor interact with human T-cell lymphotropic virus type-1 tax in a multi-histone acetyltransferase/activator-enhancer complex. *J. Biol. Chem.* **2000**, *275*, 11852–11857. [CrossRef] [PubMed]

156. Bogenberger, J.M.; Laybourn, P.J. Human T lymphotropic virus type 1 protein tax reduces histone levels. *Retrovirology* **2008**, *5*, e9. [CrossRef] [PubMed]

157. Rahman, S.; Quann, K.; Pandya, D.; Singh, S.; Khan, Z.K.; Jain, P. HTLV-1 tax mediated downregulation of mirnas associated with chromatin remodeling factors in T cells with stably integrated viral promoter. *PLoS ONE* **2012**, *7*, e34490. [CrossRef] [PubMed]

158. Lin, R.J.; Lin, Y.C.; Yu, A.L. MiR-149* induces apoptosis by inhibiting AKT1 and E2F1 in human cancer cells. *Mol. Carcinog.* **2010**, *49*, 719–727. [CrossRef] [PubMed]

159. Jin, L.; Hu, W.L.; Jiang, C.C.; Wang, J.X.; Han, C.C.; Chu, P.; Zhang, L.J.; Thorne, R.F.; Wilmott, J.; Scolyer, R.A.; *et al.* MicroRNA-149*, a p53-responsive microRNA, functions as an oncogenic regulator in human melanoma. *Proc. Natl. Acad. Sci. USA* **2011**, *108*, 15840–15845. [CrossRef] [PubMed]

160. Schaefer, A.; Jung, M.; Mollenkopf, H.J.; Wagner, I.; Stephan, C.; Jentzmik, F.; Miller, K.; Lein, M.; Kristiansen, G.; Jung, K. Diagnostic and prognostic implications of microrna profiling in prostate carcinoma. *Int. J. Cancer* **2010**, *126*, 1166–1176. [CrossRef] [PubMed]

161. Li, D.; Chen, P.; Li, X.Y.; Zhang, L.Y.; Xiong, W.; Zhou, M.; Xiao, L.; Zeng, F.; Li, X.L.; Wu, M.H.; *et al.* Grade-specific expression profiles of miRNAs/mRNAs and docking study in human grade I–III astrocytomas. *OMICS* **2011**, *15*, 673–682. [CrossRef] [PubMed]

162. Liu, H.; Brannon, A.R.; Reddy, A.R.; Alexe, G.; Seiler, M.W.; Arreola, A.; Oza, J.H.; Yao, M.; Juan, D.; Liou, L.S.; *et al.* Identifying mRNA targets of microrna dysregulated in cancer: With application to clear cell renal cell carcinoma. *BMC Syst. Biol.* **2010**, *4*, e51. [CrossRef] [PubMed]

163. Skalsky, R.L.; Cullen, B.R. Reduced expression of brain-enriched micrornas in glioblastomas permits targeted regulation of a cell death gene. *PLoS ONE* **2011**, *6*, e24248. [CrossRef] [PubMed]

164. Zhang, L.; Volinia, S.; Bonome, T.; Calin, G.A.; Greshock, J.; Yang, N.; Liu, C.G.; Giannakakis, A.; Alexiou, P.; Hasegawa, K.; *et al.* Genomic and epigenetic alterations deregulate microrna expression in human epithelial ovarian cancer. *Proc. Natl. Acad. Sci. USA* **2008**, *105*, 7004–7009. [CrossRef] [PubMed]

165. Cui, J.; Bi, M.; Overstreet, A.M.; Yang, Y.; Li, H.; Leng, Y.; Qian, K.; Huang, Q.; Zhang, C.; Lu, Z.; *et al.* MiR-873 regulates ERα transcriptional activity and tamoxifen resistance via targeting CDK3 in breast cancer cells. *Oncogene* **2014**. [CrossRef] [PubMed]

166. Vernin, C.; Thenoz, M.; Pinatel, C.; Gessain, A.; Gout, O.; Delfau-Larue, M.H.; Nazaret, N.; Legras-Lachuer, C.; Wattel, E.; Mortreux, F. HTLV-1 BZIP factor HBZ promotes cell proliferation and genetic instability by activating oncomirs. *Cancer Res.* **2014**, *74*, 6082–6093. [CrossRef] [PubMed]

167. Sayed, D.; Rane, S.; Lypowy, J.; He, M.; Chen, I.Y.; Vashistha, H.; Yan, L.; Malhotra, A.; Vatner, D.; Abdellatif, M. MicroRNA-21 targets sprouty2 and promotes cellular outgrowths. *Mol. Biol. Cell* **2008**, *19*, 3272–3282. [CrossRef] [PubMed]

168. Thum, T.; Gross, C.; Fiedler, J.; Fischer, T.; Kissler, S.; Bussen, M.; Galuppo, P.; Just, S.; Rottbauer, W.; Frantz, S.; *et al.* Microrna-21 contributes to myocardial disease by stimulating map kinase signalling in fibroblasts. *Nature* **2008**, *456*, 980–984. [CrossRef] [PubMed]

169. Asangani, I.A.; Rasheed, S.A.; Nikolova, D.A.; Leupold, J.H.; Colburn, N.H.; Post, S.; Allgayer, H. MicroRNA-21 (miR-21) post-transcriptionally downregulates tumor suppressor PDCD4 and stimulates invasion, intravasation and metastasis in colorectal cancer. *Oncogene* **2008**, *27*, 2128–2136. [CrossRef] [PubMed]

170. Meng, F.; Henson, R.; Wehbe-Janek, H.; Ghoshal, K.; Jacob, S.T.; Patel, T. Microrna-21 regulates expression of the pten tumor suppressor gene in human hepatocellular cancer. *Gastroenterology* **2007**, *133*, 647–658. [CrossRef] [PubMed]

171. Zhu, S.; Si, M.L.; Wu, H.; Mo, Y.Y. Microrna-21 targets the tumor suppressor gene tropomyosin 1 (TPM1). *J. Biol. Chem.* **2007**, *282*, 14328–14336. [CrossRef] [PubMed]

172. Volinia, S.; Calin, G.A.; Liu, C.G.; Ambs, S.; Cimmino, A.; Petrocca, F.; Visone, R.; Iorio, M.; Roldo, C.; Ferracin, M.; *et al.* A microrna expression signature of human solid tumors defines cancer gene targets. *Proc. Natl. Acad. Sci. USA* **2006**, *103*, 2257–2261. [CrossRef] [PubMed]

173. O'Donnell, K.A.; Wentzel, E.A.; Zeller, K.I.; Dang, C.V.; Mendell, J.T. c-Myc -regulated microRNAs modulate E2F1 expression. *Nature* **2005**, *435*, 839–843. [CrossRef] [PubMed]

174. Monzo, M.; Navarro, A.; Bandres, E.; Artells, R.; Moreno, I.; Gel, B.; Ibeas, R.; Moreno, J.; Martinez, F.; Diaz, T.; *et al.* Overlapping expression of micrornas in human embryonic colon and colorectal cancer. *Cell Res.* **2008**, *18*, 823–833. [CrossRef] [PubMed]

Viruses 2015, 7, 4047–4074

175. Novotny, G.W.; Sonne, S.B.; Nielsen, J.E.; Jonstrup, S.P.; Hansen, M.A.; Skakkebaek, N.E.; Rajpert-de Meyts, E.; Kjems, J.; Leffers, H. Translational repression of e2f1 mrna in carcinoma *in situ* and normal testis correlates with expression of the mir-17–92 cluster. *Cell Death Differ.* **2007**, *14*, 879–882. [CrossRef] [PubMed]

176. Xiao, C.; Srinivasan, L.; Calado, D.P.; Patterson, H.C.; Zhang, B.; Wang, J.; Henderson, J.M.; Kutok, J.L.; Rajewsky, K. Lymphoproliferative disease and autoimmunity in mice with increased miR-17–92 expression in lymphocytes. *Nat. Immunol.* **2008**, *9*, 405–414. [CrossRef] [PubMed]

177. Gandhi, N.S.; Tekade, R.K.; Chougule, M.B. Nanocarrier mediated delivery of siRNA/miRNA in combination with chemotherapeutic agents for cancer therapy: Current progress and advances. *J. Control. Release* **2014**, *194*, 238–256. [CrossRef] [PubMed]

178. Sethi, S.; Li, Y.; Sarkar, F.H. Regulating mirna by natural agents as a new strategy for cancer treatment. *Curr. Drug Targets* **2013**, *14*, 1167–1174. [CrossRef] [PubMed]

179. Ling, H.; Fabbri, M.; Calin, G.A. Micrornas and other non-coding RNAs as targets for anticancer drug development. *Nat. Rev. Drug Disc.* **2013**, *12*, 847–865. [CrossRef] [PubMed]

*Review*

# HTLV-1, Immune Response and Autoimmunity

**Juarez A S Quaresma** [1,†], **Gilberto T Yoshikawa** [2,†], **Roberta V L Koyama** [1,†], **George A S Dias** [1,†], **Satomi Fujihara** [3,†] and **Hellen T Fuzii** [3,*,†]

1   Science Center of Health and Biology. Pará State University, Rua Perebebuí, 2623, Belém, Pará 66087-670, Brazil; juarez@ufpa.br (J.A.S.Q.); robertakoyamareumato@gmail.com (R.V.L.K.); georgealbertodias@yahoo.com.br (G.A.S.D.)
2   Science Health Institute, Federal University of Pará, Praça Camilo Salgado, 1, Belém, Pará 66055-240, Brazil; gyoshikawa@uol.com.br
3   Tropical Medicine Center, Federal University of Pará, Av. Generalíssimo Deodoro, 92, Belém, Pará 66055-240, Brazil; satomifujihara@gmail.com
*   Correspondence: hellenfuzii@gmail.com; Tel.: +55-91-3201-0954
†   These authors contributed equally to this work.

Academic Editor: Louis M. Mansky
Received: 18 October 2015; Accepted: 14 December 2015; Published: 24 December 2015

**Abstract:** Human T-lymphotropic virus type-1 (HTLV-1) infection is associated with adult T-cell leukemia/lymphoma (ATL). Tropical spastic paraparesis/HTLV-1-associated myelopathy (PET/HAM) is involved in the development of autoimmune diseases including Rheumatoid Arthritis (RA), Systemic Lupus Erythematosus (SLE), and Sjögren's Syndrome (SS). The development of HTLV-1-driven autoimmunity is hypothesized to rely on molecular mimicry, because virus-like particles can trigger an inflammatory response. However, HTLV-1 modifies the behavior of CD4$^+$ T cells on infection and alters their cytokine production. A previous study showed that in patients infected with HTLV-1, the activity of regulatory CD4$^+$ T cells and their consequent expression of inflammatory and anti-inflammatory cytokines are altered. In this review, we discuss the mechanisms underlying changes in cytokine release leading to the loss of tolerance and development of autoimmunity.

**Keywords:** Human T-lymphotropic virus type-1 (HTLV-1); immune response; autoimmunity

## 1. Introduction

The etiology of autoimmune diseases is unknown, but it is clear that the interaction between genes and the environment is an important step in breaking immune tolerance to self-antigens. This can lead to inflammation and destruction of specific tissues and organs. Although host genetic background contributes to autoimmunity, research indicates that infectious agents are one of most important environmental factors responsible for the development of autoimmune diseases [1–3]. Chronic inflammatory responses to infections have been associated with the initiation and exacerbation of autoimmune diseases [1,4,5].

Human T-lymphotropic virus type-1 (HTLV-1) is associated to a number of diseases, such as HTLV-1-associated myelopathy/tropical spastic paraparesis (HAM/TSP), and autoimmune diseases such as the Sjögren's Syndrome (SS), arthropathies, and uveitis, which are often related to changes in the immune response [6–9]. The HTLV-1 virus infects CD4$^+$ T lymphocytes, and can modify the cell function. CD4$^+$ T lymphocytes are the central acquired immune response regulators. Changes in their behavior can trigger inflammatory reactions that can break immune system tolerance, leading to autoimmunity. In this review, we discuss immunological changes in HTLV-1 infection and its association with autoimmune diseases.

## 2. Human T-lymphotropic Virus Type-1 (HTLV-1)

Human T-lymphotropic virus type-1 (HTLV-1) is classified as a complex type C retrovirus belonging to the genus *Deltaretrovirus*, family *Retroviridae*, and subfamily *Orthoretrovirinae* [10,11].

The morphological structure of this virus is similar to other retroviruses; the capsid contains two simple RNA strands together with the reverse transcriptase and integrase enzymes. These enzymes are important for insertion of the virus into the host genome, resulting in a provirus [12–15].

The virus genome contains structural and functional genes, such as the *gag*, *pro/pol*, and *env* that are flanked by two long terminal repeat (LTR) regions. Additionally, the *pX* region was identified in the region between the *env* gene and the 3'-LTR region. The *pX* region codes for the Tax (*p*40), REX (*p*27), *p*12, *p*13, *p*21, and *p*30 regulatory proteins that are involved in viral infection and proliferation. Tax is a phosphoprotein that exerts an essential role in viral transcription and cell behavior transformation [16–19]. It has pleiotropic functions that occur on a very wide spectrum of interactions with cellular proteins. Tax can modify signal transduction pathways of the host–cell that induce the transcription factors NF-κB, cAMP response element binding (CREB), Serum response factor (SRF) and activator protein 1 (AP-1) [12,13,20–22].

Recently, another gene was identified, *hbz* (*HTLV-1 b-ZIP factor*), that codes for a protein involved in the pathogenesis of the virus together with Tax. HBZ can function in two different molecular forms, mRNA and protein. In mRNA form, it can promote cell proliferation by positive regulation of E2F1. HBZ protein can down-regulate Tax expression and can interact directly with c-Jun and c-Jun-B [23,24].

HTLV-1 infection is associated with several diseases, primarily with adult T-cell lymphoma (ATL) and HAM/TSP. HAM/TSP patients present a series of immunological dysfunctions, including spontaneous proliferation of HTLV-infected T CD4+ lymphocytes, an increase in the migratory capacity of circulating leukocytes, and increased production of inflammatory cytokines—particularly neurotoxic cytokines such as IFN-γ and TNF-α—in affected regions along the spinal cord [6,16,25–28]. In HAM/TSP patients, there is a predominance of Th1 cytokines (IFN-γ) and a reduction in Th2 cytokines (IL-4 and IL-10), this is likely to cause greater circulation of immune cells between peripheral blood and the central nervous system (CNS), leading to inflammation of the nervous tissue [29–31]. Among the several existing theories related to HAM/TSP development, the most widely accepted is that it is a virally induced, cytotoxic, demyelinating inflammatory process of a chronic and progressive nature. The lymphocytes are activated during spastic paraparesis; when they cross the blood–brain barrier, the inflammatory process initiates in the CNS, resulting in lesions [10,15,26,32,33]. Another theory is direct cytotoxicity mechanism, where HTLV-1-cytotoxic CD8+ T cell cross the blood-brain barrier and destroy HTLV-1 infected glia cells by cytotoxicity or cytokine production. The last theory suggests that autoimmunity mechanism can cause lesions by molecular mimicry. A host neuronal protein seems to be similar to Tax protein from the virus, which can cause immune cross-reaction, leading to CNS inflammation [6,10,34,35].

## 3. Immunological Changes in HTLV-1-Infected Patients

Patients infected with HTLV-1 may develop a number of associated diseases, such as HAM/TSP, or other autoimmune diseases such as the Sjögren's Syndrome (SS), arthropathies, and uveitis, which are often related to changes in the immune response [7–9,35]. These changes may occur due to infection of the CD4+ T lymphocytes by HTLV-1.

HTLV-1-infected CD4+ T lymphocytes exhibit altered signaling cascades and transcription factor activation, leading to changes in cell behavior [7–9,16,36,37].

Many studies have reported that the HTLV-1 Tax protein affects several transcription factors including CREB/ATF, NF-κB, AP-1, SRF, and Nuclear factor of activated T-cells (NFAT), as well as a number of signaling cascades involving PDZ domain-containing proteins such as Rho-GTPases and Janus kinase (JAK)/signal transducer and activator of transcription (STAT), thus altering the transforming growth factor-β (TGF-β) cascades. These factors are involved in cell proliferation and activation, including expression of cytokines and activation of viral proteins [17–19,38–41].

The expression of forkhead/winged helix transcription factor (FOXP3), which is an important transcription factor, has also been reported to be altered in patients infected with HTLV-1. FOXP3 is an essential transcription factor for the differentiation, function, and homeostasis of regulatory T cells (Tregs). Irregularities in the expression of FOXP3 may lead to loss of immune tolerance and the probable development of autoimmune diseases [41,42].

Previous studies have demonstrated that an increase in FOXP3 expression in patients that developed ATL leads to an exacerbated Treg function, resulting in increased production of IL-10 and TGF-β, which in turn triggers the immunosuppression phenotype observed in these patients. In contrast, studies performed with patients that developed HAM/TSP showed a decrease in FOXP3 expression and in the production of the IL-10 and TGF-β cytokines responsible for suppression of the immune response [31,36,43–45]. This loss of suppressive function may lead to an exacerbation of the disease process, since inflammation is not controlled and the inflammatory process is perpetuated. A study performed by Yamano *et al.* [31] showed that persistent activation of the immune response, induced by Tax, in patients with HAM/TSP may be associated with a decrease in the expression of CD4$^+$CD25$^+$FOXP3$^+$ T cells that possess a suppressive function and an accumulation of CD4$^+$CD25$^+$FOXP3$^-$ T cells that can exacerbate the pathogenic process of HAM/TSP. The authors demonstrated that in patients with HAM/TSP, there was an increase in the sub-population of IFN-γ-producing T cells with a CD4$^+$CD25$^+$FOXP3$^-$ phenotype and that this increase correlated with the clinical severity of HAM/TSP.

Changes in signaling pathways and transcription factor activation caused by viral proteins play an important role in modifying immune response homeostasis, resulting in a cytokine environment that may influence the immune cells phenotype. The effect of the environment may lead to autocrine and paracrine activation and thus influence T cell differentiation and homeostasis. IFN-γ can stimulate the production of Th1 cells, while the presence of IL-4 leads to the differentiation of Th2 cells. Several studies have shown that the presence of these cytokines triggers the activity of known suppressors of cytokine signaling (SOCS) family proteins, which are able to inhibit the recruitment of STATs or the activation of JAKs. SOCS play a role in the maturation, differentiation, and maintenance of T lymphocytes [46–48]. SOCS-1 was shown to inhibit the activation of pathways stimulated by IFN-γ and IL-4, whereas SOC3 is important for maintaining the Th2 phenotype. Previous studies performed with HAM/TSP patients have demonstrated increased SOC-1 and decreased SOC-3 levels, suggesting a tendency towards the Th1 response [49].

The HTLV-1-infected CD4$^+$ T lymphocytes of HAM/TSP patients exhibit spontaneous proliferation, in addition to an increased production of proinflammatory cytokines such as IFN-γ, TNF-α, IL-1, and IL-6; neurotoxic cytokines such as IFN-γ and TNF-α, in particular, are found at high concentrations in the spinal fluid of HAM/TSP patients [9,16,50,51]. These findings demonstrate that infected individuals have an immune response characteristic of the Th1 phenotype (IFN-γ and TNF-α) and a decrease in the Th2 profile (IL-4 and IL-10) [30,31,45].

Toulza *et al.* [43] observed that individuals with HAM/TSP had similar IL-10 levels as healthy individuals and that the TGF-β levels were significantly lower compared to the control group; this in turn could result in a decrease in the Treg cell-mediated suppressive function and thus contribute to the exacerbation of inflammation.

## 4. HTLV-1 and Autoimmunity

HTLV-1 infection leads to changes in the systemic immune response even in asymptomatic patients [36,52–57]. The regulation of the immune response is carefully organized so that the organism suffers no damage and that homeostasis is maintained. This is ensured via clonal selection that occurs during lymphocyte development concomitant with the destruction of the autoreactive cells [58] and the development of Treg cells. Treg cells are able to inhibit the proliferation of T cells *in vitro* and regulate the activity of CD4$^+$ and CD8$^+$ T cells *in vivo*. There are two major types of Treg cells, naturally occurring cells and cells produced in the periphery [59].

During the development of autoimmunity, there is a loss of tolerance to self-antigens, causing an inflammatory response that attacks organs and tissues of the individual. The pathogenesis of autoimmune diseases is generally studied in the context of T helper cells and the balance between the Th1 and Th2 responses. Thus, it has been observed that some diseases such as Rheumatoid Arthritis (RA) fit the Th1 profile, in which a cell-mediated response occurs, while others such as Systemic Lupus Erythematosus (SLE) fit the Th2 profile, involving antibodies and immunocomplexes in their physiopathology [60]. However, during the development of autoimmunity, there is a combination of several factors that are not only immunological, but also genetic and environmental [61,62]. Of these environmental factors, infections are of great importance, since they act as a trigger for autoimmunity [63,64].

The association between autoimmunity and HTLV-1 infection has been previously described; however, the mechanisms underlying this association are not yet fully understood. Many studies have indicated that molecular mimicry could be the trigger for the development of certain diseases. However, as previously described, HTLV-1 can result in several immune response anomalies since it infects CD4+ T lymphocytes and alters their behavior (Figure 1) [65,66].

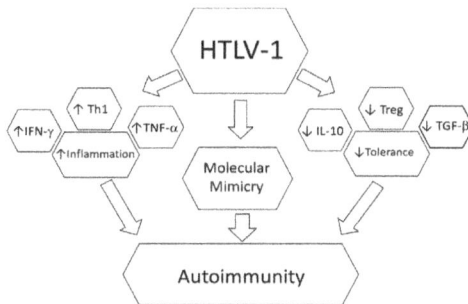

**Figure 1.** Possible mechanisms involved in HTLV-1 association with autoimmunity.

## 5. Rheumatoid Arthritis

RA is a chronic and incapacitating disease that affects 1% of the world's population. Although the etiopathology of the disease is not fully understood, it is characterized by chronic polyarthritis that may lead to the destruction of articulation if not properly treated [67–71]. During the early stages of RA, the cytokines expressed in the synovium are mainly IL-2, 4, 13, 15, and 17. Once the disease is established, expression of IFN-$\gamma$, TNF-$\alpha$, and IL-10 is observed, with low-level expression of IL-2, -4, -5, and -13. Furthermore, there is a correlation between serum cytokines and disease progression [72,73]. The development and progression of RA depends on the migration of the T lymphocytes into the synovium. Previous studies have observed proliferation of the synovium and T cell infiltration in HTLV infected patients that develop RA. Several studies reported the presence of HTLV proviral DNA in synovial liquid and tissue cells and that T cells in both the synovium and synovial cells were infected with HTLV-1. Nishioka *et al.* [74] also reported the expression of Tax mRNA in synovial cells from HTLV-I-associated arthropathy patients. Tax may induce cell proliferation, as well as the production of inflammatory cytokines. In addition, similar to what occurs in HAM/TSP, there is a migration of lymphocytes into the CNS. This migration may be associated with the viral load of the patient, as demonstrated by Yakova *et al.* [75]. These authors observed that patients infected with HTLV-1 that had RA or connective tissue disease had a higher viral load compared to asymptomatic patients; however, it was similar to the viral load of patients that developed HAM/TSP. Moreover, the viral load in the synovium was higher in RA patients [74–79].

## 6. Sjögren's Syndrome

SS is defined as a systemic autoimmune disorder, manifesting primarily as xerostomia (dry mouth) and xerophthalmia (dry keratoconjunctivitis) due to the lymphocytic infiltration of the salivary and lachrymal glands, which in turn leads to the destruction of the ducts. Furthermore, antinuclear antibodies (ANA) and other self-antibodies, such as anti-SS-A (Ro) and SS-B (La), are also found in these patients. Disease development is also associated with genetic and hormonal factors [80,81]. Several viral infections, such as HTLV-1, may also be associated with the occurrence of this disease. A number of studies have reported a high prevalence of HTLV-1 in SS patients [82,83]. Nakamura *et al.* [84] reported a high prevalence of anti-HTLV-1 IgA in the salivary glands of SS patients. Another interesting related factor is the level of mononuclear infiltrate in SS patients infected with HTLV-1; SS patients with HTLV-1 show higher infiltrate levels compared with SS patients not infected with HTLV-1 [82–84].

## 7. Systemic Lupus Erythematosus

SLE is a systemic autoimmune disease of unknown etiology. This disease progresses with polymorphic clinical manifestations and periods of exacerbation and remission. Disease development is associated with a genetic predisposition and environmental factors, such as exposure to sunlight and viral infection [85–88].

The association between HTLV-1 and SLE is still controversial [89–91]. One possible mechanism proposed for this association is a process of molecular mimicry through the endogenous sequence related to HTLV-1 (HRES-1) in the development of SLE. This could trigger the production of self-antibodies, leading to the formation of immunocomplexes that are deposited in the tissues; this in turn could cause complement fixation and inflammation, which are pathogenic characteristics of SLE [92]. Other studies have demonstrated the expression of HTLV-1 antigens in the mononuclear cells present in the peripheral blood of individuals with SLE and infected with HTLV-1, following three or more days of *in vitro* culturing. This indicates the occurrence of viral replication in SLE patients, which could explain the high seropositivity for HTLV-1 and HTLV-2 observed in these patients [93].

## 8. Conclusions

Several studies have described HTLV-1 infection in the context of autoimmune diseases, although this subject is still under debate. Molecular mimicry has been hypothesized as a possible mechanism; however, HTLV-1-promoted altered cytokine release is critically involved in breaking tolerance. HTLV-1-induced changes in the activity of regulatory CD4 T-cell molecules affect the homeostasis of cytokines, including IFN-γ, TNF-α, TGF-β and IL-10, and disrupt the balance in inflammatory and anti-inflammatory responses, leading to the loss of tolerance and the development of autoimmunity.

**Acknowledgments:** We thank Federal University of Pará and Pará State University for supporting our work.

**Author Contributions:** Juarez A S Quaresma, Gilberto T Yoshikawa, George A S Dias, Roberta V L Koyama, Satomi Fujihara and Hellen T Fuzii, wrote the review with equal contribution.

**Conflicts of Interest:** The authors declare no conflict of interest.

## References

1. Von Herrath, M.G.; Fujinami, R.S.; Whitton, J.L. Microorganisms and autoimmunity: Making the barren field fertile? *Nat. Rev. Microbiol.* **2003**, *1*, 151–157. [CrossRef] [PubMed]
2. Libbey, J.E.; Fujinami, R.S. Potential triggers of MS. Results Probl. *Cell Differ.* **2010**, *51*, 21–42.
3. Fujinami, R.S. Viruses and autoimmune disease—Two sides of the same coin? *Trends Microbiol.* **2001**, *9*, 377–381. [CrossRef]
4. McCoy, L.; Tsunoda, I.; Fujinami, R.S. Multiple sclerosis and virus induced immune responses: Autoimmunity can be primed by molecular mimicry and augmented by bystander activation. *Autoimmunity* **2006**, *39*, 9–19. [CrossRef] [PubMed]

5. Sfriso, P.; Ghirardello, A.; Botsios, C.; Tonon, M.; Zen, M.; Bassi, N.; Bassetto, F.; Doria, A. Infections and autoimmunity: The multifaceted relationship. *J. Leukoc. Biol.* **2010**, *87*, 385–395. [CrossRef] [PubMed]

6. Araujo, A.Q.; Silva, S.T. The HTLV-1 neurological complex. *Lancet Neurol.* **2006**, *5*, 1068–1076. [CrossRef]

7. Gessain, A.; Barin, F.; Vernant, J.C.; Gout, O.; Maurs, L.; Calender, A.; de Thé, G. Antibodies to human T lymphotropicvírus type I in patients with tropical spastic paraparesis. *Lancet* **1985**, *2*, 407–410. [CrossRef]

8. Román, G.C.; Osame, M. Identity of HTLV-I-associated tropical spastic paraparesis and HTLV-I-associated myelopathy. *Lancet* **1988**, *1*. [CrossRef]

9. Romanelli, L.C.F.; Caramelli, P.; Proietti, A.B.F.C. Human T-cell leukemia virus type 1 (HTLV-1): When do we suspect of infection? *Rev. Assoc. Med. Brás.* **2010**, *56*, 340–347. [CrossRef] [PubMed]

10. Cooper, S.A.; van der Loeff, M.S.; Taylor, G.P. The neurology of HTLV-1 infection. *Pract. Neurol.* **2009**, *9*, 16–26. [CrossRef] [PubMed]

11. Gessain, A.; Mahieux, R. Tropical spastic paraparesis and HTLV-1 associated myelopathy: Clinical, epidemiological, virological and therapeutic aspects. *Rev. Neurol.* **2012**, *168*, 257–269. [CrossRef] [PubMed]

12. Boxus, M.; Willems, L. Mechanisms of HTLV-1 persistence and transformation. *Br. J. Cancer* **2009**, *101*, 1497–1501. [CrossRef] [PubMed]

13. Hoshino, H. Cellular factors involved in HTLV-1 entry and pathogenicit. *Front. Microbiol.* **2012**, *3*. [CrossRef] [PubMed]

14. Lairmore, M.D.; Anupam, R.; Bowden, N.; Haines, R.; Haynes, R.A., II; Ratner, L.; Green, P.L. Molecular determinants of human T-lymphotropic virus type 1 transmission and spread. *Viruses* **2011**, *3*, 1131–1165. [CrossRef] [PubMed]

15. Matsuura, E.; Yamano, Y.; Jacobson, S. Neuroimmunity of HTLV-I Infection. *J. Neuroimmune Pharmacol.* **2010**, *5*, 310–325. [CrossRef] [PubMed]

16. Best, I.; López, G.; Verdonck, K.; González, E.; Tipismana, M.; Gotuzzo, E.; Vanham, G.; Clark, D. IFN-γ production in response to Tax 161-233, and frequency of CD4+ Foxp3+ and Lin HLA-DRhigh CD123+ cells, discriminate HAM/TSP patients from asymptomatic HTLV-1-carriers in a Peruvian population. *Immunology* **2009**, *128*, e777–e786. [CrossRef] [PubMed]

17. Marriott, S.J.; Semmes, O.J. Impact of HTLV-I Tax on cell cycle progression and the cellular DNA damage repair response. *Oncogene* **2005**, *24*, 5986–5995. [CrossRef] [PubMed]

18. Grassmann, R.; Aboud, M.; Jeang, K.T. Molecular mechanisms of cellular transformation by HTLV-1 Tax. *Oncogene* **2005**, *24*, 5976–5985. [CrossRef] [PubMed]

19. Cheng, H.; Ren, T.; Sun, S.C. New insight into the oncogenic mechanism of the retroviral oncoprotein Tax. *Protein Cell* **2012**, *3*, 581–589. [CrossRef] [PubMed]

20. Higuchi, M.; Fujii, M. Distinct functions of HTLV-1 Tax1 from HTLV-2 Tax2 contribute key roles to viral pathogenesis. *Retrovirology* **2009**, *6*. [CrossRef] [PubMed]

21. Jeang, K.T. HTLV-1 and adult T-cell leukemia: Insights into viral transformation of cells 30 years after virus discovery. *J. Formos. Med. Assoc.* **2010**, *109*, 688–693. [CrossRef]

22. Beimling, P.; Moelling, K. Direct interaction of CREB protein with 21 bp Tax-response elements of HTLV-ILTR. *Oncogene* **1992**, *7*, 257–262. [PubMed]

23. Satou, Y.; Yasunaga, J.; Yoshida, M.; Matsuoka, M. HTLV-I basic leucine zipper factor gene mRNA supports proliferation of adult T cell leukemia cells. *Proc. Natl. Acad. Sci. USA* **2006**, *103*, 720–725. [CrossRef] [PubMed]

24. Matsuoka, M. Human T-cell leukemia virus type I (HTLV-I) infection and the onset of adult T-cell leukemia (ATL). *Retrovirology* **2005**, *2*. [CrossRef] [PubMed]

25. Cabral, F.; Arruda, L.B.; de Araújo, M.L.; Montanheiro, P.; Smid, J.; de Oliveira, A.C.; Duarte, A.J.; Casseb, J. Detection of human T-cell lymphotropic virus type 1 in plasma samples. *Virus Res.* **2012**, *163*, 87–90. [CrossRef] [PubMed]

26. Journo, C.; Mahieux, R. HTLV-1 and innate immunity. *Viruses* **2011**, *3*, 1374–1394. [CrossRef] [PubMed]

27. Nagai, M.; Yamano, Y.; Brennan, M.B.; Mora, C.A.; Jacobson, S. Increased HTLV-I proviral load and preferential expansion of HTLV-I Tax-specific CD8+ T cells in cerebrospinal fluid from patients with HAM/TSP. *Ann. Neurol.* **2001**, *50*, 807–812. [CrossRef] [PubMed]

28. Gallo, R.C. Research and discovery of the first human cancer virus, HTLV-1. *Best Pract. Res. Clin. Haematol.* **2011**, *24*, 559–565. [CrossRef] [PubMed]

29. Morgan, O. HTLV-1 associated myelopathy/tropical spastic paraparesis: How far have we come? *West Indian Med. J.* **2011**, *60*, 505–512. [PubMed]

30. Ahuja, J.; Lepoutre, V.; Wigdahl, B.; Khan, Z.K.; Jain, P. Induction of pro-inflammatory cytokines by human T-cell leukemia virus type-1 Tax protein as determined by multiplexed cytokine protein array analyses of human dendritic cells. *Biomed. Pharmacother.* **2007**, *61*, 201–208. [CrossRef] [PubMed]

31. Yamano, Y.; Araya, N.; Sato, T.; Utsunomiya, A.; Azakami, K.; Hasegawa, D.; Izumi, T.; Fujita, H.; Aratani, S.; Yagishita, N.; *et al.* Abnormally high levels of virus-infected INF-γ+CCR4+CD4+CD25+ T cells in a retrovirus-associated neuroinflammatory disorder. *PLoS ONE* **2009**, *4*. [CrossRef] [PubMed]

32. Oliére, S.; Douville, R.; Sze, A.; Belgnaoui, S.M.; Hiscott, J. Modulation of innate immune responses during human T-cell leukemia virus (HTLV-1) pathogenesis. *Cytokine Growth Factor Rev.* **2011**, *22*, 197–210. [CrossRef] [PubMed]

33. Puccioni-Sohler, M.; Gasparetto, E.; Cabral-Castro, M.J.; Slatter, C.; Vidal, C.M.; Cortes, R.D.; Rosen, B.R.; Mainero, C. HAM/TSP: Association between white matter lesions on magnetic resonance imaging, clinical and cerebrospinal fluid findings. *Arq. Neuropsiquiatr.* **2012**, *70*, 246–251. [CrossRef] [PubMed]

34. Saito, M. Immunogenetics and the pathological mechanisms of human T-cell leukemia virustype 1-(HTLV-1-) associated myelopathy/tropical spastic paraparesis (HAM/TSP). *Interdiscip. Perspect. Infect. Dis.* **2010**, *2010*. [CrossRef]

35. Shoeibi, A.; Etemadi, M.; Ahmadi, A.M.; Amini, M.; Boostani, R. "HTLV-I Infection" twenty-year research in neurology department of mashhad university of medical sciences. *Iran. J. Basic Med. Sci.* **2013**, *16*, 202–207. [PubMed]

36. Satou, Y.; Matsuoka, M. HTLV-1 and the host immune system: How the virus disrupts immune regulation, leading to HTLV-1 associated diseases. *J. Clin. Exp. Hematop.* **2010**, *50*, 1–8. [CrossRef] [PubMed]

37. Fuzii, H.T.; Dias, G.A.S.; Barros, R.J.; Falcão, L.F.; Quaresma, J.A. Immunopathogenesis of HTLV-1-assoaciated myelopathy/tropical spastic paraparesis (HAM/TSP). *Life Sci.* **2014**, *104*, 9–14. [CrossRef] [PubMed]

38. Azran, I.; Schavinsky-Khrapunsky, Y.; Aboud, M. Role of Tax protein in human T-cell leukemia virus type-I leukemogenicity. *Retrovirology* **2004**, *1*. [CrossRef] [PubMed]

39. Kibler, K.V.; Jeang, K.T. CREB/ATF-Dependent repression of cyclin A by human T-cell leukemia virus Type 1 Tax protein. *J. Virol.* **2001**, *75*, 2161–2173. [CrossRef] [PubMed]

40. Matsumoto, J.; Ohshima, T.; Isono, O.; Shimotohno, K. HTLV-1 HBZ suppresses AP-1 activity by impairing both the DNA-binding ability and the stability of c-Jun protein. *Oncogene* **2005**, *24*, 1001–1010. [CrossRef] [PubMed]

41. Wildin, R.S.; Freitas, A. IPEX and FOXP3: Clinical and research perspectives. *J. Autoimmun.* **2005**, *25*, 56–62. [CrossRef] [PubMed]

42. Bacchetta, R.; Gambineri, E.; Roncarolo, M.G. Role of regulatory T cells and FOXP3 in human diseases. *J. Allergy Clin. Immunol.* **2007**, *120*, 227–235. [CrossRef] [PubMed]

43. Toulza, F.; Heaps, A.; Tanaka, Y.; Taylor, G.P.; Bangham, C.R. High frequency of CD4+FoxP3+ cells in HTLV-1 infection: Inverse correlation with HTLV-1-especific CTL response. *Blood* **2008**, *111*, 5047–5053. [CrossRef] [PubMed]

44. Brito-Melo, G.E.; Peruhype-Magalhães, V.; Teixeira-Carvalho, A.; Barbosa-Stancioli, E.F.; Carneiro-Proietti, A.B.; Catalan-Soares, B.; Ribas, J.G.; Martins-Filho, O.A. IL-10 produced by CD4+ and CD8+ T cells emerge as a putative immunoregulatory mechanism to counterbalance the monocyte-derived TNF-α and guarantee asymptomatic clinical status during chronic HTLV-I infection. *Clin. Exp. Immunol.* **2007**, *147*, 35–44. [CrossRef] [PubMed]

45. Santos, S.B.; Muniz, A.L.; Carvalho, E.M. HTLV-1-associated myelopathy Immunopathogenesis. *Gaz. Méd. Bahia* **2009**, *79*, 11–17.

46. Palmer, D.C.; Restifo, N.P. Suppressors of cytokine signaling (SOCS) in T cell differentiation, maturation, and function. *Trends Immunol.* **2009**, *30*, 592–602. [CrossRef] [PubMed]

47. Tamiya, T.; Kashiwagi, I; Takahashi, R.; Yasukawa, H.; Yoshimura, A. Suppressors of cytokine signaling (SOCS) proteins and JAK/STAT pathways: Regulation of T-cell inflammation by SOCS1 and SOCS3. *Arterioscler. Thromb. Vasc. Biol.* **2011**, *31*, 980–985. [CrossRef] [PubMed]

48. Yoshimura, A.; Suzuki, M.; Sakaguchi, R.; Hanada, T.; Yasukawa, H. SOCS, inflammation, and autoimmunity. *Front. Immunol.* **2012**, *12*. [CrossRef] [PubMed]

49. Nishiura, Y.; Nakamura, T.; Fukushima, N.; Moriuchi, R.; Katamine, S.; Eguchi, K. Increased mRNA expression of Th1-cytokine signaling molecules in patients with HTLV-I-associated myelopathy/tropical spastic paraparesis. *Tohoku J. Exp. Med.* **2004**, *204*, 289–298. [CrossRef] [PubMed]

50. Araújo, A.Q.; Leite, A.C.; Lima, M.A.; Silva, M.T. HTLV-1 and neurological conditions: When to suspect and when to order a diagnostic test for HTLV-1 infection? *Arq. Neuropsiquiatr.* **2009**, *67*, 132–138. [CrossRef] [PubMed]

51. Castro-Costa, C.M.; Araújo, A.Q.; Câmara, C.C.; Ferreira, A.S.; Santos, T.J.; Castro-Costa, S.B.; Alcântara, R.N.; Taylor, G.P. Pain in tropical spastic paraparesis/HTLV-I associated myelopathy patients. *Arq. Neuropsiquiatr.* **2009**, *67*, 866–870. [CrossRef] [PubMed]

52. Heraud, J.M.; Merien, F.; Mortreux, F.; Mahieux, R.; Kazanji, M. Immunological changes and cytokine gene expression during primary infection with human T-cell leukaemia virus type 1 in squirrel monkeys (*Saimiri sciureus*). *Virology* **2007**, *361*, 402–411. [CrossRef] [PubMed]

53. Michaëlsson, J.; Barbosa, H.M.; Jordan, K.A.; Chapman, J.M.; Brunialti, M.K.; Neto, W.K.; Nukui, Y.; Sabino, E.C.; Chieia, M.A.; Oliveira, A.S.; *et al.* The frequency of CD127$^{low}$ expressing CD4$^+$CD25$^{high}$ T regulatory cells is inversely correlated with human T lymphotrophic virus type-1 (HTLV-1) proviral load in HTLV-1-infection and HTLV-1-associated myelopathy/tropical spastic paraparesis. *BMC Immunol.* **2008**, *9*. [CrossRef] [PubMed]

54. Montes, M.; Sanchez, C.; Verdonck, K.; Lake, J.E.; Gonzalez, E.; Lopez, G.; Terashima, A.; Nolan, T.; Lewis, D.E.; Gotuzzo, E.; *et al.* Regulatory T cell expansion in HTLV-1 and strongyloidiasis co-infection is associated with reduced IL-5 responses to Strongyloides stercoralis antigen. *PLoS Negl. Trop. Dis.* **2009**, *3*. [CrossRef] [PubMed]

55. Santos, S.B.; Porto, A.F.; Muniz, A.L.; Jesus, A.R.; Magalhães, E.; Melo, A.; Dutra, W.O.; Gollob, K.J.; Carvalho, E.M. Exacerbated inflammatory cellular imune response characteristics of HAM/TSP is observed in a large proportionof HTLV-I asymptomatic carriers. *BMC Infect. Dis.* **2004**, *4*. [CrossRef]

56. Brito-Melo, G.E.; Martins-Filho, O.A.; Carneiro-Proietti, A.B.; Catalan-Soares, B.; Ribas, J.G.; Thorum, G.W.; Barbosa-Stancioli, E.F. Phenotypic study of peripheral blood leucocytes in HTLV-I-infected individuals from Minas Gerais, Brazil. *Scand. J. Immunol.* **2002**, *55*, 621–628. [CrossRef] [PubMed]

57. Coutinho, R.J.; Grassi, M.F.; Korngold, A.B.; Olavarria, V.N.; Galvão-Castro, B.; Mascarenhas, R.E. Human T lymphotropic virus type 1 (HTLV-1) proviral load induces activation of T-lymphocytes in asymptomatic carriers. *BMC Infect. Dis.* **2014**, *22*. [CrossRef] [PubMed]

58. Walker, L.S.; Abbas, A.K. The enemy within: Keeping self-reactive T cells at bay in the periphery. *Nat. Rev. Immunol.* **2002**, *2*, 11–19. [CrossRef] [PubMed]

59. O'Garra, A.; Vieira, P. Regulatory T cells and mechanisms of immune system control. *Nat. Med.* **2004**, *10*, 801–805. [CrossRef] [PubMed]

60. Moudgil, K.D.; Choubey, D. Cytokines in autoimmunity: Role in induction, regulation, and treatment. *J. Interferon Cytokine Res.* **2011**, *31*, 695–703. [CrossRef] [PubMed]

61. Parks, C.G.; Miller, F.W.; Pollard, K.M.; Selmi, C.; Germolec, D.; Joyce, K.; Rose, N.R.; Humble, M.C. Expert panel workshop consensus statement on the role of the environment in the development of autoimmune disease. *Int. J. Mol. Sci.* **2014**, *15*, 14269–14297. [CrossRef] [PubMed]

62. Cárdenas-Roldán, J.; Rojas-Villarraga, A.; Anaya, J.M. How do autoimmune diseases cluster in families? A systematic review and meta-analysis. *BMC. Med.* **2013**, *73*. [CrossRef]

63. Lossius, A.; Johansen, J.N.; Torkildsen, O.; Vartdal, F.; Holmøy, T. Epstein-Barr virus in systemic lupus erythematosus, rheumatoid arthritis and multiple sclerosis-association and causation. *Viruses* **2012**, *4*, 3701–3730. [CrossRef] [PubMed]

64. Getts, D.R.; Chastain, E.M.; Terry, R.L.; Miller, S.D. Virus infection, antiviral immunity, and autoimmunity. *Immunol. Rev.* **2013**, *255*, 197–209. [CrossRef] [PubMed]

65. Levin, M.C.; Lee, S.M.; Kalume, F.; Morcos, Y.; Dohan, F.C.; Hasty, K.A.; Callaway, J.C.; Zunt, J.; Desiderio, D.; Stuart, J.M. Autoimmunity due to molecular mimicry as a cause of neurological disease. *Nat. Med.* **2002**, *8*, 509–513. [CrossRef] [PubMed]

66. García-Vallejo, F.; Domínguez, M.C.; Tamayo, O. Autoimmunity and molecular mimicry in tropical spastic paraparesis/human T-lymphotropic virus-associated myelopathy. *Braz. J. Med. Biol. Res.* **2005**, *38*, 241–250. [PubMed]

67. Alamanos, Y.; Voulgari, P.V.; Drosos, A.A. Incidence and prevalence of rheumatoid arthritis, based on the 1987 American College of Rheumatology criteria: A systematic review. *Semin. Arthritis Rheum.* **2006**, *36*, 182–188. [CrossRef] [PubMed]
68. Mota, L.M.H.; Laurindo, I.M.M.; Santos Neto, L.L. Artrite reumatoide inicial–conceitos. *Rev. Assoc. Med. Bras.* **2010**, *56*, 227–229. [CrossRef] [PubMed]
69. Coenen, M.J.H.; Gregersen, P.K. Rheumatoid arthritis: A view of the current genetic landscape. *Genes Immun.* **2008**, *77*, 1–11. [CrossRef] [PubMed]
70. Firestein, G.S. Evolving concepts of rheumatoid arthritis. *Nature* **2003**, *5*, 356–361. [CrossRef] [PubMed]
71. Feldmann, M.; Brennan, F.M. Rheumatoid arthritis. *Cell* **1996**, *85*, 307–310. [CrossRef]
72. Meyer, P.W.; Hodkinson, B.; Ally, M.; Musenge, E.; Wadee, A.A.; Fickl, H.; Tikly, M.; Anderson, R. Circulating cytokine profiles and their relationships with autoantibodies, acute phase reactants, and disease activity in patients with rheumatoid arthritis. *Med. Inflamm.* **2010**, *2010*. [CrossRef] [PubMed]
73. Steiner, G.; Tohidast-Akrad, M.; Witzmann, G.; Vesely, M.; Studnicka-Benke, A.; Gal, A.; Kunaver, M.; Zenz, P.; Smolen, J.S. Cytokine production by synovial T cells in rheumatoid arthritis. *Rheumatology* **1999**, *38*, 202–213. [CrossRef] [PubMed]
74. Nishioka, K.; Nakajima, T.; Hasunuma, T.; Sato, K. Rheumatic manifestation of human leukemia virus infection. *Rheum. Dis. Clin. N. Am.* **1993**, *19*, 489–503.
75. Yakova, M.; Lézin, A.; Dantin, F.; Lagathu, G.; Olindo, S.; Jean-Baptiste, G.; Arfi, S.; Césaire, R. Increased proviral load in HTLV-1-infected patients with rheumatoid arthritis or connective tissue disease. *Retrovirology* **2005**, *1*. [CrossRef]
76. Firestein, G.S.; Zvaifler, N.J. Rheumatoid arthritis: A disease of disordered immunity. In *Inflammation: Basic Principles and Clinical Correlates*, 2nd ed.; Raven Press Ltd.: New York, NY, USA, 1992; pp. 959–975.
77. Eguchi, K.; Origuchi, T.; Takashima, H.; Iwata, K.; Katamine, S.; Nagataki, S. High seroprevalence of anti-HTLV-1 antibody in rheumatoid arthritis. *Arthritis Rheum.* **1996**, *39*, 463–466. [CrossRef] [PubMed]
78. Brzustewicz, E.; Bryl, E. The role of cytokines in the pathogenesis of rheumatoid arthritis—Practical and potential application of cytokines as biomarkers and targets of personalized therapy. *Cytokine* **2015**, *76*, 527–536. [CrossRef] [PubMed]
79. Ijichi, S.; Matsuda, T.; Maruyama, I.; Izumihara, T.; Kojima, K.; Niimura, T.; Maruyama, Y.; Sonoda, S.; Yoshida, A.; Osame, M. Arthritis in a human T lymphotropic virus type I (HTLV-I) carrier. *Ann. Rheum. Dis.* **1990**, *49*, 718–721. [CrossRef] [PubMed]
80. Fox, R.I.; Tornwall, J.; Maruyama, T.; Stern, M. Evolving concepts of diagnosis, pathogenesis, and therapy of Sjogren's syndrome. *Curr. Opin. Rheumatol.* **1998**, *10*, 446–456. [CrossRef] [PubMed]
81. Ambrosi, A.; Wahren-Herlenius, M. Update on the immunobiology of Sjögren's syndrome. *Curr. Opin. Rheumatol.* **2015**, *27*, 468–475. [CrossRef] [PubMed]
82. Eguchi, K.; Matsuoka, N.; Ida, H.; Nakashima, M.; Sakai, M.; Sakito, S.; Kawakami, A.; Terada, K.; Shimada, H.; Kawabe, Y. Primary Sjögren's syndrome with antibodies to HTLV-1: Clinical and laboratory features. *Ann. Rheum. Dis.* **1992**, *51*, 769–776. [CrossRef] [PubMed]
83. Vernant, J.C.; Buisson, G.; Magdeleine, J.; Thore, J.; Jouannelle, A.; Neisson-Vernant, C.; Monplaisir, N. T-lymphocyte alveolitis, tropical spastic paresis, and Sjogren syndrome. *Lancet* **1988**, *1*. [CrossRef]
84. Nakamura, H.; Takahashi, Y.; Yamamoto-Fukuda, T.; Horai, Y.; Nakashima, Y.; Arima, K.; Nakamura, T.; Koji, T.; Kawakami, A. Direct infection of primary salivary gland epithelial cells by human T lymphotropic virus type I in patients with Sjögren's syndrome. *Arthritis Rheumatol.* **2015**, *67*, 1096–1106. [CrossRef] [PubMed]
85. Scofield, L.; Alarcón, G.S.; Cooper, G.S. Employment and disability issues in systemic lupus erythematosus: A review. *Arthritis Care Res.* **2008**, *59*, 1475–1479. [CrossRef] [PubMed]
86. Cooper, G.S.; Gilbert, K.M.; Greidinger, E.L.; James, J.A.; Pfau, J.C.; Reinlib, L.; Richardson, B.C.; Rose, N.R. Recent advances and opportunities in research on lupus: Environmental influences and mechanisms of disease. *Cien. Saude Colet.* **2009**, *14*, 1865–1876. [CrossRef] [PubMed]
87. Tiffin, N.; Adeyemo, A.; Okpechi, I. A diverse array of genetic factors contribute to the pathogenesis of systemic lupus erythematosus. *Orphanet J. Rare Dis.* **2013**, *7*. [CrossRef] [PubMed]
88. Yap, D.Y.; Lai, K.N. Cytokines and their roles in the pathogenesis of systemic lupus erythematosus: From basics to recent advances. *J. Biomed. Biotechnol.* **2010**, *2010*. [CrossRef] [PubMed]

89. Shirdel, A.; Hashemzadeh, K.; Sahebari, M.; Rafatpanah, H.; Hatef, M.; Rezaieyazdi, Z.; Mirfeizi, Z.; Faridhosseini, R. Is there any association between human lymphotropic virus type I (HTLV-I) infection and systemic lupus erythematosus? *Iran. J. Basic Med. Sci.* **2013**, *16*, 252–257. [PubMed]

90. Sugimoto, T.; Okamoto, M.; Koyama, T.; Takashima, H.; Saeki, M.; Kashiwagi, A.; Horie, M. The occurrence of systemic lupus erythematosus in an asymptomatic carrier of human T-cell lymphotropic virus type I. *Clin. Rheumatol.* **2007**, *26*, 1005–1007. [CrossRef] [PubMed]

91. Akimoto, M.; Matsushita, K.; Suruga, Y.; Aoki, N.; Ozaki, A.; Uozumi, K.; Tei, C.; Arima, N. Clinical manifestations of human T lymphotropic virus type I-infected patients with systemic lupus erythematosus. *J. Rheumatol.* **2007**, *34*, 1841–1848. [PubMed]

92. Magistrelli, C.; Samoilova, E.; Agarwal, R.K.; Banki, K.; Ferrante, P.; Vladutiu, A.; Phillips, P.E.; Perl, A. Polymorphic genotypes of the HRES-1 human endogenous retrovirus locus correlate with systemic lupus erythematosus and autoreactivity. *Immunogenetics* **1999**, *49*, 829–834. [CrossRef] [PubMed]

93. Olsen, R.G.; Tarr, M.J.; Mathes, L.E.; Whisler, R.; du Plessis, D.; Schulz, E.J.; Blakeslee, J.R. Serological and virological evidence of human T-lymphotropic virus in systemic lupus erythematosus. *Med. Microbiol. Immunol.* **1987**, *176*, 53–64. [CrossRef] [PubMed]

*Review*

# Mother-to-Child Transmission of HTLV-1 Epidemiological Aspects, Mechanisms and Determinants of Mother-to-Child Transmission

**Florent Percher [1,2,3], Patricia Jeannin [1,2], Sandra Martin-Latil [4], Antoine Gessain [1,2], Philippe V. Afonso [1,2], Aurore Vidy-Roche [1,2,3] and Pierre-Emmanuel Ceccaldi [1,2,3,*]**

[1] Pasteur Institute, Virology Department, Epidemiology and Physiopathology of Oncogenic Viruses Unit, F-75015 Paris, France; florent.percher@pasteur.fr (F.P.); patricia.jeannin@pasteur.fr (P.J.); antoine.gessain@pasteur.fr (A.G.); philippe.afonso@pasteur.fr (P.V.A.); aurore.vidy@pasteur.fr (A.V.-R.)

[2] UMR CNRS 3569, Paris 75015, France

[3] Sorbonne Paris Cité, Cellule Pasteur, Université Paris Diderot, Institut Pasteur 75015, Paris

[4] ANSES, Enteric Viruses Unit, Maisons-Alfort 94706, France; sandra.martin-latil@anses.fr

\* Correspondence: pierre-emmanuel.ceccaldi@pasteur.fr; Tel.: +33-1-45-68-87-82; Fax: +33-1-40-61-34-65

Academic Editor: Louis M. Mansky

Received: 6 November 2015; Accepted: 27 January 2016; Published: 3 February 2016

**Abstract:** Human T-cell Lymphotropic Virus type 1 (HTLV-1) is a human retrovirus that infects at least 5–10 million people worldwide, and is the etiological agent of a lymphoproliferative malignancy; Adult T-cell Leukemia/Lymphoma (ATLL); and a chronic neuromyelopathy, HTLV-1 Associated Myelopathy/Tropical Spastic Paraparesis (HAM/TSP), as well as other inflammatory diseases such as infective dermatitis and uveitis. Besides sexual intercourse and intravenous transmission, HTLV-1 can also be transmitted from infected mother to child during prolonged breastfeeding. Some characteristics that are linked to mother-to-child transmission (MTCT) of HTLV-1, such as the role of proviral load, antibody titer of the infected mother, and duration of breastfeeding, have been elucidated; however, most of the mechanisms underlying HTLV-1 transmission during breast feeding remain largely unknown, such as the sites of infection and cellular targets as well as the role of milk factors. The present review focuses on the latest findings and current opinions and perspectives on MTCT of HTLV-1.

**Keywords:** human; HTLV-1; intestinal barrier; retrovirus; breastfeeding

---

## 1. Introduction

Human T-cell Leukemia Virus Type 1 (HTLV-1) infects at least 5–10 million people worldwide, mainly in highly endemic areas such as southern Japan, West/Central Africa, the Caribbean region, and parts of South America and Melanesia [1]. HTLV-1 infection is mostly associated with two distinct diseases: a lymphoproliferation, Adult T cell Leukemia/Lymphoma (ATLL), and an inflammatory neurological disease, tropical spastic paraparesis or HTLV-1 associated myelopathy (HAM/TSP). Additionally, HTLV-1 is associated with other inflammatory diseases such as infective dermatitis, some uveitis and some myositis. Although HTLV-1 preferentially infects CD4[+] T cells [2], CD8[+] T-cells may play an important role as reservoir in the host [3], and to a lesser extent, infected monocytes and B lymphocytes, dendritic cells, and endothelial cells may be found [4,5]. Different modes of transmission have been identified for HTLV-1: (1) sexual contact; (2) transfusion of contaminated blood; and (3) from mother to child (MTCT) [6]. In each case, such a transmission involves the transfer of infected body fluid (semen, blood, and milk, respectively). In the case of MTCT, cohort studies on HTLV-1 infected carriers indicate that infection during childhood is a potent risk factor for the development of ATLL [7]. It is now clear that HTLV-1 MTCT mainly involves prolonged breastfeeding, as demonstrated by epidemiological, virological and experimental data. However, the mechanisms of such a transmission remain largely unknown. For instance, the nature of the infected cells present in the milk, the anatomical sites of viral entry through the mucosa, the first cellular targets of infection, the role of anti-HTLV-1 antibodies present in breast milk, and the role of other milk factors that may influence MTCT have not been completely addressed. The present review focuses on such mechanisms, current studies and perspectives.

## 2. Evidence of HTLV-1 MTCT during Breastfeeding

First evidence of HTLV-1 transmission from infected mother to children during lactation has been brought by epidemiological studies. HTLV-1 infection was more prevalent among breastfed children than bottle-fed children in Japan [8,9], with a rate of seroconversion of 15.7% among children that had been breastfed for 12 months, compared to 3.6% for bottle-fed children for similar period [10]. Moreover, there is a correlation between MTCT rate and breastfeeding duration. Thus, in a prospective study in Jamaica, Wiktor *et al.* [11] reported that breastfeeding beyond 12 months was associated with a transmission rate of 32%, compared to a transmission rate of 9% for shorter breastfeeding durations. Similarly, Takahashi *et al.* [12] showed that a six-month duration of breastfeeding was a critical point in the rate of seroconversion, since rates of 4.4% and 14.4% were found for children that had been breastfed for periods under six months or over seven months, respectively. A major piece of evidence supporting HTLV-1 transmission through breastfeeding has been brought in the 1980s in Japan, where Hino and coworkers started a pilot study to screen pregnant women in Nagasaki Prefecture for anti-HTLV-1 antibodies (for a review, see [13]). HTLV-1 prevalence was around 4%. Interestingly, HTLV-1 prevalence among the elder children of the HTLV-1 carrier mothers was approximately 20%, and mothers of the HTLV-1 positive children were usually HTLV-1 positive (92%), thus showing evidence of MTCT. More importantly, in 1987, the ATLL Prevention Program Nagasaki, which aimed to refrain seropositive mothers from breastfeeding in the Nagasaki Prefecture, resulted in a huge reduction of HTLV-1 MTCT from 20.3% to 2.5% [14]. The major importance of breastfeeding in HTLV-1 MTCT was later confirmed in other areas [15]. Of note, this residual rate (2.5%) of MTCT in the absence of breastfeeding raised the possibility of minor secondary routes, such as contamination during delivery, or intrauterine transmission. This latter route remains controversial, since contradictory studies on the presence of HTLV-1 in cord-blood samples from seropositive babies have been reported [16–18].

From a virological point of view, it is known that many retroviruses may be transmitted via breast milk, such as Moloney murine leukemia virus [19,20], Mouse Mammary Tumor Virus [21], or Caprine Arthritis Encephalitis Virus [22]. Concerning HTLV-1, viral antigens [23], and antibodies to HTLV-1 were found in the milk of seropositive mothers [24]. The proviral load in breast milk is strongly predictive of the risk of MTCT, increasing from 4.7/1000 person-months for a provirus load in milk lower than 0.18% to 28.7/1000 person-months for a provirus load higher than 1.5% [25].

From an experimental point of view, oral inoculation of peripheral blood lymphocytes isolated from ATLL patients to adult common marmosets (*Callithrix jacus*) was able to induce seroconversion within 2.5 months, and the virus was detected in the animal peripheral blood lymphocytes [26]. This study also demonstrated that $5.6 \times 10^7$ cells from ATLL patients were sufficient to infect these animals. Similarly, oral inoculation of concentrated fresh milk from HTLV-1 seropositive mothers in the same animal model could transmit the infection [27]. Oral transmission of HTLV-1 could be likewise observed in other animal models. Oral inoculation of four rabbits for eight weeks with an HTLV-1-infected rabbit lymphoid cell line resulted in the seroconversion of one animal, and it was possible to generate from this animal a lymphoid cell line productively infected with HTLV-1 [28]. Similarly, oral inoculation of HTLV-1-infected lymphoid cell line (*i.e.*, MT-2) to rats induced a persistent HTLV-1 infection in the absence of both humoral and cellular immune responses [29].

## 3. The Mechanisms of HTLV-1 MTCT

Altogether, these data indicate that breastfeeding is a major route for HTLV-1 MTCT. However, the mechanisms of HTLV-1 passage through the digestive tract remain largely unknown.

A first point to address concerns the source of HTLV-1 infection in breast milk. It is known that cell-to-cell contact is required for efficient viral transmission *in vivo* [30] as well as *in vitro* [31,32], except for dendritic cells that can be infected directly with cell-free HTLV-1 virions [33]. Cell free virions have not been detected so far in breast milk, thus the potential source of infection in breast milk may come from infected cells, such as lymphocytes, macrophages, or breast epithelial mammary cells. Since it has been estimated that breastfed children ingest an average of $10^8$ leucocytes a day, considering prolonged breastfeeding [34,35], infected lymphocytes could provide a strong source of infection in milk [36]. HTLV-1 infected mononuclear cells can be found in the milk from seropositive mothers during early lactation [23,37]. Of note, cellular components in breast milk can be found even after long-term lactation (over 5 years) [38], even if the ratio between the different cell types varies along the time: for example, the major part of cells in mother's early milk and colostrum is constituted of macrophages [39]. It has been found that leukocytes and epithelial cells from the mammary gland are susceptible to HTLV-1 infection [38]. This observation was confirmed by the evidence that mammary basal epithelial cells can be productively infected with HTLV-1 and are able to transfer infection to peripheral blood lymphocytes [40,41]. In addition, in a case report of an ATLL male patient with pseudogynecomasty, breast biopsy revealed the presence of mammary epithelial cells productively infected with HTLV-1 [42]. Such data support the hypothesis that basal and/or luminal epithelial cells may constitute a reservoir of HTLV-1 infectivity. The importance of mammary epithelial cells in viral transmission during lactation has also been evoked for Bovine Leukemia Virus (BLV), another deltaretrovirus that is transmitted from BLV-infected cows to calves during lactation [43]. Whatever the cell types involved, this can provide a more or less continuous source of infection.

A second question concerning the mechanisms of HTLV-1 transmission through the digestive tract is the anatomical site of viral entry. Up until now, no studies have addressed this question; in fact, animal models studies (marmoset, rabbit, and rat) on oral inoculation of HTLV-1 were mainly focused on seroconversion, progression towards an ATLL-like disease, and immune status of the host. Among the different possible sites of entry, the palatine tonsils and gut seem to be of particular importance due to their enrichment in possible targets (lymphoid cells and M cells), their function in antigen sampling [44] and their structure. Moreover, although it does not constitute a proof of viral entry, HTLV-1 may be retrieved in these structures, as shown in the tonsils for HTLV-1 infected humans [45], in the intestine and mesenteric lymph nodes of squirrel monkeys inoculated intravenously with HTLV-1 [46] and in Gut-Associated-Lymphoid Tissues from orally inoculated rabbits [47]. Whatever the primary sites of HTLV-1 passage/infection, tonsils and/or gut, the virus encounters an epithelium, pluristratified or monostratified, respectively. This allows *in vitro* studies using classical models of human epithelial cell monolayers.

Very few *in vitro* studies have been published concerning the mechanisms of passage of HTLV-1 across the intestinal barrier. In 1992, Zacharopoulos *et al.* [48] indicated that a human enterocytic cell lines (*i.e.*, I407) was susceptible to HTLV-1 infection *in vitro*, as shown by electron microscopy, *in situ* hybridization, and PCR amplification. However, it was unclear if the cells were fully differentiated as no data on the epithelium integrity and tight junctions were shown. In contrast, our laboratory performed studies on the susceptibility of three different human enterocytic cell lines on compartmentalized culture devices, with assessment of the enterocyte differentiation: existence of a tight epithelial barrier was checked by electron and confocal microscopy, and the trans-epithelial resistance was measured [49]. In this study, as summarized in Figure 1, it was demonstrated that HTLV-1 infected lymphocytes were unable to disrupt the epithelial barrier integrity or infect human enterocytes, in contrast to previous studies on a blood–brain barrier model in human brain endothelial cells [5,50]. However, it was shown that HTLV-1 virions were able to cross the epithelial barrier by transcytosis mechanism, and productively infect underlying human dendritic cells [49].

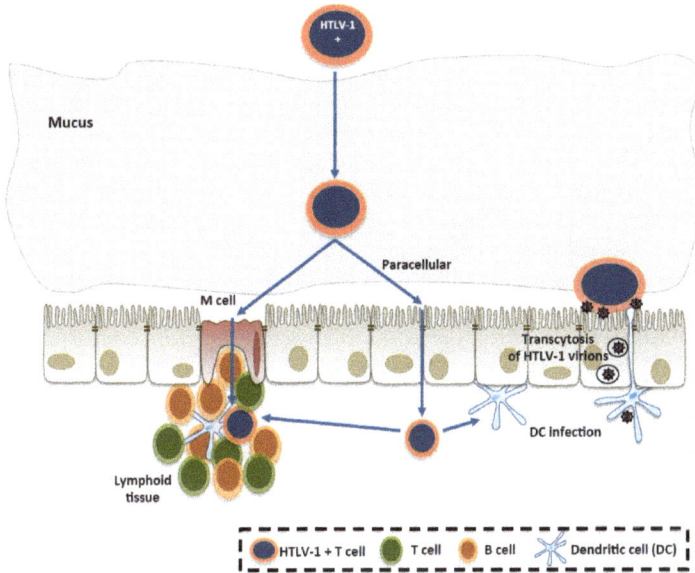

**Figure 1.** Hypothetical mechanisms of HTLV-1 (Human T-cell Lymphotropic Virus type 1 ) passage through the intestinal epithelium.

HTLV-1-infected lymphocytes, once they have crossed the mucus layer, could either pass through the epithelium via M cells, or in-between enterocytes (paracellular passage), to reach the lamina propria where potential targets of HTLV-1 infection, such as T lymphocytes (and/or dendritic cells and B lymphocytes) are located. From *in vitro* studies [49], the infection of enterocytes or the epithelial layer disruption seems to be excluded. Another possibility, suggested in the same study, could be viral transcytosis through the enterocyte, with infection of underlying dendritic cells.

Interestingly, viral transcytosis through enterocytes was observed only in the presence of HTLV-1 infected lymphocytes, and not in the case of purified virions. Such a mechanism of viral transcytosis is reminiscent of previous work showing transcytosis of HIV across an epithelial barrier, without infection of enterocytes, and subsequent infection of macrophages or CD4 lymphocytes located to the basal side of the epithelium [51]. These *in vitro* results highlight one of the potential mechanisms proposed for HTLV-1 passage, which are summarized in Figure 1.

## 4. Determinants of HTLV-1 MTCT

Studies on HTLV-1 MTCT determinants have focused mainly on genetic host factors, immunological host factors, lactation duration and milk components.

The genetic host factors that control HTLV-1 infection by breastfeeding have been investigated by Plancoulaine *et al.* [52], who began in the 1990s a large epidemiological study in endemic villages of French Guiana. The authors found a dominant major gene predisposing to HTLV-1 infection, in addition to the expected familial correlations due to the transmission routes (mother to child and spouse to spouse) [53]. Previous studies had shown that *HLA* (Human Leukocyte Antigen) genes distribution was different for ATLL or HAM/TSP patients compared to asymptomatic carriers (for example, see [54]), but the study by Plancoulaine *et al.* [53] indicated a genetic predisposition for HTLV-1 infection itself for 1.5% of the population, which concerned almost all infected children under 10 years of age, *i.e.*, infected through breastfeeding. Further studies allowed mapping a major susceptibility locus for HTLV-1 infection during childhood to chromosome 6q27 [55].

Concerning the immunological factors involved in HTLV-1 MTCT, the role of maternal anti-HTLV-1 antibodies may appear controversial. Such studies have to take into account the duration of breastfeeding, since the protective role of anti-HTLV-1 antibodies has been demonstrated in a rabbit model of infection, where passive immunization was shown to prevent milk-borne transmission of HTLV-1 to offspring [56]. Moreover, it has been shown *in vitro* that the addition of HTLV-1 serum cord blood plasma is able to prevent the infection of human neonatal lymphocytes when co-cultured with breast-milk cells of HTLV-1 carrier mothers [12]. However, it has been suggested that higher anti-HTLV-1 antibodies titer in the serum of the mother, as well as the presence of anti-Tax antibodies, is associated with a higher risk of children infection [11,52,57–59]. However, a high anti-HTLV-1 antibody titer in the serum may be correlated with a high provirus load in PBMCs, which is a risk factor for HTLV-1 MTCT [57]. In an analysis including the provirus load in maternal PBMCs, the presence of anti-Tax antibodies and the anti-HTLV-1 titers, it was found that a higher maternal proviral load and a higher anti-HTLV-1 antibody titer were independently associated with a higher risk of HTLV-1 MTCT, whereas the presence of anti-Tax antibodies was not [60].

Another point to take into account concerns the other milk components that may influence HTLV-1 MTCT transmission. As an example, it has been shown that lactoferrin, an iron-binding milk glycoprotein, was able to enhance HTLV-1 replication, by transcriptional activation of HTLV-1 Long Terminal Repeat (LTR), the viral promoter [61]. This effect on HTLV-1 infection seems to be specific since lactoferrin did not show any effect on HIV-1 LTR, and was even able to inhibit HIV-1 infection, probably by interfering with viral fusion and entry steps [61]. As an interesting example of "positive feedback", the same authors further demonstrated that lactoferrin expression was up-regulated during HTLV-1 infection, probably in a paracrine manner involving Tax-induced NF-κB activation [62].

## 5. Ongoing Research on HTLV-1 MTCT and Perspectives

One of the major remaining questions on MTCT concerns the sites of primary passage/infection of HTLV-1 in the digestive tract. The mechanisms of HTLV-1 infection after oral inoculation should be addressed *in vivo* using a humanized mouse as a model of HTLV-1 infection [63–65], in complement to the rabbit model. In particular, combination of histopathological studies and bioluminescence imaging will allow determining the preferential sites of HTLV-1 entry (palatine tonsils, and/or gut). In parallel, the use of transgenic, knock-out, and knock-in mice depleted for different cell types (M cells, dendritic cells, and macrophages) will allow assessing the role of the different cell types in the first steps of infection. These studies will also benefit from an *in vitro* approach, such as differentiation of M cells from enterocytic cell lines, on compartmentalized culture devices, as already done to show the role of these cells in virus transport across the epithelium [66]. Combination of *in vivo/in vitro* studies will also allow delineating the role of factors such as milk components (e.g., lactoperoxidase), the proviral load, and the antibody titer in HTLV-1 transport through the intestinal epithelium.

Another perspective concerns the comprehension of the role of breastfeeding duration on HTLV-1 MTCT. It seems rather clear that such a role corresponds to a combination of the cumulative viral input, the changes over time in milk composition in infected cell types, maternal antibodies, and the immune status and maturation of the neonate's gut. It is known that neonatal life (as well as prenatal) organizes and controls mucosal homeostasis through endogenous and exogenous factors that drive the development and maturation of the intestinal immune system (for a review, see [67]). Recent studies have shown the effects of intestinal microbiota in the development of the immune system and intestinal architecture [68,69]. Gut microbiota could have an impact on HTLV-1 MTCT efficiency, as reported for HIV-1 [70] or Mouse Mammary Tumor Virus [71], as well as the human milk microbiota, which participate in the neonate's gut microbiota constitution [72].

Altogether, these further studies could delineate new preventive strategies to counteract HTLV-1 MTCT, as well as provide new data on the general mechanisms of pathogenic agents in MTCT.

*Viruses* **2016**, *8*, 40

**Acknowledgments:** The authors thank Florence Buseyne, Olivier Cassar and Catherine Cecilio for helpful advices and improvement of the manuscript. Florent Percher is a recipient of a DIM Malinf (Région Ile de France) Ph.D. fellowship. Part of this work is financially supported by the Ligue contre le Cancer.

**Author Contributions:** All the authors conceived and wrote the paper.

**Conflicts of Interest:** The authors declare no conflict of interest.

## References

1. Gessain, A.; Cassar, O. Epidemiological aspects and world distribution of HTLV-1 Infection. *Front. Microbiol.* **2012**, *3*, 1–23. [CrossRef] [PubMed]

2. Richardson, J.H.; Edwards, A.J.; Cruickshank, J.K.; Rudge, P.; Dalgleish, A.G. In vivo cellular tropism of human T-cell leukemia virus type 1. *J. Virol.* **1990**, *64*, 5682–5687. [PubMed]

3. Nagai, M.; Brennan, M.B.; Sakai, J.A.; Mora, C.A.; Jacobson, S. CD8$^+$ T cells are an *in vivo* reservoir for human T-cell lymphotropic virus type I. *Blood* **2001**, *98*, 1858–1861. [CrossRef] [PubMed]

4. Koyanagi, Y.; Yoshida, T.; Suzuki, M.; Uma, A.; Ananthasubramaniam, L.; Ramajayam, S.; Yamamoto, N. Dual infection of HIV-1 and HTLV-I in south India: A study on a patient with AIDS-related complex. *Microbiol. Immunol.* **1993**, *37*, 983–986. [CrossRef] [PubMed]

5. Afonso, P.V.; Ozden, S.; Cumont, M.C.; Seilhean, D.; Cartier, L.; Rezaie, P.; Mason, S.; Lambert, S.; Huerre, M.; Gessain, A.; et al. Alteration of blood-brain barrier integrity by retroviral infection. *PLoS Pathog.* **2008**, *4*, e1000205. [CrossRef] [PubMed]

6. Carneiro-Proietti, A.B.; Amaranto-Damasio, M.S.; Leal-Horiguchi, C.F.; Bastos, R.H.; Seabra-Freitas, G.; Borowiak, D.R.; Ribeiro, M.A.; Proietti, F.A.; Ferreira, A.S.; Martins, M.L. Mother-to-Child transmission of human T-Cell lymphotropic viruses-1/2: What we know, and what are the gaps in understanding and preventing this route of infection. *J. Pediatr. Infect. Dis. Soc.* **2014**, *3* (Suppl. 1), S24–S29. [CrossRef] [PubMed]

7. Murphy, E.L.; Hanchard, B.; Figueroa, J.P.; Gibbs, W.N.; Lofters, W.S.; Campbell, M.; Goedert, J.J.; Blattner, W.A. Modelling the risk of adult T-cell leukemia/lymphoma in persons infected with human T-lymphotropic virus type I. *Int. J. Cancer* **1989**, *43*, 250–253. [CrossRef] [PubMed]

8. Ando, Y.; Nakano, S.; Saito, K.; Shimamoto, I.; Ichijo, M.; Toyama, T.; Hinuma, Y. Transmission of adult T-cell leukemia retrovirus (HTLV-I) from mother to child: Comparison of bottle- with breast-fed babies. *Jpn. J. Cancer Res.* **1987**, *78*, 322–324. [PubMed]

9. Hino, S.; Sugiyama, H.; Doi, H.; Ishimaru, T.; Yamabe, T.; Tsuji, Y.; Miyamoto, T. Breaking the cycle of HTLV-I transmission via carrier mothers' milk. *Lancet* **1987**, *2*, 158–159. [CrossRef]

10. Hino, S.; Katamine, S.; Miyata, H.; Tsuji, Y.; Yamabe, T.; Miyamoto, T. Primary prevention of HTLV-I in Japan. *J. Acquir. Immune Defic. Syndr. Hum. Retrovirol.* **1996**, *13* (Suppl. 1), S199–S203. [CrossRef] [PubMed]

11. Wiktor, S.Z.; Pate, E.J.; Rosenberg, P.S.; Barnett, M.; Palmer, P.; Medeiros, D.; Maloney, E.M.; Blattner, W.A. Mother-to-child transmission of human T-cell lymphotropic virus type I associated with prolonged breast-feeding. *J. Hum. Virol.* **1997**, *1*, 37–44. [PubMed]

12. Takahashi, K.; Takezaki, T.; Oki, T.; Kawakami, K.; Yashiki, S.; Fujiyoshi, T.; Usuku, K.; Mueller, N.; Osame, M.; Miyata, K.; et al. Inhibitory effect of maternal antibody on mother-to-child transmission of human T-lymphotropic virus type I. The Mother-to-Child Transmission Study Group. *Int. J. Cancer* **1991**, *49*, 673–677. [CrossRef] [PubMed]

13. Hino, S. Establishment of the milk-borne transmission as a key factor for the peculiar endemicity of human T-lymphotropic virus type 1 (HTLV-1): The ATL Prevention Program Nagasaki. *Proc. Jpn. Acad. Ser. B Phys. Biol. Sci.* **2011**, *87*, 152–166. [CrossRef] [PubMed]

14. Hino, S.; Katamine, S.; Miyata, H.; Tsuji, Y.; Yamabe, T.; Miyamoto, T. Primary prevention of HTLV-1 in Japan. *Leukemia* **1997**, *11* (Suppl. 3), 57–59. [CrossRef] [PubMed]

15. Ribeiro, M.A.; Martins, M.L.; Teixeira, C.; Ladeira, R.; Oliveira Mde, F.; Januario, J.N.; Proietti, F.A.; Carneiro-Proietti, A.B. Blocking vertical transmission of human T cell lymphotropic virus type 1 and 2 through breastfeeding interruption. *Pediatr. Infect. Dis. J.* **2012**, *31*, 1139–1143. [CrossRef] [PubMed]

16. Hino, S.; Yamaguchi, K.; Katamine, S.; Sugiyama, H.; Amagasaki, T.; Kinoshita, K.; Yoshida, Y.; Doi, H.; Tsuji, Y.; Miyamoto, T. Mother-to-child transmission of human T-cell leukemia virus type-I. *Jpn. J. Cancer Res.* **1985**, *76*, 474–480. [PubMed]

17. Komuro, A.; Hayami, M.; Fujii, H.; Miyahara, S.; Hirayama, M. Vertical transmission of adult T-cell leukaemia virus. *Lancet* **1983**, *1*, 240. [CrossRef]
18. Satow, Y.; Hashido, M.; Ishikawa, K.; Honda, H.; Mizuno, M.; Kawana, T.; Hayami, M. Detection of HTLV-I antigen in peripheral and cord blood lymphocytes from carrier mothers. *Lancet* **1991**, *338*, 915–916. [PubMed]
19. Chakraborty, J.; Clark, S.; Okonta, H.; Duggan, J. A small animal model for mother-to-fetus transmission of ts1, a murine retrovirus. *Viral Immunol.* **2003**, *16*, 191–201. [CrossRef] [PubMed]
20. Duggan, J.; Okonta, H.; Chakraborty, J. Transmission of Moloney murine leukemia virus (ts-1) by breast milk. *J. Gen. Virol.* **2006**, *87 Pt 9*, 2679–2684. [CrossRef] [PubMed]
21. Bittner, J.J. Relation of nursing to the extra-chromosomal theory of breast cancer in mice. *Am. J. Cancer* **1939**, *35*, 90–97. [PubMed]
22. Le Jan, C.; Bellaton, C.; Greenland, T.; Mornex, J.F. Mammary transmission of caprine arthritis encephalitis virus: A 3D model for *in vitro* study. *Reprod. Nutr. Dev.* **2005**, *45*, 513–523. [CrossRef] [PubMed]
23. Kinoshita, K.; Hino, S.; Amagaski, T.; Ikeda, S.; Yamada, Y.; Suzuyama, J.; Momita, S.; Toriya, K.; Kamihira, S.; Ichimaru, M. Demonstration of adult T-cell leukemia virus antigen in milk from three sero-positive mothers. *Gann* **1984**, *75*, 103–105. [PubMed]
24. Matsubara, F.; Haraguchi, K.; Harada, K.; Koizumi, A. Screening for antibodies to human T-cell leukemia virus type I in Japanese breast milk. *Biol. Pharm. Bull.* **2012**, *35*, 773–776. [CrossRef] [PubMed]
25. Li, H.C.; Biggar, R.J.; Miley, W.J.; Maloney, E.M.; Cranston, B.; Hanchard, B.; Hisada, M. Provirus load in breast milk and risk of mother-to-child transmission of human T lymphotropic virus type I. *J. Infect. Dis.* **2004**, *190*, 1275–1278. [CrossRef] [PubMed]
26. Yamanouchi, K.; Kinoshita, K.; Moriuchi, R.; Katamine, S.; Amagasaki, T.; Ikeda, S.; Ichimaru, M.; Miyamoto, T.; Hino, S. Oral transmission of human T-cell leukemia virus type-I into a common marmoset (*Callithrix jacchus*) as an experimental model for milk-borne transmission. *Jpn. J. Cancer Res.* **1985**, *76*, 481–487. [PubMed]
27. Kinoshita, K.; Yamanouchi, K.; Ikeda, S.; Momita, S.; Amagasaki, T.; Soda, H.; Ichimaru, M.; Moriuchi, R.; Katamine, S.; Miyamoto, T.; *et al.* Oral infection of a common marmoset with human T-cell leukemia virus type-I (HTLV-I) by inoculating fresh human milk of HTLV-I carrier mothers. *Jpn. J. Cancer Res.* **1985**, *76*, 1147–1153. [PubMed]
28. Uemura, Y.; Kotani, S.; Yoshimoto, S.; Fujishita, M.; Yano, S.; Ohtsuki, Y.; Miyoshi, I. Oral transmission of human T-cell leukemia virus type I in the rabbit. *Jpn. J. Cancer Res.* **1986**, *77*, 970–973. [PubMed]
29. Kato, H.; Koya, Y.; Ohashi, T.; Hanabuchi, S.; Takemura, F.; Fujii, M.; Tsujimoto, H.; Hasegawa, A.; Kannagi, M. Oral administration of human T-cell leukemia virus type 1 induces immune unresponsiveness with persistent infection in adult rats. *J. Virol.* **1998**, *72*, 7289–7293. [PubMed]
30. Okochi, K.; Sato, H.; Hinuma, Y. A retrospective study on transmission of adult T cell leukemia virus by blood transfusion: Seroconversion in recipients. *Vox Sang.* **1984**, *46*, 245–253. [CrossRef] [PubMed]
31. Yamamoto, N.; Okada, M.; Koyanagi, Y.; Kannagi, M.; Hinuma, Y. Transformation of human leukocytes by cocultivation with an adult T cell leukemia virus producer cell line. *Science* **1982**, *217*, 737–739. [CrossRef] [PubMed]
32. Popovic, M.; Sarin, P.S.; Robert-Gurroff, M.; Kalyanaraman, V.S.; Mann, D.; Minowada, J.; Gallo, R.C. Isolation and transmission of human retrovirus (human T-cell leukemia virus). *Science* **1983**, *219*, 856–859. [CrossRef] [PubMed]
33. Jones, K.S.; Petrow-Sadowski, C.; Huang, Y.K.; Bertolette, D.C.; Ruscetti, F.W. Cell-free HTLV-1 infects dendritic cells leading to transmission and transformation of CD4+ T cells. *Nat. Med.* **2008**, *14*, 429–436. [CrossRef] [PubMed]
34. Jarvinen, K.M.; Geller, L.; Bencharitiwong, R.; Sampson, H.A. Presence of functional, autoreactive human milk-specific IgE in infants with cow's milk allergy. *Clin. Exp. Allergy* **2012**, *42*, 238–247. [CrossRef] [PubMed]
35. Ogra, S.S.; Weintraub, D.I.; Ogra, P.L. Immunologic aspects of human colostrum and milk: Interaction with the intestinal immunity of the neonate. *Adv. Exp. Med. Biol.* **1978**, *107*, 95–107. [PubMed]
36. Proietti, F.A.; Carneiro-Proietti, A.B.; Catalan-Soares, B.C.; Murphy, E.L. Global epidemiology of HTLV-I infection and associated diseases. *Oncogene* **2005**, *24*, 6058–6068. [CrossRef] [PubMed]
37. Kinoshita, K.; Amagasaki, T.; Hino, S.; Doi, H.; Yamanouchi, K.; Ban, N.; Momita, S.; Ikeda, S.; Kamihira, S.; Ichimaru, M.; *et al.* Milk-borne transmission of HTLV-I from carrier mothers to their children. *Jpn. J. Cancer Res.* **1987**, *78*, 674–680. [PubMed]

38. Southern, S.O.; Southern, P.J. Persistent HTLV-I infection of breast luminal epithelial cells: A role in HTLV transmission? *Virology* **1998**, *241*, 200–214. [CrossRef] [PubMed]
39. Satomi, M.; Shimizu, M.; Shinya, E.; Watari, E.; Owaki, A.; Hidaka, C.; Ichikawa, M.; Takeshita, T.; Takahashi, H. Transmission of macrophage-tropic HIV-1 by breast-milk macrophages via DC-SIGN. *J. Infect. Dis.* **2005**, *191*, 174–181. [CrossRef] [PubMed]
40. LeVasseur, R.J.; Southern, S.O.; Southern, P.J. Mammary epithelial cells support and transfer productive human T-cell lymphotropic virus infections. *J. Hum. Virol.* **1998**, *1*, 214–223. [PubMed]
41. Takeuchi, H.; Takahashi, M.; Norose, Y.; Takeshita, T.; Fukunaga, Y.; Takahashi, H. Transformation of breast milk macrophages by HTLV-I: Implications for HTLV-I transmission via breastfeeding. *Biomed. Res.* **2010**, *31*, 53–61. [CrossRef] [PubMed]
42. Loureiro, P.; Southern, S.O.; Southern, P.J.; Pombo-de-Oliveira, M.S. Clinicopathological studies of a patient with adult T-cell leukemia and pseudogynecomasty. *Am. J. Hematol.* **2000**, *65*, 256–259. [CrossRef]
43. Buehring, G.C.; Kramme, P.M.; Schultz, R.D. Evidence for bovine leukemia virus in mammary epithelial cells of infected cows. *Lab. Invest.* **1994**, *71*, 359–365. [PubMed]
44. Schulz, O.; Pabst, O. Antigen sampling in the small intestine. *Trends Immunol.* **2013**, *34*, 155–161. [CrossRef] [PubMed]
45. Takenouchi, N.; Matsuoka, E.; Moritoyo, T.; Nagai, M.; Katsuta, K.; Hasui, K.; Ueno, K.; Eizuru, Y.; Usuku, K.; Osame, M.; *et al.* Molecular pathologic analysis of the tonsil in HTLV-I-infected individuals. *J. Acquir. Immune Defic. Syndr.* **1999**, *22*, 200–207. [CrossRef] [PubMed]
46. Kazanji, M. HTLV type 1 infection in squirrel monkeys (*Saimiri sciureus*): A promising animal model for HTLV type 1 human infection. *AIDS Res. Hum. Retrovir.* **2000**, *16*, 1741–1746. [CrossRef] [PubMed]
47. Haynes, R.A., 2nd; Ware, E.; Premanandan, C.; Zimmerman, B.; Yu, L.; Phipps, A.J.; Lairmore, M.D. Cyclosporine-induced immune suppression alters establishment of HTLV-1 infection in a rabbit model. *Blood* **2010**, *115*, 815–823. [CrossRef] [PubMed]
48. Zacharopoulos, V.R.; Perotti, M.E.; Phillips, D.M. Lymphocyte-facilitated infection of epithelia by human T-cell lymphotropic virus type I. *J. Virol.* **1992**, *66*, 4601–4605. [PubMed]
49. Martin-Latil, S.; Gnadig, N.F.; Mallet, A.; Desdouits, M.; Guivel-Benhassine, F.; Jeannin, P.; Prevost, M.C.; Schwartz, O.; Gessain, A.; Ozden, S.; *et al.* Transcytosis of HTLV-1 across a tight human epithelial barrier and infection of subepithelial dendritic cells. *Blood* **2012**, *120*, 572–580. [CrossRef] [PubMed]
50. Afonso, P.V.; Ozden, S.; Prevost, M.C.; Schmitt, C.; Seilhean, D.; Weksler, B.; Couraud, P.O.; Gessain, A.; Romero, I.A.; Ceccaldi, P.E. Human blood-brain barrier disruption by retroviral-infected lymphocytes: Role of myosin light chain kinase in endothelial tight-junction disorganization. *J. Immunol.* **2007**, *179*, 2576–2583. [CrossRef] [PubMed]
51. Hocini, H.; Bomsel, M. Infectious human immunodeficiency virus can rapidly penetrate a tight human epithelial barrier by transcytosis in a process impaired by mucosal immunoglobulins. *J. Infect. Dis.* **1999**, *179* (Suppl. 3), S448–S453. [CrossRef] [PubMed]
52. Plancoulaine, S.; Buigues, R.P.; Murphy, E.L.; van Beveren, M.; Pouliquen, J.F.; Joubert, M.; Remy, F.; Tuppin, P.; Tortevoye, P.; de The, G.; *et al.* Demographic and familial characteristics of HTLV-1 infection among an isolated, highly endemic population of African origin in French Guiana. *Int. J. Cancer* **1998**, *76*, 331–336. [CrossRef]
53. Plancoulaine, S.; Gessain, A.; Joubert, M.; Tortevoye, P.; Jeanne, I.; Talarmin, A.; de The, G.; Abel, L. Detection of a major gene predisposing to human T lymphotropic virus type I infection in children among an endemic population of African origin. *J. Infect. Dis.* **2000**, *182*, 405–412. [CrossRef] [PubMed]
54. Jeffery, K.J.; Usuku, K.; Hall, S.E.; Matsumoto, W.; Taylor, G.P.; Procter, J.; Bunce, M.; Ogg, G.S.; Welsh, K.I.; Weber, J.N.; *et al.* HLA alleles determine human T-lymphotropic virus-I (HTLV-I) proviral load and the risk of HTLV-I-associated myelopathy. *Proc. Natl. Acad. Sci. USA* **1999**, *96*, 3848–3853. [CrossRef] [PubMed]
55. Plancoulaine, S.; Gessain, A.; Tortevoye, P.; Boland-Auge, A.; Vasilescu, A.; Matsuda, F.; Abel, L. A major susceptibility locus for HTLV-1 infection in childhood maps to chromosome 6q27. *Hum. Mol. Genet.* **2006**, *15*, 3306–3312. [CrossRef] [PubMed]
56. Sawada, T.; Iwahara, Y.; Ishii, K.; Taguchi, H.; Hoshino, H.; Miyoshi, I. Immunoglobulin prophylaxis against milkborne transmission of human T cell leukemia virus type I in rabbits. *J. Infect. Dis.* **1991**, *164*, 1193–1196. [CrossRef] [PubMed]

57. Ureta-Vidal, A.; Angelin-Duclos, C.; Tortevoye, P.; Murphy, E.; Lepere, J.F.; Buigues, R.P.; Jolly, N.; Joubert, M.; Carles, G.; Pouliquen, J.F.; *et al.* Mother-to-child transmission of human T-cell-leukemia/lymphoma virus type I: Implication of high antiviral antibody titer and high proviral load in carrier mothers. *Int. J. Cancer* **1999**, *82*, 832–836. [CrossRef]

58. Sawada, T.; Tohmatsu, J.; Obara, T.; Koide, A.; Kamihira, S.; Ichimaru, M.; Kashiwagi, S.; Kajiyama, W.; Matsumura, N.; Kinoshita, K.; *et al.* High risk of mother-to-child transmission of HTLV-I in p40tax antibody-positive mothers. *Jpn. J. Cancer Res.* **1989**, *80*, 506–508. [CrossRef] [PubMed]

59. Hirata, M.; Hayashi, J.; Noguchi, A.; Nakashima, K.; Kajiyama, W.; Kashiwagi, S.; Sawada, T. The effects of breastfeeding and presence of antibody to p40tax protein of human T cell lymphotropic virus type-I on mother to child transmission. *Int. J. Epidemiol.* **1992**, *21*, 989–994. [CrossRef] [PubMed]

60. Hisada, M.; Maloney, E.M.; Sawada, T.; Miley, W.J.; Palmer, P.; Hanchard, B.; Goedert, J.J.; Manns, A. Virus markers associated with vertical transmission of human T lymphotropic virus type 1 in Jamaica. *Clin. Infect. Dis.* **2002**, *34*, 1551–1557. [CrossRef] [PubMed]

61. Moriuchi, M.; Moriuchi, H. A milk protein lactoferrin enhances human T cell leukemia virus type I and suppresses HIV-1 infection. *J. Immunol.* **2001**, *166*, 4231–4236. [CrossRef] [PubMed]

62. Moriuchi, M.; Moriuchi, H. Induction of lactoferrin gene expression in myeloid or mammary gland cells by human T-cell leukemia virus type 1 (HTLV-1) tax: Implications for milk-borne transmission of HTLV-1. *J. Virol.* **2006**, *80*, 7118–7126. [CrossRef] [PubMed]

63. Banerjee, P.; Tripp, A.; Lairmore, M.D.; Crawford, L.; Sieburg, M.; Ramos, J.C.; Harrington, W., Jr.; Beilke, M.A.; Feuer, G. Adult T-cell leukemia/lymphoma development in HTLV-1-infected humanized SCID mice. *Blood* **2010**, *115*, 2640–2648. [CrossRef] [PubMed]

64. Villaudy, J.; Wencker, M.; Gadot, N.; Gillet, N.A.; Scoazec, J.Y.; Gazzolo, L.; Manz, M.G.; Bangham, C.R.; Dodon, M.D. HTLV-1 propels thymic human T cell development in "human immune system" Rag2$^{-/-}$ $\lambda$c$^{-/-}$ mice. *PLoS Pathog.* **2011**, *7*, e1002231. [CrossRef] [PubMed]

65. Dodon, M.D.; Villaudy, J.; Gazzolo, L.; Haines, R.; Lairmore, M. What we are learning on HTLV-1 pathogenesis from animal models. *Front. Microbiol.* **2012**, *3*, 320. [CrossRef] [PubMed]

66. Ouzilou, L.; Caliot, E.; Pelletier, I.; Prevost, M.C.; Pringault, E.; Colbere-Garapin, F. Poliovirus transcytosis through M-like cells. *J. Gen. Virol.* **2002**, *83 Pt 9*, 2177–2182. [CrossRef] [PubMed]

67. Renz, H.; Brandtzaeg, P.; Hornef, M. The impact of perinatal immune development on mucosal homeostasis and chronic inflammation. *Nat. Rev. Immunol.* **2012**, *12*, 9–23. [CrossRef] [PubMed]

68. Van de Pavert, S.A.; Mebius, R.E. New insights into the development of lymphoid tissues. *Nat. Rev. Immunol.* **2010**, *10*, 664–674. [CrossRef] [PubMed]

69. Maranduba, C.M.; De Castro, S.B.; de Souza, G.T.; Rossato, C.; da Guia, F.C.; Valente, M.A.; Rettore, J.V.; Maranduba, C.P.; de Souza, C.M.; do Carmo, A.M.; *et al.* Intestinal microbiota as modulators of the immune system and neuroimmune system: Impact on the host health and homeostasis. *J. Immunol. Res.* **2015**, *2015*. [CrossRef] [PubMed]

70. Shu, Z.; Ma, J.; Tuerhong, D.; Yang, C.; Upur, H. How intestinal bacteria can promote HIV replication. *AIDS Rev.* **2013**, *15*, 32–37. [PubMed]

71. Kane, M.; Case, L.K.; Kopaskie, K.; Kozlova, A.; MacDearmid, C.; Chervonsky, A.V.; Golovkina, T.V. Successful transmission of a retrovirus depends on the commensal microbiota. *Science* **2011**, *334*, 245–249. [CrossRef] [PubMed]

72. Fernandez, L.; Langa, S.; Martin, V.; Jimenez, E.; Martin, R.; Rodriguez, J.M. The microbiota of human milk in healthy women. *Cell Mol. Biol.* **2013**, *59*, 31–42. [PubMed]

*viruses*

MDPI

*Review*

# Genetic Markers of the Host in Persons Living with HTLV-1, HIV and HCV Infections

Tatiane Assone [1,2,*], Arthur Paiva [2], Luiz Augusto M. Fonseca [3] and Jorge Casseb [1,2,*]

1   Laboratory of Dermatology and Immune deficiencies, Department of Dermatology, University of São Paulo Medical School, LIM56, Av. Dr. Eneas de Carvalho Aguiar 500, 3rd Floor, Building II, São Paulo, SP, Brazil
2   Institute of Tropical Medicine of São Paulo, São Paulo, Brazil; arthurmpaiva@usp.br
3   Department of Preventive Medicine, University of São Paulo Medical School, São Paulo, Brazil; augustom@hcnet.usp.br
*   Correspondence: tatianeassone@usp.br (T.A.); jcasseb@usp.br (J.C.); Tel.: +55-11-3061-7194 (T.A. & J.C.); Fax: +55-11-3081-7190 (T.A. & J.C.)

Academic Editor: Louis M. Mansky
Received: 24 September 2015; Accepted: 15 January 2016; Published: 3 February 2016

**Abstract:** Human T-cell leukemia virus type 1 (HTLV-1), hepatitis C virus (HCV) and human immunodeficiency virus type 1 (HIV-1) are prevalent worldwide, and share similar means of transmission. These infections may influence each other in evolution and outcome, including cancer or immunodeficiency. Many studies have reported the influence of genetic markers on the host immune response against different persistent viral infections, such as HTLV-1 infection, pointing to the importance of the individual genetic background on their outcomes. However, despite recent advances on the knowledge of the pathogenesis of HTLV-1 infection, gaps in the understanding of the role of the individual genetic background on the progress to disease clinically manifested still remain. In this scenario, much less is known regarding the influence of genetic factors in the context of dual or triple infections or their influence on the underlying mechanisms that lead to outcomes that differ from those observed in monoinfection. This review describes the main factors involved in the virus–host balance, especially for some particular human leukocyte antigen (HLA) haplotypes, and other important genetic markers in the development of HTLV-1-associated myelopathy/tropical spastic paraparesis (HAM/TSP) and other persistent viruses, such as HIV and HCV.

**Keywords:** HTLV-1; HIV-1; HCV; genetic factors

## 1. Introduction

Human T-cell leukemia virus type 1 (HTLV-1), hepatitis C virus (HCV) and human immunodeficiency virus type 1 (HIV-1) have been reviewed, since these viruses are prevalent worldwide, and share similar means of transmission, and superposition of infected populations, such as intravenous drug users (IDU) and commercial sex workers, especially in endemic areas for HTLVs. In addition, these infections may influence each other in evolution and outcome, including cancer or immunodeficiency.

Many studies have reported the influence of genetic markers on the host immune response against different persistent viral infections, such as HTLV-1 infection, pointing to the importance of the individual genetic background on their outcomes. However, despite recent advances on the knowledge of the pathogenesis of HTLV-1 infection, gaps in the understanding of the role of the individual genetic background on the progress to clinically manifested disease still remain. In this scenario, much less is known about the influence of genetic factors in the context of dual or triple infections, or their influence on the underlying mechanisms that lead to outcomes that differ from those observed in monoinfection. HTLV-1 is an ancient infection, with better adaptation to its host than HIV or HCV. HIV often complicates the evolution of HCV and HTLV-1, but the reverse may not be true, with a higher spontaneous clearance rate of HCV in those triple co-infected with HCV/HIV/HTLV than in those with HCV/HIV, or monoinfected with HCV.

This review describes the main factors involved in the virus–host balance, especially for some particular human leukocyte antigen(HLA) haplotypes, and other important genetic markers in the development of HTLV-1-associated myelopathy/tropical spastic paraparesis (HAM/TSP) and other persistent viruses, such as HIV and HCV.

## 2. HLA Risk for HAM/TSP

HAM/TSP occurs in only 1%–2% of those infected by HTLV-1,thus, host genetic factors interacting with viral factors have been suggested as determinant to the outcome; that is, whether the individual will develop an effective immune response to HTLV-1 or will progress to clinically manifested HAM/TSP. Among those genetic factors, the most frequently studied have been the haplotypes of HLA Class 1 (HLA-A, HLA-B, HLA-C) of the major histocompatibility complex (MHC), molecules that encode glycoproteins expressed on the surface of almost all nucleated cells, and of which the major function is the presentation of antigenic peptides to CD8+ T lymphocytes. The effectiveness of a specific immune response to HTLV-1, especially the cytotoxic CD8+ T lymphocytes' response (CTL), has been shown to be the key to control the provirus load (PVL) of HTLV-1 [1]. The most compelling evidence for a role of host CTLs came from the observation of a population in Southern Japan, where the presence of two genes of the HLA class 1 (*HLA-A\*02* or *HLA-Cw\*08*) was associated with a lower PVL and decreased prevalence of HAM/TSP [2,3].

Some specific HLA alleles have been associated with protection, while others have been associated with an increased risk of HAM/TSP. However, unlike the expressions of both *HLA-A\*02* and *HLA-Cw\*08*, which are associated with a protective effect, *HLA-DRB1\*0101* and *HLA-B\*5401* have been linked with an increased susceptibility to HAM/TSP [2,3]. The association of *HLA-DRB1\*0101* with disease susceptibility only becomes evident in the absence of the *HLA-A\*02* protective effect [2], whereas *HLA-B\*5401* is independently associated with susceptibility to disease; moreover, among patients with HAM/TSP, *HLA-B\*5401* is associated with a significant increase in PVL. *HLA-A\*02*, *HLA-Cw\*08* and *HLA-DR1* are also found in the population of Southern Japan, where they are associated with a higher risk for HAM/TSP [4].

There may be differences in the frequency of HLA alleles in different populations, and changes in the protective effect of certain HLA alleles according to ethnicity (Table 1). The same protective effect of *HLA-A\*02* in HAM/TSP, seen in Japanese, has been reported in a small sample of 29 individuals from London, 27 of whom had a Caribbean origin [2], a finding also observed in Brazil [5,6] but not in other populations, such as Afro-Caribbean individuals from Martinique [7], Jamaica [8], Spain [9] and Iran [10,11].

HTLV-1 PVL is controlled by the host immune response, with a dominant role for an effective CTL response [8]. Long-term studies have shown that CTL is determinant of the outcome; for example, in Japanese cohorts, *HLA-A\*02* and *HLA-C\*08* play a protective role against the development of HAM/TSP, whereas *HLA-B\*54* is associated with a higher risk. This outcome is possibly related to the killing of infected cells with *HLA-A2* or *-C\*08* restricted HTLV-1 epitopes, resulting in decreased PVLs [9]. In fact, such HLAs have the ability to present peptides derived from viral proteins to CD8+ T cells, which are mostly protective during HTLV-1 infection [10].

**Table 1.** Distribution of human leukocyte antigen (HLA) haplotypes according to risk of HTLV-1-associated myelopathy/tropical spastic paraparesis (HAM/TSP) development.

| HLA Allele | Japanese | Brazilians | Iranians | Spanish | Afro-Caribbean (Martinique) | Afro-Caribbean (London) | Jamaicans |
|---|---|---|---|---|---|---|---|
| A*02 | ++ | + | 0 | 0 | 0 | ++ | 0 |
| Cw*08 | ++ | 0 | 0 | 0 | | | |
| A*24 | - | | 0 | | 0 | | |
| B*07 | - | ± | 0 | - | 0 | | 0 |
| B*5401 | - | ⊗ | ⊗ | ⊗ | ⊗ | ⊗ | ⊗ |
| DRB1*0101 | ± | 0 | ± | - | 0 | | 0 |
| DRB1*11 | - | - | - | | - | 0 | 0 |

++ protective effect; + tendency to protective effect; - susceptibility; ± susceptibility only in negative *HLA-A*02*; 0 no associated effect; ⊗ HLA not prevalent.

In Iran, alleles *HLA-A*02*, *HLA-Cw*08* and *HLA-A*24* were not associated with a lower risk of HAM/TSP or lower provirus load [10,12]. In Brazil, *HLA-Cw*08* showed no protective effect, and, among *HLA-B*07* individuals, only those negative for *HLA-A*02* [6] were susceptible to HAM/TSP [6]. In Spain, no association between the presence of protective alleles (*HLA-A*02* and/or *HLA-Cw08*) and HAM/TSP could be demonstrated nor were there significant differences in PVLs; however, HAM/TSP was significantly associated with *HLA-B*07* and *HLA-DRB1*0101* [9]. The *HLA-B*5401* allele was not found in the populations of Iran, Brazil and Spain (Table 1), and has been described almost exclusively in East Asian individuals [13].

Alleles associated with a higher risk, such as *HLA-DRB1*0101*, were also associated with susceptibility to HAM/TSP in Iran's population, as well as in Japanese, an effect was observed only among *HLA-A*02*-negative individuals, and not occurring in *HLA-A*02*-positive individuals [10]. The association of *HLA-DR*11* with HAM/TSP, previously described in Japanese patients, was observed only in Brazilian patients [5]. Among Brazilian individuals, *HLA-Cw*07* was associated with HAM/TSP only in the absence of *HLA-A*02* [6].

The protective ability of HLA class 1 allele correlates with the affinity to bind antigenic peptides derived from the HTLV-1 proteins [14]. However, contrary to what was expected, HTLV-1 antigen, which is recognized by the protective immune response class 1 immune dominant Tax, was not associated with this protein, but rather with the regulatory protein encoded hemoglobin subunit zeta (HBZ) on the negative strand of the provirus. A combination of theoretical methods for the prediction of epitopes [15] and cellular laboratory experiments demonstrated that the binding to epitopes of the protective *HLA-A*02* and *HLA-Cw*08* alleles are stronger than that of the detrimental *HLA-B*54* [14]. In that study, HLA class 1 molecules that bind strongly to HBZ epitopes were significantly associated with the asymptomatic state, an association remaining even after patients with *HLA-A*02*, *HLA-A*08* and *HLA-B*54* were excluded from the analysis, demonstrating that the protective effect of binding HBZ is common to several HLA alleles and not just a feature of particular alleles. Moreover, among both asymptomatic subjects and HAM/TSP patients who carry protective alleles, epitopes that could bind HBZ were strongly associated with a significant reduction in HTLV-1 PVL [14].

## 3. Interferon Lambda 3 (IFN-λ3)

IFN lambda 3 (IFN-λ3) is an important cytokine that is responsible for an unspecific antiviral response by interacting with the HLA class II receptor, inducing intracellular signaling by janus kinase/signal transducers and activators of transcription (JAK/STAT) and mitogen-activated protein kinases (MAPK). Host genetic background in HLA class II, encoded by single nucleotide polymorphisms (SNPs), can lead to a spatial conformation in the receptor, modifying the attachment that avoids interaction between IFN-λ3 and its receptor, inducing a genetic by stand interaction [16].

The first reports about the role of *IL28B* (coding for IFN-λ3) on HAM/TSP outcomes could not clearly show the connection [17,18]. However, it was noted that HAM/TSP patients presented an independent association with the polymorphism in *IL28B* SNP rs8099917 (GG), when compared to asymptomatic HTLV-1 carriers [19]; such a finding has not been reported for other infections, such as HIV and HBV infections [20], except for patients with acute HIV infection, whose response to antiretrovirals was related to SNP rs12979860 [21].

In recent years, an association between IFN-λ3 polymorphisms and anti-HCV treatment with pegylated interferon (PEG-IFN) outcome was described [22]. The correlation between the polymorphism of IFN-λ3 was noted in two positions (rs12979860 and rs8099917) [16,22]. Co-infection with HIV or HTLV is widely present among HCV subjects, making interaction a possibility, and potentially changing both the pathogenesis of the disease and/or the response to treatment [23,24].

It is noteworthy that the immune response seems to be a crucial factor in the pathogenesis of HAM/TSP. For example, a study showed that patients with HAM/TSP had higher levels of IFN-gamma compared to asymptomatic patients [25]. Furthermore, the polymorphism of rs12979860 SNP profile induces the production of IFN-λ3 and an immune response to HTLV-1, leading to neuronal injury in the spinal cord [26,27]. As reported, the interferon stimulated genes likely regulate the expression of cytokines and this regulation may differ in the infected tissue and between cell types within the liver and spinal cord [28]. It is known that IFN-λ3 attenuates interleukin (IL)-13 production, leading to a protective effect and decreased inhibition between killer-cell immunoglobulin-like receptor (*KIR*) and *HLA-C* [29]. Thus, it is possible to infer that IFN-λ3 and two other interferons, IL-28A and IL-29, can activate the JAK-STAT cascade, which is similar and probably synergistic with type 1 interferons (e.g., interferon alpha), although using different receptors, and contributing to the immune pathogenesis of HAM/TSP.

## 4. miRNA

The importance of microRNA (miRNA) in the replicative cycle of several other viruses, as well as to the progression of associated pathologies, was established over the past decade. Furthermore, the involvement of miRNAs in altering the life cycle of HTLV-1 and progression to neurodegenerative diseases and related oncogene has been investigated [30]. Various miRNA-derived transcript proteins can change the features of HTLV-1, either interacting with the restructuring of chromatin or manipulating components of the RNA interference (RNAi), by providing multiple routes through which miRNA expression, *etc.*, can be down-regulated in the cell host. Furthermore, the mechanism of action by which deregulation of host miRNAs can affect cells infected with HTLV-1 can vary substantially through the silencer, including miRNA-induced silencing complex of RNA (RISC), gene transcription, inhibition of components RNAi, and chromatin remodeling. These changes induced by miRNA can lead to increased cell survival, invasiveness, proliferation and differentiation; they can also allow viral latency. Recent studies have shown the involvement of successful miRNAs in the life cycle and pathogenesis of HTLV-1, but there are still significant issues to be addressed. In Table 2, a summary of the already identified miRNAs and their biological effects on HTLV-1 are presented. In HAM/TSP, the miRNA involved in the pathogenesis is the miR-132, while, in adult T-cell leukemia/lymphoma (ATL), miR-223 is responsible for promoting oncogenesis, an important biological marker [31].

HTLV-1 can change the miRNA profiles of infected cells, contributing to cell transformation and leading to the development of ATL and/or HAM/TSP [30]. Moreover, the modification of chromatin by viral proteins and host cell miRNAs may contribute to the deregulation of cells expressing miRNA and may possibly serve as a key mechanism by which the virus manipulates the status of miRNA host. Recent discoveries attempt to validate the importance of variation in the levels of miRNA mediated by HTLV-1, but there is a gap on this recent field of study. Moreover, a better understanding of the molecular mechanisms may contribute to a better understanding of viral regulation and cellular regulatory pathways. Taken together, this knowledge can identify potential therapeutic intervention points in the future [31].

**Table 2.** Identified miRNAs and their biological effects on HTLV-1.

| MiRNA | Regulation | miRNA Target | Function |
|-------|-----------|--------------|----------|
| miR-21 | Upregulated | PTEN | Antiapoptotic |
| miR-93 | Upregulated | p21 (WAF1/CIP1); MICB | Antiapoptotic |
| miR-132 | Downregulated | p300 | Immune evasion |
| miR-143-p3 | Upregulated | AChE; PKA; GRα | Increase of viral transcription |
| miR-146 a | Upregulated | Unknown | Pro-inflammatory |
| miR-149 | Downregulated | p300 | Proliferation |
| miR-155 | Upregulated | TP53INP1; Unknown | Proliferation |
| miR-873 | Downregulated | p300 | Proliferation |

PTEN: Phosphatase and tensin homolog; MICB: MHC class I polypeptide-related sequence B; AChE: *Acetylcholinesterase*; PKA: Protein kinase A; GRα: glucocorticoid receptor α; TP53INP1: Tumor protein P53 inducible nuclear protein 1.

## 5. Killer Cell Immunoglobulin-Like Receptors (KIRs)

KIR present a high tendency to suffer genetic mutations, indicating a high polymorphic capacity. HLA-C molecules present ligands for KIR2DL receptors, with a functionally relevant indistinct to determine KIR specificity, like *HLA-C* group 1 (*HLA-C1*) alleles, where the HLA-C alpha 1 domain is ligand for the inhibitory receptors KIR2DL2 and KIR2DL3 and the activating receptor KIR2DS2 [32,33]. *HLA-C* group 2 (*HLA-C2*) alleles are involved in inhibitory KIR2DL1 and in the activation of KIR2DS1 [32–35]. KIR2DL3 and its ligand HLA-C1 have been associated with an increased likelihood of spontaneous [36,37] and treatment-induced HCV clearance [37,38]. This association is attributed to differential natural killer (NK) cell activation and function in the context of this KIR/HLA interaction [39]. SNPs from the *HLA-C* coding regions show weak associations with sustained virologic response(SVR) [16,40] for chronic hepatitis C.

New evidence has shown that the pathogenic mechanism of disease-associated HTLV-1 infection is an impairment of the immunity [41]. The KIR genotype influences CTL efficiency, by affecting HLA class I-mediated HTLV-1 immunity [42], and KIRs influence both innate and adaptive immunity (Figure 1) [1,42], however KIR2DL2 gene is associated with an enhancement of the effect of known protective or detrimental HLA class I alleles on PVL and HAM/TSP risk, for multiple HLA-A, -B and -C molecules. Surprisingly, KIR2DL2 also exhibits the same behavior in HCV infection, another unrelated virus. In HTLV-1 infection, KIR2DL2 enhanced the protective and detrimental effects of *HLA-C\*08* and *B\*54*, respectively, on disease status and enhanced the association between *B\*54* and high PVL in HAM/TSP patients [42]. In HCV infection, KIR2DL2 enhanced the protective effect of *B\*57* on its spontaneous clearance and the association between *B\*57* and low viral load in chronic carriers [42]. These observations suggest that KIR2DL2 enhancement of the HLA class I-restricted response may be a general mechanism.

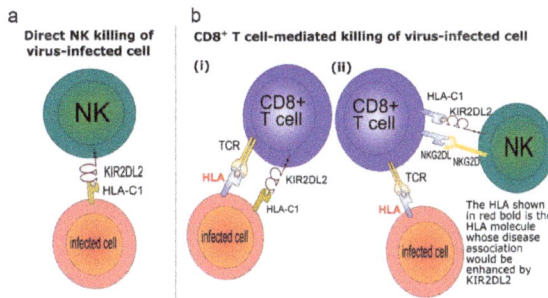

**Figure 1.** Mechanisms explaining inhibitory killer-cell immunoglobulin-like receptor (KIR) enhancement of human leukocyte antigen (HLA) class I associations.

## 6. Genes and Susceptibility to HIV Infection

Susceptibility to HIV infection and the clinical course after infection are both influenced by the complex interaction of factors related to the human host, the virus and the surrounding environment, resulting in large epidemiological and clinical heterogeneity among infected individuals. Host genetic factors play an important role in this variability and pathogenesis. For example, the gene encoding the CCR5 (chemokine (C-C motif) receptor 5) co-receptor necessary for infection of R5 strains of HIV, which usually initiate infection, and may influence both the acquisition of infection and the rate of progression to disease.

The *CCR5* gene is located on chromosome 3 and individuals with *CCR5 Δ32* deletion acquire protection against virus to produce a defective protein, which is not expressed on the cell surface, preventing the virus from binding to the CCR5 co-receptor to penetrate the cell. While homozygous for *CCR5* deletion has Δ32-protection against HIV [43–45], heterozygous for the mutation has a delay of two to four years of progression to acquired immune deficiency syndrome (AIDS) [46–48]. However, homozygous for the *CCR5Δ32* deletion mutation is the only genotype identified as being capable of protecting against HIV infection. The *CCR5Δ32* allele occurs at a frequency of 4% to 15% in the Caucasian population, with a higher frequency in the Northern European populations [44,49]. Other genetic potential protective effects against infection caused by deletions seem to involve more complex interactions between two or more gene variants. Apparently, Asians and Africans with *CCL5-403A/A* genotype (chemotactic chemokine ligand 5), also called RANTES (regulated on activation, normal, T cell expressed and secreted), could be resistant to HIV-1 but controlled studies are still lacking to confirm this observation [50]. In China, a genetic variant of SDF-1, the primary ligand of CXCR4 ((C-X-C motif) receptor 4), was associated with resistance to HIV infection in intravenous drug users [51].

ZNRD1 (zinc ribbon domain-containing protein 1) is a DNA-dependent RNA polymerase catalyzing the transcription of DNA into RNA required to complete the life cycle of HIV, which was subsequently identified as SNPs associated with depletion of CD4+ T cells. A haplotype in ZNRD1 gene was associated with a 35% reduction in the risk of HIV acquisition in Euro-Americans (Americans with European ancestry) and ZNRD1 variants also affect the progression of HIV-1 infection to disease in European, American, and African cohorts [52].

Langerin, also known as CD207, is a transmembrane receptor encoded by the *CD207* gene and expressed in Langerhans cells, scattered throughout the genital mucosal epithelium where the transmission of HIV occurs. Although Langerhans cells are considered dendritic cells (DC), and immature DCs are involved in the transmission of HIV to lymphocyte T [53], there is evidence suggesting that Langerin prevents HIV transmission. The HIV particles captured by Langerin are internalized and degraded in the Birbeck granules [54]. However, a mutation in the gene of persons deficient in the Langerin Birbeck granules has been described [55].

Defensins are small cationic proteins rich in cysteine produced by leukocytes and epithelial cells that are active against bacteria, viruses, and fungi. They play a role in immunity penetration through the cell membrane and pore formation flue material [56–58]. Mammalian defensins are classified as alpha, beta, and theta defensins [58]. Alpha-defensin bound to receptor CD4 and gp120 glycoprotein of the viral envelope may negatively modulate CD4+ T cells and inactivate viral particles through disruption of membrane [59,60]. Accordingly, they could block HIV entry directly by inactivating it or by blocking or eliminating viral receptor on the cell surface. Beta-defensins have mechanism of action similar to that of alpha-defensin, blocking virus entry of both the tropic virus strains macrophage (R5 viruses), as well as tropic strains for T cells (X4 viruses), achieving its effect by direct inactivation of viral particles or negative modulation of CXCR420 [61]. Six human beta-defensins were identified in epithelial cells, although they can be present in up to 28 different human genes [62]. Theta-defensins in humans and chimpanzees are only found as inactive pseudogenes, which are transcribed into mRNA, but are home to premature stop codons that prevent expression of functional products [63]. However, by reconstituting the human putative ancestral gene for theta-defensin, the presence of potent activity against strains X422 and R5 was observed *in vitro*. The reconstituted product, called human retrociclin-1, binds to CD4 molecule and gp120, preventing viral entry into target cells [64].

TREX-1 (three prime repair exonuclease), which degrades cytosolic DNA, preventing unnecessary immune response against free nucleic acids, is a limiting factor for HIV-1 and polymorphism of a single nucleotide rs3135945 and was associated with susceptibility to HIV infection, emphasizing the participation of TREX-1 in anti-HIV response [65].

### 6.1. Genes that Influence the Dynamic Progression of AIDS

Apart from individuals uninfected by HIV, despite repeated sexual exposure to the virus in high risk situations, known as exposed uninfected, there is a small proportion of HIV-infected individuals who remain clinically and immunologically healthy for more than one or two decades after being seroconverted, while in others infection may be characterized by an extremely rapid progression to AIDS within one year [66]. Host genetic factors possibly contribute to this heterogeneity, as demonstrated by many studies in which genetic polymorphisms in human genes are able to influence the risk of HIV infection and progression to AIDS [66].

The type of HLA is one of the host genetic factors associated with the course of HIV infection. The *HLA-B* alleles are considered the primary genetic determinants of disease progression, according to the categorization in rapid progressors, slow progressors and long-term non-progressors [67]. While *HLA-B35* is associated with rapid progression to AIDS [67–69], *HLA-B*5701* and *HLA-B27* are more prevalent among long-term non-progressors (LTNP) [70–75], of which 1% are the elite controllers (EC), who are mainly characterized by maintaining persistently undetectable viral loads without antiretroviral treatment.

Associations among the SNPs of MHC class I, MHC class III and LTNP phenotype are observed [76] is also noted in other factors, such as the co-expression of multiple HLA protectors, SNPs of HLA-C, and stronger T cell responses against the HIV proteins in elite controllers individuals [77]. These different genetic variant combinations may have addictive, synergistic or inhibitory effects determining the course of HIV infection. HLA class II also contributes to the immune response in the control of the viral load of HIV patients and distinct stratifications of *HLA-DRB1* effect on HIV viremia between controllers and progressors are associated with different subsets of HLA-DRB1 alleles, with *DRB1*15:02* significantly associated with low viremia and *DRB1*03:01* with high viremia [78,79].

Unlike individuals with two copies of the mutation *CCR5Δ32* who are protected from HIV infection by non-functionality of *CCR5* [43–45], heterozygous individuals for this mutation can be infected by HIV R5 strains but exhibit an altered activity of the chemokine receptor, resulting in delayed progression to AIDS [46–48]. Occurring at a frequency of up to 15% in the Caucasian population, especially in the Northern European population, the allele *CCR5Δ32* is virtually absent among natives of Africa and its marginal presence in Asian populations may be due to gene flow from Caucasian populations [80–82].

Polymorphism in otherwise apparently normal chemokine receptors also has some degree of influence on disease progression. For example, CCR2 chemokine receptor can function as a HIV co-receptor in some situations [83]. Person with homozygous or heterozygous for *CCR2-64I* which results mutation from valine to isoleucine changed at amino acid position 64, progress more slowly to AIDS than those homozygous for the wild type variant, although not all studies confirm this association [84]. There is also controversy as to CX3CR1, fractalkine, a receptor for chemokine, for which the initial studies showed an association with more rapid progression to AIDS [85–87].

The beta-chemokines MIP-1α (CCL3), MIP-1β (CCL4) and RANTES (CCL5) are natural ligands of CCR543. Two natural variants have been described and named CCL3L1 and CCL4L147. CCL3L1 (also known as MIP-1αP) is the most potent CCR5 agonist and is a strong inhibitor of infection by R5 strains of HIV-1 [88,89]. RANTES (CCL5), through promoters 28G-403A, could also slow progression to AIDS, whereas RANTES has another variant, named 1.1.C, which accelerates AIDS development [84].

The LTNP condition probably results from a complex association of various genetic factors rather than only one variant of a single gene, as confirmed by the later discovery of two new association between allelic variants of *TNF-a-238* genes and *PDCD1-7209* and LTNP situation [90].

### 6.2. Genes Important to Anti-HIV Treatment

Abacavir is a reverse transcriptase inhibitor antiretroviral used in current clinical practice with other antiretroviral agents with few interactions with other drugs and favorable long-term toxicity profile. However, it has the most important adverse effect on the immune-mediated hypersensitivity reaction, affecting 5% to 8% of patients in the first weeks of treatment, requiring the immediate cessation of treatment [91]. Their subsequent reintroduction is contraindicated due to the risk of recurrence of the reaction with greater severity, speed and risk of death [91]. A hypersensitivity reaction to abacavir is strongly related to the presence of the *HLA-B*5701* [92–96] and avoiding abacavir in patients with *HLA-B*5701* reduces the incidence of that reaction [97]. The effectiveness of screening for *HLA-B*5701* in preventing hypersensitivity reaction to abacavir has been established, although its cost-effectiveness depends on factors that vary between populations and health care settings, and the availability of test [98].

The observation that individuals homozygous for the *CCR5-Δ32* deletion show protection against HIV led to the development of drugs antagonist of the CCR5 chemokine receptor, blocking this receptor and inhibiting HIV entry that uses CCR5. As the CCR5 antagonists administration carries the risk of selection of viral variants able to use alternative CXCR4 co-receptor, tropism for co-receptor should be assessed prior to clinical use of inhibitor [99]. Maraviroc was the first antiretroviral drug with this mechanism of action; other drugs of this group include vicriviroc, cenicriviroc, adaptavir, INCB-9471 and PRO-140.

In 2007, an HIV-infected adult patient living in Berlin developed acute myelogenous leukemia and was treated with a transplant from an allogeneic hematopoietic stem cell donor who was homozygous for the *CCR5-Δ32* deletion and, after stopping antiretroviral therapy following transplantation, his viral load remained undetectable, becoming the first confirmed case of cure of HIV infection [100]. Since then, several efforts are underway in an attempt to reproduce the Berlin patient's condition through the engineering of autologous T cells or hematopoietic stem cells resistant to HIV [100].

### 6.3. Co-Infection HIV–HTLV

HTLV-1 and -2 have the same modes of transmission as HIV, resulting in common risk factors and overlapping of populations exposed, so that in individuals infected with HIV, HTLV is 100 to 500 times more frequent than in the general population. *In vitro*, *Tax* gene products of HTLV-1 increment the release of free viral particles of HIV-1 [101] and an accelerated course of HIV-1 infection has been reported in patients co-infected with HIV-1 and HTLV-1 [24,102,103].

On the other hand, the Tax protein of HTLV-2 may have an immunomodulatory effect, increasing IFN-y synthesis by cells infected with HIV-1 [104], and HTLV-2 induces the production of CCL3 chemokine, CCL4 and CCL5, which can have a protective effect on disease progression by HIV [105,106]. Lymphocytes T CD8+ recovered from patients infected with HTLV-2 spontaneously produce high levels of chemokines [107] and these chemokine molecules are natural ligands for the CCR5, the most important co-receptor for input of HIV in cells, suppressing the infection with HIV strains with tropism for macrophages [108,109]. Thus, an association between increased production of chemokine and slower disease progression by HIV [110,111], and between increased production of chemokines and reduction of HIV [112] levels has been reported, and slower depletion of CD4+ T cells in individuals with co-infection HIV-1/HTLV-2 [105,113–115], as well as lower plasma HIV RNA than for the mono HIV-infected subjects [116].

### 7. HCV and Genetic Susceptibility

Approximately 20%–30% of individuals infected with HCV will clear the virus spontaneously [117], and around of 185 million are infected with this virus worldwide [118]. Chronic hepatitis C is the major cause of hepatocellular carcinoma, end-stage liver disease, and liver transplantation in the USA [117]. Genome-wide association studies (GWAs) identified variants located in the interferon lambda region, and it is associated with HCV clearance spontaneously in subjects under treatment with interferon alfa [26], and interferon lambda 4 (*IFNL4*) plays a fundamental role in these associations [119]. Interferon lambda 4 protein (IFN-λ4) was found in persons who have the ΔG allele of the ss469415590 variant (*IFNL4-ΔG*), another important variant is rs12979860, which was associated with spontaneous HCV clearance [120,121] and is a SNP located in IL28B.

Indeed, the linkage disequilibrium is strong among *IFNL4-ΔG* allele and *rs12979860-T* allele is unfavorable in individuals of European or Asian ancestry; nevertheless, this linkage disequilibrium is moderate in individuals of African ancestry [119]. Among black participants, *IFNL4-ΔG* genotype was associated with spontaneous HCV clearance more strongly than *rs12979860* genotype [120,121]. On the other hand, the linkage disequilibrium in the SNP of IL28B in *rs8099917-T* was described as an important predictor of sustained virologic response in chronic HCV subjects undergoing treatment with pegylated alfa interferon and ribavirin over 48 weeks [16].

Other mutations in the innate immune response regulator genes have important roles in the response to treatment in patients with chronic HCV infection, which emphasizes the important role of *KIR* and *HLA* genotypes [40,122]. The variant *KIR2DS3* gene was described as the principal gene associated with chronic HCV infection, whereas the reduction of *HLA-Bw4+ KIR3DS1+* was associated with an increased risk of developing hepatocellular carcinoma. Therefore, they have a role in the innate system in developing HCV-related disorders, of which *KIR2DS3* and *KIR2D* genes stand out as related to HCV disease progression, and lymphoproliferative disorders [122]. This was observed in Brazil, and the most important *KIR* variants associated with SRV was *KIR2DS5*, whereas *KIR2DL2* was associated with chronic hepatitis C [123].

### 8. Concluding Remarks

There are several studies on the immune response against persistent viral infections. In contrast, only a small number of studies on genetic markers have been published recently. This review highlights some specific *HLA* alleles that have been associated with protection or with increased risk of HAM/TSP development; *KIR2DL2* and IFN-λ3 *rs8099917* (GG) polymorphism, associated with HLA class I-restricted, may be involved in the pathogenic mechanism of HIV and HCV infections. All of these polymorphisms should be studied in the future as potential markers in HTLV-1, as well for HIV and HCV infected subjects.

**Acknowledgments:** To all participants who contributed to this study. To Jerusa Smid, Augusto Penalva de Oliveira and Philip J. Norris for helpful discussions. Support: CNPq: 234058/2014-5; FAPESP: 2012/23397-0; FAPESP: 2014/22827-7.

**Author Contributions:** TA conceived and wrote the main text; AP discussed and wrote the main text; LAM revised and discussion of text; JC conceived and read the final version of this manuscript.

**Conflicts of Interest:** The authors declare no conflict of interest.

### References

1. Cook, L.B.; Elemans, M.; Rowan, A.G.; Asquith, B. HTLV-1: Persistence and pathogenesis. *Virology* **2013**, *435*, 131–140. [CrossRef] [PubMed]
2. Jeffery, K.J.; Usuku, K.; Hall, S.E.; Matsumoto, W.; Taylor, G.P.; Procter, J.; Bunce, M.; Ogg, G.S.; Welsh, K.I.; Weber, J.N.; *et al.* HLA alleles determine human T-lymphotropic virus-I (HTLV-I) proviral load and the risk of HTLV-I-associated myelopathy. *Proc. Natl. Acad. Sci. USA* **1999**, *96*, 3848–3853. [CrossRef] [PubMed]

3. Jeffery, K.J.; Siddiqui, A.A.; Bunce, M.; Lloyd, A.L.; Vine, A.M.; Witkover, A.D.; Izumo, S.; Usuku, K.; Welsh, K.I.; Osame, M.; *et al.* The influence of HLA class I alleles and heterozygosity on the outcome of human T cell lymphotropic virus type I infection. *J.Immunol.* **2000**, *165*, 7278–7284. [CrossRef] [PubMed]

4. Bangham, C.R.; Hall, S.E.; Jeffery, K.J.; Vine, A.M.; Witkover, A.; Nowak, M.A.; Wodarz, D.; Usuku, K.; Osame, M.; *et al.* Genetic control and dynamics of the cellular immune response to the human T-cell leukaemia virus, HTLV-I. *Philos. Trans. R. Soc. Lond. B Biol. Sci.* **1999**, *354*, 691–700. [CrossRef] [PubMed]

5. Borducchi, D.M.; Gerbase-DeLima, M.; Morgun, A.; Shulzhenko, N.; Pombo-de-Oliveira, M.S.; Kerbauy, J.; Rodrigues de Oliveira, J.S.; *et al.* Human leucocyte antigen and human T-cell lymphotropic virus type 1 associated diseases in Brazil. *Br. J. Haematol.* **2003**, *123*, 954–955. [CrossRef] [PubMed]

6. Catalan-Soares, B.C.; Carneiro-Proietti, A.B.; Da Fonseca, F.G.; Correa-Oliveira, R.; Peralva-Lima, D.; Portela, R.; Ribas, J.G.; Goncalves, D.U.; Interdisciplinary, H.R.G.; Proietti, F.A.; *et al.* HLA class I alleles in HTLV-1-associated myelopathy and asymptomatic carriers from the Brazilian cohort GIPH. *Med. Microbiol. Immunol.* **2009**, *198*, 1–3. [CrossRef] [PubMed]

7. Deschamps, R.; Bera, O.; Belrose, G.; Lezin, A.; Bellance, R.; Signate, A.; Cabre, P.; Smadja, D.; Cesaire, R.; Olindo, S.; *et al.* Absence of consistent association between human leukocyte antigen-I and -II alleles and human T-lymphotropic virus type 1 (HTLV-1)-associated myelopathy/tropical spastic paraparesis risk in an HTLV-1 French Afro-Caribbean population. *Int. J. Infect. Dis.* **2010**, *14*, e986–e990. [CrossRef] [PubMed]

8. Goedert, J.J.; Li, H.C.; Gao, X.J.; Chatterjee, N.; Sonoda, S.; Biggar, R.J.; Cranston, B.; Kim, N.; Carrington, M.; Morgan, O.; *et al.* Risk of human T-lymphotropic virus type I-associated diseases in Jamaica with common HLA types. *Int. J. Cancer* **2007**, *121*, 1092–1097. [CrossRef] [PubMed]

9. Trevino, A.; Vicario, J.L.; Lopez, M.; Parra, P.; Benito, R.; Ortiz de Lejarazu, R.; Ramos, J.M.; Del Romero, J.; de Mendoza, C.; Soriano, V.; *et al.* Association between HLA alleles and HAM/TSP in individuals infected with HTLV-1. *J. Neurol.* **2013**, *260*, 2551–2555. [CrossRef] [PubMed]

10. Sabouri, A.H.; Saito, M.; Usuku, K.; Bajestan, S.N.; Mahmoudi, M.; Forughipour, M.; Sabouri, Z.; Abbaspour, Z.; Goharjoo, M.E.; Khayami, E.; *et al.* Differences in viral and host genetic risk factors for development of human T-cell lymphotropic virus type 1 (HTLV-1)-associated myelopathy/tropical spastic paraparesis between Iranian and Japanese HTLV-1-infected individuals. *J. Gen. Virol.* **2005**, *86 Pt 3*, 773–781. [CrossRef] [PubMed]

11. Rafatpanah, H.; Pravica, V.; Faridhosseini, R.; Tabatabaei, A.; Ollier, W.; Poulton, K.; Thomson, W.; Hutchinson, I.; *et al.* Association between HLA-DRB1*01 and HLA-Cw*08 and outcome following HTLV-I infection. *Iran. J. Immunol.* **2007**, *4*, 94–100. [PubMed]

12. Taghaddosi, M.; Rezaee, S.A.; Rafatpanah, H.; Rajaei, T.; Farid Hosseini, R.; Narges, V. Association between HLA Class I Alleles and Proviral Load in HTLV-I Associated Myelopathy/Tropical Spastic Paraperesis (HAM/TSP) Patients in Iranian Population. *Iran. J. Basic Med. Sci.* **2013**, *16*, 264–267. [PubMed]

13. Imanishi, T.; Akaza, T.; Kimura, A.; Tokunaga, K.; Gojobori, T. Allele and haplotype frequencies for HLA and complement loci in various ethnic groups. In *HLA. 1*; Tsuji, K., Aizawa, M., Sasazuki, T., Eds.; Oxford University Press: New York, NY, USA, 1992; p. 1065.

14. Macnamara, A.; Rowan, A.; Hilburn, S.; Kadolsky, U.; Fujiwara, H.; Suemori, K.; Yasukawa, M.; Taylor, G.; Bangham, C.R.; Asquith, B.; *et al.* HLA class I binding of HBZ determines outcome in HTLV-1 infection. *PLoS Pathog.* **2010**, *6*, e1001117. [CrossRef] [PubMed]

15. MacNamara, A.; Kadolsky, U.; Bangham, C.R.; Asquith, B. T-cell epitope prediction: Rescaling can mask biological variation between MHC molecules. *PLoS Comput. Biol.* **2009**, *5*, e1000327. [CrossRef] [PubMed]

16. Suppiah, V.; Moldovan, M.; Ahlenstiel, G.; Berg, T.; Weltman, M.; Abate, M.L.; Bassendine, M.; Spengler, U.; Dore, G.J.; Powell, E.; *et al.* IL28B is associated with response to chronic hepatitis C interferon-alpha and ribavirin therapy. *Nat. Genet.* **2009**, *41*, 1100–1104. [CrossRef] [PubMed]

17. Sanabani, S.S.; Nukui, Y.; Pereira, J.; da Costa, A.C.; de Oliveira, A.C.; Pessoa, R.; Leal, F.E.; Segurado, A.C.; Kallas, E.G.; Sabino, E.C.; *et al.* Lack of evidence to support the association of a single IL28B genotype SNP rs12979860 with the HTLV-1 clinical outcomes and proviral load. *BMC Infect. Dis.* **2012**, *12*, 374. [CrossRef] [PubMed]

18. Trevino, A.; Lopez, M.; Vispo, E.; Aguilera, A.; Ramos, J.M.; Benito, R.; Roc, L.; Eiros, J.M.; de Mendoza, C.; Soriano, V.; *et al.* Development of tropical spastic paraparesis in human T-lymphotropic virus type 1 carriers is influenced by interleukin 28B gene polymorphisms. *Clin. Infect. Dis.* **2012**, *55*, e1–e4. [CrossRef] [PubMed]

19. Assone, T.; de Souza, F.V.; Gaester, K.O.; Fonseca, L.A.; Luiz Odo, C.; Malta, F.; Pinho, J.R.; Goncalves Fde, T.; Duarte, A.J.; de Oliveira, A.C.; *et al.* IL28B gene polymorphism SNP rs8099917 genotype GG is associated with HTLV-1-associated myelopathy/tropical spastic paraparesis (HAM/TSP) in HTLV-1 carriers. *PLoS Negl. Trop. Dis.* **2014**, *8*, e3199. [CrossRef] [PubMed]

20. Martin, M.P.; Qi, Y.; Goedert, J.J.; Hussain, S.K.; Kirk, G.D.; Hoots, W.K.; Buchbinder, S.; Carrington, M.; Thio, C.L.; *et al.* IL28B polymorphism does not determine outcomes of hepatitis B virus or HIV infection. *J. Infect. Dis.* **2010**, *202*, 1749–1753. [CrossRef] [PubMed]

21. Machmach, K.; Abad-Molina, C.; Romero-Sanchez, M.C.; Abad, M.A.; Ferrando-Martinez, S.; Genebat, M.; Pulido, I.; Viciana, P.; Gonzalez-Escribano, M.F.; Leal, M.; *et al.* IL28B single-nucleotide polymorphism rs12979860 is associated with spontaneous HIV control in white subjects. *J. Infect. Diseases.* **2013**, *207*, 651–655. [CrossRef] [PubMed]

22. Ge, D.; Fellay, J.; Thompson, A.J.; Simon, J.S.; Shianna, K.V.; Urban, T.J.; Heinzen, E.L.; Qiu, P.; Bertelsen, A.H.; Muir, A.J.; *et al.* Genetic variation in IL28B predicts hepatitis C treatment-induced viral clearance. *Nature* **2009**, *461*, 399–401. [CrossRef] [PubMed]

23. Cardoso, D.F.; Souza, F.V.; Fonseca, L.A.; Duarte, A.J.; Casseb, J. Influence of human T-cell lymphotropic virus type 1 (HTLV-1) Infection on laboratory parameters of patients with chronic hepatitis C virus. *Rev. Inst. Med. Trop. Sao Paulo* **2009**, *51*, 325–329. [CrossRef] [PubMed]

24. Brites, C.; Sampalo, J.; Oliveira, A. HIV/human T-cell lymphotropic virus coinfection revisited: Impact on AIDS progression. *AIDS Rev.* **2009**, *11*, 8–16. [PubMed]

25. Montanheiro, P.A.; Penalva de Oliveira, A.C.; Smid, J.; Fukumori, L.M.; Olah, I.; Da, S.D.A.J.; Casseb, J.; *et al.* The elevated interferon gamma production is an important immunological marker in HAM/TSP pathogenesis. *Scand. J. Immunol.* **2009**, *70*, 403–407. [CrossRef] [PubMed]

26. Balagopal, A.; Thomas, D.L.; Thio, C.L. IL28B and the control of hepatitis C virus infection. *Gastroenterology* **2010**, *139*, 1865–1876. [CrossRef] [PubMed]

27. Jacobson, S.; Krichavsky, M.; Flerlage, N.; Levin, M. Immunopathogenesis of HTLV-I associated neurologic disease: Massive latent HTLV-I infection in bone marrow of HAM/TSP patients. *Leukemia* **1997**, *11* (Suppl. 3), 73–75. [PubMed]

28. Ferri, C.; Monti, M.; La Civita, L.; Careccia, G.; Mazzaro, C.; Longombardo, G.; Lombardini, F.; Greco, F.; Pasero, G.; Bombardieri, S.; *et al.* Hepatitis C virus infection in non-Hodgkin's B-cell lymphoma complicating mixed cryoglobulinaemia. *Eur. J. Clin. Investig.* **1994**, *24*, 781–784. [CrossRef]

29. Koziel, M.J.; Dudley, D.; Wong, J.T.; Dienstag, J.; Houghton, M.; Ralston, R.; Walker, B.D. Intrahepatic cytotoxic T lymphocytes specific for hepatitis C virus in persons with chronic hepatitis. *J.Immunol.* **1992**, *149*, 3339–3344. [PubMed]

30. Bellon, M.; Lepelletier, Y.; Hermine, O.; Nicot, C. Deregulation of microRNA involved in hematopoiesis and the immune response in HTLV-I adult T-cell leukemia. *Blood* **2009**, *113*, 4914–4917. [CrossRef] [PubMed]

31. Ruggero, K.; Corradin, A.; Zanovello, P.; Amadori, A.; Bronte, V.; Ciminale, V.; D'Agostino, D.M. Role of microRNAs in HTLV-1 infection and transformation. *Mol. Aspects Med.* **2010**, *31*, 367–382. [CrossRef] [PubMed]

32. Colonna, M.; Borsellino, G.; Falco, M.; Ferrara, G.B.; Strominger, J.L. HLA-C is the inhibitory ligand that determines dominant resistance to lysis by NK1- and NK2-specific natural killer cells. *Proc. Natl. Acad. Sci. USA* **1993**, *90*, 12000–12004. [CrossRef] [PubMed]

33. Wagtmann, N.; Rajagopalan, S.; Winter, C.C.; Peruzzi, M.; Long, E.O. Killer cell inhibitory receptors specific for HLA-C and HLA-B identified by direct binding and by functional transfer. *Immunity* **1995**, *3*, 801–809. [CrossRef]

34. Numasaki, M.; Tagawa, M.; Iwata, F.; Suzuki, T.; Nakamura, A.; Okada, M.; Iwakura, Y.; Aiba, S.; Yamaya, M. IL-28 elicits antitumor responses against murine fibrosarcoma. *J. Immunol.* **2007**, *178*, 5086–5098. [CrossRef] [PubMed]

35. Sato, A.; Ohtsuki, M.; Hata, M.; Kobayashi, E.; Murakami, T. Antitumor activity of IFN-lambda in murine tumor models. *J. Immunol.* **2006**, *176*, 7686–7694. [CrossRef] [PubMed]

36. Khakoo, S.I.; Thio, C.L.; Martin, M.P.; Brooks, C.R.; Gao, X.; Astemborski, J.; Cheng, J.; Goedert, J.J.; Vlahov, D.; Hilgartner, M.; *et al.* HLA and NK cell inhibitory receptor genes in resolving hepatitis C virus infection. *Science* **2004**, *305*, 872–874. [CrossRef] [PubMed]

37. Knapp, S.; Warshow, U.; Hegazy, D.; Brackenbury, L.; Guha, I.N.; Fowell, A.; Little, A.M.; Alexander, G.J.; Rosenberg, W.M.; Cramp, M.E.; *et al.* Consistent beneficial effects of killer cell immunoglobulin-like receptor 2DL3 and group 1 human leukocyte antigen-C following exposure to hepatitis C virus. *Hepatology* **2010**, *51*, 1168–1175. [CrossRef] [PubMed]

38. Vidal-Castineira, J.R.; Lopez-Vazquez, A.; Martinez-Borra, J.; Martinez-Camblor, P.; Prieto, J.; Lopez-Rodriguez, R.; Sanz-Cameno, P.; de la Vega, J.; Rodrigo, L.; Perez-Lopez, R.; *et al.* Diversity of killer cell immunoglobulin-like receptor (KIR) genotypes and KIR2DL2/3 variants in HCV treatment outcome. *PLoS ONE* **2014**, *9*, e99426.

39. Ahlenstiel, G.; Martin, M.P.; Gao, X.; Carrington, M.; Rehermann, B. Distinct KIR/HLA compound genotypes affect the kinetics of human antiviral natural killer cell responses. *J. Clin. Investig.* **2008**, *118*, 1017–1026. [CrossRef] [PubMed]

40. Suppiah, V.; Gaudieri, S.; Armstrong, N.J.; O'Connor, K.S.; Berg, T.; Weltman, M.; Abate, M.L.; Spengler, U.; Bassendine, M.; Dore, G.J.; *et al.* IL28B, HLA-C, and KIR variants additively predict response to therapy in chronic hepatitis C virus infection in a European Cohort: A cross-sectional study. *PLoS Med.* **2011**, *8*, e1001092. [CrossRef] [PubMed]

41. Kamihira, S.; Usui, T.; Ichikawa, T.; Uno, N.; Morinaga, Y.; Mori, S.; Nagai, K.; Sasaki, D.; Hasegawa, H.; Yanagihara, K.; *et al.* Paradoxical expression of IL-28B mRNA in peripheral blood in human T-cell leukemia virus type-1 mono-infection and co-infection with hepatitis C virus. *Virol. J.* **2012**, *9*, 40. [CrossRef] [PubMed]

42. Seich Al Basatena, N.K.; Macnamara, A.; Vine, A.M.; Thio, C.L.; Astemborski, J.; Usuku, K.; Osame, M.; Kirk, G.D.; Donfield, S.M.; Goedert, J.J.; *et al.* KIR2DL2 enhances protective and detrimental HLA class I-mediated immunity in chronic viral infection. *PLoS Pathog.* **2011**, *7*, e1002270. [CrossRef] [PubMed]

43. Liu, R.; Paxton, W.A.; Choe, S.; Ceradini, D.; Martin, S.R.; Horuk, R.; MacDonald, M.E.; Stuhlmann, H.; Koup, R.A.; Landau, N.R. Homozygous defect in HIV-1 coreceptor accounts for resistance of some multiply-exposed individuals to HIV-1 infection. *Cell* **1996**, *86*, 367–377. [CrossRef]

44. Dean, M.; Carrington, M.; Winkler, C.; Huttley, G.A.; Smith, M.W.; Allikmets, R.; Goedert, J.J.; Buchbinder, S.P.; Vittinghoff, E.; Gomperts, E.; *et al.* Genetic restriction of HIV-1 infection and progression to AIDS by a deletion allele of the CKR5 structural gene. Hemophilia Growth and Development Study, Multicenter AIDS Cohort Study, Multicenter Hemophilia Cohort Study, San Francisco City Cohort, ALIVE Study. *Science* **1996**, *273*, 1856–1862. [PubMed]

45. Samson, M.; Libert, F.; Doranz, B.J.; Rucker, J.; Liesnard, C.; Farber, C.M.; Saragosti, S.; Lapoumeroulie, C.; Cognaux, J.; Forceille, C.; *et al.* Resistance to HIV-1 infection in caucasian individuals bearing mutant alleles of the CCR-5 chemokine receptor gene. *Nature* **1996**, *382*, 722–725. [CrossRef] [PubMed]

46. Schinkel, J.; Langendam, M.W.; Coutinho, R.A.; Krol, A.; Brouwer, M.; Schuitemaker, H. No evidence for an effect of the CCR5 delta32/+ and CCR2b 64I/+ mutations on human immunodeficiency virus (HIV)-1 disease progression among HIV-1-infected injecting drug users. *J. Infect. Dis.* **1999**, *179*, 825–831. [CrossRef] [PubMed]

47. Stewart, G.J.; Ashton, L.J.; Biti, R.A.; Ffrench, R.A.; Bennetts, B.H.; Newcombe, N.R.; Benson, E.M.; Carr, A.; Cooper, D.A.; Kaldor, J.M. Increased frequency of CCR-5 delta 32 heterozygotes among long-term non-progressors with HIV-1 infection. The Australian Long-Term Non-Progressor Study Group. *AIDS* **1997**, *11*, 1833–1838. [CrossRef] [PubMed]

48. Michael, N.L.; Louie, L.G.; Sheppard, H.W. CCR5-delta 32 gene deletion in HIV-1 infected patients. *Lancet* **1997**, *350*, 741–742. [CrossRef]

49. Martinson, J.J.; Chapman, N.H.; Rees, D.C.; Liu, Y.T.; Clegg, J.B. Global distribution of the CCR5 gene 32-basepair deletion. *Nat. Genet.* **1997**, *16*, 100–103. [CrossRef] [PubMed]

50. He, J.; Li, X.; Tang, J.; Jin, T.; Liao, Q.; Hu, G. Association between chemotactic chemokine ligand 5 -403G/A polymorphism and risk of human immunodeficiency virus-1 infection: A meta-analysis. *OncoTargets Ther.* **2015**, *8*, 727–734.

51. Gong, X.; Liu, Y.; Liu, F.L.; Jin, L.; Wang, H.; Zheng, Y.T. A SDF1 genetic variant confers resistance to HIV-1 infection in intravenous drug users in China. *Infect. Genet. Evol.* **2015**, *34*, 137–142. [CrossRef] [PubMed]

52. An, P.; Goedert, J.J.; Donfield, S.; Buchbinder, S.; Kirk, G.D.; Detels, R.; Winkler, C.A. Regulatory variation in HIV-1 dependency factor ZNRD1 associates with host resistance to HIV-1 acquisition. *J. Infect. Dis.* **2014**, *210*, 1539–1548. [CrossRef] [PubMed]

53. Fahrbach, K.M.; Barry, S.M.; Ayehunie, S.; Lamore, S.; Klausner, M.; Hope, T.J. Activated CD34-derived Langerhans cells mediate transinfection with human immunodeficiency virus. *J. Virol.* **2007**, *81*, 6858–6868. [CrossRef] [PubMed]

54. de Witte, L.; Nabatov, A.; Pion, M.; Fluitsma, D.; de Jong, M.A.; de Gruijl, T.; Piguet, V.; van Kooyk, Y.; Geijtenbeek, T.B. Langerin is a natural barrier to HIV-1 transmission by Langerhans cells. *Nat. Med.* **2007**, *13*, 367–371. [CrossRef] [PubMed]

55. Schwartz, O. Langerhans cells lap up HIV-1. *Nat. Med.* **2007**, *13*, 245–246. [CrossRef] [PubMed]

56. Cole, A.M.; Ganz, T. Human antimicrobial peptides: Analysis and application. *BioTechniques* **2000**, *29*, 822–826, 828,830–831. [PubMed]

57. Jenssen, H.; Hamill, P.; Hancock, R.E. Peptide antimicrobial agents. *Clin. Microbiol. Rev.* **2006**, *19*, 491–511. [CrossRef] [PubMed]

58. Klotman, M.E.; Chang, T.L. Defensins in innate antiviral immunity. *Nat. Rev. Immunol.* **2006**, *6*, 447–456. [CrossRef] [PubMed]

59. Furci, L.; Sironi, F.; Tolazzi, M.; Vassena, L.; Lusso, P. Alpha-defensins block the early steps of HIV-1 infection: Interference with the binding of gp120 to CD4. *Blood* **2007**, *109*, 2928–2935. [PubMed]

60. Chang, T.L.; Vargas, J., Jr.; DelPortillo, A.; Klotman, M.E. Dual role of alpha-defensin-1 in anti-HIV-1 innate immunity. *J. Clin. Investig.* **2005**, *115*, 765–773. [CrossRef] [PubMed]

61. Weinberg, A.; Quinones-Mateu, M.E.; Lederman, M.M. Role of human beta-defensins in HIV infection. *Adv. Dent. Res.* **2006**, *19*, 42–48. [CrossRef] [PubMed]

62. Schutte, B.C.; Mitros, J.P.; Bartlett, J.A.; Walters, J.D.; Jia, H.P.; Welsh, M.J.; Casavant, T.L.; McCray, P.B. Discovery of five conserved beta-defensin gene clusters using a computational search strategy. *Proc. Natl. Acad. Sci. USA* **2002**, *99*, 2129–2133. [CrossRef] [PubMed]

63. Cole, A.M.; Hong, T.; Boo, L.M.; Nguyen, T.; Zhao, C.; Bristol, G.; Zack, J.A.; Waring, A.J.; Yang, O.O.; Lehrer, R.I. Retrocyclin: A primate peptide that protects cells from infection by T- and M-tropic strains of HIV-1. *Proc. Natl. Acad. Sci. USA* **2002**, *99*, 1813–1818. [CrossRef] [PubMed]

64. Gallo, S.A.; Wang, W.; Rawat, S.S.; Jung, G.; Waring, A.J.; Cole, A.M.; Lu, H.; Yan, X.; Daly, N.L.; Craik, D.J.; et al. Theta-defensins prevent HIV-1 Env-mediated fusion by binding gp41 and blocking 6-helix bundle formation. *J. Biol. Chem.* **2006**, *281*, 18787–18792. [CrossRef] [PubMed]

65. Cole, A.M.; Lehrer, R.I. Minidefensins: Antimicrobial peptides with activity against HIV-1. *Curr. Pharm. Des.* **2003**, *9*, 1463–1473. [CrossRef] [PubMed]

66. Teixeira, S.L.; de Sa, N.B.; Campos, D.P.; Coelho, A.B.; Guimaraes, M.L.; Leite, T.C.; Veloso, V.G.; Morgado, M.G. Association of the HLA-B*52 allele with non-progression to AIDS in Brazilian HIV-1-infected individuals. *Genes Immun.* **2014**, *15*, 256–262. [CrossRef] [PubMed]

67. Gao, X.; O'Brien, T.R.; Welzel, T.M.; Marti, D.; Qi, Y.; Goedert, J.J.; Phair, J.; Pfeiffer, R.; Carrington, M. HLA-B alleles associate consistently with HIV heterosexual transmission, viral load, and progression to AIDS, but not susceptibility to infection. *AIDS* **2010**, *24*, 1835–1840. [CrossRef] [PubMed]

68. Klein, M.R.; Keet, I.P.; D'Amaro, J.; Bende, R.J.; Hekman, A.; Mesman, B.; Koot, M.; de Waal, L.P.; Coutinho, R.A.; Miedema, F. Associations between HLA frequencies and pathogenic features of human immunodeficiency virus type 1 infection in seroconverters from the Amsterdam cohort of homosexual men. *J. Infect. Dis.* **1994**, *169*, 1244–1249. [CrossRef] [PubMed]

69. Gao, X.; Nelson, G.W.; Karacki, P.; Martin, M.P.; Phair, J.; Kaslow, R.; Goedert, J.J.; Buchbinder, S.; Hoots, K.; Vlahov, D.; et al. Effect of a single amino acid change in MHC class I molecules on the rate of progression to AIDS. *N. Engl. J. Med.* **2001**, *344*, 1668–1675. [CrossRef] [PubMed]

70. Migueles, S.A.; Sabbaghian, M.S.; Shupert, W.L.; Bettinotti, M.P.; Marincola, F.M.; Martino, L.; Hallahan, C.W.; Selig, S.M.; Schwartz, D.; Sullivan, J.; et al. HLA B*5701 is highly associated with restriction of virus replication in a subgroup of HIV-infected long term nonprogressors. *Proc. Natl. Acad. Sci. USA* **2000**, *97*, 2709–2714. [CrossRef] [PubMed]

71. Kaslow, R.A.; Carrington, M.; Apple, R.; Park, L.; Munoz, A.; Saah, A.J.; Goedert, J.J.; Winkler, C.; O'Brien, S.J.; Rinaldo, C.; et al. Influence of combinations of human major histocompatibility complex genes on the course of HIV-1 infection. *Nat. Med.* **1996**, *2*, 405–411. [CrossRef] [PubMed]

72. Hendel, H.; Caillat-Zucman, S.; Lebuanec, H.; Carrington, M.; O'Brien, S.; Andrieu, J.M.; Schachter, F.; Zagury, D.; Rappaport, J.; Winkler, C.; et al. New class I and II HLA alleles strongly associated with opposite patterns of progression to AIDS. *J. Immunol.* **1999**, *162*, 6942–6946. [PubMed]

73. Altfeld, M.; Addo, M.M.; Rosenberg, E.S.; Hecht, F.M.; Lee, P.K.; Vogel, M.; Yu, X.G.; Draenert, R.; Johnston, M.N.; Strick, D.; *et al.* Influence of HLA-B57 on clinical presentation and viral control during acute HIV-1 infection. *AIDS* **2003**, *17*, 2581–2591. [CrossRef] [PubMed]

74. Catano, G.; Kulkarni, H.; He, W.; Marconi, V.C.; Agan, B.K.; Landrum, M.; Anderson, S.; Delmar, J.; Telles, V.; Song, L.; *et al.* HIV-1 disease-influencing effects associated with ZNRD1, HCP5 and HLA-C alleles are attributable mainly to either HLA-A10 or HLA-B*57 alleles. *PLoS ONE* **2008**, *3*, e3636. [CrossRef] [PubMed]

75. Antoni, G.; Guergnon, J.; Meaudre, C.; Samri, A.; Boufassa, F.; Goujard, C.; Lambotte, O.; Autran, B.; Rouzioux, C.; Costagliola, D.; *et al.* MHC-driven HIV-1 control on the long run is not systematically determined at early times post-HIV-1 infection. *AIDS* **2013**, *27*, 1707–1716. [CrossRef] [PubMed]

76. Moroni, M.; Ghezzi, S.; Baroli, P.; Heltai, S.; De Battista, D.; Pensieroso, S.; Cavarelli, M.; Dispinseri, S.; Vanni, I.; Pastori, C.; *et al.* Spontaneous control of HIV-1 viremia in a subject with protective HLA-B plus HLA-C alleles and HLA-C associated single nucleotide polymorphisms. *J. Transl. Med.* **2014**, *12*, 335. [CrossRef] [PubMed]

77. Ballana, E.; Ruiz-de Andres, A.; Mothe, B.; Ramirez de Arellano, E.; Aguilar, F.; Badia, R.; Grau, E.; Clotet, B.; del Val, M.; Brander, C.; *et al.* Differential prevalence of the HLA-C—35 CC genotype among viremic long term non-progressor and elite controller HIV+ individuals. *Immunobiology* **2012**, *217*, 889–894. [CrossRef] [PubMed]

78. Julg, B.; Moodley, E.S.; Qi, Y.; Ramduth, D.; Reddy, S.; Mncube, Z.; Gao, X.; Goulder, P.J.; Detels, R.; Ndung'u, T.; Walker, B.D.; Carrington, M. Possession of HLA class II DRB1*1303 associates with reduced viral loads in chronic HIV-1 clade C and B infection. *J. Infect. Dis.* **2011**, *203*, 803–809. [CrossRef] [PubMed]

79. Ranasinghe, S.; Cutler, S.; Davis, I.; Lu, R.; Soghoian, D.Z.; Qi, Y.; Sidney, J.; Kranias, G.; Flanders, M.D.; Lindqvist, M.; *et al.* Association of HLA-DRB1-restricted CD4(+) T cell responses with HIV immune control. *Nat. Med.* **2013**, *19*, 930–933. [CrossRef] [PubMed]

80. Ramana, G.V.; Vasanthi, A.; Khaja, M.; Su, B.; Govindaiah, V.; Jin, L.; Singh, L.; Chakraborty, R. Distribution of HIV-1 resistance-conferring polymorphic alleles SDF-1-3'A, CCR2-64I and CCR5-Delta32 in diverse populations of Andhra Pradesh, South India. *J. Genet.* **2001**, *80*, 137–140. [CrossRef] [PubMed]

81. Voevodin, A.; Samilchuk, E.; Dashti, S. Frequencies of SDF-1 chemokine, CCR-5, and CCR-2 chemokine receptor gene alleles conferring resistance to human immunodeficiency virus type 1 and AIDS in Kuwaitis. *J. Med. Virol.* **1999**, *58*, 54–58. [CrossRef]

82. Chatterjee, A.; Rathore, A.; Vidyant, S.; Kakkar, K.; Dhole, T.N. Chemokines and chemokine receptors in susceptibility to HIV-1 infection and progression to AIDS. *Dis. Markers* **2012**, *32*, 143–151. [CrossRef] [PubMed]

83. Smith, M.W.; Carrington, M.; Winkler, C.; Lomb, D.; Dean, M.; Huttley, G.; O'Brien, S.J. CCR2 chemokine receptor and AIDS progression. *Nat. Med.* **1997**, *3*, 1052–1053. [CrossRef] [PubMed]

84. Lama, J.; Planelles, V. Host factors influencing susceptibility to HIV infection and AIDS progression. *Retrovirology* **2007**, *4*, 52. [CrossRef] [PubMed]

85. Faure, S.; Meyer, L.; Costagliola, D.; Vaneensberghe, C.; Genin, E.; Autran, B.; Delfraissy, J.F.; McDermott, D.H.; Murphy, P.M.; Debre, P.; *et al.* Rapid progression to AIDS in HIV+ individuals with a structural variant of the chemokine receptor CX3CR1. *Science* **2000**, *287*, 2274–2277. [CrossRef] [PubMed]

86. Singh, K.K.; Hughes, M.D.; Chen, J.; Spector, S.A. Genetic polymorphisms in CX3CR1 predict HIV-1 disease progression in children independently of CD4+ lymphocyte count and HIV-1 RNA load. *J. Infect. Dis.* **2005**, *191*, 1971–1980. [CrossRef] [PubMed]

87. Vidal, F.; Vilades, C.; Domingo, P.; Broch, M.; Pedrol, E.; Dalmau, D.; Knobel, H.; Peraire, J.; Gutierrez, C.; Sambeat, M.A.; *et al.* Spanish HIV-1-infected long-term nonprogressors of more than 15 years have an increased frequency of the CX3CR1 249I variant allele. *J. Acquir. Immune Defic. Syndr.* **2005**, *40*, 527–531. [CrossRef] [PubMed]

88. Menten, P.; Wuyts, A.; van Damme, J. Macrophage inflammatory protein-1. *Cytokine Growth Factor Rev.* **2002**, *13*, 455–481. [CrossRef]

89. Irving, S.G.; Zipfel, P.F.; Balke, J.; McBride, O.W.; Morton, C.C.; Burd, P.R.; Siebenlist, U.; Kelly, K. Two inflammatory mediator cytokine genes are closely linked and variably amplified on chromosome 17q. *Nucleic Acids Res.* **1990**, *18*, 3261–3270. [CrossRef] [PubMed]

90. Nasi, M.; Riva, A.; Borghi, V.; D'Amico, R.; Del Giovane, C.; Casoli, C.; Galli, M.; Vicenzi, E.; Gibellini, L.; De Biasi, S.; *et al.* Novel genetic association of TNF-alpha-238 and PDCD1-7209 polymorphisms with long-term non-progressive HIV-1 infection. *Int. J. Infect. Dis.* **2013**, *17*, e845–e850. [CrossRef] [PubMed]
91. Hetherington, S.; McGuirk, S.; Powell, G.; Cutrell, A.; Naderer, O.; Spreen, B.; Lafon, S.; Pearce, G.; Steel, H. Hypersensitivity reactions during therapy with the nucleoside reverse transcriptase inhibitor abacavir. *Clin. Ther.* **2001**, *23*, 1603–1614. [CrossRef]
92. Mallal, S.; Nolan, D.; Witt, C.; Masel, G.; Martin, A.M.; Moore, C.; Sayer, D.; Castley, A.; Mamotte, C.; Maxwell, D.; *et al.* Association between presence of HLA-B*5701, HLA-DR7, and HLA-DQ3 and hypersensitivity to HIV-1 reverse-transcriptase inhibitor abacavir. *Lancet* **2002**, *359*, 727–732. [CrossRef]
93. Hetherington, S.; Hughes, A.R.; Mosteller, M.; Shortino, D.; Baker, K.L.; Spreen, W.; Lai, E.; Davies, K.; Handley, A.; Dow, D.J.; *et al.* Genetic variations in HLA-B region and hypersensitivity reactions to abacavir. *Lancet* **2002**, *359*, 1121–1122. [CrossRef]
94. Martin, A.M.; Nolan, D.; Gaudieri, S.; Almeida, C.A.; Nolan, R.; James, I.; Carvalho, F.; Phillips, E.; Christiansen, F.T.; Purcell, A.W.; *et al.* Predisposition to abacavir hypersensitivity conferred by HLA-B*5701 and a haplotypic Hsp70-Hom variant. *Proc. Natl. Acad. Sci. USA* **2004**, *101*, 4180–4185. [CrossRef] [PubMed]
95. Hughes, A.R.; Mosteller, M.; Bansal, A.T.; Davies, K.; Haneline, S.A.; Lai, E.H.; Nangle, K.; Scott, T.; Spreen, W.R.; Warren, L.L.; *et al.* Association of genetic variations in HLA-B region with hypersensitivity to abacavir in some, but not all, populations. *Pharmacogenomics* **2004**, *5*, 203–211. [CrossRef] [PubMed]
96. Phillips, E.J.; Wong, G.A.; Kaul, R.; Shahabi, K.; Nolan, D.A.; Knowles, S.R.; Martin, A.M.; Mallal, S.A.; Shear, N.H. Clinical and immunogenetic correlates of abacavir hypersensitivity. *AIDS* **2005**, *19*, 979–981. [CrossRef] [PubMed]
97. Rauch, A.; Nolan, D.; Martin, A.; McKinnon, E.; Almeida, C.; Mallal, S. Prospective genetic screening decreases the incidence of abacavir hypersensitivity reactions in the Western Australian HIV cohort study. *Clin. Infect. Dis.* **2006**, *43*, 99–102. [CrossRef] [PubMed]
98. Mallal, S.; Phillips, E.; Carosi, G.; Molina, J.M.; Workman, C.; Tomazic, J.; Jagel-Guedes, E.; Rugina, S.; Kozyrev, O.; Cid, J.F.; *et al.* HLA-B*5701 screening for hypersensitivity to abacavir. *N. Engl. J. Med.* **2008**, *358*, 568–579. [CrossRef] [PubMed]
99. Gilliam, B.L.; Riedel, D.J.; Redfield, R.R. Clinical use of CCR5 inhibitors in HIV and beyond. *J. Transl. Med.* **2011**, *9* (Suppl. 1), S9. [CrossRef] [PubMed]
100. Yukl, S.A.; Boritz, E.; Busch, M.; Bentsen, C.; Chun, T.W.; Douek, D.; Eisele, E.; Haase, A.; Ho, Y.C.; Hutter, G.; *et al.* Challenges in detecting HIV persistence during potentially curative interventions: A study of the Berlin patient. *PLoS Pathog.* **2013**, *9*, e1003347. [CrossRef] [PubMed]
101. Zack, J.A.; Cann, A.J.; Lugo, J.P.; Chen, I.S. HIV-1 production from infected peripheral blood T cells after HTLV-I induced mitogenic stimulation. *Science* **1988**, *240*, 1026–1029. [CrossRef] [PubMed]
102. Bartholomew, C.; Blattner, W.; Cleghorn, F. Progression to AIDS in homosexual men co-infected with HIV and HTLV-I in Trinidad. *Lancet* **1987**, *2*, 1469. [CrossRef]
103. Sobesky, M.; Couppie, P.; Pradinaud, R.; Godard, M.C.; Alvarez, F.; Benoit, B.; Carme, B.; Lebeux, P.; *et al.* Coinfection with HIV and HTLV-I infection and survival in AIDS stage. French Guiana Study. GECVIG (Clinical HIV Study Group in Guiana). *Presse Med.* **2000**, *29*, 413–416. [PubMed]
104. Bovolenta, C.; Pilotti, E.; Mauri, M.; Turci, M.; Ciancianaini, P.; Fisicaro, P.; Bertazzoni, U.; Poli, G.; Casoli, C. Human T-cell leukemia virus type 2 induces survival and proliferation of CD34(+) TF-1 cells through activation of STAT1 and STAT5 by secretion of interferon-gamma and granulocyte macrophage-colony-stimulating factor. *Blood* **2002**, *99*, 224–231. [CrossRef] [PubMed]
105. Casoli, C.; Vicenzi, E.; Cimarelli, A.; Magnani, G.; Ciancianaini, P.; Cattaneo, E.; Dall'Aglio, P.; Poli, G.; Bertazzoni, U. HTLV-II down-regulates HIV-1 replication in IL-2-stimulated primary PBMC of coinfected individuals through expression of MIP-1alpha. *Blood* **2000**, *95*, 2760–2769. [PubMed]
106. Balistrieri, G.; Barrios, C.; Castillo, L.; Umunakwe, T.C.; Giam, C.Z.; Zhi, H.; Beilke, M.A.; *et al.* Induction of CC-chemokines with antiviral function in macrophages by the human T lymphotropic virus type 2 transactivating protein, Tax2. *Viral Immunol.* **2013**, *26*, 3–12. [CrossRef] [PubMed]

107. Lewis, M.J.; Gautier, V.W.; Wang, X.P.; Kaplan, M.H.; Hall, W.W. Spontaneous production of C-C chemokines by individuals infected with human T lymphotropic virus type II (HTLV-II) alone and HTLV-II/HIV-1 coinfected individuals. *J. Immunol.* **2000**, *165*, 4127–4132. [CrossRef] [PubMed]

108. Dragic, T.; Litwin, V.; Allaway, G.P.; Martin, S.R.; Huang, Y.; Nagashima, K.A.; Cayanan, C.; Maddon, P.J.; Koup, R.A.; Moore, J.P.; *et al.* HIV-1 entry into CD4+ cells is mediated by the chemokine receptor CC-CKR-5. *Nature* **1996**, *381*, 667–673. [CrossRef] [PubMed]

109. Connor, R.I.; Sheridan, K.E.; Ceradini, D.; Choe, S.; Landau, N.R. Change in coreceptor use correlates with disease progression in HIV-1–infected individuals. *J. Exp. Med.* **1997**, *185*, 621–628. [CrossRef] [PubMed]

110. Ullum, H.; Cozzi Lepri, A.; Victor, J.; Aladdin, H.; Phillips, A.N.; Gerstoft, J.; Skinhoj, P.; Pedersen, B.K. Production of beta-chemokines in human immunodeficiency virus (HIV) infection: Evidence that high levels of macrophage inflammatory protein-1beta are associated with a decreased risk of HIV disease progression. *J. Infect. Dis.* **1998**, *177*, 331–336. [CrossRef] [PubMed]

111. Cocchi, F.; DeVico, A.L.; Yarchoan, R.; Redfield, R.; Cleghorn, F.; Blattner, W.A.; Garzino-Demo, A.; Colombini-Hatch, S.; Margolis, D.; Gallo, R.C. Higher macrophage inflammatory protein (MIP)-1alpha and MIP-1beta levels from CD8+ T cells are associated with asymptomatic HIV-1 infection. *Proc. Natl. Acad. Sci. USA* **2000**, *97*, 13812–13817. [CrossRef] [PubMed]

112. Ferbas, J.; Giorgi, J.V.; Amini, S.; Grovit-Ferbas, K.; Wiley, D.J.; Detels, R.; Plaeger, S. Antigen-specific production of RANTES, macrophage inflammatory protein (MIP)-1alpha, and MIP-1beta *in vitro* is a correlate of reduced human immunodeficiency virus burden *in vivo*. *J. Infect. Dis.* **2000**, *182*, 1247–1250. [CrossRef] [PubMed]

113. Magnani, G.; Elia, G.; Casoli, C.; Calzetti, C.; Degli Antoni, A.; Fiaccadori, F. HTLV-II does not adversely affect the natural history of HIV-1 infection in intravenous drug users. *Infection* **1995**, *23*, 63. [CrossRef] [PubMed]

114. Cimarelli, A.; Duclos, C.A.; Gessain, A.; Cattaneo, E.; Casoli, C.; Biglione, M.; Mauclere, P.; Bertazzoni, U. Quantification of HTLV-II proviral copies by competitive polymerase chain reaction in peripheral blood mononuclear cells of Italian injecting drug users, central Africans, and Amerindians. *J. Acquir. Immune Defic. Syndr. Hum. Retrovirol.* **1995**, *10*, 198–204. [CrossRef] [PubMed]

115. Beilke, M.A.; Theall, K.P.; O'Brien, M.; Clayton, J.L.; Benjamin, S.M.; Winsor, E.L.; Kissinger, P.J. Clinical outcomes and disease progression among patients coinfected with HIV and human T lymphotropic virus types 1 and 2. *Clin. Infect. Dis.* **2004**, *39*, 256–263. [CrossRef] [PubMed]

116. Bassani, S.; Lopez, M.; Toro, C.; Jimenez, V.; Sempere, J.M.; Soriano, V.; Benito, J.M. Influence of human T cell lymphotropic virus type 2 coinfection on virological and immunological parameters in HIV type 1-infected patients. *Clin. Infect. Dis.* **2007**, *44*, 105–110. [CrossRef] [PubMed]

117. National Institutes of Health. Consensus Development Conference Statement: Management of hepatitis C 2002 (June 10–12, 2002). *Gastroenterology* **2002**, *123*, 2082–2099.

118. Mohd-Hanafiah, K.; Groeger, J.; Flaxman, A.D.; Wiersma, S.T. Global epidemiology of hepatitis C virus infection: New estimates of age-specific antibody to HCV seroprevalence. *Hepatology* **2013**, *57*, 1333–1342. [CrossRef] [PubMed]

119. Shebl, F.M.; Pfeiffer, R.M.; Buckett, D.; Muchmore, B.; Chen, S.; Dotrang, M.; Prokunina-Olsson, L.; Edlin, B.R.; O'Brien, T.R. IL28B rs12979860 genotype and spontaneous clearance of hepatitis C virus in a multi-ethnic cohort of injection drug users: Evidence for a supra-additive association. *J. Infect. Dis.* **2011**, *204*, 1843–1847. [CrossRef] [PubMed]

120. Thomas, D.L.; Thio, C.L.; Martin, M.P.; Qi, Y.; Ge, D.; O'Huigin, C.; Kidd, J.; Kidd, K.; Khakoo, S.I.; Alexander, G.; *et al.* Genetic variation in IL28B and spontaneous clearance of hepatitis C virus. *Nature* **2009**, *461*, 798–801. [CrossRef] [PubMed]

121. Aka, P.V.; Kuniholm, M.H.; Pfeiffer, R.M.; Wang, A.S.; Tang, W.; Chen, S.; Astemborski, J.; Plankey, M.; Villacres, M.C.; Peters, M.G.; *et al.* Association of the IFNL4-DeltaG Allele With Impaired Spontaneous Clearance of Hepatitis C Virus. *J. Infect. Dis.* **2014**, *209*, 350–354. [CrossRef] [PubMed]

122. De Re, V.; Caggiari, L.; De Zorzi, M.; Repetto, O.; Zignego, A.L.; Izzo, F.; Tornesello, M.L.; Buonaguro, F.M.; Mangia, A.; Sansonno, D.; *et al.* Genetic diversity of the KIR/HLA system and susceptibility to hepatitis C virus-related diseases. *PLoS ONE* **2015**, *10*, e0117420.

123. de Vasconcelos, J.M.; de Jesus Maues Pereira Moia, L.; Amaral Ido, S.; Miranda, E.C.; Cicalisetakeshita, L.Y.; de Oliveira, L.F.; de Araujo Melo Mendes, L.; Sastre, D.; Tamegao-Lopes, B.P.; de Aquino Pedroza, L.S.; *et al.* Association of killer cell immunoglobulin-like receptor polymorphisms with chronic hepatitis C and responses to therapy in Brazil. *Genet. Mol. Biol.* **2013**, *36*, 22–27. [CrossRef] [PubMed]

# Section 2:
# Research Articles

*viruses*

MDPI

Article

# PRMT5 Is Upregulated in HTLV-1-Mediated T-Cell Transformation and Selective Inhibition Alters Viral Gene Expression and Infected Cell Survival

Amanda R. Panfil [1,2], Jacob Al-Saleem [1,2], Cory M. Howard [1,2], Jessica M. Mates [3], Jesse J. Kwiek [1,3,4], Robert A. Baiocchi [5,6] and Patrick L. Green [1,2,6,*]

[1]   Center for Retrovirus Research, The Ohio State University, Columbus, OH 43210, USA;
      panfil.6@osu.edu (A.R.P.); al-saleem.1@osu.edu (J.A.-S.); howard.937@osu.edu (C.M.H.);
      kwiek.2@osu.edu (J.J.K.)
[2]   Department of Veterinary Biosciences, The Ohio State University, Columbus, OH 43210, USA
[3]   Center for Microbial Interface Biology, The Ohio State University, Columbus, OH 43210, USA;
      mates.6@osu.edu
[4]   Department of Microbiology and Department of Microbial Infection and Immunity,
      The Ohio State University, Columbus, OH 43210, USA
[5]   Division of Hematology, Department of Internal Medicine, The Ohio State University, Columbus, OH 43210,
      USA; baiocchi.1@osu.edu
[6]   Comprehensive Cancer Center and Solove Research Institute, The Ohio State University, Columbus,
      OH 43210, USA
*    Correspondence: green.466@osu.edu; Tel.: +1-614-688-4899

Academic Editor: Louis M. Mansky
Received: 6 October 2015; Accepted: 18 December 2015; Published: 30 December 2015

**Abstract:** Human T-cell leukemia virus type-1 (HTLV-1) is a tumorigenic retrovirus responsible for development of adult T-cell leukemia/lymphoma (ATLL). This disease manifests after a long clinical latency period of up to 2–3 decades. Two viral gene products, Tax and HBZ, have transforming properties and play a role in the pathogenic process. Genetic and epigenetic cellular changes also occur in HTLV-1-infected cells, which contribute to transformation and disease development. However, the role of cellular factors in transformation is not completely understood. Herein, we examined the role of protein arginine methyltransferase 5 (PRMT5) on HTLV-1-mediated cellular transformation and viral gene expression. We found PRMT5 expression was upregulated during HTLV-1-mediated T-cell transformation, as well as in established lymphocytic leukemia/lymphoma cell lines and ATLL patient PBMCs. shRNA-mediated reduction in PRMT5 protein levels or its inhibition by a small molecule inhibitor (PRMT5i) in HTLV-1-infected lymphocytes resulted in increased viral gene expression and decreased cellular proliferation. PRMT5i also had selective toxicity in HTLV-1-transformed T-cells. Finally, we demonstrated that PRMT5 and the HTLV-1 p30 protein had an additive inhibitory effect on HTLV-1 gene expression. Our study provides evidence for PRMT5 as a host cell factor important in HTLV-1-mediated T-cell transformation, and a potential target for ATLL treatment.

**Keywords:** HTLV-1; PRMT5; transformation; ATLL; Tax; HBZ; p30; lymphoma

---

## 1. Introduction

Human T-cell leukemia virus type-1 (HTLV-1) is a tumorigenic retrovirus that infects an estimated 15–20 million people worldwide [1]. This blood-borne pathogen is the causative infectious agent of adult T-cell leukemia/lymphoma (ATLL), a disease of CD4+ T-cells [2–4]. HTLV-1 is also associated with inflammatory disorders such as HTLV-1-associated myelopathy/tropical spastic paraparesis (HAM/TSP) [5,6]. The likelihood of developing ATLL is between 2%–6% during the lifetime of an infected individual [7], with symptoms taking up to 20–30 years to present. Despite the long clinical latency period, diseases such as ATLL are extremely aggressive and usually fatal. ATLL is highly chemotherapy-resistant, and while many current therapies (e.g., antivirals AZT/IFN-α, proteasome inhibitors, anti-CCR4 monoclonal antibody) improve ATLL patient survival (reviewed in [8]), the patients consistently relapse.

As a complex retrovirus, HTLV-1 has a genome that encodes structural and enzymatic proteins (Gag, Pro, Pol, Env), regulatory proteins (Tax and Rex), and several accessory proteins (p30, p12/p8, HBZ). Studies have shown that at least two viral gene products, Tax and HBZ, have transforming properties and play a role in the pathogenic process [9,10]. Tax acts as a viral transcriptional activator of HTLV-1 gene expression through activation of the viral long terminal repeat (LTR) and various cellular signaling pathways such as CREB, NF-κB, and AP-1 [11,12]. Tax also causes deregulation of the cell cycle by silencing cellular checkpoints that guard against DNA structural damage and abnormal chromosomal segregation, thus leading to the accumulation of mutations in HTLV-1 infected cells [13]. However, Tax expression is lost in greater than 70% of ATLL cells due to genetic and/or epigenetic changes in the HTLV-1 provirus, which include deletion or methylation of the viral 5' LTR. These changes abolish expression of other viral genes with the exception of HBZ. In fact, HBZ is the only viral gene that is intact and expressed in all ATLL cases [14,15]. HBZ protein is expressed from a promoter located in the viral 3' LTR; current data indicates that HBZ promotes proliferation of ATLL cells through both its mRNA and protein forms [15,16]. HBZ protein has also been shown to interact with several cellular transcription factors such as CREB and CBP, p300, JunD, JunB, and c-Jun and to act as a negative regulator of Tax-mediated HTLV-1 transcription [17–22].

Although Tax is indispensable for viral transformation, the mechanisms by which the virus persists *in vivo* and transforms CD4+ T-cells are not completely understood. The requirement for Tax and other viral proteins *in vivo* suggests that expression of viral proteins early in infection plays a major role in viral replication, infected cell survival, and disease development. A favored theory within the field is that the virus is critically dependent on Tax early in infection to initiate transformation, but then Tax expression is highly regulated and often times silenced to prevent immune detection. HBZ is hypothesized to provide the maintenance or cell survival signal necessary for the transformation process. Over time, the combination of genetic and epigenetic changes in an HTLV-1-infected cell can lead to transformation and potentially, disease development [23]. While we know that the viral proteins Tax and HBZ are intimately involved in the cell transformation process, neither is sufficient, which suggests the involvement of cellular factors.

Chromatin remodeling complexes and associated co-repressors such as histone deacetylases (HDAC), DNA methyltransferases (DNMT), and protein arginine methyltransferase 5 (PRMT5) participate in silencing tumor suppressor gene expression and contribute to cellular transformation [24–26]. Recent reports have indicated PRMT5 over-expression as relevant to the pathogenesis of many cancers, including lymphomas, melanomas, and astrocytomas [27–32]. PRMT5 is a type II PRMT enzyme that silences the transcription of key regulatory genes by symmetric di-methylation (S2Me) of arginine (R) residues on histone proteins (H4R3 and H3R8) [33]. PRMT5 is also involved in a wide variety of cellular processes, including RNA processing, transcriptional regulation, and signal transduction pathway regulation that are highly relevant to the pathogenesis of cancer [34–36]. Recently, PRMT5 was found to play a critical role in Epstein-Barr virus (EBV)—driven B-cell transformation [31].

Our group previously identified PRMT5 as a binding partner of the HTLV-1 accessory protein p30, using mass spectrometry [37]. p30 is encoded from a doubly spliced mRNA and is dispensable for viral infection and T-cell transformation *in vitro*, but is required for establishment of viral persistence in an *in vivo* rabbit model of infection [38,39]. p30 negatively regulates viral gene transcription at both the transcriptional and post-transcriptional levels by competing with Tax for binding to CBP/p300 and retaining the *tax/rex* mRNA in the nucleus, respectively [40–42].

Currently, there have been no studies investigating the role of PRMT5 in T-cell malignancies, including HTLV-1-associated disease. Therefore, we sought to determine if PRMT5 plays a role in

HTLV-1 transformation/malignancy. Indeed, we found PRMT5 levels were upregulated during T-cell transformation and in established lymphocytic leukemia/lymphoma cell lines. Our data suggested that PRMT5 negatively regulated HTLV-1 viral gene expression, which indicated that PRMT5 could be an important cellular regulator of the viral transformation process. Furthermore, selective inhibition of PRMT5 by a novel small molecule inhibitor (PRMT5i) in HTLV-1-positive cell lines reduced cell survival; therefore, PRMT5 may represent an important therapeutic target for ATLL.

## 2. Materials and Methods

### 2.1. Cell Lines and Culture

HEK293T and pA-18G-BHK-21 cells were maintained in Dulbecco's modified Eagle's medium (DMEM) supplemented with 10% fetal bovine serum (FBS) (Gemini Bio-Products, Broderick, CA, USA), 2 mM glutamine, penicillin (100 U/mL), and streptomycin (100 µg/mL). PBL-ACH and ACH.2 cells (early passage HTLV-1-immortalized human T-cells) were maintained in RPMI 1640 supplemented with 20% FBS, 10 U/mL recombinant human interleukin-2 (rhIL-2; Roche Applied Biosciences, Indianapolis, IN, USA), 2 mM glutamine, 100 U/mL penicillin, and 100 µg/mL streptomycin. SLB-1 cells (HTLV-1-transformed T-cell line) were maintained in Iscove's medium supplemented with 10% FBS, 2 mM glutamine, 100 U/mL penicillin, and 100 µg/mL streptomycin. C8166, MT-1, MT-2, Hut-102 (HTLV-1-transformed T-cell lines), Hut-78 and Jurkat cells (HTLV-1-negative transformed T-cell lines) were maintained in RPMI 1640 supplemented with 10% FBS, 2 mM glutamine, 100 U/mL penicillin, and 100 µg/mL streptomycin. TL-Om1, ATL-43T, and ATL-ED cells (ATL-derived T-cell lines) were maintained in RPMI 1640 supplemented with 10% FBS, 2 mM glutamine, 100 U/mL penicillin, and 100 µg/mL streptomycin. ATL-55T cells (ATL-derived T-cell line) were maintained in RPMI 1640 supplemented with 20% FBS, 20 U/mL rhIL-2, 2 mM glutamine, 100 U/mL penicillin, and 100 µg/mL streptomycin. The parental 729.B (uninfected) and derivative 729.ACH (HTLV-1 producing) cell lines were maintained in Iscove's medium supplemented with 10% FBS, 2 mM glutamine, 100 U/mL penicillin, and 100 µg/mL streptomycin. ACH-2 (HIV-1$_{LAV}$) cells were obtained through the AIDS Research and Reference Reagent Program, Division of AIDS, NIAID, NIH. ACH-2 cells (HIV-1$_{LAV}$) were maintained in RPMI 1640 supplemented with 10% FBS, 2 mM glutamine, 100 U/mL penicillin, and 100 µg/mL streptomycin. All cells were grown at 37 °C in a humidified atmosphere of 5% $CO_2$ and air. The SeAx cell line was derived from peripheral blood of a patient with Sezary syndrome (kind gift from Dr. Henry Wong, The Ohio State University). Human PBMCs were isolated using Ficoll-Paque PLUS (GE Healthcare Life Sciences, Pittsburgh, PA, USA) and naïve T-cells were enriched using a Pan T-Cell Isolation Kit (Miltenyi Biotec, Inc., Gaithersburg, MD, USA). Whole blood samples obtained from ATLL patients were a generous gift from Dr. Lee Ratner (Washington University, St. Louis, MO, USA).

### 2.2. Plasmids and Cloning

Plasmid DNA was purified on maxi-prep columns according to the manufacturer's protocol (Qiagen, Valencia, CA, USA). The flag-tagged PRMT5 expression vector was described previously [37]. p30 cDNA was cloned into the pcDNA3.1 expression vector (Invitrogen, Grand Island, NY, USA) to create pcDNA3.1-p30. The S-tagged Tax and HBZ expression vectors contained Tax or HBZ cDNA inserted into a pTriEx™-4 Neo vector (Novagen, Madison, WI, USA) for mammalian cell expression of S-tagged Tax and HBZ proteins. The plasmid containing the wild-type HTLV-1 infectious proviral clone, ACHneo, was described previously [43]. The LTR-1-luciferase reporter plasmid and transfection efficiency control plasmid TK-renilla were described previously [41].

### 2.3. Immunoblotting

Cell lysates from luciferase assays were harvested in Passive Lysis Buffer (Promega, Madison, WI, USA) containing protease inhibitor cocktail (Roche) and quantitated using an ND-1000 Nanodrop spectrophotometer (ThermoFisher, Waltham, MA, USA). All other cell lysates were harvested in NP-40 lysis buffer containing protease inhibitor cocktail, and were quantitated using the Pierce™ BCA Protein Assay Kit (ThermoFisher) and a FilterMax F5 Multi-Mode Microplate Reader (Molecular Devices, Sunnyvale, CA, USA). Equivalent amounts of protein were separated in Mini-PROTEAN® TGX™ Precast 4%–20% Gels (Bio-Rad, Hercules, CA, USA) and transferred to nitrocellulose membranes. Membranes were blocked in PBS containing 5% milk and 0.1% Tween-20 and incubated with primary antibody. The following antibodies were used: anti-PRMT5 (ab31751, 1:1000; Abcam, Cambridge, MA, USA), anti-p19 (patient anti-sera specific for gag proteins), anti-Flag clone M2 (1:1000; Agilent, Wilmington, DE, USA), anti-p30 (rabbit anti-sera), anti-Tax (rabbit anti-sera), anti-HBZ (rabbit anti-sera), anti-p27 (1:250; Santa Cruz Biotechnology, Dallas, TX, USA), anti-p21 (1:250; Santa Cruz Biotechnology), anti-cyclin B1 (1:250; Santa Cruz Biotechnology), and anti-β-actin 1:5000 (Sigma, St. Louis, MO, USA). The secondary antibodies used were HRP goat-anti-rabbit and goat-anti-mouse (1:5000; Santa Cruz Biotechnology). Blots were developed using Immunocruz Luminol Reagent (Santa Cruz Biotechnology). Images were taken using the Amersham Imager 600 (GE Healthcare Life Sciences) and densitometric data was calculated using the ImageQuantTL program (GE Healthcare Life Sciences).

### 2.4. Quantitative RT-PCR

Total RNA was isolated from $10^6$ cells per condition using the RNeasy Mini Kit (Qiagen) according to the manufacturer's instructions. Isolated RNA was quantitated and DNase-treated using recombinant DNase I (Roche). Reverse transcription was performed using the Omniscript RT Kit (Qiagen) according to the manufacturer's instructions. The instrumentation and general principles of the CFX96 Touch™ Real-Time PCR Detection System (Bio-Rad) are described in detail in the operator's manual. PCR amplification was carried out in 96-well plates with optical caps. The final reaction volume was 20 µL consisting of 10 µL iQ™ SYBR® Green Supermix (Bio-Rad), 300 nM of each specific primer, and 2 µL of cDNA template. For each run, sample cDNA and a no-template control were assayed in triplicate. The reaction conditions were 95 °C for 5 min, followed by 40 cycles of 94 °C for 30 s, 56 °C for 30 s, and 72 °C for 45 s. Primer pairs to specifically detect viral mRNA species (tax, hbz), prmt5, st7, and gapdh were described previously [28,44]. Data are presented in histogram form as means with standard deviations from triplicate experiments.

### 2.5. Co-Culture Immortalization Assays

Long-term immortalization assays were performed as detailed previously [45]. Briefly, $2 \times 10^6$ freshly isolated human PBMCs were co-cultivated at a 2:1 ratio with lethally irradiated cells (729.B uninfected parental; 729.ACH HTLV-1-producing) in 24-well culture plates (media was supplemented with 10 U/mL rhIL-2). HTLV-1 gene expression was confirmed by the detection of p19 Gag protein in the culture supernatant, and was measured weekly by p19 ELISA (Zeptometrix, Buffalo, NY, USA). Viable cells were counted weekly by Trypan blue dye exclusion.

### 2.6. Packaging and Infection of Lentivirus Vectors

Lentiviral vectors expressing five different PRMT5-directed shRNAs (target set RHS4533-EG10419), and the universal negative control, pLKO.1 (RHS4080) were purchased from Open Biosystems (Dharmacon, Lafayette, CO, USA) and propagated according to the manufacturer's instructions. HEK293T cells were transfected with lentiviral vector(s) expressing shRNAs, plus DNA vectors encoding HIV Gag/Pol and VSV-G in 10-cm dishes with Lipofectamine®2000 according to the manufacturer's instructions. Media containing lentiviral particles was collected 72 h later and filtered through 0.45 µm pore size filters. Lentiviral particles were then concentrated using ultracentrifugation in a Sorvall SW-41 swinging bucket rotor. Lymphoid cell lines were infected with the concentrated lentivirus in 8 µg/mL polybrene by spinoculation at 2000 x*g* for 2 h at room temperature. HEK293T cells were infected with the concentrated lentivirus in 8 µg/mL polybrene. After 72 h, stable cell lines were selected by treatment with 1–2 µg/mL puromycin for 7 days.

## 2.7. PRMT5i Treatment

A selective PRMT5 inhibitor (PRMT5i) drug was recently described by Alinari *et al.* [31]. Lymphoid cells were seeded into a 12-well plate at $0.5 \times 10^6$ cells/mL. Indicated concentrations of the PRMT5i were added to duplicate wells. Cells were incubated at 37 °C for 48 h. After incubation, cell viability and proliferation were measured. Cell viability was determined using Trypan blue dye exclusion. Cellular proliferation was measured in duplicate in each condition using the CellTiter 96® AQ$_{ueous}$ One Solution Cell Proliferation Assay (Promega). Cells were collected by slow centrifugation (5 min, 800 x$g$) for downstream qRT-PCR analysis as described above.

## 2.8. HIV-1 Gene Expression in Chronically Infected Cells

ACH-2 (HIV-1$_{LAV}$) cells were seeded into a 12-well plate at $1.6 \times 10^6$ cells/mL. Indicated concentrations of PRMT5i were added in triplicate wells. Cells were incubated at 37 °C for 48 h. After incubation, 10 μL of culture supernatant was removed and freeze-thawed once for reverse transcriptase (RT) assays. Briefly, 10 μL of culture supernatant was incubated overnight at 37 °C with 25 μL buffer (50 mM Tris-HCl pH 7.8, 75 mM KCl, 2 mM DTT, 5 mM MgCl$_2$, 5 μg/mL Poly$_{(dA:dT)}$, and 0.5% NP-40) containing 10 μCi/mL [α$^{32}$P]-labeled dTTP (Perkin Elmer, Waltham, MA, USA). A volume of 10 μL was spotted onto a DEAE filtermat (Perkin Elmer), air dried at room temperature, then washed 5× with 1× saline-sodium citrate buffer (SSC) and 2× with 85% ethanol. Filtermats were air dried and exposed to a phosphorimaging screen for 2.5 h at room temperature. Density, counts/mm$^2$, was determined using the Typhoon Scanner (GE Healthcare Life Sciences) and Quantity One software (Bio-Rad).

## 2.9. Transient Transfections, Reporter Assays, and p19 Gag ELISA

HEK293T cells were transfected using Lipofectamine®2000 Transfection Reagent according to the manufacturer's instructions. Each transfection experiment was performed in triplicate and presented as means plus standard deviations. In general, HEK293T cells in a 6-well dish were transfected with approximately 1–2 μg total DNA consisting of 20 ng TK-renilla (transfection control), 100 ng LTR-1-luciferase, 1 μg ACHneo, 500 ng Flag-PRMT5, 500 ng pcDNA3.1-p30, 100 ng S-tag-Tax, or 500 ng S-tag-HBZ, where indicated. HEK293T cells were harvested 48 h post-transfection in Passive Lysis Buffer (Promega). Relative firefly and Renilla luciferase units were measured in a FilterMax F5 Multi-Mode Microplate Reader using the Dual-Luciferase® Reporter Assay System (Promega) according to the manufacturer's instructions. Each condition was performed in duplicate. Extracts also were subjected to immunoblotting to verify equivalent protein levels. Cell supernatants (48 h) were used for p19 ELISA (Zeptometrix).

## 2.10. Annexin V Staining

Lymphoid cells were seeded into a 6-well plate at $1 \times 10^6$ cells/mL. Indicated concentrations of the PRMT5i were added to the appropriate wells. Cells were incubated at 37 °C for 24 h. After incubation, cells were collected by slow centrifugation (5 min, 800 x$g$) for apoptosis analysis via flow cytometry. Collected cells were stained using the FITC Annexin V Apoptosis Detection Kit I (BD Biosciences; San Diego, CA, USA) according to the manufacturer's instructions.

## 2.11. ChIP Assays

pA-18G-BHK-21 cells are a Syrian Hamster kidney cell line stably transfected with a plasmid vector containing the lacZ bacterial gene under the control of a HTLV-1-LTR promoter as previously described [46]. pA-18G-BHK-21 cells were transfected in 10-cm dishes (1 μg ACHneo and 5 μg flag-PRMT5) using Lipofectamine®2000 according to the manufacturer's instructions. Cells were cross-linked in fresh 1% paraformaldehyde for 10 min at room temperature. The cross-linking reaction was quenched using 125 mM glycine. Following cell lysis and DNA fragmentation by sonication,

DNA-protein complexes were immunoprecipitated with anti-PRMT5 (Santa Cruz) and control anti-IgG (Santa Cruz) antibodies. Immunoprecipitated DNA-protein complexes were washed using sequential low-salt, high-salt, lithium chloride, and Tris-EDTA (TE) buffers. DNA was purified using the Qiagen Gel Extraction Kit (Qiagen). The presence of specific DNA fragments in each precipitate was detected using PCR. Primers used for amplifying the HTLV-1 LTR were 5'-CCACAGGCGGGAGGCGGCAGAA-3' and 5'-TCATAAGCTCAGACCTCCGGGAAG-3' and LacZ-coding region were 5'-AAAATGGTCTGCTGCTG-3' and 5'-TGGCTTCATCCACCACA-3'. Quantification of each ChIP experiment was performed using ImageJ software.

## 3. Results

### 3.1. PRMT5 Was Upregulated in T-Cell Leukemia/Lymphoma Cells

Recently, PRMT5 over-expression was identified to be involved in the pathogenesis of hematologic (lymphoma) and solid tumors (melanoma, astrocytomas) [27–32]. To determine whether PRMT5 is important to HTLV-1 biology and pathogenesis, we first examined the levels of PRMT5 protein (Figure 1A) and RNA (Figure 1B) in a wide variety of T lymphocytic leukemia/lymphoma cells, including HTLV-1-transformed T-cell lines (PBL-ACH, ACH.2, SLB-1, Hut-102, MT-1, MT-2, C8166), ATL-derived T-cell lines (TL-Om1, ATL-43T, ATL-55T, ATL-ED), HTLV-1-negative T-cell lines (SeAx, Jurkat), and naïve primary T-cells. Both PRMT5 protein and RNA were upregulated in all T-cell leukemia/lymphoma cell lines compared to naïve T-cells. Interestingly, although protein and RNA levels were upregulated, PRMT5 RNA was not directly correlated to PRMT5 protein levels, which suggested a post-transcriptional method of regulation. PRMT5 protein (Figure 1C) and RNA (Figure 1D) were also upregulated in 3 of 4 and 4 of 4 PBMC samples from ATLL patients, respectively, relative to HTLV-1-negative naïve T-cells.

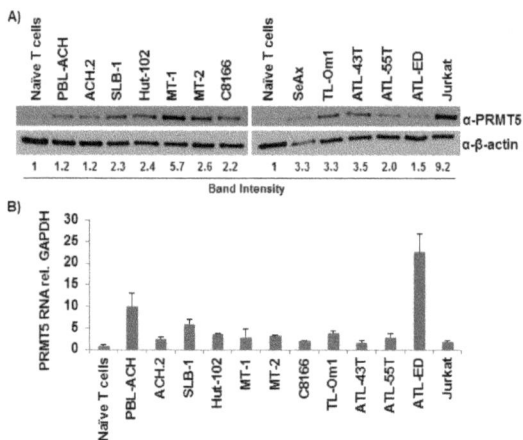

**Figure 1.** *Cont.*

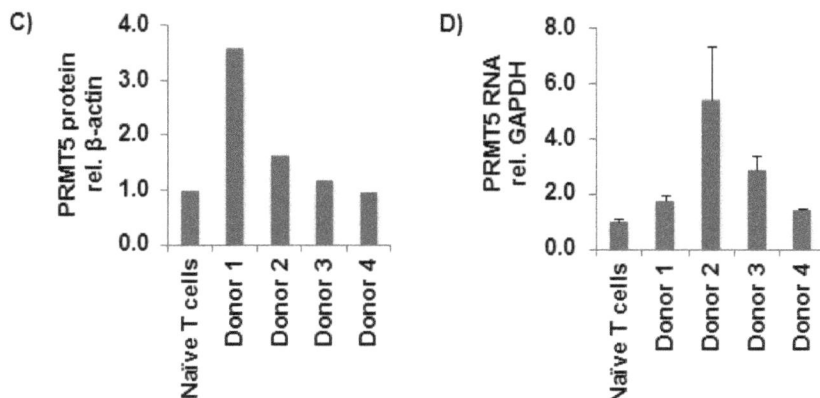

**Figure 1.** PRMT5 was upregulated in T-cell leukemia/lymphoma cells. (**A**) Total cell lysates of HTLV-1-transformed cell lines (PBL-ACH, ACH.2, SLB-1, Hut-102, MT-1, MT-2, C8166), ATL-derived cell lines (TL-Om1, ATL-43T, ATL-55T, ATL-ED), HTLV-1-negative transformed cell lines (SeAx, Jurkat), and naïve T-cells were subjected to immunoblot analysis to compare the levels of endogenous PRMT5 expression. β-Actin expression was used as a loading control. The amount of PRMT5 in each cell line was measured relative to β-actin; the level of PRMT5 expression obtained with naïve T-cells was set at 1; (**B**) Quantitative RT-PCR for PRMT5 and GAPDH was performed on mRNA isolated from cells in panel A. Total PRMT5 mRNA level was determined using the ΔΔCt method [47] and normalized to relative GAPDH levels. Data are presented in histogram form with means and standard deviations from triplicate experiments; (**C**) Lysates of ATLL PBMCs from four independent donors and naïve T-cells were subjected to immunoblot analysis to compare endogenous PRMT5 protein expression levels. β-Actin expression was used as a loading control. The amount of PRMT5 in each condition was measured relative to β-actin and depicted in histogram form; the level of expression obtained with naïve T-cells was set to 1. Each sample was measured in duplicate; (**D**) Quantitative RT-PCR for PRMT5 and GAPDH was performed on mRNA isolated from cells in panel C. Total PRMT5 mRNA level was determined using the ΔΔCt method and normalized to relative GAPDH levels. Data are presented in histogram form with means and standard deviations from triplicate experiments.

### 3.2. PRMT5 Levels Were Elevated during HTLV-1-Mediated Cellular Transformation

We next determined whether PRMT5 becomes dysregulated and over-expressed during HTLV-1-driven T-cell transformation. Freshly isolated human PBMCs co-cultured with lethally irradiated HTLV-1 producer cells (729.ACHi) in the presence of 10 U/mL of human IL-2 showed progressive growth consistent with HTLV-1 immortalization (Figure 2A, left panel). As a control, PBMCs co-cultured with lethally irradiated 729.B (HTLV-1-negative parental line) were unable to sustain progressive growth. We also detected continuous accumulation of p19 Gag in the culture supernatants of PBMCs co-cultured with 729.ACHi cells, which indicated viral replication and virion production; as expected, irradiated HTLV-1 producer cells alone failed to grow or produce p19 over time (Figure 2A, right panel). We examined PRMT5 protein (Figure 2B) and RNA (Figure 2C) levels throughout the 10-week *in vitro* transformation assay. Protein and RNA were isolated from two independent wells of cells at weekly time points. Our data revealed that PRMT5 protein and RNA were upregulated at each time point, to varying degrees, throughout the transformation assay.

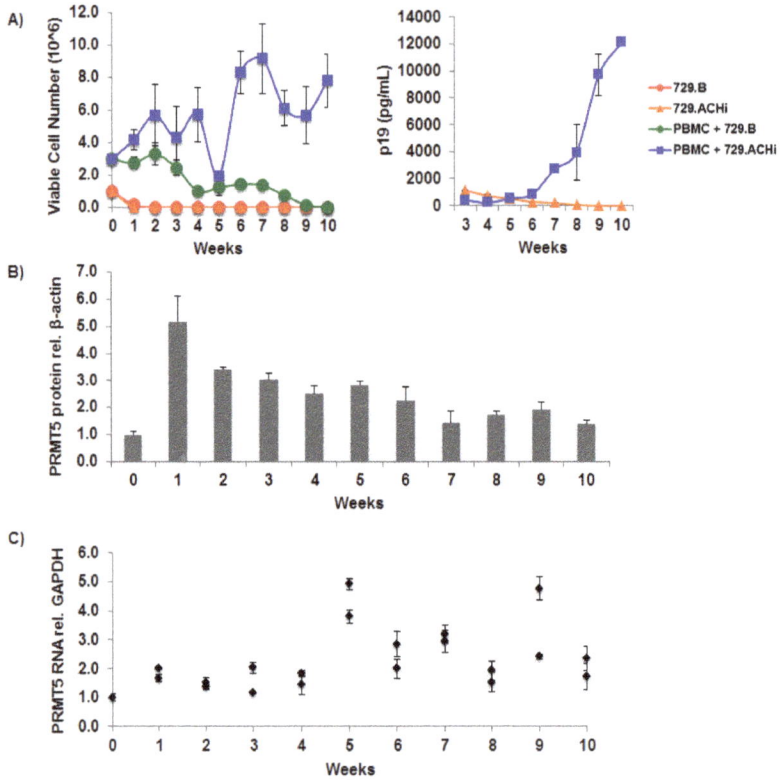

**Figure 2.** PRMT5 levels were elevated during HTLV-1-mediated cellular transformation. Freshly isolated PBMCs ($2 \times 10^6$ cells) were co-cultivated with $10^6$ lethally irradiated 729.B uninfected parental or 729.ACH HTLV-1-producer cells in 24-well plates. (**A**) A growth curve showing the immortalization process as measured by T-cell number and ELISA data showing p19 Gag protein production at weekly intervals are presented. Means and standard deviations of data from each time point were determined from four random independent wells; Cells were also collected at weekly intervals and analyzed by immunoblotting for PRMT5 protein expression (**B**); and qRT-PCR analysis for PRMT5 RNA levels (**C**). PRMT5 levels are shown relative to an internal control (β-actin, GAPDH) for each time point. Resting PBMC PRMT5 levels were set to 1. Means and standard deviations of data from each time point were determined from two random independent wells.

### 3.3. Loss of Endogenous PRMT5 Increased HTLV-1 Gene Expression

To determine whether PRMT5 over-expression is a marker for T-cell transformation and/or contributes to the process of HTLV-1 transformation, we utilized shRNA vectors to knockdown PRMT5 expression in three different HTLV-1-transformed cell lines. As shown in Figure 3A, shRNA-mediated knockdown of endogenous PRMT5 expression in the early passage HTLV-1 immortalized T-cell line, PBL-ACH, resulted in a significant increase of viral p19 Gag protein production (left and right panel) and a significant increase in the levels of Tax and HBZ gene expression (middle panel). Knockdown of PRMT5 expression likewise significantly increased Tax and HBZ transcript levels in another HTLV-1-transformed cell line, SLB-1 (Figure 3B). Finally, knockdown of PRMT5 protein expression in the Tax-negative ATL-derived T-cell line, TL-Om1, significantly increased HBZ transcript levels (Figure 3C).

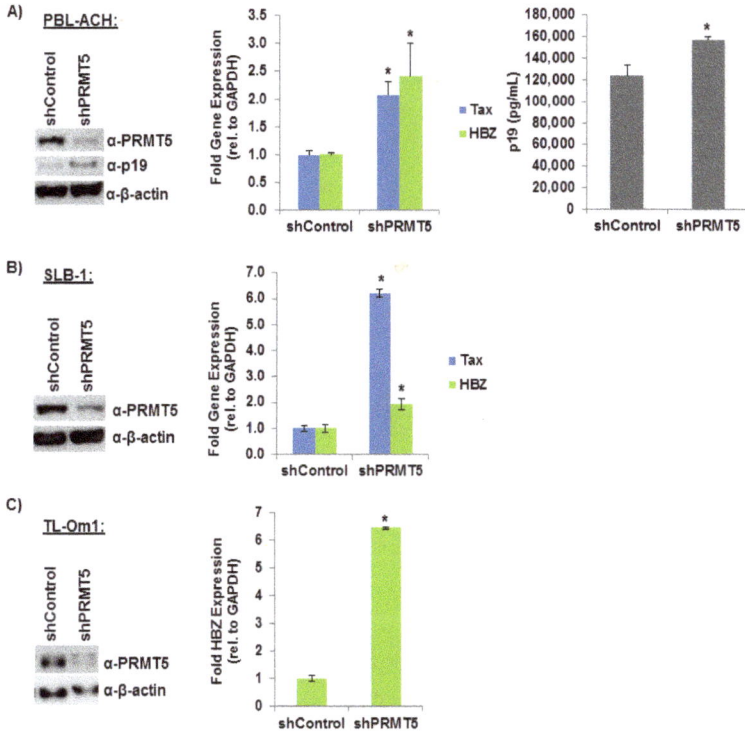

**Figure 3.** Loss of endogenous PRMT5 increased HTLV-1 gene expression. (**A**) PBL-ACH cells (HTLV-1-transformed) were infected with a pool of five different lentiviral vectors directed against PRMT5, or control shRNAs. The cells were selected for 7 days using puromycin. Immunoblot analysis was performed to compare the levels of PRMT5, p19 (Gag), and β-actin (loading control) in each condition (left panel). Quantitative RT-PCR (middle panel) for Tax, HBZ, and GAPDH was performed on mRNA isolated from shControl and shPRMT5 cells. Transcript levels were determined using the ΔΔCt method and normalized to relative GAPDH levels. Levels of Tax and HBZ relative to GAPDH in shControl cells were set to 1. Data are presented in histogram form with means and standard deviations from triplicate experiments. HTLV-1 gene expression was quantified by the detection of the p19 Gag protein in the culture supernatant using ELISA (right panel); (**B**) SLB-1 cells (HTLV-1-transformed) were infected with a pool of five different lentiviral vectors directed against PRMT5 or control shRNAs. The cells were selected for 7 days using puromycin. Immunoblot analysis was performed to compare the levels of PRMT5 and β-actin protein (loading control) in each condition (left panel). Quantitative RT-PCR for Tax, HBZ, and GAPDH was performed on mRNA isolated from shControl and shPRMT5 cells (right panel). Transcript levels were determined using the ΔΔCt method and normalized to relative GAPDH levels. Data are presented in histogram form with means and standard deviations from triplicate experiments. Levels of Tax and HBZ relative to GAPDH in shControl cells were set to 1; (**C**) TL-Om1 cells (ATL-derived, Tax-negative) were infected with a pool of five different lentiviral vectors directed against PRMT5, or control shRNAs. The cells were selected for 7 days using puromycin. Immunoblot analysis was performed to compare the levels of PRMT5 and β-actin protein (loading control) in each condition (left panel). Quantitative RT-PCR for HBZ and GAPDH was performed on mRNA isolated from shControl and shPRMT5 cells (right panel). HBZ transcript level was determined using the ΔΔCt method and normalized to relative GAPDH levels. Data are presented in histogram form with means and standard deviations from triplicate experiments. Levels of HBZ relative to GAPDH in shControl cells were set to 1. Student's *t* test was performed to determine significant differences in viral transcript levels between shControl and shPRMT5 cells; $p < 0.05$ (*).

### 3.4. Selective Inhibition of PRMT5 Enhanced HTLV-1 Gene Expression

Recently, a first-in-class, small-molecule PRMT5 inhibitor (PRMT5i) was developed [31]. This novel inhibitor selectively blocks S2Me-H4R3 (symmetric di-methylation of H4R3) but is inactive against other type I and type II PRMT enzymes, which highlights its specificity for PRMT5. To evaluate whether PRMT5 over-expression is a marker for T-cell transformation or contributes to the process of HTLV-1 transformation, we treated six different HTLV-1-transformed cell lines with titrating amounts of PRMT5i ranging from 10 μM to 50 μM. As shown in Figure 4A–D, inhibition of PRMT5 resulted in a significant increase in Tax and HBZ viral gene expression in the HTLV-1-transformed T-cell lines PBL-ACH, ACH.2 SLB-1, and Hut-102. ST7 transcript levels were measured as a control to ensure successful PRMT5 inhibition because PRMT5 was reported to repress the tumor suppressor ST7 in MCL [28]. Treatment with PRMT5i also resulted in a significant increase in HBZ transcript levels in the Tax-negative, ATL-derived T-cell lines TL-Om1 and ATL-ED (Figure 4E,F). HTLV-1-negative cell lines, Jurkat and Hut-78, were also treated with titrating amounts of PRMT5i. ST7 transcript levels were measured to ensure successful PRMT5 inhibition (Figure 4G). PRMT5 expression levels were examined and found to be relatively unchanged in all PRMT5i treated cells tested (western blot lower panels; Figure 4A–G). ACH-2 cells (HIV-1$_{LAV}$) were also treated with titrating amounts of PRMT5i (Figure 4H) to examine if PRMT5 might regulate viral gene expression of another human retrovirus. Treatment with PRMT5i did not significantly alter the expression of HIV-1 as measured by RT activity in the cell supernatant.

### 3.5. Selective Inhibition of PRMT5 Decreased Cell Proliferation and Viability

In a recent report, PRMT5 was linked to proliferation in B-cell lines and MCL because knockdown of PRMT5 expression reduced cell proliferation [28]. Treatment of HTLV-1-positive T-cell lines with titrating amounts of PRMT5i or shRNA-mediated knockdown of PRMT5 resulted in a significant decrease in cellular proliferation (Figure 5A,B) and cell viability (Figure 5C). Surprisingly, the same dose of PRMT5i had minimal effects on proliferation and viability of Jurkat and Hut-78 cells (HTLV-1-negative). The level of cellular apoptosis (Figure 5D) and senescence (Figure 5E–H) was also measured in response to titrating amounts of PRMT5i treatment. The number of apoptotic cells was increased in all cell lines in response to PRMT5i treatment; however, the amount of apoptotic cells in HTLV-1-transformed lines was higher. We found a slight increase in the level of cellular senescence (as measured by increased p21 and p27 levels and decreased cyclin B1 levels) in response to PRMT5i in the HTLV-1-positive cell lines examined.

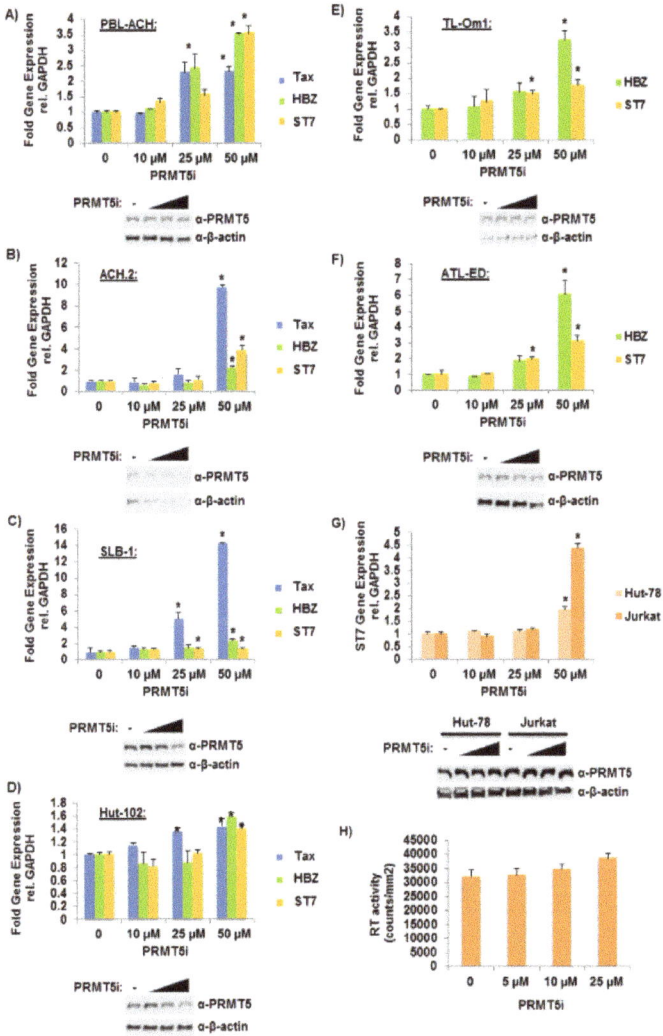

**Figure 4.** Selective inhibition of PRMT5 enhanced HTLV-1 gene expression. (**A**) PBL-ACH, (**B**) ACH.2, (**C**) SLB-1, (**D**) Hut-102, (**E**) TL-Om1, (**F**) ATL-ED, (**G**) Jurkat, and Hut-78 cells were treated with titrating amounts of PRMT5i for 48 h. (PBL-ACH, ACH.2, SLB-1, and Hut-102 cells are HTLV-1-tranformed lines; TL-Om1 and ATL-ED are ATL-derived Tax-negative lines; Jurkat and Hut-78 are HTLV-1 negative lines.) Cells were collected by slow centrifugation at 800 x*g* for 5 min. Quantitative RT-PCR for Tax (where indicated), HBZ (where indicated), ST7, and GAPDH was performed on mRNA isolated from cells in each condition. Transcript level was determined using the $\Delta\Delta$Ct method and normalized to relative GAPDH levels. Data are presented in histogram form with means and standard deviations from duplicate experiments; vehicle-treated cells were set to 1 (upper panel). Student's *t* test was performed to determine significant differences in viral transcript levels between vehicle and inhibitor treated cells; $p < 0.05$ (*). Inhibitor-treated cells were also subjected to immunoblot analysis to compare the levels of endogenous PRMT5 expression. $\beta$-actin expression was used as a loading control (lower panel); (**H**) ACH-2 cells were treated with titrating amounts of PRMT5i for 48 h. Average reverse transcriptase (RT) activity of triplicate experiments is depicted.

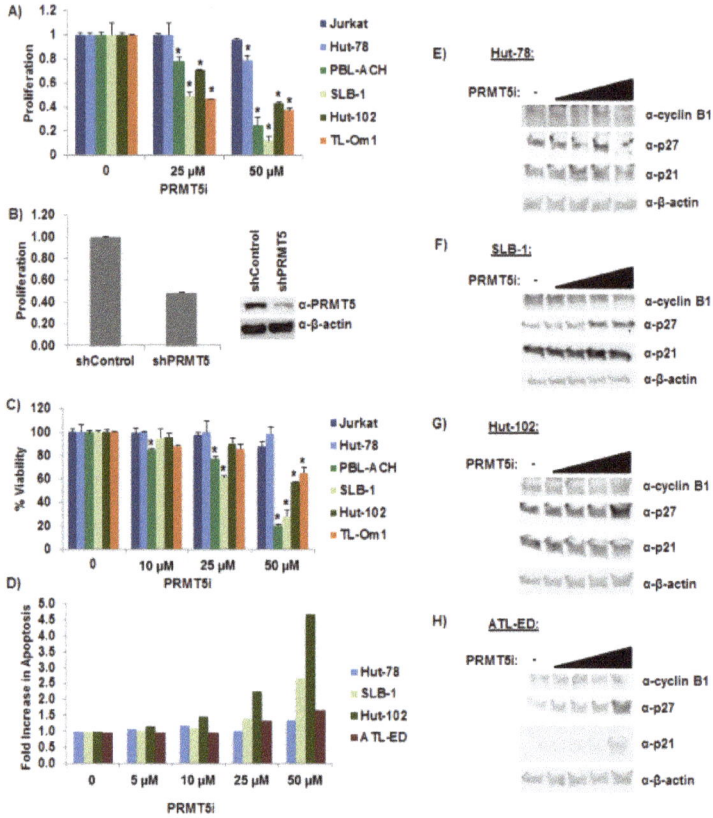

**Figure 5.** Selective inhibition of PRMT5 decreased cell proliferation and viability. (**A**) Cellular proliferation was measured using an MTS cellular proliferation assay in Jurkat, Hut-78, PBL-ACH, SLB-1, Hut-102, and TL-Om1 cell lines after 48 h of PRMT5i treatment. (Jurkat and Hut-78 are HTLV-1 negative lines; PBL-ACH, SLB-1, and Hut-102 cells are HTLV-1-tranformed lines; TL-Om1 is an ATL-derived Tax-negative line.) Data are presented in histogram form with mean and standard deviations from duplicate experiments. Vehicle treatment for each cell type was set at 1. Student's *t* test was performed to determine significant differences in cellular proliferation between vehicle and inhibitor-treated cells; $p < 0.05$ (*); (**B**) SLB-1 cells were infected with a pool of five different lentiviral vectors directed against PRMT5 or control shRNAs. The cells were selected for 7 days using puromycin. Immunoblot analysis was performed to compare the levels of PRMT5 and β-actin protein (loading control) in each condition (right panel). Cellular proliferation was measured using an MTS cellular proliferation assay (left panel); (**C**) Cellular viability was measured using Trypan blue exclusion method in Jurkat, Hut-78, PBL-ACH, SLB-1, Hut-102, and TL-Om1 after 48 h of PRMT5i treatment. Data are presented in histogram form with means and standard deviations from duplicate experiments. Vehicle treatment for each cell type was set at 100% viability. Student's *t* test was performed to determine significant differences in cellular viability between vehicle and inhibitor treated cells; $p < 0.05$ (*); (**D**) Cellular apoptosis was measured using a FITC Annexin V Apoptosis Detection Kit as described in the Materials and Methods. The percentage of cells undergoing apoptosis in vehicle (DMSO) treated cells was set at 1. The fold increase in apoptosis was measured in Hut-78, SLB-1, Hut-102, and ATL-ED cells in response to titrating amounts of PRMT5i after 24 h of treatment; (ATL-ED is an ATL-derived Tax-negative line.) (**E–H**) Immunoblot analysis was performed in Hut-78, SLB-1, Hut-102, and ATL-ED cells after 24 h of PRMT5i treatment to compare the levels of p21, p27, cyclin B1, and β-actin expression (loading control) in each condition (right panel).

## 3.6. PRMT5 Negatively Regulated HTLV-1 Gene Expression

A recent report identified PRMT5 as a binding partner of the HTLV-1 accessory protein p30, a known negative regulator of HTLV-1 gene expression [37]. To investigate the effect(s) of exogenous PRMT5 on p30, HEK293T cells were transfected with a LTR-1-luciferase reporter vector (LTR-1-luc), TK-renilla control, the ACHneo proviral clone, a flag-tagged PRMT5 expression vector, and a p30 expression vector as indicated (Figure 6A). Luciferase activity was measured after 48 h. The luciferase activity of the empty control reflected the amount of Tax and therefore, was a measure of transcription from the provirus. In the presence of either exogenous p30 or PRMT5, LTR-1-luciferase reporter was significantly repressed (left panel). However, in the presence of both p30 and PRMT5, there was an additive effect on LTR-1-luciferase repression. The amount of viral p19 Gag protein present in the culture supernatant of each condition also was examined using ELISA, which provided another method to quantify HTLV-1 gene expression (middle panel). Similar to the luciferase results, p30 and PRMT5 individually repressed viral p19 Gag production, and the presence of both p30 and PRMT5 repressed viral gene transcription further. We next asked whether PRMT5 was required for p30 function by transducing HEK293T cells with shRNA vectors directed against PRMT5 or a scramble control (Figure 6B). Scramble and shPRMT5 HEK293T cell lines were selected for 7 days using puromycin to ensure sufficient knockdown of endogenous PRMT5. After selection, each cell line was transfected with LTR-1-luc, TK-renilla control, the ACHneo proviral clone, and a p30 expression vector as indicated. Luciferase activity was measured after 48 h. Knockdown of endogenous PRMT5 significantly enhanced viral transcription, as measured by LTR-1-luciferase activity (left panel) and p19 Gag ELISA (middle panel). In addition, reduced levels of PRMT5 did not significantly affect the ability of p30 to repress viral transcription.

## 3.7. PRMT5 Did Not Affect Tax Transcriptional Function

We next asked what effect PRMT5 had on Tax and HBZ transcriptional activity. HEK293T cells were transfected with LTR-1-luc, TK-renilla control, the ACHneo proviral clone, flag-PRMT5 expression vector, S-tag-Tax expression vector, and S-tag-HBZ expression vector as indicated (Figure 7A). Tax activated transcription while HBZ repressed Tax-mediated transcriptional activation (left and middle panels), as expected [17–19]. In the presence of exogenous PRMT5, Tax transcriptional activity and HBZ-mediated repression of Tax transcriptional activity were decreased. To determine if PRMT5 was able to specifically repress Tax in the absence of other viral genes, HEK293T cells were transfected with LTR-1-luc, TK-renilla control, flag-PRMT5 expression vector, and an S-tag-Tax expression vector (Figure 7B). Tax was able to activate LTR-1-luciferase activity, while PRMT5 had no effect in the presence or absence of Tax on LTR-1 luciferase activity. Our results suggested that PRMT5 requires the HTLV-1 proviral DNA to suppress Tax-induced LTR activation. To examine if viral factors other than Tax are implicated in suppression of viral transcription by PRMT5, we examined the effect of HBZ or p30 with PRMT5 on LTR-1 luciferase activity. HEK293T cells were transfected with LTR-1-luc, TK-renilla control, flag-PRMT5 expression vector, and an S-tag-HBZ or p30 expression vector (Figure 7C,D). Neither HBZ nor p30 activated or repressed LTR-1-luciferase activity in the absence of Tax, as expected [17–22,41,42]. Also, PRMT5 had no effect in the presence or absence of HBZ or p30 on LTR-1-luciferase activity. To determine if PRMT5 associated with the viral LTR promoters *in vivo*, we performed ChIP assays using pA-18G-BHK-21 cells. The Syrian Hamster kidney cell line was stably transfected with a plasmid vector containing the HTLV-1-LTR promoter that drives expression of lacZ [46]. pA-18G-BHK-21 cells were transfected with ACHneo proviral DNA and flag-PRMT5 expression vector (Figure 7E). PRMT5 was associated with the viral LTR, but not the downstream LacZ-coding region (negative control).

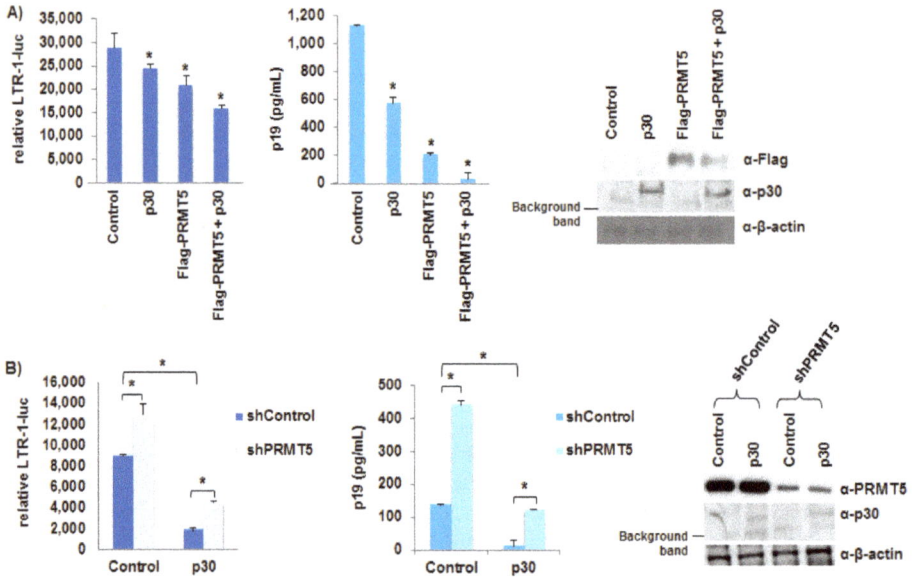

**Figure 6.** PRMT5 negatively regulated HTLV-1 gene expression. (**A**) HEK293T cells were transfected with 20 ng TK-renilla, 100 ng LTR-1-luciferase reporter, 1 μg ACHneo, 500 ng flag-PRMT5, and 500 ng p30 expression plasmid as indicated. At 48 h post-transfection, cell lysates were collected and luciferase levels measured; relative luciferase activity for each condition is shown (left panel). The decreases in relative LTR-1-luc activity compared to control were significant ($p < 0.05$ (*)). HTLV-1 gene expression was quantified by the detection of the p19 Gag protein in the culture supernatant of each condition using ELISA (middle panel). The decreases in p19 Gag levels compared to control were significant ($p < 0.05$ (*)). Immunoblot analysis was performed to compare the levels of PRMT5 (Flag), p30, and β-actin (loading control) in each condition (right panel); (**B**) HEK293T cells were infected with a pool of five different lentiviral vectors directed against PRMT5 or control shRNAs. The cells were selected for 7 days using puromycin. HEK293T shControl and shPRMT5 cells were then transfected with 20 ng TK-renilla, 100 ng LTR-1-luciferase reporter, 1 μg ACHneo, and 500 ng p30 expression plasmid as indicated. Forty-eight hours post-transfection, cell lysates were collected, and luciferase levels measured; relative luciferase activity for each condition is shown (left panel). The differences in relative LTR-1-luc activity were significant ($p < 0.05$ (*)). HTLV-1 gene expression was quantified by the detection of the p19 Gag protein in the culture supernatant of each condition using ELISA (middle panel). The differences in p19 Gag levels were significant ($p < 0.05$ (*)). Immunoblot analysis was performed to compare the levels of endogenous PRMT5, p30, and β-actin (loading control) in each condition (right panel).

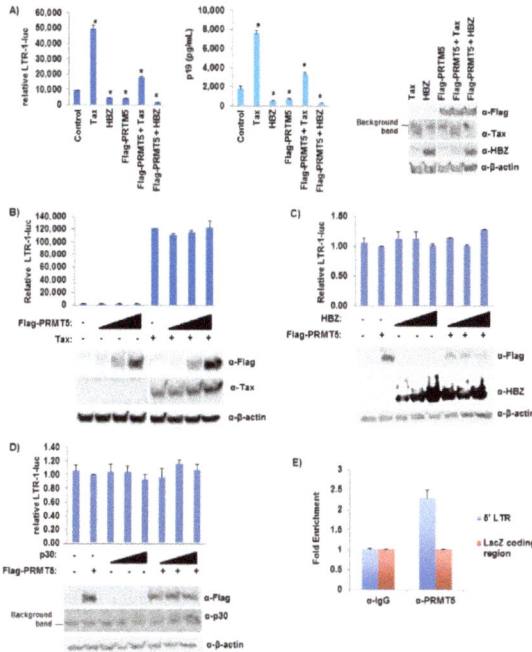

**Figure 7.** PRMT5 did not affect Tax transcriptional function. (**A**) HEK293T cells were transfected with 20 ng TK-renilla, 100 ng LTR-1-luciferase reporter, 1 µg ACHneo, 100 ng S-tag-Tax, 500 ng S-tag-HBZ, and 500 ng flag-PRMT5 as indicated. Forty-eight hours post-transfection, cell lysates were collected and luciferase levels measured; relative luciferase activity for each condition is shown (left panel). The differences in relative LTR-1-luc activity compared to control were significant ($p < 0.05$ (*)). HTLV-1 gene expression was quantified by the detection of the p19 Gag protein in the culture supernatant of each condition using ELISA (middle panel). The differences in p19 Gag levels compared to control were significant ($p < 0.05$ (*)). Immunoblot analysis was performed to compare the levels of PRMT5 (Flag), Tax, HBZ, and β-actin (loading control) in each condition (right panel). All samples were analyzed on the same nitrocellulose membrane; (**B**) HEK293T cells were transfected with 20 ng TK-renilla, 100 ng LTR-1-luciferase reporter, 100 ng S-tag-Tax, and titrating amounts of flag-PRMT5 as indicated. Forty-eight hours post-transfection, cell lysates were collected, and luciferase levels measured; relative luciferase activity for each condition is shown (upper panel). Immunoblot analysis was performed to compare the levels of PRMT5 (Flag), Tax, and β-actin (loading control) in each condition (lower panel). All samples were analyzed on the same nitrocellulose membrane; (**C**) HEK293T cells were transfected with 20 ng TK-renilla, 100 ng LTR-1-luciferase reporter, 500 ng flag-PRMT5, and titrating amounts of HBZ as indicated. Forty-eight hours post-transfection, cell lysates were collected and luciferase levels measured; relative luciferase activity for each condition is shown (upper panel). Immunoblot analysis was performed to compare the levels of PRMT5 (Flag), HBZ, and β-actin (loading control) in each condition (lower panel); (**D**) HEK293T cells were transfected with 20 ng TK-renilla, 100 ng LTR-1-luciferase reporter, 500 ng flag-PRMT5, and titrating amounts of p30 as indicated. Forty-eight hours post-transfection, cell lysates were collected and luciferase levels measured; relative luciferase activity for each condition is shown (upper panel). Immunoblot analysis was performed to compare the levels of PRMT5 (Flag), p30, and β-actin (loading control) in each condition (lower panel); (**E**) ChIP assays were performed on cross-linked chromatin from pA-18G-BHK-21 cells using either IgG or PRMT5 antibodies, and the retained DNA was amplified using LTR or LacZ coding region-specific primers. Fold-enrichment with each antibody was calculated relative to the IgG sample. Each ChIP experiment was repeated twice in duplicate.

## 4. Discussion

HTLV-1 is a tumorigenic retrovirus and the causative infectious agent of ATLL, an extremely aggressive and fatal disease of CD4+ T-cells [2–4]. In culture, HTLV-1 can effectively immortalize and eventually transform primary human T-cells. However, in infected individuals, the incidence of disease is only 2%–6% [7] after an extensive clinical latency period. Evidence suggests both genetic and epigenetic changes in the cellular environment that accumulate over time contribute to the development of ATLL [23]. While many aspects of HTLV-1 biology have been revealed, the detailed mechanisms of ATLL development remain poorly defined. Recently, the cellular protein PRMT5 has been shown to play a critical role in EBV-driven B-cell transformation as well as the pathogenesis of many types of hematologic and solid tumors [27–32]. We hypothesized that PRMT5 could be important in HTLV-1-mediated cellular transformation and in regulation of viral replication. Given the development of a novel small molecular inhibitor (PRMT5i) [31], identification of PRMT5 as a factor during HTLV-1 transformation will provide valuable insights into improved strategies to treat patients with ATLL.

To determine the importance of PRMT5 in HTLV-1-infected cells, we first examined the expression level of endogenous PRMT5 in a variety of HTLV-1-transformed, ATLL-derived, and HTLV-1-negative T-cells lines (Figure 1A). PRMT5 proteins levels were upregulated in HTLV-1-positive cells, but also in all transformed T-cell lines, regardless of origin. This is not surprising given that PRMT5 over-expression has recently been identified in lung carcinoma, glioblastoma, B-cell lymphoma, mantle cell lymphoma, and melanoma, to name just a few [28,31,48–50]. It appears that PRMT5 over-expression is a hallmark of most transformed cells, not specifically HTLV-1-transformed cells. Previous work by Pal *et al.* found decreased PRMT5 mRNA levels in mantle cell lymphoma cell lines despite abundant PRMT5 protein over-expression [28]. In this instance, the increase in PRMT5 protein was not due to an increase in mRNA levels, but instead was due to a decrease in the inhibitory miRNAs miR-92b and miR-96, which allowed for enhanced PRMT5 translation. Conversely, a recent report by Shilo *et al.* found both PRMT5 protein and mRNA were upregulated in lung tumors [48]. We examined the level of PRMT5 mRNA in a variety of HTLV-1-transformed, ATLL-derived, and HTLV-1-negative T-cell lines and found that the PRMT5 mRNA level was increased in every transformed cell line relative to naïve T-cells (Figure 1B). However, the increase in PRMT5 mRNA did not directly correlate with the level of PRMT5 protein expression, which suggested some degree of post-transcriptional regulation. Because these experiments were conducted in cell lines grown *in vitro*, we also examined the level of PRMT5 protein and RNA in total PBMCs isolated from ATLL patients (Figure 1C,D). Both PRMT5 protein and RNA were upregulated in a majority of ATLL patient samples. The increased level of PRMT5 RNA and protein expression in patient PBMCs was not as prominent as what was found in transformed cell lines, likely due to the use of total PBMCs, which contain a mixture of normal and leukemic cells.

HTLV-1 infection of CD4+ T-cells does not always lead to transformation. A delicate balance must be achieved between viral gene expression and certain genetic and epigenetic events to result in transformation. Using a long-term immortalization co-culture assay, we found both PRMT5 protein and RNA were upregulated throughout the immortalization process (Figure 2). It is important to note the producer cells were lethally irradiated, and although there was some residual p19 Gag detected in the supernatant, the producer cells were dead by week 1. Since only a portion of the total co-culture assay was tested, the levels of both protein and RNA fluctuated from week to week; however, the overall trend showed that PRMT5 was upregulated.

Regulation of viral gene expression early after infection is highly relevant for successful transformation; for example, too much Tax expression can cause a phenomenon known as Tax-induced senescence (TIS) [51]. Using shRNA vectors directed against PRMT5, we found that knockdown of PRMT5 enhanced HTLV-1 viral gene expression and decreased cellular proliferation in HTLV-1-infected cell lines (Figures 3 and 5B). Given the importance of PRMT5 in cellular proliferation, long-term stable cell lines were difficult to create. Thus, we transduced cells and selected with drug for less than two weeks, which provided the added benefit of less antigenic drift within the cell population over time. Similar results were obtained using a novel, small molecule inhibitor of PRMT5 (Figure 4A–F). Because Tax expression is lost in a majority of ATLL-transformed cells and only HBZ is expressed in every cell, we included the Tax-negative ATLL transformed cell lines, TL-Om1 and ATL-ED, in our studies. PRMT5 knockdown and inhibition enhanced HBZ expression in ATLL transformed cell lines. Of interest, PRMT5i did not affect HIV-1 gene expression, which suggested that PRMT5 was not a global repressor of all retrovirus gene expression (Figure 4G). Because Tax and HBZ are driven from separate viral promoters (5' LTR and 3' LTR opposite strand, respectively), this finding would suggest that PRMT5 is a global repressor of HTLV-1 transcription. In support of this hypothesis, we found PRMT5 associated with the viral LTR using ChIP analysis (Figure 7E).

Using reporter gene assays, we found PRMT5 inhibited HTLV-1 gene transcription, but not Tax protein specifically (Figures 6 and 7A,B). We also found LTR promoter activation was unaffected by PRMT5 (with or without viral accessory proteins HBZ or p30) in the absence of the proviral genome (Figure 7B–D). This result was not surprising since one of the functions of HBZ is to repress Tax-mediated transcriptional activation of the viral LTR and the function of p30 is to retain unspliced *tax/rex* mRNA in the nucleus. Taken together, these results suggested that other viral proteins were required for the repressive effects of PRMT5, and/or PRMT5 affected a cellular transcription factor responsible for activating viral transcription. Although not required for the repressive effects of p30 on viral gene expression, we did find PRMT5 and p30 had additive repressive effects on viral transcription, which adds yet another level of regulation to HTLV-1 gene expression. The roles of additional PRMT5 interacting partners, such as MEP50, in PRMT5-mediated HTLV-1 gene regulation are also a possibility to explore in the future. MEP50 is a WD-40 repeat protein and a common PRMT5 cofactor, likely present in most PRMT5-containing complexes *in vivo* [52]. Phosphorylation of MEP50 by Cdk4 alters the activity and targeting of the PRMT5 protein in cells [53].

Importantly, we found PRMT5i treatment or shRNA-mediated knockdown of PRMT5 in HTLV-1-positive cell lines caused a decrease in cell proliferation compared to HTLV-1-negative cell lines (Figure 5A,B). Furthermore, PRMT5i was selectively toxic to HTLV-1-positive cell lines (Figure 5C). These results suggested that HTLV-1-positive cells rely strongly on PRMT5 for cellular growth and survival. Treatment with PRMT5i induced cellular apoptosis to some degree in all cell lines (Figure 5D). Interestingly, HTLV-1-transformed cell lines underwent noticeably more apoptosis than either the HTLV-1-negative or the ATL-derived cell lines. Previous reports have found that aberrant expression of Tax protein can lead to TIS in cells [51]. We did observe a slight increase in cellular senescence in response to PRMT5i in the HTLV-1-positive cell lines tested, including Tax-expressing HTLV-1-transformed cells and Tax-negative ATL-derived cells (Figure 5E–H). We would predict an increase in cellular senescence in the HTLV-1-transformed cell lines, as they are the only Tax-expressing lines. However, these cell lines also express HBZ, which has been reported to repress TIS. Another possibility is the level of Tax expression induced in response to PRMT5i in our cell lines was not substantial enough to elicit a measurable increase in cellular senescence. In summary, our study highlighted the significance of PRMT5 in HTLV-1-mediated cellular transformation and its importance as a target for the newly developed PRMT5i, presenting a viable strategy for treatment of ATLL.

**Acknowledgments:** We thank Kate Hayes-Ozello for editorial comments on the manuscript. This work was supported by grants from the National Institutes of Health (AI111125 and CA100730) to Patrick L. Green, Leukemia Lymphoma Society translational research project to Robert A. Baiocchi, and Pelotonia Idea Grant to Patrick L. Green and Robert A. Baiocchi.

**Author Contributions:** Amanda R. Panfil and Patrick L. Green conceived and designed the experiments; Amanda R. Panfil, Jacob Al-Saleem, Cory M. Howard, and Jessica M. Mates performed the experiments; Amanda R. Panfil, Jacob Al-Saleem, Cory M. Howard, Jessica M. Mates, and Patrick L. Green analyzed the data; Jesse J. Kwiek, Robert A. Baiocchi, and Patrick L. Green contributed reagents/materials/analysis tools; Amanda R. Panfil and Patrick L. Green wrote the paper.

**Conflicts of Interest:** The authors declare no conflict of interest.

## References

1. Proietti, F.A.; Carneiro-Proietti, A.B.; Catalan-Soares, B.C.; Murphy, E.L. Global epidemiology of HTLV-I infection and associated diseases. *Oncogene* **2005**, *24*, 6058–6068. [CrossRef] [PubMed]
2. Poiesz, B.J.; Ruscetti, F.W.; Gazdar, A.F.; Bunn, P.A.; Minna, J.D.; Gallo, R.C. Detection and isolation of type C retrovirus particles from fresh and cultured lymphocytes of a patient with cutaneous T-cell lymphoma. *Proc. Natl. Acad. Sci. USA* **1980**, *77*, 7415–7419. [CrossRef] [PubMed]
3. Yoshida, M.; Miyoshi, I.; Hinuma, Y. Isolation and characterization of retrovirus from cell lines of human adult T-cell leukemia and its implication in the disease. *Proc. Natl. Acad. Sci. USA* **1982**, *79*, 2031–2035. [CrossRef] [PubMed]
4. Hinuma, Y.; Nagata, K.; Hanaoka, M.; Nakai, M.; Matsumoto, T.; Kinoshita, K.-I.; Shirakawa, S.; Miyoshi, I. Adult T-cell leukemia: Antigen in an ATL cell line and detection of antibodies to the antigen in human sera. *Proc. Natl. Acad. Sci. USA* **1981**, *78*, 6476–6480. [CrossRef] [PubMed]
5. Osame, M.; Izumo, S.; Igata, A.; Matsumoto, M.; Matsumoto, T.; Sonoda, S.; Tara, M.; Shibata, Y. Blood transfusion with HTLV-I associated myelopathy. *Lancet* **1986**, *2*, 104–105. [CrossRef]
6. Gessain, A.; Barin, F.; Vernant, J.C.; Gout, O.; Maurs, L.; Calender, A.; de The, G. Antibodies to human T-lymphotropic virus type-I in patients with tropical spastic paraparesis. *Lancet* **1985**, *2*, 407–410. [CrossRef]
7. Taylor, G.P. Editorial commentary: Human T-cell lymphotropic virus type 1 (HTLV-1) and HTLV-1-associated myelopathy/tropical spastic paraparesis. *Clin. Infect. Dis.* **2015**, *61*, 57–58. [CrossRef] [PubMed]
8. Utsunomiya, A.; Choi, I.; Chihara, D.; Seto, M. Recent advances in the treatment of adult T-cell leukemia-lymphomas. *Cancer Sci.* **2015**, *106*, 344–351. [CrossRef] [PubMed]
9. Cheng, H.; Ren, T.; Sun, S.C. New insight into the oncogenic mechanism of the retroviral oncoprotein Tax. *Protein Cell* **2012**, *3*, 581–589. [CrossRef] [PubMed]
10. Matsuoka, M.; Green, P.L. The HBZ gene, a key player in HTLV-1 pathogenesis. *Retrovirology* **2009**, *6*. [CrossRef] [PubMed]
11. Grassmann, R.; Aboud, M.; Jeang, K.T. Molecular mechanisms of cellular transformation by HTLV-1 Tax. *Oncogene* **2005**, *24*, 5976–5985. [CrossRef] [PubMed]
12. Bex, F.; Gaynor, R.B. Regulation of gene expression by HTLV-I Tax protein. *Methods* **1998**, *16*, 83–94. [CrossRef] [PubMed]
13. Arima, N.; Tei, C. HTLV-I Tax related dysfunction of cell cycle regulators and oncogenesis of adult T cell leukemia. *Leuk. Lymphoma* **2001**, *40*, 267–278. [CrossRef] [PubMed]
14. Satou, Y.; Yasunaga, J.; Yoshida, M.; Matsuoka, M. HTLV-I basic leucine zipper factor gene mRNA supports proliferation of adult t cell leukemia cells. *Proc. Natl. Acad. Sci. USA* **2006**, *103*, 720–725. [CrossRef] [PubMed]
15. Arnold, J.; Zimmerman, B.; Li, M.; Lairmore, M.D.; Green, P.L. Human T-cell leukemia virus type-1 antisense-encoded gene, HBZ, promotes T lymphocyte proliferation. *Blood* **2008**, *112*, 3788–3797. [CrossRef] [PubMed]
16. Arnold, J.; Yamamoto, B.; Li, M.; Phipps, A.J.; Younis, I.; Lairmore, M.D.; Green, P.L. Enhancement of infectivity and persistence *in vivo* by HBZ, a natural antisense coded protein of HTLV-1. *Blood* **2006**, *107*, 3976–3982. [CrossRef] [PubMed]
17. Clerc, I.; Polakowski, N.; Andre-Arpin, C.; Cook, P.; Barbeau, B.; Mesnard, J.M.; Lemasson, I. An interaction between the human T cell leukemia virus type 1 basic leucine zipper factor (HBZ) and the KIX domain of p300/CBP contributes to the down-regulation of Tax-dependent viral transcription by HBZ. *J. Biol. Chem.* **2008**, *283*, 23903–23913. [CrossRef] [PubMed]
18. Gaudray, G.; Gachon, F.; Basbous, J.; Biard-Piechaczyk, M.; Devaux, C.; Mesnard, J. The complementary strand of the human T-cell leukemia virus type 1 RNA genome encodes a bZIP transcription factor that down-regulates viral transcription. *J. Virol.* **2002**, *76*, 12813–12822. [CrossRef] [PubMed]
19. Lemasson, I.; Lewis, M.R.; Polakowski, N.; Hivin, P.; Cavanagh, M.H.; Thebault, S.; Barbeau, B.; Nyborg, J.K.; Mesnard, J.M. Human T-cell leukemia virus type 1 (HTLV-1) bZIP protein interacts with the cellular transcription factor CREB to inhibit HTLV-1 transcription. *J. Virol.* **2007**, *81*, 1543–1553. [CrossRef] [PubMed]
20. Thebault, S.; Basbous, J.; Hivin, P.; Devaux, C.; Mesnard, J.M. HBZ interacts with JunD and stimulates its transcriptional activity. *FEBS Lett.* **2004**, *562*, 165–170. [CrossRef]

21. Basbous, J.; Arpin, C.; Gaudray, G.; Piechaczyk, M.; Devaux, C.; Mesnard, J. HBZ factor of HTLV-1 dimerizes with transcription factors JunB and c-Jun and modulates their transcriptional activity. *J. Biol. Chem.* **2003**, *278*, 43620–43627. [CrossRef] [PubMed]

22. Matsumoto, J.; Ohshima, T.; Isono, O.; Shimotohno, K. HTLV-1 HBZ suppresses AP-1 activity by impairing both the DNA-binding ability and the stability of c-Jun protein. *Oncogene* **2005**, *24*, 1001–1010. [CrossRef] [PubMed]

23. Matsuoka, M.; Jeang, K.T. Human T-cell leukaemia virus type 1 (HTLV-1) infectivity and cellular transformation. *Nat. Rev. Cancer* **2007**, *7*, 270–280. [CrossRef] [PubMed]

24. Egger, G.; Liang, G.; Aparicio, A.; Jones, P.A. Epigenetics in human disease and prospects for epigenetic therapy. *Nature* **2004**, *429*, 457–463. [CrossRef] [PubMed]

25. Ganesan, A.; Nolan, L.; Crabb, S.J.; Packham, G. Epigenetic therapy: Histone acetylation, DNA methylation and anti-cancer drug discovery. *Curr. Cancer Drug Targets* **2009**, *9*, 963–981. [CrossRef] [PubMed]

26. Poke, F.S.; Qadi, A.; Holloway, A.F. Reversing aberrant methylation patterns in cancer. *Curr. Med. Chem.* **2010**, *17*, 1246–1254. [CrossRef] [PubMed]

27. Majumder, S.; Alinari, L.; Roy, S.; Miller, T.; Datta, J.; Sif, S.; Baiocchi, R.; Jacob, S.T. Methylation of histone H3 and H4 by PRMT5 regulates ribosomal RNA gene transcription. *J. Cell. Biochem.* **2010**, *109*, 553–563. [CrossRef] [PubMed]

28. Pal, S.; Baiocchi, R.A.; Byrd, J.C.; Grever, M.R.; Jacob, S.T.; Sif, S. Low levels of miR-92b/96 induce PRMT5 translation and H3R8/H4R3 methylation in mantle cell lymphoma. *EMBO J.* **2007**, *26*, 3558–3569. [CrossRef] [PubMed]

29. Pal, S.; Vishwanath, S.N.; Erdjument-Bromage, H.; Tempst, P.; Sif, S. Human SWI/SNF-associated PRMT5 methylates histone H3 arginine 8 and negatively regulates expression of ST7 and NM23 tumor suppressor genes. *Mol. Cell. Biol.* **2004**, *24*, 9630–9645. [CrossRef] [PubMed]

30. Wang, L.; Pal, S.; Sif, S. Protein arginine methyltransferase 5 suppresses the transcription of the RB family of tumor suppressors in leukemia and lymphoma cells. *Mol. Cell. Biol.* **2008**, *28*, 6262–6277. [CrossRef] [PubMed]

31. Alinari, L.; Mahasenan, K.V.; Yan, F.; Karkhanis, V.; Chung, J.H.; Smith, E.M.; Quinion, C.; Smith, P.L.; Kim, L.; Patton, J.T.; *et al.* Selective inhibition of protein arginine methyltransferase 5 blocks initiation and maintenance of B-cell transformation. *Blood* **2015**, *125*, 2530–2543. [CrossRef] [PubMed]

32. Chung, J.; Karkhanis, V.; Tae, S.; Yan, F.; Smith, P.; Ayers, L.W.; Agostinelli, C.; Pileri, S.; Denis, G.V.; Baiocchi, R.A.; *et al.* Protein arginine methyltransferase 5 (PRMT5) inhibition induces lymphoma cell death through reactivation of the retinoblastoma tumor suppressor pathway and polycomb repressor complex 2 (PRC2) silencing. *J. Biol. Chem.* **2013**, *288*, 35534–35547. [CrossRef] [PubMed]

33. Karkhanis, V.; Hu, Y.J.; Baiocchi, R.A.; Imbalzano, A.N.; Sif, S. Versatility of PRMT5-induced methylation in growth control and development. *Trends Biochem. Sci.* **2011**, *36*, 633–641. [CrossRef] [PubMed]

34. Rank, G.; Cerruti, L.; Simpson, R.J.; Moritz, R.L.; Jane, S.M.; Zhao, Q. Identification of a PRMT5-dependent repressor complex linked to silencing of human fetal globin gene expression. *Blood* **2010**, *116*, 1585–1592. [CrossRef] [PubMed]

35. Kim, C.; Lim, Y.; Yoo, B.C.; Won, N.H.; Kim, S.; Kim, G. Regulation of post-translational protein arginine methylation during HeLa cell cycle. *Biochim. Biophys. Acta* **2010**, *1800*, 977–985. [CrossRef] [PubMed]

36. Ancelin, K.; Lange, U.C.; Hajkova, P.; Schneider, R.; Bannister, A.J.; Kouzarides, T.; Surani, M.A. Blimp1 associates with Prmt5 and directs histone arginine methylation in mouse germ cells. *Nat. Cell Biol.* **2006**, *8*, 623–630. [CrossRef] [PubMed]

37. Doueiri, R.; Anupam, R.; Kvaratskhelia, M.; Green, K.B.; Lairmore, M.D.; Green, P.L. Comparative host protein interactions with HTLV-1 p30 and HTLV-2 p28: Insights into difference in pathobiology of human retroviruses. *Retrovirology* **2012**, *9*. [CrossRef] [PubMed]

38. Koralnik, I.J.; Gessain, A.; Klotman, M.E.; lo Monico, A.; Berneman, Z.N.; Franchini, G. Protein isoforms encoded by the pX region of human T-cell leukemia/lymphotropic virus type I. *Proc. Natl. Acad. Sci. USA* **1992**, *89*, 8813–8817. [CrossRef] [PubMed]

39. Lairmore, M.D.; Albrecht, B.; D'Souza, C.; Nisbet, J.W.; Ding, W.; Bartoe, J.T.; Green, P.L.; Zhang, W. *In vitro* and *in vivo* functional analysis of human T cell lymphotropic virus type 1 pX open reading frames I and II. *AIDS Res. Hum. Retrovir.* **2000**, *16*, 1757–1764. [CrossRef] [PubMed]

40. Zhang, W.; Nisbet, J.W.; Albrecht, B.; Ding, W.; Kashanchi, F.; Bartoe, J.T.; Lairmore, M.D. Human T-lymphotropic virus type 1 p30$^{II}$ regulates gene transcription by binding CREB binding protein/p300. *J. Virol.* **2001**, *75*, 9885–9895. [CrossRef] [PubMed]

41. Younis, I.; Khair, L.; Dundr, M.; Lairmore, M.D.; Franchini, G.; Green, P.L. Repression of human T-cell leukemia virus type 1 and 2 replication by a viral mRNA-encoded posttranscriptional regulator. *J. Virol.* **2004**, *78*, 11077–11083. [CrossRef] [PubMed]

42. Nicot, C.; Dundr, J.M.; Johnson, J.R.; Fullen, J.R.; Alonzo, N.; Fukumoto, R.; Princler, G.L.; Derse, D.; Misteli, T.; Franchini, G. HTLV-1-encoded p30$^{II}$ is a post-transcriptional negative regulator of viral replication. *Nat. Med.* **2004**, *10*, 197–201. [CrossRef] [PubMed]

43. Anderson, M.D.; Ye, J.; Xie, L.; Green, P.L. Transformation studies with a human T-cell leukemia virus type 1 molecular clone. *J. Virol. Methods* **2004**, *116*, 195–202. [CrossRef] [PubMed]

44. Li, M.; Green, P.L. Detection and quantitation of HTLV-1 and HTLV-2 mRNA species by real-time RT-PCR. *J. Virol. Methods* **2007**, *142*, 159–168. [CrossRef] [PubMed]

45. Green, P.L.; Ross, T.M.; Chen, I.S.Y.; Pettiford, S. Human T-cell leukemia virus type II nucleotide sequences between *env* and the last exon of *tax/rex* are not required for viral replication or cellular transformation. *J. Virol.* **1995**, *69*, 387–394. [PubMed]

46. Astier-Gin, T.; Portail, J.P.; Lafond, F.; Guillemain, B. Identification of HTLV-I- or HTLV-II-producing cells by cocultivation with BHK-21 cells stably transfected with a LTR-lacZ gene construct. *J. Virol. Methods* **1995**, *51*, 19–29. [CrossRef]

47. Livak, K.J.; Schmittgen, T.D. Analysis of relative gene expression data using real-time quantitative PCR and the $2^{-\Delta\Delta C_T}$ method. *Methods* **2001**, *25*, 402–408. [CrossRef] [PubMed]

48. Shilo, K.; Wu, X.; Sharma, S.; Welliver, M.; Duan, W.; Villalona-Calero, M.; Fukuoka, J.; Sif, S.; Baiocchi, R.; Hitchcock, C.L.; *et al.* Cellular localization of protein arginine methyltransferase-5 correlates with grade of lung tumors. *Diagn. Pathol.* **2013**, *8*. [CrossRef] [PubMed]

49. Yan, F.; Alinari, L.; Lustberg, M.E.; Martin, L.K.; Cordero-Nieves, H.M.; Banasavadi-Siddegowda, Y.; Virk, S.; Barnholtz-Sloan, J.; Bell, E.H.; Wojton, J.; *et al.* Genetic validation of the protein arginine methyltransferase PRMT5 as a candidate therapeutic target in glioblastoma. *Cancer Res.* **2014**, *74*, 1752–1765. [CrossRef] [PubMed]

50. Nicholas, C.; Yang, J.; Peters, S.B.; Bill, M.A.; Baiocchi, R.A.; Yan, F.; Sif, S.; Tae, S.; Gaudio, E.; Wu, X.; *et al.* PRMT5 is upregulated in malignant and metastatic melanoma and regulates expression of MITF and p27(Kip1.). *PLoS ONE* **2013**, *8*, e74710. [CrossRef] [PubMed]

51. Zhi, H.; Yang, L.; Kuo, Y.L.; Ho, Y.K.; Shih, H.M.; Giam, C.Z. Nf-κb hyper-activation by HTLV-1 Tax induces cellular senescence, but can be alleviated by the viral anti-sense protein HBZ. *PLoS Pathog.* **2011**, *7*, e1002025. [CrossRef] [PubMed]

52. Krause, C.D.; Yang, Z.H.; Kim, Y.S.; Lee, J.H.; Cook, J.R.; Pestka, S. Protein arginine methyltransferases: Evolution and assessment of their pharmacological and therapeutic potential. *Pharmacol. Ther.* **2007**, *113*, 50–87. [CrossRef] [PubMed]

53. Aggarwal, P.; Vaites, L.P.; Kim, J.K.; Mellert, H.; Gurung, B.; Nakagawa, H.; Herlyn, M.; Hua, X.; Rustgi, A.K.; McMahon, S.B.; *et al.* Nuclear cyclin D1/CDK4 kinase regulates CUL4 expression and triggers neoplastic growth via activation of the PRMT5 methyltransferase. *Cancer Cell* **2010**, *18*, 329–340. [CrossRef] [PubMed]

# viruses

MDPI

*Article*

# HTLV-1 Rex Tunes the Cellular Environment Favorable for Viral Replication

Kazumi Nakano * and Toshiki Watanabe

Department of Computational Biology and Medical Sciences, Graduate School of Frontier Sciences,
The University of Tokyo, 4-6-1, Shirokanedai, Minatoku, Tokyo 108-8639, Japan; tnabe@ims.u-tokyo.ac.jp
* Correspondence: nakanokz@ims.u-tokyo.ac.jp; Tel.: +81-3-5449-5295

Academic Editor: Louis M. Mansky
Received: 9 November 2015; Accepted: 9 February 2016; Published: 24 February 2016

**Abstract:** Human T-cell leukemia virus type-1 (HTLV-1) Rex is a viral RNA binding protein. The most important and well-known function of Rex is stabilizing and exporting viral mRNAs from the nucleus, particularly for unspliced/partially-spliced mRNAs encoding the structural proteins essential for viral replication. Without Rex, these unspliced viral mRNAs would otherwise be completely spliced. Therefore, Rex is vital for the translation of structural proteins and the stabilization of viral genomic RNA and, thus, for viral replication. Rex schedules the period of extensive viral replication and suppression to enter latency. Although the importance of Rex in the viral life-cycle is well understood, the underlying molecular mechanism of how Rex achieves its function has not been clarified. For example, how does Rex protect unspliced/partially-spliced viral mRNAs from the host cellular splicing machinery? How does Rex protect viral mRNAs, antigenic to eukaryotic cells, from cellular mRNA surveillance mechanisms? Here we will discuss these mechanisms, which explain the function of Rex as an organizer of HTLV-1 expression based on previously and recently discovered aspects of Rex. We also focus on the potential influence of Rex on the homeostasis of the infected cell and how it can exert its function.

**Keywords:** HTLV-1 Rex; pro-viral expression; unspliced RNA; NMD; alternative splicing; cell cycle regulation

---

## 1. Molecular Events in the Host Cell Caused by HTLV-1 Infection

Infection of T-cells with human T-cell leukemia virus type 1 (HTLV-1) causes adult T-cell leukemia (ATL), HTLV-I associated myelopathy/tropical spastic paraparesis (HAM/TSP) and HTLV-1 uveitis (HU) [1,2], although the molecular basis of such variations in the pathogenesis of HTLV-1 has not been fully elucidated. The structure of the genomic HTLV-1 RNA and the molecular events triggered by HTLV-1 infection have been thoroughly investigated [3–7]. Briefly, the genomic RNA of HTLV-1 is composed of 8685 nucleotides with two long terminal repeats (LTRs), which function as the viral promoter, at the both 5′ and 3′ ends. Although the genomic RNA is compact, HTLV-1 has various RNA signals to obtain the most out of its coding potential. By utilizing (1) three overlapped reading frames with two -1 programmed ribosomal frameshift signal (-1PRF), (2) two alternative splicing sites, (3) and multiple start and stop codons, HTLV-1 genomic RNA encodes more than 10 viral proteins [8]. HTLV-1 has three alternatively-spliced forms of viral mRNAs, which are unspliced, singly (partially)-spliced and doubly (fully)-spliced. The unspliced HTLV-1 mRNA encodes Gag, Pro, and Pol proteins, while singly-spliced RNA encodes Env. The doubly-spliced HTLV-1 mRNA encodes functional accessory proteins, such as Tax, Rex, P30II, p12, p13 in sense open reading frames (ORFs) and HBZ (HTLV-1 basic leucine zipper factor protein) in an anti-sense ORF.

After integration to the human genome, transcription and translation from the HTLV-1 provirus rely entirely on the host cell machinery. The translated viral accessory proteins then function in a precise schedule for effective viral replication [9,10] (Figure 1). The viral mRNA from the provirus for the first round of transcription is completely spliced to *tax/rex* mRNA by the cellular splicing machinery. Tax is more effectively translated from *tax/rex* mRNA because of its stronger Kozak sequence compared with that of Rex [11]. Then, Tax stimulates the transactivation of LTRs for enhanced *tax/rex* mRNA transcription. Such feed-forward activation of the HTLV-1 provirus results in the gradual accumulation of Rex in the infected cell. Subsequently, accumulation of sufficient Rex permits Rex-mediated nuclear export of unspliced and partially spliced viral RNA. The active export of these viral mRNAs to the cytoplasm by Rex results in enhanced translation of the viral structural proteins, Gag, Pro, Pol, and Env and, thereby, enhances viral replication. Inversely with the active nuclear-export of unspliced and partially-spliced viral mRNA by Rex, that of *tax/rex* mRNA is reduced; thus, cellular concentrations of Tax and Rex proteins are also decreased. Moreover, p30II from the minor doubly-spliced viral mRNA binds and retains *tax/rex* mRNA in the nucleoli by its strong nucleolar localization signal (NoLS). In combination, the cellular levels and activities of Tax and Rex proteins are gradually reduced, and both viral expression and replication are diminished to enter the latency. Rende *et al.* [12] mathematically analyzed the molecular events in early-phase HTLV-1 infection and confirmed that viral expression was indeed divided into two phases. The first phase was Tax/Rex expression, and the second phase was structural protein expression, which were both controlled by the functions of Tax and Rex. Furthermore, they concluded that the two-phase kinetics of HTLV-1 expression was strictly regulated by Rex, indicating that Rex is the major conductor of HTLV-1 expression.

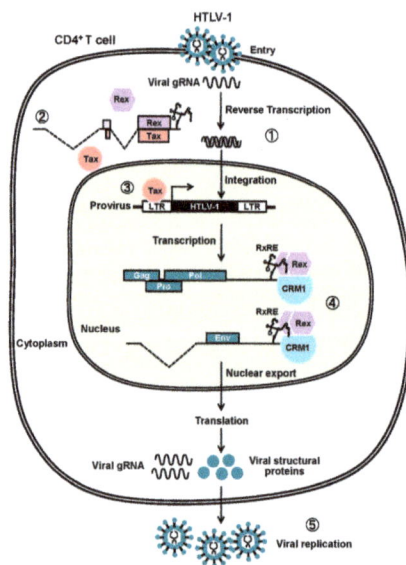

**Figure 1.** After HTLV-1 entry, the viral genomic RNA is reverse-transcribed and integrated into the host human genome (1). The viral mRNA from the provirus for the first-round of transcription is completely spliced to *tax/rex* mRNA by the cellular splicing machinery (2). Tax stimulates the transactivation of LTRs for further viral transcription, resulting in the gradual accumulation of Rex in the infected cell (3). Rex then starts exporting the unspliced and partially spliced viral mRNAs, encoding Gag, Pro, Pol, and Env, to the cytoplasm by binding to RxRE of viral mRNA (4),resulting in active viral replication (5). Due to active nuclear export of unspliced and partially spliced viral mRNA by Rex, that of *tax/rex* mRNA is eventually reduced to enter the latency.

## 2. Canonical Rex Function as a Post-Transcriptional Regulator of Viral Expression

### 2.1. Rex-Dependent Nuclear Export of Viral mRNAs

Rex binds to the Rex Responsive Element (RxRE) of the HTLV-1 mRNAs to form Rex-viral mRNA complex for selective nuclear-export. Unlike the Rev Responsive Element (RRE) in human immunodeficiency virus type-1 (HIV-1) mRNAs, RxRE is in all HTLV-1 mRNAs [5,13]. The RxRE of HTLV-1 mRNA maps to the region of 255 nucleotides (nt) from the U3 to the R region of the 3'-LTR and forms a stable secondary structure with four stem loops. Such a unique structure of RxRE is considered to function as the landmark for Rex to selectively bind to the viral mRNAs [14,15]. Although, all HTLV-1 derived mRNAs have RxRE, the nuclear export efficiency by Rex is different among HTLV-1 mRNAs. It has been widely accepted that cytoplasmic accumulations of unspliced and partially-spliced HTLV-1 mRNAs are Rex dependent, while that of fully spliced *tax/rex* mRNA is suppressed by Rex [12,16]. Subsequently, Bai *et al.* demonstrated that Rex also stimulated the nuclear-export of *tax/rex* mRNA and Tax expression, at least partially [17]. Most recently, Cavallari *et al.* elegantly demonstrated that not only unspliced *gag/pol* mRNA and singly spliced *env* mRNA, but also some of singly- and fully-spliced mRNAs encoding viral accessory proteins were also nuclear-exported in Rex-dependent manner. They showed that *p30II*, *p12/p8*, and *p13* mRNAs were Rex-dependent, while *tax/rex* and *p21rex* mRNAs were Rex-independent. Interestingly, all Rex-dependent viral mRNAs contain 75 nt intronic regions, which control Rex-dependency as a *cis*-acting sequence [18]. Another study group demonstrated that HBZ, the antisense protein of HTLV-1, inhibited the nuclear-export of intron-containing mRNA by Rex, thus inhibited active viral replication and induced latency [19]. These reports suggest that Rex-mediated nuclear exports of HTLV-1 mRNAs are finely-tuned by RxRE, inherent *cis*-acting viral sequence, and viral proteins.

Rex binds to Chromosomal Maintenance 1 (CRM1), also known as Exportin 1 (XPO1), via its nuclear export signal (NES) for nuclear export. CRM1 is a cellular nuclear export protein which is responsible for the translocation of various cellular proteins with NES. Thus, the export of HTLV-1 mRNAs to the cytoplasm is dependent on CRM1, which is separated from the bulk cellular mRNAs exported in an Aly/Ref export-factor-dependent manner. The molecular mechanism of the RxRE-Rex-CRM1 complex formation has been extensively studied by Hakata *et al.* [20,21]. The authors revealed that the Rex has to be multimerized to bind to RxRE. They also propose a possibility that CRM1 is involved not only in the translocation of Rex but also in its multimerization. Therefore, one may speculate that Rex initially forms a complex with CRM1, which assists oligomerization of Rex on CRM1 before binding to RxRE. Nevertheless, the detailed order of the complex formation has never been investigated.

### 2.2. Primary Structure of Rex and Its Function

HTLV-1 Rex protein consists of 189 amino acids with its molecular weight of approximately 27 kDa. The HTLV-1 Rex protein contains several functional domains essential for its function. The primary structure of Rex has been well-described [22,23]. The N-terminal arginine-rich RNA-binding domain (aa 1–19) is required for binding to RxRE. Additionally, this domain overlaps with the nuclear localization signal (NLS), which is essential for Rex to shuttle-back to nucleus with importin-β, and with the p30II-binding domain. Rex interacts with CRM1 through the NES (aa 66–118) for nuclear export. Rex has two multimerization domains (aa 57–66 and 106–124) with NES in between and both of them are considered to be necessary for stable oligomerization of Rex. Most recently, a stability domain was identified at the C-terminal region of Rex (aa 170–189) [24–26]. The authors demonstrated that deletion of the stability domain destabilized Rex significantly but did not influence the function of Rex.

### 2.3. Rex Activity and Phosphorylation

It has been described that the activity Rex is finely regulated through its phosphorylation [27] at several serine(Ser)/threonine(Thr) residues [24]. The treatment of HUT102 (an HTLV-1-infected cell line) with a protein kinase C inhibitor, H-7 [1-(5-isoquinolinyl-sulfonyl)-2-methylpiperazine] destabilized unspliced viral mRNA and reduced the expression level of Gag-p19 protein [27]. To date, seven phosphorylation sites of Rex have been identified at Thr-22, Ser-36, Thr-37, Ser-70, Ser-97, Ser-106, and Thr-174 [24,28]. Kesic *et al.* [24] evaluated the importance of phosphorylation sites and demonstrated that Rex phosphorylation at Ser-97 and Thr-174 was the most critical for the efficiency of RxRE-dependent nuclear export by Rex.

*2.4. Regulation of Rex by Other HTLV-1 Viral Proteins*

As described above, Rex plays a central role in selective expression of HTLV-1 viral structural proteins and is, thus, in active viral reproduction. The Rex activity is critical to switch from the early productive period to late latent period. Therefore Rex activity has to be finely tuned during HTLV-1 infection. HTLV-1 has an elegant auto-regulatory mechanism to regulate the activity of Rex, *i.e.*, by two-phased HTLV-1 expression kinetics, which is described above (see Section 1), and by the function of other viral proteins.

The suppressive function of p30II for the Rex has been well investigated, following extensive reproduction of HTLV-1 virus by Tax and Rex, p30II is expressed from the minor doubly-spliced HTLV-1 mRNA. P30II selectively binds to *tax/rex* mRNA. Then, p30II, with a strong nucleolar localization signal (NoLS), localizes and retained *tax/rex* mRNA in nucleoli, thus preventing their expression and functions. Such time-lagged operations of the positive (Tax and Rex) and negative (p30II) regulators of HTLV-1 promotes the early infectious phase followed by the late infectious phase with a rapid shutdown to escape from the host immune surveillance against pathogens [29–32]. On the other hand, Rex binds to p30II and rescues *tax/rex* mRNA [32]. The timing of Rex-p30II interaction is considered to regulate switching from the early active-viral-reproduction phase to the late rapid-shutdown phase to escape from the host immune system. Most recently, HBZ, the antisense protein of HTLV-1, was demonstrated to inhibit the nuclear-export of intron-containing mRNA by Rex, thus inhibiting active viral replication and induced latency [19].

Since p21Rex is constitutively expressed in primary peripheral blood mononuclear cells from HTLV-1 carriers and ATL patients [33–35], it has been expected that p21Rex plays a role in the HTLV-1 life cycle, such as p27Rex suppressor as a dominant-negative isoform, although, a clear biological function of p21Rex has not been elucidated, yet. P21Rex is expressed from a defective HTLV-1 mRNA without the exon 2, and lacks the N-terminus 78 amino acids of p27Rex, ranging from NLS to the N'-multimerization domain [36,37]. Without NLS, p21Rex localizes to the cytoplasm; thus, the functional importance of this isoform has not yet been elucidated. More recently, Bai *et al.* demonstrated that p21Rex neither nuclear-exported the viral mRNA, nor influenced the p27Rex function [17]. Therefore, p21Rex seems not to be involved in HTLV-1 lifecycle as an isoform of Rex, although possible roles of this short Rex isoform in the cellular biological pathways are required to be elucidated in the future.

## 3. Non-Canonical Functions of Rex: Exploring New Aspects of Rex

*3.1. NMD Inhibition by Rex*

3.1.1. Rex Stabilizes HTLV-1 Genomic RNA by Inhibition of NMD

For viruses, stabilization of viral mRNAs in the host cells is a major issue to overcome for self-replication [38]. Recently, a lot of attention has been directed towards one of the host mRNA decay mechanisms in the host−pathogen interaction, Nonsense-mediated mRNA decay (NMD), and has revealed how viruses evade NMD and protect viral mRNAs [39–42]. These reports have shown that each virus has its own strategy to stabilize viral mRNAs; for example, by an inherent viral RNA stabilization mechanism, utilizing host RNA stability factors, inhibition of the host mRNA decay machinery, or hijacking the host cell RNA metabolism with viral nucleases [38,43].

NMD is an essential and evolutionarily-conserved cellular mRNA quality control mechanism. The principal function of NMD is to prevent the expression of harmful truncated proteins by selective elimination of aberrant mRNAs containing premature termination codons (PTCs) (see reviews [43,44]) (Figure 2A). As indicated above, the major Rex function is the stabilization and export of the viral unspliced and partially-spliced mRNAs to cytoplasm. However, unspliced HTLV-1 mRNA (*i.e.*, viral genomic RNA) contains various RNA signals, such as multiple start and stop codons, overlapping ORFs, programmed ribosomal frameshift signals, and a long 3′-untranslated region (>1000 nt). These RNA signals are unusual for eukaryotic cells and have the potential to initiate NMD (Figure 2B). However, it is not clear how HTLV-1 evades NMD to protect its genomic RNA. Our laboratory has demonstrated that full-length HTLV-1 transcripts exhibit enhanced turnover in NMD-activated cells that overexpress UPF1, while knockdown of UPF1 by small interfering (si) RNA promotes enhanced stability of HTLV-1 genomic mRNA [45]. By confirming that the genomic and full-length mRNAs of HTLV-1 are sensitive to NMD, we further demonstrated that Rex inhibited NMD. We suggest that through the inhibition of NMD, Rex stabilizes viral transcripts in the cytoplasm to secure translation of viral structural proteins. In contrast, it is highly probable that Rex also perturbs cellular mRNA metabolism and host cell homeostasis by inhibition of the global NMD activity. It is noteworthy that Rex-mediated inhibition of NMD is not RNA- or sequence-specific, but Rex establishes a general blockage of NMD. Thus, not only the viral transcripts, but also natural host-encoded NMD substrates are stabilized in the presence of Rex. We demonstrated that Rex stabilized well-known NMD target mRNAs, such as *IL-6*, *MAP3K14*, and *FYN* mRNAs.

It has been reported that *IL-2Rα* mRNA was stabilized up to a five-fold level in Rex-overexpressing cells compared with the control cells without Rex [46,47], although the underlying mechanism has not been clarified. Since *IL-2Rα* mRNA can be a NMD target because of its upstream (u)-ORF structure, we speculate that this mRNA is stabilized by Rex through NMD inhibition. Indeed, UPF1 knockdown by siRNA in HeLa cells resulted in a significant increase in the *IL-2Rα* mRNA expression (Nakano unpublished data).

### 3.1.2. How Does Rex Protect Viral mRNAs from NMD in the Cytoplasm?

NMD is a complex mechanism coupled with splicing and translation. Briefly, the core components of NMD are UPF1, UPF2, and UPF3, which detect the PTC-containing mRNA to be degraded via NMD, and SMG1, SMG5, SMG6, and SMG7, which phosphorylate/dephosphorylate UPF1 (Figure 2B). This results in the regulation of the activity of UPF1, the key molecule of NMD. UPF1 is a component of the termination complex, assembled when the ribosome reaches the termination codon of mRNA, while UPF2 and UPF3 are components of the exon junction complex (EJC) formed at the exon-exon boundary and are removed by the ribosome while it moves through. In normal mRNA, the termination codon is in the last exon, thus EJC does not remain at the end of translation. In contrast, PTC is located upstream of EJC, thus UPF1 on PTC comes into contact with UPF2 and UPF3, which triggers the phosphorylation of UPF1 by SMG1, and the onset of NMD. Phosphorylated UPF1 is dephosphorylated by the SMG5/SMG7 complex and recycled, while SMG7 completes the NMD process in mRNA processing bodies (p-bodies) of the cytoplasm. Recently, it was reported that SMG6 functioned as an endonuclease in the degradation of PTC-containing mRNA (see review [43]).

Since NMD machinery is coupled with splicing and translation apparatus, we speculate that Rex may influence the overall NMD activity directly via interaction with NMD core-components, and indirectly via interaction with splicing and translational machinery. Comprehensive protein-protein interactome analysis between Rex and the host-cellular proteins by a high-resolution mass spectrometry (MS) will be fruitful to understand the overall molecular landscape of NMD inhibition by Rex.

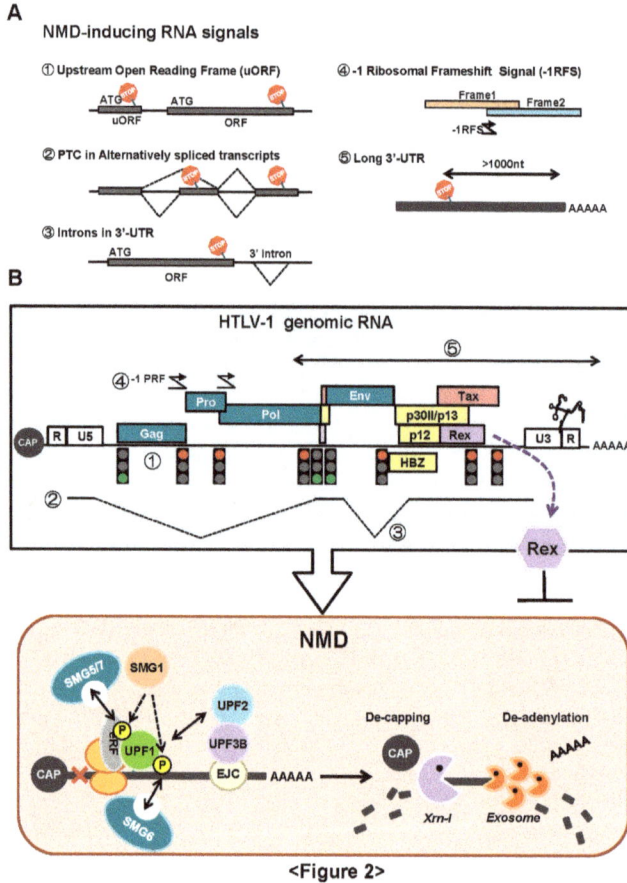

<Figure 2>

**Figure 2.** (**A**) NMD is an essential and evolutionarily-conserved cellular mRNA quality control mechanism. The principal function of NMD is to prevent the expression of harmful truncated proteins by selective elimination of aberrant mRNAs containing premature termination codons (PTCs). It has been reported that NMD also regulates the expression levels of normal mRNAs, which inherently contain RNA signals to generate PTC: 1. Upstream (U)-ORF, 2. alternative splicing producing PTC, 3. intron in 3'-UTR, 4. -1 programmed ribosomal frameshift signal (-1PRF) and 5. long 3'-UTR more than 1000 nt; (**B**) The genomic RNA of HTLV-1 contains various RNA signals potentially initiate NMD, *i.e.*, 1. multiple start and stop codons, 2. two alternative splicing sites, 3. intron in the 3'-UTR region of *gag/pro/pol* mRNA, 4. overlapping ORFs, and -1 PRF and 5. a long 3'-UTR (>1000 nt). We reported that HTLV-1 genomic RNA was indeed destabilized by NMD, and HTLV-1 Rex had a new function to inhibit NMD [32].

### 3.2. Regulation of mRNA Splicing Machinery by Rex

It is well known that mRNA splicing is coupled with transcription, and virtually all primary (unspliced) mRNAs are spliced at the site of transcription (see review [48]). Thus, HTLV-1 unspliced and partially-spliced mRNAs cannot evade splicing only through selective nuclear export by Rex. Consequently, we may speculate that Rex has a function to inhibit cellular splicing activity against the viral mRNAs, which is independent of the well-known CRM1-dependent nuclear export mechanism of Rex. Gröne *et al.* [49] demonstrated that Rex increased the nuclear quantity of unspliced viral mRNAs and reduced the number of spliced viral mRNAs. Since Rex did not influence the total quantity of viral transcripts, the authors conclude that Rex has a function by which it reduces splicing activity.

SF2/ASF regulates splicing activity and plays an important role in the splice-site selection [48,50]. Splicing patterns of HTLV-1 mRNA are governed by SF2/ASF *i.e.*, the differential pX splice site utilization of HTLV-1 mRNA is dependent on the expression level of SF2/ASF [51], although the viral mechanism to regulate the splicing activity through SF2/ASF has not been fully investigated. Interestingly, Powell *et al.* [52] demonstrated that HIV-1 Rev suppressed cellular splicing activity by recruiting SF2/ASF to the Rev-RRE RNP complex. Tange *et al.* [53] also showed the interaction between Rev and p32, the ASF/SF2-associated protein. The authors speculated that the interaction might function as a bridge between Rev and the host cellular splicing machinery. Considering the homologous function and molecular mechanism of HTLV-1 Rex and HIV-Rev, it is possible that Rex has a similar mechanism to suppress splicing machinery by binding and inhibiting the function of SF2/ASF.

Heterogeneous Nuclear Ribonucleoprotein A1 (hnRNPA1) is another cellular protein which is known to interact with Rex. hnRNPA1 associates with mRNA as a component of the RNP complex in the nucleus and influences the transcription, maturation and transport of mRNA [54]. HnRNPA1 also plays a crucial role in the regulation of alternative splicing, mainly as a splicing suppressor [48]. A recent study clearly showed that the expression level of hnRNPA1 had strong implications for the determination of exon-inclusion/skipping [55]. Hamaia *et al.* [56] first demonstrated that Rex function was impaired in a T cell line not infected by HTLV-1, Jurkat, and speculated that Rex was unable to bind to RxRE in the cell line. Later, the same study group found that hnRNPA1 bound to RxRE in competition with Rex, thus influencing the function of Rex [57]. Subsequently, Kress *et al.* [58] demonstrated that hnRNPA1 suppressed the Rex activity in a dose-dependent manner, while the suppression of hnRNPA1 in C91/PL, a HTLV-1-infected cell line, increased the Rex-dependent nuclear export of unspliced and partially-spliced mRNA. The authors proposed the possibility that hnRNPA1 enhances the splicing processes of viral mRNA. Indeed, hnRNPA1 caused enhanced exon 2 skipping in HTLV-1 mRNA [51]. On the other hand, the basal hnRNPA1 level was lower in HTLV-1-infected T cell lines (C91/PL, MT2, and HUT102) compared with other T cell lines without HTLV-1 infection (CBL and Jurkat) [57]. The authors concluded that HTLV-1 may have a mechanism to downregulate hnRNPA1, which is not advantageous for viral replication. Glutathione S-transferase (GST)-Rex pulldown assays conducted in our laboratory showed that Rex physically interacted with hnRNPA1 (Nakano, unpublished data). The underlying mechanism of how Rex is involved in the downregulation/inhibition of hnRNPA1 requires further investigation.

It has been shown that Rex changes the preference for exon usage during *FYN* mRNA splicing/maturation from exon7B to exon7A, resulting in enhanced production of the brain-type Fyn-B instead of the T cell-type Fyn-T [59]. Fyn is a proto-oncogene, belonging to the membrane-associated tyrosine kinase family. Its overexpression/disorder has been implicated to the tumorigenesis of several malignancies. Fyn has two major isoforms of distinct functions, Fyn-B expressed in the brain and Fyn-T expressed exclusively in hematopoietic cells, which are derived from exon7A and exon7B, respectively. Picard *et al.* [60] reported that the expression level of *FYN-B* mRNA was significantly increased in acute lymphoblastic leukemia or chronic lymphocytic leukemia. As indicated above, hnRNPA1, the regulator of exon usage, is downregulated in HTLV-1-infected cells. Moreover, we found that Rex interacts with hnRNPA1. If Rex itself is involved in the downregulation and/or suppression of hnRNPA1, such deregulation of hnRNPA1 function by Rex may have implications to alterations in the exon usage during mRNA maturation, such as observed in *FYN* mRNA. Aberrant overexpression of various splicing variants caused by genetic lesions in the splicing machinery may have implication to the HTLV-1 pathogenesis.

Taken together, the biological significance of the molecular interactions between Rex and the splicing-regulatory proteins in the regulation of splicing activity and splicing patterns requires elucidation in the future.

### 3.3. Cell-Cycle Regulation: Does Rex Interfere the Host Cell-Cycle Regulation?

A wealth of evidence has indicated that a number of viruses have mechanisms to modify cellular cell-cycle regulation for the promotion of viral replication. It has been well documented and reviewed that HIV-1 Vpr induces G2 arrest of the host cell-cycle [61–63]. The G2/M check point or DNA damage checkpoint is regulated by the activity of the Cdc2 (Cdk1) and CyclynB complexes, which are finely tuned by various kinases and phosphatases. Cdc2 undergoes inhibitory phosphorylation by Wee1 and Myt1, or Chk1/2, which are activated by the ATM/ATR DNA damage response pathway. Cdc25s are phosphatases and activate Cdc2 by dephosphorylation. When the cell senses DNA damage, Cdc2 is inhibited by the ATM/ATR pathway, and the cell cycle is arrested at G2. At the G2/M transition, PLK1 phosphorylates Cdc25s and Wee1 for activation and inhibition, respectively. Thus, Cdc2 is activated to enter the M phase (see review [64]). Furthermore, PLK1 is phosphorylated and activated by Aurora kinase A (AURKA) and its co-factor, Bora [65]. For the molecular mechanism of G2 arrest by Vpr, Zhao and Elder [61] indicated the importance of the interaction between Vpr and IκB kinase-associated serine/threonine protein phosphatase 2A (PP2A) in the induction of G2 arrest, although the detailed mechanism has yet to be clarified. Goh *et al.* [66] demonstrated that Vpr interacts with and inhibits Cdc25C. However, because Vpr is also known to activate ATR and Chk1 [67], it has not been fully elucidated whether Vpr directly inhibits Cdc25C or does so through the ATR pathway. Noronha *et al.* [68] investigated the influence of Vpr using a different methodology and showed that Vpr altered the subcellular localization of CyclinB1, Wee1, and Cdc25C. These authors also found that Vpr-induced herniations of the nuclear envelope and speculated that such disrupted nuclear architecture might interrupt normal cell-cycle progression.

Why does HIV-1 Vpr induce G2 arrest? For viral replication, transcription from the provirus and translation of viral proteins are dependent on the host machinery. It is thought that G2 arrest by Vpr is beneficial for the selective translation of viral proteins (Figure 3). The m$^7$G-Cap structure of transcribed mRNA is first recognized by the Cap binding complex (CBC) and subjected to the pioneer round of translation for the quality check of mRNA. Then, CBC is replaced by the eIF4F complex for the steady-state translational procedure, which is regulated by eIF4E within the eIF4F complex. It is thought that the translation of a viral protein from HIV-1 mRNA relies on CBC-dependent pioneer-round translation, which is cell-cycle independent. In contrast, the major eIF4E-dependent translation is inhibited during G2 phase. Therefore, G2-arrest by Vpr can enhance the translation of viral proteins (see review [69]). Furthermore, Sharma *et al.* [70] elegantly demonstrated that Vpr abrogates activated (phosphorylated) eIF4E levels. They also showed that CBC was retained at the Cap structure of unspliced and partially-spliced HIV-1 mRNAs in the cytoplasm. Taken together, these findings indicate that Vpr suppresses eIF4E activity by the reduction of its active form, as well as by the induction of G2-arrest. Thus, only CBC-bound HIV-1 mRNAs can be effectively subjected to cellular translational machinery (Figure 3). Most recently, it has been demonstrated that Vpr interacts with and activates the SLX4 endonuclease complex, which activates the DNA damage/repair response through the ATR/Chk1 pathway, resulting in G2 arrest [71].

HTLV-1, with a similar life-cycle to HIV-1, may have a similar strategy to enhance self-reproduction. However, the HIV-1 Vpr homologue has not been identified among the HTLV-1-encoded proteins. In the review by Zhao and Elder [61], they mention that HTLV-1 Tax showed similar characteristics to HIV-1 Vpr, such as binding to PP2A and the induction G2 arrest. Haoudi *et al.* [72] first indicated that Tax bound to and activated Chk2 in the DNA damage response, resulting in G2 arrest. Moreover, the interaction between Tax and Chk2 was further investigated by the same group, and they subsequently concluded that Tax inhibits the Chk2-induced DNA damage response through its retention in chromatin in order to evade the cellular DNA damage response to Tax-induced DNA instability [73]. Another study group also demonstrated that Tax bound to and inhibited the activity of Chk1, which is also involved in the ATM/ATR-mediated DNA damage response [74]. Fu *et al.* [75] showed that Tax interacted with PP2A to activate I kappa B kinase (IKK), thus influencing the nuclear factor (NF)-κB pathway. Together, these previous reports indicate that Tax inhibits the ATM/ATR-dependent

DNA damage response and, thus, is not likely to induce G2 arrest. Anupam *et al.* [76] conducted a protein-interactome analysis for p30II and demonstrated that p30II interacted with ATM and modulated the activity of the G2/M checkpoint. There have been no reports implicating Rex to the cell-cycle regulation. Yet, we observed G2 arrest in CEM (ALL patient-derived human T cell line) overexpressing Rex (Nakano, unpublished data). Rex has arginine-rich NLS at the N-terminus similar to Vpr-NLS2, which is essential for G2 arrest. It has been demonstrated that eIF4E specifically binds to the mRNAs of cell-cycle promoting proteins in nucleus and is exported by CRM1. This mechanism is separated from TAP/NXF1 and REF/Aly-dependent export of bulk mRNAs [77–80]. Since Rex is also nuclear-exported by CRM1, we speculate that Rex may compete for CRM1 with eIF4E. Consequently, Rex may suppress the eIF4E-CRM1-dependent nuclear export of mRNAs encoding cell-cycle promoting proteins and, therefore, may induce cell-cycle arrest. The interaction between Rex and eIF4E and other cell-cycle regulating proteins should be investigated in the future.

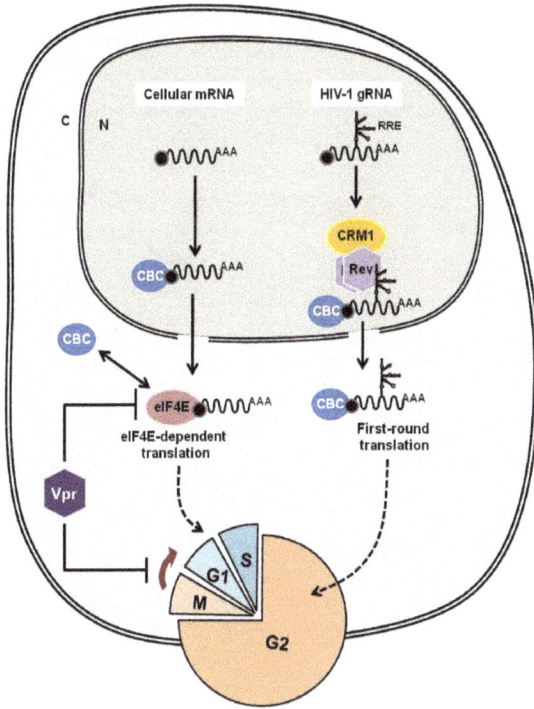

**Figure 3.** The G2-arrest by Vpr is beneficial for the selective translation of viral proteins. In the regular mRNA translation, the $m^7$G-Cap structure of transcribed mRNA is first recognized by the Cap binding complex (CBC) and subjected to the pioneer round of translation for the quality check of mRNA. Then, CBC is replaced by the eIF4F complex for the steady-state translational procedure, which is regulated by eIF4E within the eIF4F complex. The majority of eIF4E-dependent translation is inhibited during G2 phase. On the other hand, CBC-dependent pioneer-round translation is cell-cycle independent. Since the translation of a viral protein from HIV-1 mRNA relies on CBC-dependent pioneer-round translation, G2-arrest by Vpr can enhance the translation of viral proteins by suppressing the eIF4E-dependent translation. Additionally, Vpr reduces the activated (phosphorylated) eIF4E level. Taken together, Vpr assists selective translation of HIV-1 mRNAs by induction of G2-arrest, as well as by suppression of eIF4E activity.

## 4. Function of Rex and the Viral Pathogenesis

### 4.1. Do Rex-1/Rex-2 Functions Relate to the Pathogenesities of HTLV-1/HTLV-2?

Comparative analysis between Rex-1 (Rex) from HTLV-1 and Rex-2 from HTLV-2 can be helpful to understand the relationship between the function of Rex and the viral pathogenesity [13]. Both HTLV-1 and HTLV-2 belong to the same genus [81] and infect human T cells. Both viruses encode a similar set of viral proteins, including Tax and Rex and, thus, reproduce through a similar pathway. Yet, only HTLV-1 causes ATL and HAM/TSP in infected T cells, but not HTLV-2. The primary structures of Rex-1 and Rex-2 show 60% homology with common functional domains, such as RNA binding domain (RBD)/NLS, two multimerization domains, nuclear export signal (NES), and stability domain (SD) [24–26] and, thus, function as the viral RNA binding/transporting proteins through the common cellular pathways. On the other hand, the position of RxRE in the viral mRNA is different between HTLV-1 and HTLV-2, which may modulate impacts of Rex-1 and Rex-2 functions in their respective viral life cycles [13]. It has been clarified that all HTLV-1 mRNAs have RxRE , which is located in the U3/R region, while only unspliced HTLV-2 mRNA has RxRE, which is located in the R/U5 region [82]. Thus, it can be speculated that Rex-1 nuclear-exports all HTLV-1 mRNAs, including *tax/rex* mRNA, which enhances Tax/Rex expression and, thus, viral reproduction, whereas Rex-2 does not, resulting in a low viral production. Indeed, Bai *et al.* demonstrated that the nuclear export of the doubly spliced *tax/rex* mRNA of HTLV-1 was also enhanced by Rex-1 in a RxRE-1/CRM1-dependent manner [17]. Differences in nuclear export efficiencies of viral mRNA/Rex/CRM1 complex between HTLV-1 and HTLV-2 may influence viral replications and activities and, thus on pathogenesities of these viruses in infected T cells.

### 4.2. HTLV-1 Rex and HIV-1 Rev: Are They Similar or Different?

#### 4.2.1. HIV-1 Rev, the Molecular Counterpart of HTLV-1 Rex

Rev protein of HIV-1 (Human Immunodeficiency Virus type-I) is the molecular counterpart of HTLV-1 Rex. HTLV-1 and HIV-1 both belong to the family of *Retroviridae*, and are further specified to the genuses of *Deltaretrovirus* and *Lentivirus*, respectively. In addition, the major tropism of both viruses is human CD4$^+$ T cells. HTLV-1 and HIV-1 have genomic RNA of a similar size, *i.e.*, about 8.5 knt and 9.75 knt, respectively, which encodes viral proteins with considerably homologous functions. HTLV-1 Rex and HIV-1 Rev bind viral mRNAs and shuttle between the nucleus and cytoplasm for nuclear export of viral transcripts, through quite similar mechanisms yet, interestingly, the homology between their amino acid sequences is very low [22,83]. Messenger-RNAs of HTLV-1 and HIV-1 have regions to form complex secondary structures called RxRE and RRE. Rex and Rev bind to viral mRNAs thorough RxRE and RRE as highly-specific landmarks, respectively. In terms of primary functional domains, both Rex and Rev have arginine-rich RNA binding domains for selective binding to their respective responsive elements (Figure 4A). It has been well documented that Rex and Rev stabilize unspliced and partially-spliced viral mRNAs, encoding viral structural proteins, and actively transport them to the cytoplasm for selective translation (Figure 4B). For shuttling between nucleus and cytoplasm, both proteins have NLSs for binding to importin-β and NESs for binding to CRM1 [23,83]. Furthermore, they bind to B-23 via NLSs to be translocated to the nucleolus [84–87] (Figure 4B). While the uniform role and mechanism of Rex and Rev are extensively discussed, some differences in the detailed molecular mechanism between Rex and Rev have been also described. It is now accepted that multimerization is essential for Rex to interact with RxRE, yet the monomer Rex is still able to bind to CRM1 for translocation to the cytoplasm [20–22]. Quite the opposite, the monomer Rev is known to bind to RRE, however multimerization of Rev up to 12 molecules is necessary for stable binding to CRM1 and for effective cytoplasmic-translocation [88,89]. These differences between these two viral RNA binding proteins may be closely related to the nuclear export efficiency of viral mRNAs, thus, viral replication and, consequently, to different disease associations, *i.e.*, ATL and HAM/TSP with HTLV-1, and AIDS with HIV-1.

**A**

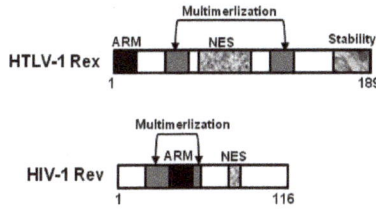

**B**

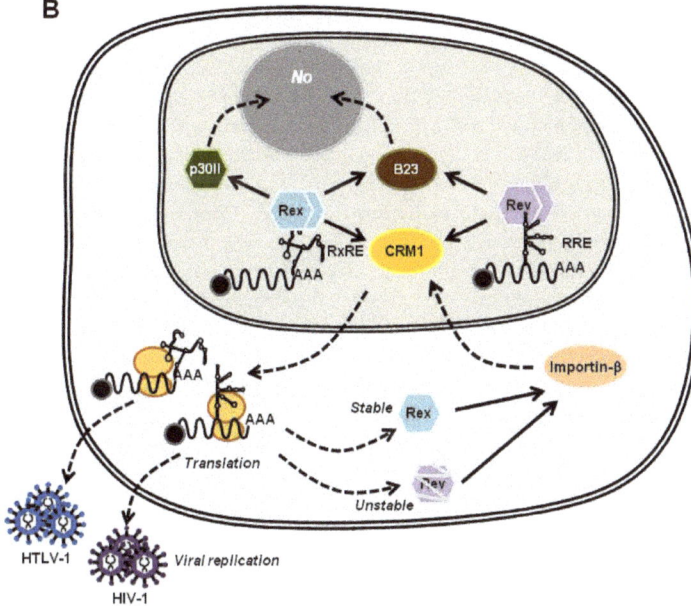

**Figure 4.** (**A**) The primary structures of HTLV-1 Rex and HIV-1 Rev. The homology in the primary sequences of HTLV-1 Rex and HIV-1 Rev is low, but they share most of the functional domains critical for their function, *i.e.*, arginine-rich RNA binding domain, NLS, NES, and two multimerization domains; (**B**) HTLV-1 Rex and HIV-1 Rev play similar functions through similar mechanisms. In the nucleolus, Rex and Rev specifically bind to the respective viral mRNAs through Rex responsive element (RxRE) for Rex and the Rev responsive element (RRE) for Rev. They stabilize unspliced or partially-spliced viral mRNA and actively transport these to the cytoplasm for selective translation of viral structural proteins by CRM1 binding through their NES. Rex and Rev return to the nucleus by binding to Importin-β, and further translocated to the nucleolus (No) by binding to B-23 via NLS.

### 4.2.2. The Structural Biology of HTLV-1 Rex; Learning from that of HIV-1 Rev

Along with accumulation of knowledge in the molecular characteristics and functions, both Rex and Rev propose a common question; how these small viral proteins bind to RxRE/RRE of viral mRNAs and CRM1 simultaneously, and how they form a large, but stable, RNP complex for the nuclear export of viral mRNAs. To answer these questions, information from the structural biology may contribute significantly. In terms of structural biology, investigations into Rev have progressed far more than those of Rex. In contrast, our knowledge of the structure of Rex has not been updated from that of the N-terminal arginine-rich domain (aa 1–16) of the Rex peptide solved by nuclear magnetic resonance [90]. On the contrary, great efforts have been made to obtain the structural information on the Rev−Rev dimer interface or the Rev−RRE interaction since the early 2000′s. Daugherty *et al.* [91] clarified for the first time that Rev formed a homo-oligomer on the RRE via an oligomerization domain. In 2010, two different groups reported the partial structure of Rev in a dimer form [92,93]. Daugherty *et al.* [93] proposed the "jellyfish model" in which a rigid Rev dimer(s) forms an oligomeric structure, similar to the head of a jelly fish, while an unstructured NES region extended from each Rev molecule to form a tentacle-like alignment in binding to CRM1. They also mentioned that at the RNA-binding surfaces, "the two ARMs are arranged to reach out from the body of the Rev dimer to grasp RNA, much as two human arms are positioned to grip objects".

Based on the Rev structure, the detailed molecular mechanism of HIV-1 viral mRNA nuclear export by Rev has since been clarified greatly. Fang *et al.* [94] investigated the structure of RRE thoroughly and demonstrated that RRE functioned as the topological landmark for Rev by forming an unusual structure, to which only Rev was able to specifically bind with high affinity. Finally, it was demonstrated that the RRE-Rev-CRM1 "HIV-1 export complex" was assembled co-transcriptionally at the transcription site, thus unspliced HIV-1 mRNAs were stably exported from the nucleus [95]. Moreover, it has been demonstrated that RRE-Rev assembly starts with changes in the RRE structure so that binding to the first two dimer complexes of Rev leads to further conformational changes of RRE, triggering oligomerization of Rev [96]. This enables stable binding to the CRM1 dimer [97]. Structural analysis of the Rev-dimer and RRE complex revealed that the Rev-dimer architecture can be flexibly altered depending on the structure of RRE [98]. The authors speculated that such changes in RRE structure and in the Rev-dimer architecture might alter the whole architecture of the "jellyfish complex" *i.e.*, the RRE-Rev-CRM1 complex. Overall, the structural biology of Rev indeed provided tremendous information to clarify the regulatory mechanism of the nuclear export efficiency of the HIV-1 export complex, thus resulting in pathogenesis of HIV-1. At present, the same questions still remain for HTLV-1 Rex. The structural biology of Rex is promising to provide significant information to answer these questions in the future.

## 5. Conclusions

Accumulating data on the analysis of the Rex interactome shows that Rex has a significantly high potential to interact with a wide variety of cellular proteins. These cellular proteins are crucial for the maintenance of the cellular homeostasis by playing essential roles in mRNA surveillance and metabolism, nucleo-cytoplasmic shuttling, tumor growth regulation and in post-translational modification of proteins, such as SUMOylation [13,99,100]. These data strongly suggest that Rex modifies a wide range of cellular pathways in order to organize the host cellular environment suitable for the stabilization and translocation of viral mRNAs, as well as for selective translation of viral proteins for effective self-replication (Figure 5A). Such Rex-oriented tuning of the host cell environment can alter cellular homeostasis, and thus may provide a basis for the pathogenesis of HTLV-1 (Figure 5B).

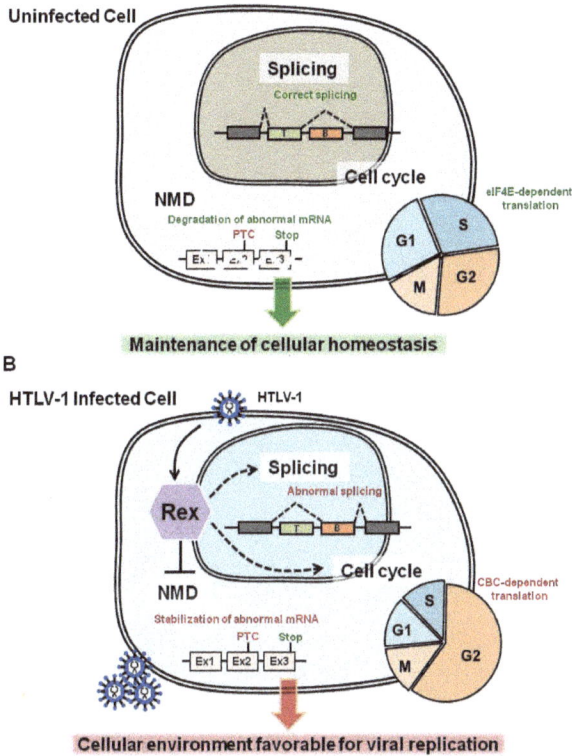

**Figure 5.** In this review, we focused on three cellular pathways, NMD, splicing machinery, and cell-cycle regulation, since we may expect that Rex interacts with these pathways to adjust the cellular environment suitable for the viral replication. (**A**) In the normal cells, the activities of NMD, splicing, and cell-cycle regulation are optimized to maintain the cellular homeostasis by eliminating PTC-containing harmful mRNAs, production of correctly spliced mRNAs encoding functional proteins, and adjusting the cell-cycle for n optimal cell proliferation rate and for effective eIF4E-dependent RNA translations, respectively; (**B**) In HTLV-1 infected cells, Rex inhibits NMD for stabilization of the viral genomic mRNA [32]. Additionally, based on previous reports and newly discovered aspects of Rex in our laboratory, we assume that Rex may suppress the activity of splicing machinery and may induce G2 arrest. These adjustments of host-cell mechanisms are favorable for the stabilization and translocation of viral mRNAs, as well as for selective translation of viral proteins for effective self-replication. On the other hand, suppression/alteration in these pathways may cause accumulation of abnormal PTC-containing mRNAs, thus harmful proteins; abnormal splicing patterns; deregulated cell proliferation; and suppression of eIF4E-dependent translation. Therefore, Rex-oriented tuning of the host cell environment may alter cellular homeostasis, and provide a basis for the pathogenesis of HTLV-1.

**Acknowledgments:** This work was supported by Grants-in-Aid for Scientific Research from the Ministry of Education, Culture, Sports, Science, and Technology of Japan, to TW (No. 19659241) and to KN (No. 22700863, No. 24501304 and No. 15K06827).

**Author Contributions:** K.N. and T.W. conceived and designed the experiments; K.N. performed the experiments and analyzed the data; T.W. contributed reagents/materials/analysis tools; K.N. and T.W. wrote the paper.

**Conflicts of Interest:** The authors declare no conflicts of interest.

## References

1. Gallo, R.C. The discovery of the first human retrovirus: HTLV-1 and HTLV-2. *Retrovirology* **2005**, *7*, 1–7.
2. Takatsuki, K. Discovery of adult T-cell leukemia. *Retrovirology* **2005**, *3*, 1–3.
3. Seiki, M.; Hattori, S.; Hirayama, Y.; Yoshida, M. Human adult T-cell leukemia virus: Complete nucleotide sequence of the provirus genome integrated in leukemia cell DNA. *Proc. Natl. Acad. Sci. USA* **1983**, *80*, 3618–3622. [CrossRef] [PubMed]
4. Johnson, J.M.; Harrod, R.; Franchini, G. Molecular biology and pathogenesis of the human T-cell leukaemia/lymphotropic virus Type-1 (HTLV-1). *Int. J. Exp. Pathol.* **2001**, *82*, 135–147. [CrossRef] [PubMed]
5. Franchini, G.; Fukumoto, R.; Fullen, J.R. T-cell control by human T-cell leukemia/lymphoma virus type 1. *Int. J. Hematol.* **2003**, *78*, 280–296. [CrossRef] [PubMed]
6. Kashanchi, F.; Brady, J.N. Transcriptional and post-transcriptional gene regulation of HTLV-1. *Oncogene* **2005**, *24*, 5938–5951. [CrossRef] [PubMed]
7. Kannian, P.; Green, P.L. Human T Lymphotropic Virus Type 1 (HTLV-1): Molecular Biology and Oncogenesis. *Viruses* **2010**, *2*, 2037–2077. [CrossRef] [PubMed]
8. Balvay, L.; Lastra, M.L.; Sargueil, B.; Darlix, J.-L.; Ohlmann, T. Translational control of retroviruses. *Nat. Rev. Microbiol.* **2007**, *5*, 128–140. [CrossRef] [PubMed]
9. Corradin, A.; DI Camillo, B.; Rende, F.; Ciminale, V.; Toffolo, G.M.; Cobelli, C. Retrovirus HTLV-1 gene circuit: A potential oscillator for eukaryotes. *Pac. Symp. Biocomput.* **2010**, *432*, 421–432.
10. Li, M.; Kesic, M.; Yin, H.; Yu, L.; Green, P.L. Kinetic analysis of human T-cell leukemia virus type 1 gene expression in cell culture and infected animals. *J. Virol.* **2009**, *83*, 3788–3797. [CrossRef] [PubMed]
11. Green, P.L.; Chen, I.S.Y. Regulation of human T cell leukemia virus expression. *FASEB J.* **1990**, *4*, 169–174. [PubMed]
12. Rende, F.; Cavallari, I.; Corradin, A.; Silic-Benussi, M.; Toulza, F.; Toffolo, G.M.; Tanaka, Y.; Jacobson, S.; Taylor, G.P.; D'Agostino, D.M.; *et al.* Kinetics and intracellular compartmentalization of HTLV-1 gene expression: Nuclear retention of HBZ mRNAs. *Blood* **2011**, *117*, 4855–4859. [CrossRef] [PubMed]
13. Nakano, K.; Watanabe, T. HTLV-1 Rex: The courier of viral messages making use of the host vehicle. *Front. Microbiol.* **2012**, *3*, 330. [CrossRef] [PubMed]
14. Ahmed, Y.F.; Gilmartin, G.M.; Hanly, S.M.; Nevins, J.R.; Greene, W.C. The HTLV-I Rex response element mediates a novel form of mRNA polyadenylation. *Cell* **1991**, *64*, 727–737. [CrossRef]
15. Ahmed, Y.F.; Hanly, S.M.; Malim, M.H.; Cullen, B.R.; Greene, W.C. Structure-function analyses of the HTLV-I Rex and HIV-1 Rev RNA response elements: Insights into the mechanism of Rex and Rev action. *Genes Dev.* **1990**, *4*, 1014–1022. [CrossRef] [PubMed]
16. Hidaka, M.; Inoue, J.; Yoshida, M.; Seiki, M. Post-transcriptional regulator (rex) of HTLV-1 initiates expression of viral structural proteins but suppresses expression of regulatory proteins. *EMBO J.* **1988**, *7*, 519–523. [PubMed]
17. Bai, X.T.; Sinha-Datta, U.; Ko, N.L.; Bellon, M.; Nicot, C. Nuclear export and expression of HTLV-I tax/rex mRNA is RxRE/Rex-dependent. *J. Virol.* **2012**, *86*, 4559–4565. [CrossRef] [PubMed]
18. Cavallari, I.; Rende, F.; Bona, M.K.; Sztuba-solinska, J.; Silic-benussi, M.; Tognon, M.; Legrice, S.F.J.; Franchini, G.; Ciminale, V. Expression of alternatively spliced human T-cell leukemia virus type 1 mRNAs is influenced by mitosis and by a novel *cis*-acting regulatory sequence. *J. Virol.* **2016**, *90*, 1486–1498.
19. Philip, S.; Zahoor, M.A.; Zhi, H.; Ho, Y.-K.; Giam, C.-Z. Regulation of human T-lymphotropic virus type I latency and reactivation by HBZ and Rex. *PLoS Pathog.* **2014**, *10*, e1004040. [CrossRef] [PubMed]
20. Hakata, Y.; Umemoto, T.; Matsushita, S.; Shida, H. Involvement of human CRM1 (exportin 1) in the export and multimerization of the Rex protein of human T-cell leukemia virus type 1. *J. Virol.* **1998**, *72*, 6602–6607. [PubMed]
21. Hakata, Y.; Yamada, M.; Shida, H. Rat CRM1 is responsible for the poor activity of human T-cell leukemia virus type 1 Rex protein in rat cells. *J. Virol.* **2001**, *75*, 11515–11525. [CrossRef] [PubMed]
22. Baydoun, H.H.; Bellon, M.; Nicot, C. HTLV-1 Yin and Yang: Rex and p30 master regulators of viral mRNA trafficking. *AIDS Rev.* **2008**, *10*, 195–204. [PubMed]
23. Younis, I.; Green, P.L. The human T-cell leukemia virus Rex protein. *Front. Biosci.* **2005**, *1*, 431–445. [CrossRef]
24. Kesic, M.; Doueiri, R.; Ward, M.; Semmes, O.J.; Green, P.L. Phosphorylation regulates human T-cell leukemia virus type 1 Rex function. *Retrovirology* **2009**, *6*, 105. [CrossRef] [PubMed]

25. Kesic, M.; Ward, M.; Semmes, O.J.; Green, P.L. Site-specific phosphorylation regulates human T-cell leukemia virus type 2 Rex function *in vivo*. *J. Virol.* **2009**, *83*, 8859–8868. [CrossRef] [PubMed]

26. Xie, L.; Kesic, M.; Yamamoto, B.; Li, M.; Younis, I.; Lairmore, M.D.; Green, P.L. Human T-cell leukemia virus type 2 Rex carboxy terminus is an inhibitory/stability domain that regulates Rex functional activity and viral replication. *J. Virol.* **2009**, *83*, 5232–5243. [CrossRef] [PubMed]

27. Adachi, Y.; Nosaki, T.; Hatanaka, M. Protein kinase inhibitor H-7 blocks accumulation of unspliced mRNA of human T-cell leukemia virus type I (HTLV-I). *Biochem. Biophys. Res. Commun.* **1990**, *169*, 469–475. [CrossRef]

28. Adachi, Y.; Copeland, T.D.; Takahashi, C.; Nosaka, T.; Ahmed, A.; Oroszlan, S.; Hatanaka, M. Phosphorylation of the Rex protein of human T-cell leukemia virus type I. *J. Biol. Chem.* **1992**, *267*, 21977–21981. [PubMed]

29. Bai, X.T.; Baydoun, H.H.; Nicot, C. HTLV-I p30: A versatile protein modulating virus replication and pathogenesis. *Mol. Aspects Med.* **2010**, *31*, 344–349. [CrossRef] [PubMed]

30. Ghorbel, S.; Sinha-Datta, U.; Dundr, M.; Brown, M.; Franchini, G.; Nicot, C. Human T-cell leukemia virus type I p30 nuclear/nucleolar retention is mediated through interactions with RNA and a constituent of the 60 S ribosomal subunit. *J. Biol. Chem.* **2006**, *281*, 37150–37158. [CrossRef] [PubMed]

31. Nicot, C.; Dundr, M.; Johnson, J.M.; Fullen, J.R.; Alonzo, N.; Fukumoto, R.; Princler, G.L.; Derse, D.; Misteli, T.; Franchini, G. HTLV-1-encoded p30II is a post-transcriptional negative regulator of viral replication. *Nat. Med.* **2004**, *10*, 197–201. [CrossRef] [PubMed]

32. Sinha-Datta, U.; Datta, A.; Ghorbel, S.; Dodon, M.D.; Nicot, C. Human T-cell lymphotrophic virus type I rex and p30 interactions govern the switch between virus latency and replication. *J. Biol. Chem.* **2007**, *282*, 14608–14615. [CrossRef] [PubMed]

33. Berneman, Z.N.; Gartenhaus, R.B.; Reitz, M.S.; Blattner, W.A.; Manns, A.; Hanchard, B.; Ikehara, O.; Gallo, R.C.; Klotman, M.E. Expression of alternatively spliced human T-lymphotropic virus type I pX mRNA in infected cell lines and in primary uncultured cells from patients with adult T-cell leukemia/lymphoma and healthy carriers. *Proc. Natl. Acad. Sci. USA* **1992**, *89*, 3005–3009. [CrossRef] [PubMed]

34. Orita, S.; Takagi, S.; Saiga, A.; Minoura, N.; Araki, K.; Kinoshita, K.; Kondo, T.; Hinuma, Y.; Igarashi, H. Human T cell leukaemia virus type 1 p21X mRNA: Constitutive expression in peripheral blood mononuclear cells of patients with adult T cell leukaemia. *J. Gen. Virol.* **1992**, *73 Pt 9*, 2283–2289. [CrossRef] [PubMed]

35. Saiga, A.; Aono, Y.; Imai, J.; Kinoshita, K.; Orita, S.; Igarashi, H. Presence of antibodies to p21X and/or p27rex proteins in sera from human T-cell leukemia virus type I-infected individuals. *J. Virol. Methods* **1996**, *57*, 157–168. [CrossRef]

36. Kiyokawa, T.; Seiki, M.; Iwashita, S.; Imagawa, K.; Shimizu, F.; Yoshida, M. T-cell leukemia virus type I p27X-III and p21x-III, proteins encoded by the pX sequence of human T-cell leukemia virus type I. *Proc. Natl. Acad. Sci. USA* **1985**, *82*, 8359–8363. [CrossRef] [PubMed]

37. Orita, S.; Kobayashi, H.; Aono, Y.; Saiga, A.; Maeda, M.; Igarashi, H. p21X mRNA is expressed as a singly spliced pX transcript from defective provirus genomes having a partial delection of the pol-env region in human T-cell leukemia virus type 1-infected cells. *Nucleic Acids Res.* **1993**, *21*, 3799–3807. [CrossRef] [PubMed]

38. Dickson, A.M.; Wilusz, J. Strategies for viral RNA stability: Live long and prosper. *Trends Genet.* **2011**, *27*, 286–293. [CrossRef] [PubMed]

39. Garcia, D.; Garcia, S.; Voinnet, O. Nonsense-Mediated Decay Serves as a General Viral Restriction Mechanism in Plants. *Cell Host Microbe* **2014**, *16*, 391–402. [CrossRef] [PubMed]

40. Balistreri, G.; Horvath, P.; Schweingruber, C.; Zünd, D.; McInerney, G.; Merits, A.; Mühlemann, O.; Azzalin, C.; Helenius, A. The host nonsense-mediated mRNA decay pathway restricts mammalian RNA virus replication. *Cell Host Microbe* **2014**, *16*, 403–411. [CrossRef] [PubMed]

41. Quek, B.L.; Beemon, K. Retroviral strategy to stabilize viral RNA. *Curr. Opin. Microbiol.* **2014**, *18*, 78–82. [CrossRef] [PubMed]

42. Ramage, H.R.; Kumar, G.R.; Verschueren, E.; Johnson, J.R.; Von Dollen, J.; Johnson, T.; Newton, B.; Shah, P.; Horner, J.; Krogan, N.J.; *et al.* A combined proteomics/genomics approach links hepatitis C virus infection with nonsense-mediated mRNA decay. *Mol. Cell* **2015**, *57*, 329–340. [CrossRef] [PubMed]

43. Popp, M.; Maquat, L. Organizing principles of mammalian nonsense-mediated mRNA decay. *Annu. Rev. Genet.* **2013**, *47*, 139–165. [CrossRef] [PubMed]

44. Kervestin, S.; Jacobson, A. NMD: A multifaceted response to premature translational termination. *Nat. Rev. Mol. Cell Biol.* **2012**, *13*, 700–712. [CrossRef] [PubMed]

45. Nakano, K.; Ando, T.; Yamagishi, M.; Yokoyama, K.; Ishida, T.; Ohsugi, T.; Tanaka, Y.; Brighty, D.W.; Watanabe, T. Viral interference with host mRNA surveillance, the nonsense-mediated mRNA decay (NMD) pathway, through a new function of HTLV-1 Rex: Implications for retroviral replication. *Microbes Infect.* **2013**, *15*, 491–505. [CrossRef] [PubMed]

46. Kanamori, H.; Kodama, T.; Matsumoto, A.; Itakura, H.; Yazaki, Y. Stabilization od interleukin-2 receptor a chain mRNA by HTLV-1 Rex in mouse L cells: Lower amounts of Rex do not stabilize the mRNA. *Biochem. Biophys. Res. Commun.* **1994**, *198*, 243–250.

47. White, K.N.; Nosaka, T.; Kanamori, H.; Hatanaka, M.; Honjo, T. The nucleolar localisation signal of the HTLV-1 protein p27Rex is important for stabilisation of IL-2 receptor α subunit mRNA by p27Rex. *Biochem. Biophys. Res. Commun.* **1991**, *175*, 98–103. [CrossRef]

48. Kornblihtt, A.R.; Schor, I.E.; Alló, M.; Dujardin, G.; Petrillo, E.; Muñoz, M.J. Alternative splicing: A pivotal step between eukaryotic transcription and translation. *Nat. Rev. Mol. Cell Biol.* **2013**, *14*, 153–165. [CrossRef] [PubMed]

49. Gröne, M.; Koch, C.; Grassmann, R. The HTLV-1 Rex protein induces nuclear accumulation of unspliced viral RNA by avoiding intron excision and degradation. *Virology* **1996**, *218*, 316–325. [CrossRef] [PubMed]

50. Karni, R.; de Stanchina, E.; Lowe, S.W.; Sinha, R.; Mu, D.; Krainer, A.R. The gene encoding the splicing factor SF2/ASF is a proto-oncogene. *Nat. Struct. Mol. Biol.* **2007**, *14*, 185–193. [CrossRef] [PubMed]

51. Princler, G.; Julias, J.G.; Hughes, S.H.; Derse, D. Roles of viral and cellular proteins in the expression of alternatively spliced HTLV-1 pX mRNAs. *Virology* **2003**, *317*, 136–145. [CrossRef] [PubMed]

52. Powell, D.M.; Amaral, M.C.; Wu, J.Y.; Maniatis, T.; Greene, W.C. HIV Rev-dependent binding of SF2/ASF to the Rev response element: Possible role in Rev-mediated inhibition of HIV RNA splicing. *Proc. Natl. Acad. Sci. USA* **1997**, *94*, 973–978. [CrossRef] [PubMed]

53. Tange, T.O.; Jensen, T.H.; Kjems, J. In vitro interaction between human immunodeficiency virus type 1 Rev protein and splicing factor ASF/SF2-associated protein, p32. *J. Biol. Chem.* **1996**, *271*, 10066–10072. [PubMed]

54. Jean-Philippe, J.; Paz, S.; Caputi, M. hnRNP A1: The Swiss army knife of gene expression. *Int. J. Mol. Sci.* **2013**, *14*, 18999–19024. [CrossRef] [PubMed]

55. Huelga, S.C.; Vu, A.Q.; Arnold, J.D.; Liang, T.Y.; Liu, P.P.; Yan, B.Y.; Donohue, J.P.; Shiue, L.; Hoon, S.; Brenner, S.; *et al.* Integrative genome-wide analysis reveals cooperative regulation of alternative splicing by hnRNP proteins. *Cell Rep.* **2012**, *1*, 167–178. [CrossRef] [PubMed]

56. Hamaia, S.; Casse, H.; Gazzolo, L.; Duc Dodon, M. The human T-cell leukemia virus type 1 Rex regulatory protein exhibits an impaired functionality in human lymphoblastoid Jurkat T cells. *J. Virol.* **1997**, *71*, 8514–8521. [PubMed]

57. Dodon, M.D.; Hamaia, S.; Martin, J.; Gazzolo, L. Heterogeneous nuclear ribonucleoprotein A1 interferes with the binding of the human T cell leukemia virus type 1 Rex regulatory protein to its response element. *J. Biol. Chem.* **2002**, *277*, 18744–18752. [CrossRef] [PubMed]

58. Kress, E.; Baydoun, H.H.; Bex, F.; Gazzolo, L.; Dodon, M.D. Critical role of hnRNP A1 in HTLV-1 replication in human transformed T lymphocytes. *Retrovirology* **2005**, *2*, 8. [CrossRef] [PubMed]

59. Weil, R.; Levraud, J.P.; Dodon, M.D.; Bessia, C.; Hazan, U.; Kourilsky, P.; Israël, A. Altered expression of tyrosine kinases of the Src and Syk families in human T-cell leukemia virus type 1-infected T-cell lines. *J. Virol.* **1999**, *73*, 3709–3717. [PubMed]

60. Picard, C.; Gabert, J.; Olive, D.; Collette, Y. Altered splicing in hematological malignancies reveals a tissue-specific translational block of the Src-family tyrosine kinase fyn brain isoform expression. *Leukemia* **2004**, *18*, 1737–1739. [CrossRef] [PubMed]

61. Zhao, R.Y.; Elder, R.T. Viral infections and cell cycle G2/M regulation. *Cell Res.* **2005**, *15*, 143–149. [CrossRef] [PubMed]

62. Andersen, J.L.; Le Rouzic, E.; Planelles, V. HIV-1 Vpr: Mechanisms of G2 arrest and apoptosis. *Exp. Mol. Pathol.* **2008**, *85*, 2–10. [CrossRef] [PubMed]

63. Guenzel, C.A.; Hérate, C.; Benichou, S. HIV-1 Vpr-a still "enigmatic multitasker". *Front. Microbiol.* **2014**, *5*, 127. [CrossRef] [PubMed]

64. Boutros, R.; Lobjois, V.; Ducommun, B. CDC25 phosphatases in cancer cells: Key players? Good targets? *Nat. Rev. Cancer* **2007**, *7*, 495–507. [CrossRef] [PubMed]

65. Bruinsma, W.; Macurek, L.; Freire, R.; Lindqvist, A.; Medema, R.H. Bora and Aurora-A continue to activate Plk1 in mitosis. *J. Cell Sci.* **2014**, *127*, 801–811. [CrossRef] [PubMed]

66. Goh, W.C.; Manel, N.; Emerman, M. The human immunodeficiency virus Vpr protein binds Cdc25C: Implications for G2 arrest. *Virology* **2004**, *318*, 337–349. [CrossRef] [PubMed]

67. Lai, M.; Zimmerman, E.S.; Planelles, V.; Chen, J.; Irol, J.V. Activation of the ATR pathway by human immunodeficiency virus type 1 Vpr involves its direct binding to chromatin *in vivo*. *J. Virol.* **2005**, *79*, 15443–15451. [CrossRef] [PubMed]

68. De Noronha, C.M.C.; Sherman, M.P.; Lin, H.W.; Cavrois, M.V.; Moir, R.D.; Goldman, R.D.; Greene, W.C. Dynamic disruptions in nuclear envelope architecture and integrity induced by HIV-1 Vpr. *Science* **2001**, *294*, 1105–1109. [CrossRef] [PubMed]

69. Guerrero, S.; Batisse, J.; Libre, C.; Bernacchi, S.; Marquet, R.; Paillart, J.-C. HIV-1 replication and the cellular eukaryotic translation apparatus. *Viruses* **2015**, *7*, 199–218. [CrossRef] [PubMed]

70. Sharma, A.; Yilmaz, A.; Marsh, K.; Cochrane, A.; Boris-Lawrie, K. Thriving under stress: Selective translation of HIV-1 structural protein mRNA during Vpr-mediated impairment of eIF4E translation activity. *PLoS Pathog.* **2012**, *8*, e1002612. [CrossRef] [PubMed]

71. Brégnard, C.; Benkirane, M.; Laguette, N. DNA damage repair machinery and HIV escape from innate immune sensing. *Front. Microbiol.* **2014**, *5*, 176. [PubMed]

72. Haoudi, A.; Daniels, R.C.; Wong, E.; Kupfer, G.; Semmes, O.J. Human T-cell leukemia virus-I tax oncoprotein functionally targets a subnuclear complex involved in cellular DNA damage-response. *J. Biol. Chem.* **2003**, *278*, 37736–37744. [CrossRef] [PubMed]

73. Gupta, S.K.; Guo, X.; Durkin, S.S.; Fryrear, K.F.; Ward, M.D.; Semmes, O.J. Human T-cell leukemia virus type 1 Tax oncoprotein prevents DNA damage-induced chromatin egress of hyperphosphorylated Chk2. *J. Biol. Chem.* **2007**, *282*, 29431–29440. [CrossRef] [PubMed]

74. Park, H.U.; Jeong, J.-H.; Chung, J.H.; Brady, J.N. Human T-cell leukemia virus type 1 Tax interacts with Chk1 and attenuates DNA-damage induced G2 arrest mediated by Chk1. *Oncogene* **2004**, *23*, 4966–4974. [CrossRef] [PubMed]

75. Fu, D.-X.; Kuo, Y.-L.; Liu, B.-Y.; Jeang, K.-T.; Giam, C.-Z. Human T-lymphotropic virus type I tax activates I-kappa B kinase by inhibiting I-kappa B kinase-associated serine/threonine protein phosphatase 2A. *J. Biol. Chem.* **2003**, *278*, 1487–1493. [CrossRef] [PubMed]

76. Anupam, R.; Datta, A.; Kesic, M.; Green-Church, K.; Shkriabai, N.; Kvaratskhelia, M.; Lairmore, M.D. Human T-lymphotropic virus type 1 p30 interacts with REGgamma and modulates ATM (ataxia telangiectasia mutated) to promote cell survival. *J. Biol. Chem.* **2011**, *286*, 7661–7668. [CrossRef] [PubMed]

77. Wickramasinghe, V.O.; Laskey, R. a Control of mammalian gene expression by selective mRNA export. *Nat. Rev. Mol. Cell Biol.* **2015**, *16*, 431–442. [CrossRef] [PubMed]

78. Topisirovic, I.; Siddiqui, N.; Lapointe, V.L.; Trost, M.; Thibault, P.; Bangeranye, C.; Piñol-Roma, S.; Borden, K.L.B. Molecular dissection of the eukaryotic initiation factor 4E (eIF4E) export-competent RNP. *EMBO J.* **2009**, *28*, 1087–1098. [CrossRef] [PubMed]

79. Culjkovic, B.; Topisirovic, I.; Skrabanek, L.; Ruiz-Gutierrez, M.; Borden, K.L.B. eIF4E promotes nuclear export of cyclin D1 mRNAs via an element in the 3'UTR. *J. Cell Biol.* **2005**, *169*, 245–256. [CrossRef] [PubMed]

80. Culjkovic, B.; Topisirovic, I.; Skrabanek, L.; Ruiz-Gutierrez, M.; Borden, K.L.B. eIF4E is a central node of an RNA regulon that governs cellular proliferation. *J. Cell Biol.* **2006**, *175*, 415–426. [CrossRef] [PubMed]

81. Vandamme, A.-M.; Salemi, M.; Desmyter, J. The simian origins of the pathogenic human T-cell lymphotropic virus type I. *Trends Microbiol.* **1998**, *6*, 477–483. [CrossRef]

82. Rende, F.; Cavallari, I.; Romanelli, M.G.; Diani, E.; Bertazzoni, U.; Ciminale, V. Comparison of the genetic organization, expression strategies and oncogenic potential of HTLV-1 and HTLV-2. *Leuk. Res. Treat.* **2012**, *2012*, 1–14. [CrossRef] [PubMed]

83. Suhasini, M.; Reddy, T.R. Cellular proteins and HIV-1 Rev function. *Curr. HIV Res.* **2009**, *7*, 91–100. [CrossRef] [PubMed]

84. Adachi, Y.; Copeland, T.D.; Hatanakall, M.; Oroszlansii, S. Nucleolar targeting signal of Rex protein of human T-cell leukemia virus type I specifically binds to nucleolar shuttle protein B-23. *J. Biol. Chem.* **1993**, *268*, 13930–13934. [PubMed]

85. Palmeri, D.; Malim, M.H. Importin β can mediate the nuclear import of an arginine-rich nuclear localization signal in the absence of importin α. *Mol. Cell. Biol.* **1999**, *19*, 1218–1225. [CrossRef] [PubMed]

86. Truant, R.; Cullen, B.R. The Arginine-rich domains present in human immunodeficiency virus type 1 Tat and Rev function as direct importin β -dependent nuclear localization signals. *Mol. Cell. Biol.* **1999**, *19*, 1210–1217. [CrossRef] [PubMed]

87. Yoneda, Y. Nucleocytoplasmic protein trafic and its significance to cell function. *Genes Cells* **2000**, *5*, 777–787. [CrossRef] [PubMed]

88. Zapp, M.L.; Hope, T.J.; Parslow, T.G.; Green, M.R. Oligomerization and RNA binding domains of the type 1 human immunodeficiency virus Rev protein: A dual function for an arginine-rich binding motif. *Proc. Natl. Acad. Sci. USA* **1991**, *88*, 7734–7738. [CrossRef] [PubMed]

89. Zemmel, R.W.; Kelley, A.C.; Karn, J.; Butler, P.J. Flexible regions of RNA structure facilitate co-operative Rev assembly on the Rev-response element. *J. Mol. Biol.* **1996**, *258*, 763–777. [CrossRef] [PubMed]

90. Jiang, F.; Gorin, A.; Hu, W.; Majumdar, A.; Baskerville, S.; Xu, W.; Ellington, A.; Patel, D.J. Anchoring an extended HTLV-1 Rex peptide within an RNA major groove containing junctional base triples. *Structure* **1999**, *7*, 1461–1472. [CrossRef]

91. Daugherty, M.D.; D'Orso, I.; Frankel, A.D. A Solution to limited genomic capacity: Using adaptable binding surfaces to assemble the functional HIV Rev oligomer on RNA. *Mol. Cell* **2008**, *31*, 824–834. [CrossRef] [PubMed]

92. DiMattiaa, M.A.; Wattsc, N.R.; Stahlc, S.J.; Raderd, C.; Wingfieldc, P.T.; Stuarta, D.I.; Stevenb, A.C.; Grimes, J.M. Implications of the HIV-1 Rev dimer structure at 3.2 Å resolution for multimeric binding to the Rev response element. *Proc. Natl. Acad. Sci. USA* **2010**, *107*, 5810–5814. [CrossRef] [PubMed]

93. Daugherty, M.; Liu, B.; Frankel, A. Structural basis for cooperative RNA binding and export complex assembly by HIV Rev. *Nat. Struct. Mol. Biol.* **2010**, *17*, 1337–1343. [CrossRef] [PubMed]

94. Fang, X.; Wang, J.; O'Carroll, I.P.; Mitchell, M.; Zuo, X.; Wang, Y.; Yu, P.; Liu, Y.; Rausch, J.W.; Dyba, M.A.; *et al.* An unusual topological structure of the HIV-1 rev response element. *Cell* **2013**, *155*, 594–605. [CrossRef] [PubMed]

95. Nawroth, I.; Mueller, F.; Basyuk, E.; Beerens, N.; Rahbek, U.L.; Darzacq, X.; Bertrand, E.; Kjems, J.; Schmidt, U. Stable assembly of HIV-1 export complexes occurs cotranscriptionally. *RNA* **2014**, *20*, 1–8. [CrossRef] [PubMed]

96. Bai, Y.; Tambe, A.; Zhou, K.; Doudna, J.A.; States, U.; Division, P.B.; Berkeley, L. RNA-guided assembly of Rev-RRE nuclear export complexes. *Elife* **2014**, *3*, e03656. [CrossRef] [PubMed]

97. Booth, D.S.; Cheng, Y.; Frankel, A.D. The export receptor Crm1 forms a dimer to promote nuclear export of HIV RNA. *Elife* **2014**, *3*, e04121. [CrossRef] [PubMed]

98. Jayaraman, B.; Crosby, D.C.; Homer, C.; Ribeiro, I.; Mavor, D.; Frankel, A.D. RNA-directed remodeling of the HIV-1 protein Rev orchestrates assembly of the Rev-Rev response element complex. *Elife* **2014**, *3*, e04120. [CrossRef] [PubMed]

99. Simonis, N.; Rual, J.-F.; Lemmens, I.; Boxus, M.; Hirozane-Kishikawa, T.; Gatot, J.-S.; Dricot, A.; Hao, T.; Vertommen, D.; Legros, S.; *et al.* Host-pathogen interactome mapping for HTLV-1 and 2 retroviruses. *Retrovirology* **2012**, *9*, 26. [CrossRef] [PubMed]

100. Abe, M.; Suzuki, H.; Nishitsuji, H.; Shida, H.; Takaku, H. Interaction of human T-cell lymphotropic virus type I Rex protein with Dicer suppresses RNAi silencing. *FEBS Lett.* **2010**, *584*, 4313–4318. [CrossRef] [PubMed]

*viruses*

MDPI

Article

# Distinct Morphology of Human T-Cell Leukemia Virus Type 1-Like Particles

José O. Maldonado [1,†], Sheng Cao [2,†,‡], Wei Zhang [3,*] and Louis M. Mansky [2,*]

[1] Institute for Molecular Virology & DDS-PhD Dual Degree Program, 18-242 Moos Tower, 515 Delaware Street SE, Minneapolis, MN 55455, USA; jmaldo@umn.edu

[2] Institute for Molecular Virology, 18-242 Moos Tower, 515 Delaware Street SE, Minneapolis, MN 55455, USA; caosheng@wh.iov.cn

[3] Institute for Molecular Virology & Characterization Facility, 18-242 Moos Tower, 515 Delaware Street SE, Minneapolis, MN 55455, USA

* Correspondence: zhangwei@umn.edu (W.Z.); mansky@umn.edu (L.M.M.); Tel.: +1-612-624-1996 (W.Z.); +1-612-626-5525 (L.M.M.)

† These authors contributed equally to this work.

‡ Current address: Wuhan Institute of Virology, Chinese Academy of Science, Wuhan, China

Academic Editor: Eric O. Freed

Received: 9 March 2016; Accepted: 21 April 2016; Published: 11 May 2016

**Abstract:** The Gag polyprotein is the main retroviral structural protein and is essential for the assembly and release of virus particles. In this study, we have analyzed the morphology and Gag stoichiometry of human T-cell leukemia virus type 1 (HTLV-1)-like particles and authentic, mature HTLV-1 particles by using cryogenic transmission electron microscopy (cryo-TEM) and scanning transmission electron microscopy (STEM). HTLV-1-like particles mimicked the morphology of immature authentic HTLV-1 virions. Importantly, we have observed for the first time that the morphology of these virus-like particles (VLPs) has the unique local feature of a flat Gag lattice that does not follow the curvature of the viral membrane, resulting in an enlarged distance between the Gag lattice and the viral membrane. Other morphological features that have been previously observed with other retroviruses include: (1) a Gag lattice with multiple discontinuities; (2) membrane regions associated with the Gag lattice that exhibited a string of bead-like densities at the inner leaflet; and (3) an arrangement of the Gag lattice resembling a railroad track. Measurement of the average size and mass of VLPs and authentic HTLV-1 particles suggested a consistent range of size and Gag copy numbers in these two groups of particles. The unique local flat Gag lattice morphological feature observed suggests that HTLV-1 Gag could be arranged in a lattice structure that is distinct from that of other retroviruses characterized to date.

**Keywords:** deltaretrovirus; lentivirus; virus assembly

## 1. Introduction

Approximately 10–20 million people are infected with human T-cell leukemia virus type 1 (HTLV-1) worldwide [1,2]. HTLV-1 is a deltaretrovirus and is associated with adult T-cell leukemia/lymphoma, tropical spastic paraparesis, as well as HTLV-1-associated myelopathy [3,4]. These diseases are prevalent in places highly endemic for HTLV-1 infection such as southwestern Japan, central Africa, South America and the Caribbean. Despite the association of HTLV-1 with cancer and its significant impact on human health and well-being, the molecular mechanisms of viral replication, virus particle assembly and morphology remain poorly understood due to difficulties in propagating the virus in tissue culture.

Like other retroviruses, the assembly and budding of HTLV-1 particles is directed by the viral Gag polyprotein (recently reviewed by Maldonado *et al.* [5]). Briefly, HTLV-1 Gag molecules translocate to the plasma membrane (PM) soon after the protein is synthesized [6]. A previous study with human immunodeficiency virus type 1 (HIV-1) suggested that the viral RNA is recruited to the PM by Gag and serves as a platform to promote Gag-Gag interactions, allowing Gag to form higher order oligomers in immature particles [7]. Infectious virions are produced via a maturation process that occurs either concomitantly with or after budding of the immature virus. During virus maturation, the viral protease (PR) cleaves the Gag polyprotein into three structural proteins: matrix (MA), which remains associated with the inner leaflet of the viral membrane; capsid (CA), which organizes into a closed protein shell to package the genomic RNA; and nucleocapsid (NC) which is in complex with the viral genome.

The diameters of retrovirus particles are typically variable and commonly appear to form a normal distribution [8–16]. Calculations of the average Gag copy number per virus particle vary somewhat depending on the methods used for the measurement as well as on the type of retrovirus being analyzed. Scanning transmission electron microscopy (STEM) has previously been used successfully to determine the average Gag copy number per particle [9,10,15–18]. This method estimates the mass of the whole virus particle. Since the majority of the virus particle mass is contributed by Gag, the mass of the entire particle has been used for calculating the Gag stoichiometry. Multiple studies have reported varying Gag copy numbers, ranging from approximately 750 to 5000, which coincide with varying virus particle size distributions [10,12,13,15–19]. To date, there are no reported studies on Gag stoichiometry that would be present in the immature precursors of authentic, mature HTLV-1 particles. Determination of Gag stoichiometry is critical to understanding the mechanisms of HTLV-1 replication, for this information assists in the interpretation of HTLV-1 particle structures, and helps in determining the copy number of other viral proteins in the virus particle (e.g., Pol).

In this study, a comparative analysis of HTLV-1-like particles and authentic, mature HTLV-1 particles was performed by cryogenic transmission electron microscopy (cryo-TEM) and scanning transmission electron microscopy (STEM). These findings provide the first demonstration of the morphology of these virus-like particles (VLPs) having the unique feature of local flat Gag lattice regions that did not follow the curvature of the viral membrane and had an enlarged distance toward the membrane. Morphological features similar to that observed with other retroviruses [20] include (1) a Gag lattice with multiple discontinuities; (2) a string of bead-like densities at the inner leaflet that is associated with the Gag lattice; and (3) a Gag lattice resembling a railroad track. We also demonstrate that HTLV-1-like particles and authentic mature HTLV-1 particles possess a consistent size and Gag stoichiometry.

## 2. Materials and Methods

### 2.1. Transfection and HTLV-1-Like Particle Production

A codon-optimized HTLV-1 *gag* gene expression construct (pN3 HTLV-1 Gag, Figure 1A) was created in a similar manner to that of a previously described construct in which the yellow fluorescence protein (YFP) was fused to the carboxy-terminus of Gag (pEYFP-N3 HTLV-1 Gag) [21]. The new *gag* gene which does not have a YFP tag was synthesized with an optimal Kozak consensus sequence at the 5′ end of the gene: GCCACC**ATG**G (start codon in bold and underlined) (Figure 1A). In order to produce VLPs, six 10 cm tissue culture dishes each containing $2.2 \times 10^6$ human embryonic kidney 293T cells in 6 mL of Dulbecco's Modified Eagle Medium (DMEM) supplemented with 10% Fetal Clone III were co-transfected with the pN3-HTLV-1 Gag expression construct along with an HTLV-1 envelope protein expression construct (ratio of 10:1) using GeneJet (SignaGen, Gaithersburg, MD, USA) following the manufacturer's instructions. Twenty-four hours post-transfection, 2 mL of fresh media were added to each plate and incubated for additional 24 h at 37 °C in 5% $CO_2$. To harvest VLPs, cell culture supernatants from transfected cells were centrifuged at $3000 \times g$ for 5 min to remove large cellular debris and then filtered through a 0.2 μm filter. The samples were then concentrated and purified in the same manner as with authentic particles.

A.#    B.#    C.#

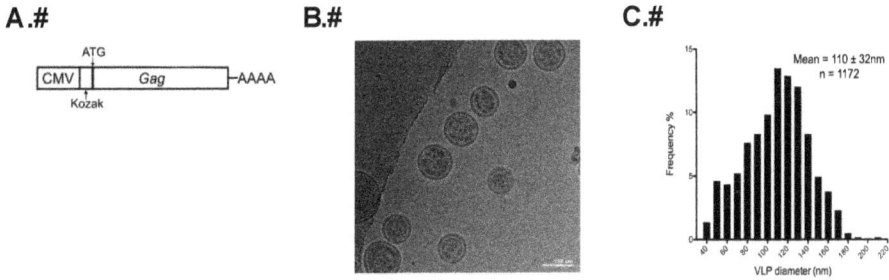

**Figure 1.** Analysis of the diameter and morphology of human T-cell leukemia virus type 1 (HTLV-1) virus-like particles (VLPs) by transmission electron microscopy (TEM). (**A**) HTLV-1-like particle expression construct. A codon-optimized Gag expression construct (pN3 HTLV-1 Gag) with a Kozak sequence was used to produce HTLV-1 VLPs; (**B**) Representative micrograph of HTLV-1-like particles of different sizes and morphology; (**C**) Size distribution of HTLV-1-like VLPs.

## 2.2. Gradient Purification of Authentic Virus Particles and VLPs

Authentic HTLV-1 particles were produced from MT-2 cells, a T-cell line chronically infected with HTLV-1, which was obtained from Dr. Douglas Richman through the NIH AIDS Reagent Program, Division of Acquired Immune Deficiency Syndrome (AIDS), National Institute of Allergy and Infectious Diseases (NIAID), National Institutes of Health (NIH) [22,23]. MT-2 cells were grown in two T-75 flasks with up to 60 mL of Roswell Park Memorial Institute (RPMI) 1640 medium supplemented with 10% Fetal Clone III, for ~10 days. After the cells reached about 90% confluency, which was indicative by the formation of large cell clumps, virus particles were harvested and centrifuged at $3000\times g$ for 5 min to remove large cellular debris and then filtered through a 0.2 μm filter.

The concentrated particles (*i.e.*, authentic virus particles or VLPs) were then ultracentrifuged through an 8% OptiPrep (60% iodixanol in water with a density of 1.32 g/mL, (Sigma-Aldrich, St. Louis, MO, USA) cushion at $109,000\times g$ for 1.5 h in a 50.1 Ti rotor (Beckman, Brea, CA, USA) at 4 °C. The particle pellet was resuspended in 0.5 mL of $1\times$ STE buffer (100 mM NaCl , 10 mM Tris-Cl, pH 7.4, 1 mM sodium chloride-Tris-ethylenediaminetetraacetic acid (EDTA), and overlaid onto a 4 mL 10%–40% OptiPrep gradient and centrifuged to equilibrium in a SW55 Ti rotor (Beckman) at $250,000\times g$ for 3 h at 4 °C. The virus- or VLP-containing fraction, at about 20% OptiPrep, was removed from the gradient using a hypodermic needle. The collected virus particles were diluted 10 fold in $1\times$ STE and pelleted at $195,000\times g$ for 1 h in a SW55 Ti rotor at 4 °C. Following centrifugation, the pellet was re-suspended in ~15 μL of $1\times$ STE at 4 °C overnight and then analyzed by cryo-TEM or STEM. The compound 2, 2′-dithiodipyridine (aldrithiol-2; AT-2) was used to inactivate authentic HTLV-1 infectivity prior to cryo-TEM or STEM analysis as previously described [24].

## 2.3. Cryo-TEM of HTLV-1-Like Particles and Authentic Virus Particles

Virus and VLP samples were prepared for cryo-TEM as previously described [21]. Briefly, 3 μL concentrated virus or VLP sample was applied to a glow-discharged c-flat holey carbon grid (Ted Pella, Redding, CA, USA) and then blotted with filter paper to remove the sample excess. The grid was then plunge frozen into liquid ethane [25] with a FEI MarkIII Vitrobot system (FEI Company, Hillsboro, OR, USA). The frozen grids were then transferred to a FEI TF30 field emission gun transmission electron microscope at liquid nitrogen temperature (FEI Company). Images were then recorded at a nominal magnification of 39,000_x and 59,000_x at low-dose (~30 electrons/$\text{Å}^2$) and 1 to 5 μm under focus conditions using a Gatan 4 k by 4 k CCD camera (Gatan Inc., Pleasanton, CA, USA).

*2.4. Determination of Particle Size*

Cryo-TEM images were analyzed by using ImageJ software (Version 1.49c, NIH, Bethesda, MD, USA). For each virus particle or VLP analyzed, two perpendicular diameters were used to calculate the average diameter [21]. Histograms of particle diameters were generated by using GraphPad Prism 6 software (Version 6.0c, GraphPad, La Jolla, CA, USA).

*2.5. Determination of Particle Mass by STEM*

The mass of virus particles or VLPs was determined by quantitative dark-field STEM, which was developed at the Brookhaven National Laboratory (BNL, Upton, NY, USA) [26]. This method allows for the study of individual unstained virus particles with minimal radiation damage. The particle sample was first mixed with tobacco mosaic virus (TMV) particles that were used as an internal control. Then the mixture was applied onto a thin-carbon transmission electronic microscopy (TEM) grid, extensively washed, blotted and freeze-dried overnight. The TEM grid was imaged under a 40 keV electron beam at $-150\,^\circ$C. The grid was first scanned with a low dose electron beam and areas with clean background were used for the final scan. Prior to the final scanning, the electron beam was focused in a nearby area to minimize radiation damage to the specimen. The low temperature and low dose imaging technique ($<500$ electrons/nm$^2$) was used to help to reduce mass loss (less than 1%) caused by electron radiation as well as to eliminate contamination from mobile hydrocarbons. Each point in the STEM image corresponds to an area of 0.625 nm$^2$ over the specimen. The whole image corresponds to 512 by 512 nm in the specimen with the center of the points separated by 1 nm. A large- and a small-angle annular dark-field detector were used to digitally record the number of scattered electrons in each scanning point. The number of scattered electrons at any scanning point is proportional to the sample mass in that local region.

STEM images were analyzed by using the PCMass software developed by the BNL STEM facility (Version 32, Brookhaven National Laboratory, Upton, New York, USA). Each virus particle or VLP in the STEM micrograph was first masked by a density profile model (*i.e.*, a sphere) in order to mimic the virus density profile. The diameter of the sphere was based on the dimension of the measured particle. The mass of each virus particle was then calculated using the sum of the electron densities within the mask and a scale factor, which was determined using the image of TMV and its mass per unit length (*i.e.*, 13.1 kDa/Å) [26,27]. The resulting histograms and graphs of particle mass distribution were generated by using GraphPad Prism 6 software (Version 6.0c, GraphPad, La Jolla, CA, USA).

## 3. Results

*3.1. Analysis of the Morphology of HTLV-1-Like Particles*

HTLV-1-like particles produced using the HTLV-1 Gag-only expression construct (Figure 1A) were observed to be spherical in shape with a mean diameter of $110 \pm 32$ nm measured from 1172 particles (Figure 1B,C). This is in contrast to a previous study using a Gag-YFP expression construct in which a mean particle diameter of $71 \pm 20$ nm was determined by cryo-TEM [21]. The electron density adjacent to the inner viral membrane was interpreted as being the immature Gag lattice. All particles with this electron density pattern were counted as Gag-containing particles. Intriguingly, many local regions of Gag assembly were observed to exhibit flat electron density features that did not strictly follow the curvature of the membrane and showed enlarged distance toward the viral membrane (4 *vs.* 8 nm) (Figure 2A,B). About 20% of particles had this morphological feature. This unique structural feature has not been reported for other retrovirus immature particles. The flat Gag density feature observed with HTLV-1-like particles suggests that HTLV-1 Gag could be arranged in a lattice structure that is distinct from that of other retroviruses characterized to date (*i.e.*, HIV-1, Mason-Pfizer monkey virus (MPMV) and Rous sarcoma virus (RSV)) [13,28,29].

Other morphological features have commonalities with other retroviruses. First, the Gag densities in the cryo-TEM images were not always continuous. In particular, smaller VLPs were observed to have multiple discontinuities in the immature Gag lattice (Figure 2A,B). The membrane regions that associate with organized Gag lattices appear to be wider and more pronounced. Between the Gag lattice and the viral membrane, a string of bead-like densities is sometimes observed (*i.e.*, in ~10% of particles analyzed) lining along the inner leaflet of the viral membrane (Figure 2E,G). These density features are likely due to the association of MA with the inner membrane of the virus particle. The MA lattice has been observed in a large membrane-enclosed multi-core structure in supernatants of HIV-1-infected cells [30]. In the cryo-TEM images (Figure 2A-G), the arrangement of the Gag molecules within the lattice is similar but not identical to that in other retrovirus immature particles such as HIV-1. The cryo-TEM image (Figure 2H) and cut-away view of the HIV-1 Gag lattice assembly in a three-dimensional (3D) reconstruction map [31,32] has two density layers: closer to the viral membrane is the CA protein layer showing an array of rod-like densities, while towards the center of the particle is the NC layer that has a continuous density. In contrast, the HTLV-1-like particles consistently display a continuous density at the region closer to the membrane inner leaflet. Underneath the continuous density layer, closer to the center of the virus particle, is an array of densities that resembled a railroad track (Figure 2H).

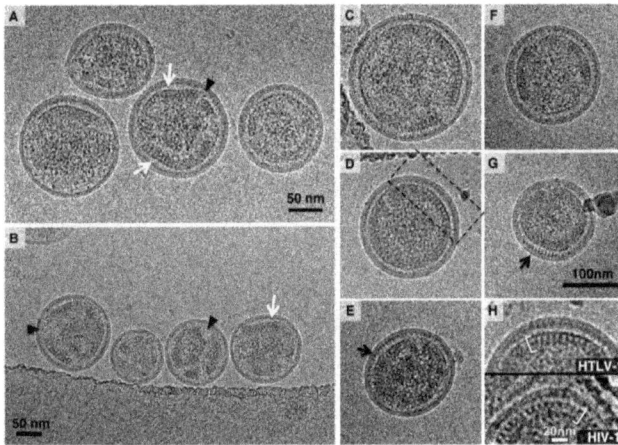

**Figure 2.** Cryogenic transmission electron microscopy (Cryo-TEM) images of HTLV-1-like particles and comparison of Gag lattice between HTLV-1 and human immunodeficiency virus type 1 (HIV-1). (**A–G**) Cryo-TEM images of HTLV-1-like particles. The white arrows indicate regions of the Gag lattice that appear flat in contrast to the curvature observed with the viral membrane. The black arrows show the membrane regions that are associated with Gag lattice and exhibit a string of bead-like densities in the inner membrane leaflet. The black arrowheads demark discontinuity of the Gag lattice. The black dash-lined box in D shows a region displayed in the top panel of H. The scale bar in G is applicable to the panels C–G; (**H**) Comparison of Gag lattice morphology between HTLV-1-like and HIV-1-like particles. The electron densities representing the Gag lattice structure are indicated by the left and right bracket, respectively.

### 3.2. Morphology of Authentic HTLV-1 Mature Particles Produced from MT-2 Cells

As a control and to confirm our previous studies with authentic HTLV-1 particles by cryo-electron tomography [33], the morphology of authentic HTLV-1 particles, harvested and purified from MT-2 cells [22,23], was also studied by cryo-TEM (Figure 3A). Mature virus particles were identified by either the presence of readily observable electron-dense cores, or by the presence of significant electron density within the particle. The lack of HTLV-1 protease inhibitors prevented the production of large numbers of authentic immature particles. Vesicles were identified by the absence of core structures or significant internal electron density. The particles were primarily spherical and heterogeneous in size. The particle diameter was determined by averaging the longest and shortest measurements of each particle. A total of 1074 authentic particles were measured and had a mean diameter of $113 \pm 23$ nm (Figure 3A,B). This measurement based on two-dimensional (2D) cryo-TEM images was in good agreement with our previous analyses using the cryo-electron tomography method [33].

**A.#**         **B.#**         **C.#**

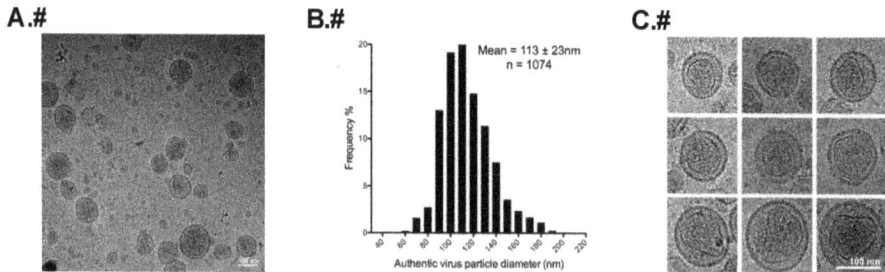

**Figure 3.** Analysis of the diameter of authentic mature HTLV-1 virus particles. (**A**) Authentic mature HTLV-1 particles produced from MT-2 cells; (**B**) Size distribution of authentic mature HTLV-1 particles; (**C**) Magnified images of authentic mature HTLV-1 particles showing irregular polyhedral-like core structures. The scale bars in A and C are 100 nm.

A gallery of cryo-TEM images of authentic HTLV-1 particles (Figure 3C) revealed that the particles contained an unordered polyhedral-like capsid core structure, which is different in each particle regardless of particle size. The core size varied by particle, with some regions of the protein capsid of the cores following the curvature of the inner leaflet of the viral lipid bilayer, while other parts of the capsid appeared completely separated from the viral membrane.

### 3.3. STEM Analyses of HTLV-1-Like Particles and Authentic Mature HTLV-1 Particles

STEM analysis was used to determine the total molecular mass of HTLV-1 particles as previously described [26]. Representative dark-field electron micrographs of HTLV-1-like particles and authentic mature HTLV-1 particles are shown in Figures 4A and 5A. Only isolated intact particles that were of the expected particle diameter range, as determined by cryo-TEM imaging, were used for mass measurements. Some smaller randomly distributed contaminants are visible in the background. Using the known mass of TMV as an internal control, we are able to obtain the average masses of HTLV-1 VLPs and authentic particles.

Both HTLV-1-like particles and authentic particles showed a wide distribution of mass diversity, which correlates with the wide particle size distribution (Figures 4B and 5B). The TMV-corrected masses of HTLV-1-like particles (Figure 4B) and HTLV-1 authentic particles (Figure 5B) were determined to be $174 \pm 96$ MDa and $204 \pm 67$ MDa, respectively (Table 1). The average masses were used for estimating the Gag copy numbers in HTLV-1-like particles and inferred immature precursors of the authentic HTLV-1 particles. AT-2 was used to inactivate the particles. It is formally possible that AT-2 treatment could affect particle morphology.

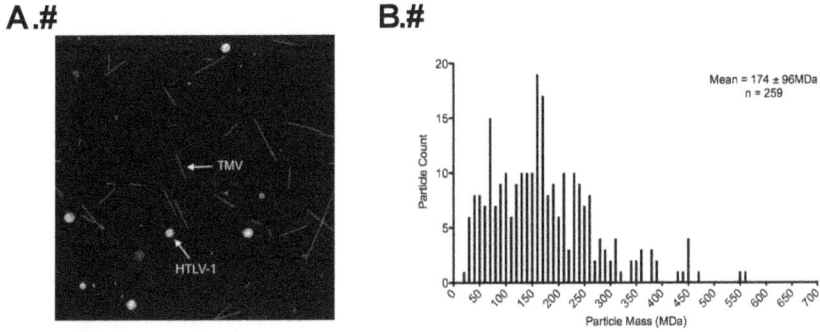

**Figure 4.** Scanning transmission electron microscopy (STEM) analysis of HTLV-1-like particles. (**A**) A STEM micrograph of HTLV-1-like particles mixed with tobacco mosaic virus (TMV); (**B**) TMV-corrected mass measurement distribution in MDa of purified HTLV-1-like particles, which was determined based on the known TMV mass per unit length of 13.1 kDa/Å.

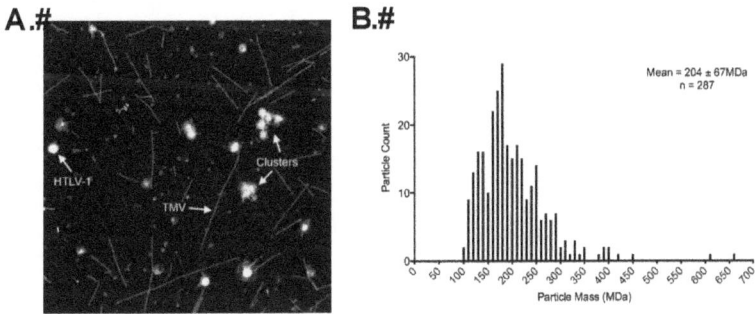

**Figure 5.** STEM analysis of authentic mature HTLV-1 virus particles. (**A**) A STEM micrograph of authentic HTLV-1 particles mixed with TMV. The region labeled as "clusters" represents closely associated viral particles and is excluded from the calculation; (**B**) The TMV-corrected measurement of mass distribution in MDa of purified authentic mature HTLV-1 particles was determined. The TMV-corrected particle mass determination was based on the known TMV mass per unit length of 13.1 kDa/Å.

**Table 1.** Summary of the mass determinations and the calculated Gag copy number per particle in human T-cell leukemia virus type 1 (HTLV-1)-like particles and authentic HTLV-1 particles.

| Measurement | | HTLV-1 Particle Sample | |
|---|---|---|---|
| | | Virus-Like Particle | Authentic Particle |
| Average Diameter (nm) [a] | | 110 | 113 |
| Average Particle Mass (MDa) [b] | | 174 | 204 |
| Mass of RNA, Lipid and Protein (MDa) | RNA [c] | 7 | 7 |
| | Lipid [d] | 70 | 80 |
| | Total protein [e] | 97 | 118 |
| Mass of Gag Molecules (MDa) | Total Gag polyprotein [f] | 70–87 | 82–106 |
| | Gag | 70–87 | 70–90 |
| | Gag-Pro | N/A | 10–13 |
| | Gag-Pro-Pol | N/A | 2.5–3 |
| Gag polyprotein copy number [g] | | 1300–1600 | 1500–1900 |

[a] As determined by cryogenic transmission electron microscopy (Cryo-TEM); [b] As determined by scanning transmission electron microscopy (STEM); [c] Mass contributed by RNA in virus-like particles was estimated experimentally as described in the Materials and Methods. The RNA mass contribution for authentic HTLV-1 particles was estimated based upon the genome size; [d] Mass contributed by lipids was estimated from average particle size and membrane thickness; [e] Mass of total protein was determined by subtraction of the RNA and lipid mass from the total particle mass as determined by STEM; [f] Total Gag polyprotein was estimated based upon the assumption that Gag contributes ~70%–90% of the total protein mass; [g] The Gag polyprotein copy number represents the range of Gag copy number in a particle that has both average mass and dimensions.

### 3.4. Calculation of Gag Stoichiometry in HTLV-1-Like Particles

The average mass of HTLV-1-like particles determined by STEM was used to estimate the average Gag copy number per virus particle. The viral RNA mass contribution from total particle mass was determined by extracting the RNA from particle lysates with RNA columns, using Roche's High Pure Viral RNA Kit (Roche Diagnostics, Indianapolis, IN, USA). The extracted viral RNA was quantified by determining the ultraviolet (UV) absorption at 260 nm and a conversion factor of 40 µg/mL × A260 optical density unit x dilution factor using a Beckman DU-65 spectrophotometer (Beckman Coulter, Brea, CA, USA). The Thermo Scientific Pierce BCA Protein Assay Kit (Thermo Fisher Scientific, Waltham, MA, USA) was used to estimate the protein content of the same sample used to determine the VLPs' RNA content. The VLPs' Gag/RNA mass ratio was determined to be 14.4:1, equivalent to about 4% of the averaged molecular mass of the VLPs measured by STEM (Table 1). Based upon the average size of the VLPs, which is 110 nm, and estimating the average thickness of the viral membrane to be 5 nm, an estimate of the number of lipid molecules in the virus envelope was made. Assuming that the distance between lipid molecules in the same leaflet was 0.85 nm [34], and the average molecular weight of lipids was 750, the mass of lipids in an averaged size particle was determined to be approximately 70 MDa (*i.e.*, ~40% of the total mass of an averaged sized particle) (Table 1).

Assuming that the mass of the Gag protein in HTLV-1-like particles is similar to that of other retroviruses, approximately 70%–90% of the total protein mass [17,35], the mass contribution of Gag in an HTLV-1-like particle with an average size of 110 nm would be 70–87 MDa. Given the molecular weight for HTLV-1 Gag is ~53 kDa [36,37], it was estimated that HTLV-1-like particles contain approximately 1300–1600 Gag polyproteins per VLP with a mass and diameter of 174 MDa and 110 nm, respectively (Table 1).

### 3.5. Estimating Gag Stoichiometry in Authentic Immature HTLV-1 Particles by Calculating Gag Copy Number in Authentic Mature HTLV-1 Particles

The same methodology used to calculate the Gag copy number in HTLV-1-like particles was used to estimate the Gag stoichiometry in authentic immature HTLV-1 particles by calculating the Gag copy number in authentic mature HTLV-1 particles. Based upon the average size of the authentic mature HTLV-1 particles, which is 113 nm, we estimated the lipid mass to be approximately 80 MDa (Table 1). The viral RNA in authentic HTLV-1 particles was calculated by assuming that each authentic particle contains two copies of the 8.5 kb genomic RNA, plus tRNA and other small RNAs comprising approximately 30% of the genomic RNA by mass. The molecular weight of RNA was estimated to be 7 MDa, equivalent to about 3.5% of the averaged molecular mass of the particle measured by STEM. Assuming that the mass of the Gag protein in retroviruses is about 70%–90% of the total protein, the Gag protein of authentic HTLV-1 particles would contribute 82–106 MDa to the total particle mass for a particle of an average size of 113 nm (Table 1). An authentic HTLV-1 particle contains three forms of the Gag polyprotein: Gag, Gag-Pro and Gag-Pro-Pol, with molecular weights of 53 kDa, 76 kDa and 180 kDa, respectively [36,37]. Given that the estimated molar ratio of the Gag, Gag-Pro, and Gag-Pro-Pol is 100:10:1 based on *in vitro* translation of viral RNA [38], it was calculated that the immature precursor of authentic HTLV-1 particles contains approximately 1500 to 1900 copies of Gag, which would result in a particle with a mass of 204 MDa and a size of 113 nm (Table 1) [34,38].

## 4. Discussion

Although HTLV-1 was the first human retrovirus to be discovered [39,40], the morphological details of HTLV-1 particles have been poorly characterized, including that of Gag stoichiometry. To combat the technical difficulties in working with HTLV-1 in cell culture, a HTLV-1 Gag-only expression model system was used to produce and purify HTLV-1-like particles. A key technical advantage of this HTLV-1 Gag model system is that it is a highly robust system that results in highly efficient production of VLPs from mammalian cells. In the absence of methodologies to efficiently produce authentic immature HTLV-1 particles, and given the absence of HTLV-1 PR inhibitors [41–44], this construct was used as a surrogate to study immature particle morphology.

The electron density of the HTLV-1 Gag lattice appears more compact than what has been previously observed for Gag lattices from HIV-1, MPMV or RSV [9,28,45]. The most intriguing morphological feature of the HTLV-1 immature Gag lattice is that about 20% of the HTLV-1-like particles had regions that appeared to be flat and did not follow the curvature of the viral membrane in multiple regions. The maximum separation between these 'flat' regions and the viral membrane was approximately 8 nm. This is the first time this observation has been made regarding the structure of an immature retroviral Gag lattice. One intriguing possibility is that this morphological feature is indicative of a more rigid lattice structure compared to that of other previously reported HIV-1 immature Gag lattice structures. The addition of a fluorophore tag on the carboxy terminus of the HTLV-1 Gag protein did affect the diameter of the VLPs (*i.e.*, average diameter of 110 nm without tag *versus* 75 nm with tag) as well as the Gag-Gag interactions, given the distinct morphological differences in the presence and absence of the fluorophore tag [21]. This is in contrast to that observed with HIV-1 Gag, where particles produced from a Gag-YFP expression construct did not influence particle size [19]. Taken together, these results imply distinct differences in the Gag assemblies in HTLV-1 immature particles compared to that of other retroviruses, particularly HIV-1.

STEM analysis led to the observation that the HTLV-1 Gag copy number distribution per particle spanned a wide range for both HTLV-1-like particles and HTLV-1 authentic particles (Figures 4B and 5B), which corresponded to the diverse particle size population (Figures 1C and 3B). HTLV-1-like particles and mature particles were found to have Gag copy numbers of 1300–1600 and 1500–1900 Gag molecules/particle, respectively, which is in the general range of Gag copy numbers observed for other retroviruses including MPMV and RSV [17,18].

The observations made by this study emphasize both unique and common morphological features of the HTLV-1-like particles in comparison to other retrovirus immature VLPs. Future studies will include a detailed determination of the immature Gag lattice, which should provide important new insights into the unique aspects of HTLV-1 particle assembly, in particular, and new insights into retroviral assembly, in general.

**Acknowledgments:** We thank Joe Wall (Brookhaven National Laboratory, NY, USA) for assistance with the STEM analyses. The cryo-TEM images are recorded using a Tecnai TF30 TEM maintained by the Characterization Facility, College of Science and Engineering, University of Minnesota. This work was supported by NIH grant RO1 GM098550. JOM was supported by NIH grants F30 DE22286 and T32 AI083196 (Institute for Molecular Virology Training Program).

**Author Contributions:** J.O.M. and S.C. performed experiments, analyzed data, and assisted in writing the manuscript. W.Z. and L.M.M. designed the study, helped analyze and interpret data, and assisted in writing the manuscript. All authors read and approved the final manuscript.

**Conflicts of Interest:** The authors declare no conflict of interest.

## Abbreviations

The following abbreviations are used in this manuscript:

| | |
|---|---|
| AT-2 | 2,2'-dithiodipyridine |
| BNL | Brookhaven National Laboratory |
| Cryo-TEM | cryogenic transmission electron microscopy |
| HIV-1 | human immunodeficiency virus type 1 |
| HTLV-1 | human T-cell leukemia virus type 1 |
| MPMV | Mason-Pfizer monkey virus |
| RSV | sarcoma virus |
| STEM | scanning transmission electron microscopy |
| TMV | tobacco mosaic virus |
| VLP | virus-like particle |

## References

1.  De The, G.; Bomford, R. An HTLV-1 vaccine: Why, how, for whom? *AIDS Res. Hum. Retrovir.* **1993**, *9*, 381–386. [CrossRef] [PubMed]
2.  Gessain, A.; Cassar, O. Epidemiological aspects and world distribution of HTLV-1 infection. *Front. Microbiol.* **2012**, *3*, 388. [CrossRef] [PubMed]
3.  Gessain, A.; Barin, F.; Vernant, J.C.; Gout, O.; Maurs, L.; Calender, A.; de The, G. Antibodies to human T-lymphotropic virus type-1 in patients with tropical spastic paraparesis. *Lancet* **1985**, *2*, 407–410. [CrossRef]
4.  Osame, M.; Usuku, K.; Izumo, S.; Ijichi, N.; Amitani, H.; Igata, A.; Matsumoto, M.; Tara, M. HTLV-1 associated myelopathy, a new clinical entity. *Lancet* **1986**, *1*, 1031–1032. [CrossRef]
5.  Maldonado, J.; Martin, J.; Mueller, J.; Zhang, W.; Mansky, L. New insights into retroviral Gag-Gag and Gag-membrane interactions. *Front. Microbiol.* **2014**, *5*. [CrossRef] [PubMed]
6.  Fogarty, K.H.; Chen, Y.; Grigsby, I.F.; Macdonald, P.J.; Smith, E.M.; Johnson, J.L.; Rawson, J.M.; Mansky, L.M.; Mueller, J.D. Characterization of cytoplasmic Gag-Gag interactions by dual-color z-scan fluorescence fluctuation spectroscopy. *Biophys. J.* **2011**, *100*, 1587–1595. [CrossRef] [PubMed]
7.  Jouvenet, N.; Simon, S.M.; Bieniasz, P.D. Imaging the interaction of HIV-1 genomes and Gag during assembly of individual viral particles. *Proc. Natl. Acad. Sci. USA* **2009**, *106*, 19114–19119. [CrossRef] [PubMed]
8.  Kingston, R.L.; Olson, N.H.; Vogt, V.M. The organization of mature Rous sarcoma virus as studied by cryoelectron microscopy. *J. Struct. Biol.* **2001**, *136*, 67–80. [CrossRef] [PubMed]
9.  Briggs, J.A.; Johnson, M.C.; Simon, M.N.; Fuller, S.D.; Vogt, V.M. Cryo-electron microscopy reveals conserved and divergent features of Gag packing in immature particles of Rous sarcoma virus and human immunodeficiency virus. *J. Mol. Biol.* **2006**, *355*, 157–168. [CrossRef] [PubMed]
10. Briggs, J.A.; Simon, M.N.; Gross, I.; Krausslich, H.G.; Fuller, S.D.; Vogt, V.M.; Johnson, M.C. The stoichiometry of Gag protein in HIV-1. *Nat. Struct. Mol. Biol.* **2004**, *11*, 672–675. [CrossRef] [PubMed]
11. Briggs, J.A.; Watson, B.E.; Gowen, B.E.; Fuller, S.D. Cryoelectron microscopy of mouse mammary tumor virus. *J. Virol.* **2004**, *78*, 2606–2608. [CrossRef] [PubMed]
12. Yeager, M.; Wilson-Kubalek, E.M.; Weiner, S.G.; Brown, P.O.; Rein, A. Supramolecular organization of immature and mature murine leukemia virus revealed by electron cryo-microscopy: Implications for retroviral assembly mechanisms. *Proc. Natl. Acad. Sci. USA* **1998**, *95*, 7299–7304. [CrossRef] [PubMed]
13. Fuller, S.D.; Wilk, T.; Gowen, B.E.; Krausslich, H.G.; Vogt, V.M. Cryo-electron microscopy reveals ordered domains in the immature HIV-1 particle. *Curr. Biol.* **1997**, *7*, 729–738. [CrossRef]

14. Butan, C.; Winkler, D.C.; Heymann, J.B.; Craven, R.C.; Steven, A.C. RSV capsid polymorphism correlates with polymerization efficiency and envelope glycoprotein content: Implications that nucleation controls morphogenesis. *J. Mol. Biol.* **2008**, *376*, 1168–1181. [CrossRef] [PubMed]

15. Yu, F.; Joshi, S.M.; Ma, Y.M.; Kingston, R.L.; Simon, M.N.; Vogt, V.M. Characterization of Rous sarcoma virus Gag particles assembled *in vitro*. *J. Virol.* **2001**, *75*, 2753–2764. [CrossRef] [PubMed]

16. Carlson, L.A.; Briggs, J.A.; Glass, B.; Riches, J.D.; Simon, M.N.; Johnson, M.C.; Muller, B.; Grunewald, K.; Krausslich, H.G. Three-dimensional analysis of budding sites and released virus suggests a revised model for HIV-1 morphogenesis. *Cell Host Microbe* **2008**, *4*, 592–599. [CrossRef] [PubMed]

17. Vogt, V.M.; Simon, M.N. Mass determination of Rous sarcoma virus virions by scanning transmission electron microscopy. *J. Virol.* **1999**, *73*, 7050–7055. [PubMed]

18. Parker, S.D.; Wall, J.S.; Hunter, E. Analysis of Mason-Pfizer monkey virus Gag particles by scanning transmission electron microscopy. *J. Virol.* **2001**, *75*, 9543–9548. [CrossRef] [PubMed]

19. Chen, Y.; Wu, B.; Musier-Forsyth, K.; Mansky, L.M.; Mueller, J.D. Fluorescence fluctuation spectroscopy on viral-like particles reveals variable gag stoichiometry. *Biophys. J.* **2009**, *96*, 1961–1969. [CrossRef] [PubMed]

20. Zhang, W.; Cao, S.; Martin, J.L.; Mueller, J.D.; Mansky, L.M. Morphology and ultrastructure of retrovirus particles. *AIMS Biophys.* **2015**, *2*, 343–369. [CrossRef] [PubMed]

21. Grigsby, I.F.; Zhang, W.; Johnson, J.L.; Fogarty, K.H.; Chen, Y.; Rawson, J.M.; Crosby, A.J.; Mueller, J.D.; Mansky, L.M. Biophysical analysis of HTLV-1 particles reveals novel insights into particle morphology and Gag stochiometry. *Retrovirology* **2010**, *7*. [CrossRef] [PubMed]

22. Haertle, T.; Carrera, C.J.; Wasson, D.B.; Sowers, L.C.; Richman, D.D.; Carson, D.A. Metabolism and anti-human immunodeficiency virus-1 activity of 2-halo-2′, 3′-dideoxyadenosine derivatives. *J. Biol. Chem.* **1988**, *263*, 5870–5875. [PubMed]

23. Harada, S.; Koyanagi, Y.; Yamamoto, N. Infection of HTLV-III/LAV in HTLV-I-carrying cells MT-2 and MT-4 and application in a plaque assay. *Science* **1985**, *229*, 563–566. [CrossRef] [PubMed]

24. Rossio, J.L.; Esser, M.T.; Suryanarayana, K.; Schneider, D.K.; Bess, J.W., Jr.; Vasquez, G.M.; Wiltrout, T.A.; Chertova, E.; Grimes, M.K.; Sattentau, Q.; et al. Inactivation of human immunodeficiency virus type 1 infectivity with preservation of conformational and functional integrity of virion surface proteins. *J. Virol.* **1998**, *72*, 7992–8001. [PubMed]

25. Baker, T.S.; Olson, N.H.; Fuller, S.D. Adding the third dimension to virus life cycles: Three-dimensional reconstruction of icosahedral viruses from cryo-electron micrographs. *Microbiol. Mol. Biol. Rev.* **1999**, *63*, 862–922. [CrossRef] [PubMed]

26. Wall, J.S.; Hainfeld, J.F.; Simon, M.N. Scanning transmission electron microscopy of nuclear structures. *Methods Cell Biol.* **1998**, *53*, 139–164. [PubMed]

27. Namba, K.; Stubbs, G. Structure of tobacco mosaic virus at 3.6 Å resolution: Implications for assembly. *Science* **1986**, *231*, 1401–1406. [CrossRef] [PubMed]

28. Bharat, T.A.; Davey, N.E.; Ulbrich, P.; Riches, J.D.; de Marco, A.; Rumlova, M.; Sachse, C.; Ruml, T.; Briggs, J.A. Structure of the immature retroviral capsid at 8 Å resolution by cryo-electron microscopy. *Nature* **2012**, *487*, 385–389. [CrossRef] [PubMed]

29. de Marco, A.; Davey, N.E.; Ulbrich, P.; Phillips, J.M.; Lux, V.; Riches, J.D.; Fuzik, T.; Ruml, T.; Krausslich, H.G.; Vogt, V.M.; et al. Conserved and variable features of Gag structure and arrangement in immature retrovirus particles. *J. Virol.* **2010**, *84*, 11729–11736. [CrossRef] [PubMed]

30. Frank, G.A.; Narayan, K.; Bess, J.W., Jr.; Del Prete, G.Q.; Wu, X.; Moran, A.; Hartnell, L.M.; Earl, L.A.; Lifson, J.D.; Subramaniam, S. Maturation of the HIV-1 core by a non-diffusional phase transition. *Nat. Commun.* **2015**, *6*, 5854. [CrossRef] [PubMed]

31. Briggs, J.A.; Riches, J.D.; Glass, B.; Bartonova, V.; Zanetti, G.; Krausslich, H.G. Structure and assembly of immature HIV. *Proc. Natl. Acad. Sci. USA* **2009**, *106*, 11090–11095. [CrossRef] [PubMed]

32. Wright, E.R.; Schooler, J.B.; Ding, H.J.; Kieffer, C.; Fillmore, C.; Sundquist, W.I.; Jensen, G.J. Electron cryotomography of immature HIV-1 virions reveals the structure of the CA and SP1 Gag shells. *EMBO J.* **2007**, *26*, 2218–2226. [CrossRef] [PubMed]

33. Cao, S.; Maldonado, J.O.; Grigsby, I.F.; Mansky, L.M.; Zhang, W. Analysis of human T-cell leukemia virus type 1 particles by using cryo-electron tomography. *J. Virol.* **2015**, *89*, 2430–2435. [CrossRef] [PubMed]

34. Cockburn, J.J.; Abrescia, N.G.; Grimes, J.M.; Sutton, G.C.; Diprose, J.M.; Benevides, J.M.; Thomas, G.J., Jr.; Bamford, J.K.; Bamford, D.H.; Stuart, D.I. Membrane structure and interactions with protein and DNA in bacteriophage PRD1. *Nature* **2004**, *432*, 122–125. [CrossRef] [PubMed]

35. Fleissner, E. Chromatographic separation and antigenic analysis of proteins of the oncornaviruses. I. Avian leukemia-sarcoma viruses. *J. Virol.* **1971**, *8*, 778–785. [PubMed]

36. Nam, S.H.; Kidokoro, M.; Shida, H.; Hatanaka, M. Processing of Gag precursor polyprotein of human T-cell leukemia virus type I by virus-encoded protease. *J. Virol.* **1988**, *62*, 3718–3728. [PubMed]

37. Nam, S.H.; Copeland, T.D.; Hatanaka, M.; Oroszlan, S. Characterization of ribosomal frameshifting for expression of pol gene products of human T-cell leukemia virus type I. *J. Virol.* **1993**, *67*, 196–203. [PubMed]

38. Mador, N.; Panet, A.; Honigman, A. Translation of gag, pro, and pol gene products of human T-cell leukemia virus type 2. *J. Virol.* **1989**, *63*, 2400–2404. [PubMed]

39. Poiesz, B.J.; Ruscetti, F.W.; Gazdar, A.F.; Bunn, P.A.; Minna, J.D.; Gallo, R.C. Detection and isolation of type C retrovirus particles from fresh and cultured lymphocytes of a patient with cutaneous T-cell lymphoma. *Proc. Natl. Acad. Sci. USA* **1980**, *77*, 7415–7419. [CrossRef] [PubMed]

40. Poiesz, B.J.; Ruscetti, F.W.; Reitz, M.S.; Kalyanaraman, V.S.; Gallo, R.C. Isolation of a new type C retrovirus (HTLV) in primary uncultured cells of a patient with sezary T-cell leukaemia. *Nature* **1981**, *294*, 268–271. [CrossRef] [PubMed]

41. Ding, Y.S.; Rich, D.H.; Ikeda, R.A. Substrates and inhibitors of human T-cell leukemia virus type I protease. *Biochemistry* **1998**, *37*, 17514–17518. [CrossRef] [PubMed]

42. Pettit, S.C.; Sanchez, R.; Smith, T.; Wehbie, R.; Derse, D.; Swanstrom, R. HIV type 1 protease inhibitors fail to inhibit HTLV-I Gag processing in infected cells. *AIDS Res. Hum. Retrovir.* **1998**, *14*, 1007–1014. [CrossRef] [PubMed]

43. Louis, J.M.; Oroszlan, S.; Tozser, J. Stabilization from autoproteolysis and kinetic characterization of the human T-cell leukemia virus type 1 proteinase. *J. Biol. Chem.* **1999**, *274*, 6660–6666. [CrossRef] [PubMed]

44. Tozser, J.; Weber, I.T. The protease of human T-cell leukemia virus type-1 is a potential therapeutic target. *Curr. Pharm. Des.* **2007**, *13*, 1285–1294. [CrossRef] [PubMed]

45. Schur, F.K.; Hagen, W.J.; Rumlova, M.; Ruml, T.; Muller, B.; Krausslich, H.G.; Briggs, J.A. Structure of the immature HIV-1 capsid in intact virus particles at 8.8 Å resolution. *Nature* **2015**, *517*, 505–508. [CrossRef] [PubMed]

*viruses*

MDPI

*Article*

# A Potential of an Anti-HTLV-I gp46 Neutralizing Monoclonal Antibody (LAT-27) for Passive Immunization against Both Horizontal and Mother-to-Child Vertical Infection with Human T Cell Leukemia Virus Type-I

**Hideki Fujii [1,\*], Mamoru Shimizu [2], Takuya Miyagi [1], Marie Kunihiro [1], Reiko Tanaka [1], Yoshiaki Takahashi [1] and Yuetsu Tanaka [1]**

[1]   Department of Immunology, Graduate School of Medicine, University of the Ryukyus, Uehara 207, Nishihara-cho, Okinawa 903-0215, Japan; miya_skywalker2008@yahoo.co.jp (T.M.); k138751@eve.u-ryukyu.ac.jp (M.K.); reiko_tanaka@s5.dion.ne.jp (R.T.); ytakah3@med.u-ryukyu.ac.jp (Y.T.); yuetsu@s4.dion.ne.jp (Y.T.)

[2]   IBL (Immuno-Biological Laboratories Co., Ltd.), Naka 1091-1, Fujioka, Gunma 375-0005, Japan; do-shimizu@ibl-japan.co.jp

\*   Correspondence: hfujii@med.u-ryukyu.ac.jp; Tel.: +81-98-895-1200; Fax: +81-98-895-1446

Academic Editor: Louis M. Mansky
Received: 9 November 2015; Accepted: 28 January 2016; Published: 3 February 2016

**Abstract:** Although the number of human T-cell leukemia virus type-I (HTLV-I)-infected individuals in the world has been estimated at over 10 million, no prophylaxis vaccines against HTLV-I infection are available. In this study, we took a new approach for establishing the basis of protective vaccines against HTLV-I. We show here the potential of a passively administered HTLV-I neutralizing monoclonal antibody of rat origin (LAT-27) that recognizes epitopes consisting of the HTLV-I gp46 amino acids 191–196. LAT-27 completely blocked HTLV-I infection *in vitro* at a minimum concentration of 5 µg/mL. Neonatal rats born to mother rats pre-infused with LAT-27 were shown to have acquired a large quantity of LAT-27, and these newborns showed complete resistance against intraperitoneal infection with HTLV-I. On the other hand, when humanized immunodeficient mice were pre-infused intravenously with humanized LAT-27 (hu-LAT-27), all the mice completely resisted HTLV-I infection. These results indicate that hu-LAT-27 may have a potential for passive immunization against both horizontal and mother-to-child vertical infection with HTLV-I.

**Keywords:** HTLV-I; NOG mice; neutralizing monoclonal antibody; envelope gp46; passive immunity

## 1. Introduction

Human T-cell leukemia virus type I (HTLV-I) [1,2] causes both neoplastic and inflammatory diseases, including adult T-cell leukemia (ATL) [3,4] and HTLV-I-associated myelopathy/tropical spastic paraparesis (HAM/TSP) [5,6]. The number of HTLV-I-infected individuals in the world has been estimated at over 10 million [7]. However, no prophylaxis vaccines or drugs against HTLV-I infection are available. HTLV-I is transmitted through contact with bodily fluids containing infected cells, most often from mother to child through breast milk or via blood transfusion. It was demonstrated that HTLV-I efficiently spreads from cell-to-cell via virological synapses [8]. The HTLV-I envelope spike consists of two glycoproteins, cell surface gp46 and trans-membrane gp21 [9], both of which are essential for HTLV-I entry into cells [10].

Since there are little or no genetic mutations in these envelope antigens among HTLV-I strains [11], it is clear that these antigens are the right targets for prophylactic vaccines. Accordingly, a line of evidence showed that a recombinant vaccinia virus (RVV) expressing gp46 and synthetic peptides corresponding to several regions of gp46 conferred immunity against HTLV-I challenge, showing a possibility of active vaccination [12–15]. However, there are a lot of hurdles before the invention of safe and effective active vaccines. As it has been demonstrated that humanized or human antibodies are safe and effective in various areas of medicine, passive immunization of anti-HTLV-I gp46 neutralizing antibodies may provide a choice for prevention of the spread of HTLV-I. Although antibodies against gp46 antigen and their neutralizing capacity are commonly demonstrated in the sera of HTLV-I-infected individuals, little is known about whether these polyclonal anti-gp46 antibodies can control human-to-human infection of HTLV-I [16]. On the way to establishing a basis for vaccine development against HTLV-I, we previously reported that our anti-gp46 neutralizing mAb (LAT-27) was capable of blocking of HTLV-I infection by direct neutralization and eradicating HTLV-I-infected cells via antibody-dependent-cellular-cytotoxicity (ADCC) *in vitro* [17]. Recently, we showed that LAT-27 is also capable of blocking primary HTLV-I infection in a humanized mouse model [18].

Here, we show that maternally transferred LAT-27 is capable of protecting newborn rats against HTLV-I infection, and suggest that humanized LAT-27 is able to block horizontal infection of humanized mice with HTLV-I. Therefore, humanized LAT-27 may be one of the candidates for passive vaccines against HTLV-I.

## 2. Materials and Methods

### 2.1. Reagents

The medium used throughout was RPMI 1640 medium (Sigma-Aldrich Inc., St. Louis, MO, USA) supplemented with 10% fetal calf serum (FCS), 100 U/mL penicillin and 100 µg/mL streptomycin (hereafter called RPMI medium). Rat and mouse monoclonal antibodies (mAbs) were purified in our laboratory from ascites fluids of CB.17-SCID mice carrying the appropriate hybridomas as described previously [17]. These antibodies were rat IgG2b mAbs anti-gp46 (clones LAT-27), rat IgG2b anti-HIV-1 p24 (clone WAP-24), mouse IgG3 anti-HTLV-I Tax (clone Lt-4). mAbs were labeled with HiLyte Fluor™ 647 using commercial labeling kits (Dojindo, Kumamoto, Japan) according to the manufacturer's instructions. PE-labeled mouse mAbs against human CD4 were purchased from BioLegend (Tokyo, Japan). Humanized-LAT-27 (hu-LAT-27) and human-mouse chimeric antibody consisting of human IgG1 Fc and a part of mouse anti-CEA were generated in collaboration with IBL (Gunma, Japan) and the information of hu-LAT-27 will be reported elsewhere.

### 2.2. Cell Culture and Syncytium Inhibition Assay

The IL-2-dependent CD4$^-$CD8$^+$ ILT-M1 cell line derived from a HAM patient was used as a source of HTLV-I (kindly provided by Kannagi of Tokyo medical and dental university) [17]. These cells were maintained in culture using RPMI medium containing 20 U/mL IL-2. Syncytium inhibition assay was carried out using a combination of ILT-M1 and HTLV-I negative Jurkat T-cell lines as reported previously [17]. ILT-M1 cell line was used because of its superiority in inducing syncytia. Briefly, a volume of 25 µL ILT-M1 cell suspension at $2 \times 10^6$ cells/mL in 20 U/mL IL-2 containing RPMI media was mixed with 50 µL of serially diluted antibody in a flat-bottom 96-well micro-titer plate for 5 min followed by the addition of a volume of 25 µL Jurkat cell suspension at $2 \times 10^6$ cells/mL. After cultivation for 16 h at 37 °C in a 5% $CO_2$ humidified incubator, syncytium formation was microscopically observed using an inverted microscope and the concentration of antibody that showed complete blocking of syncytium formation was determined.

### 2.3. ELISA

ELISA was used to quantitate rat and humanized LAT-27 in sera of rats and NOD-SCID/γc null (NOG) mice, respectively. Briefly, HTLV-I gp46 synthetic peptide [19] was coated onto 96-well

ELISA plates (Nunc) as an antigen, and the bindings of rat and humanized LAT-27 were detected with HRP-labeled anti-rat and human IgG, respectively.

*2.4. Animal Experiments*

This research was approved by the institutional review boards of the authors' institutions and written informed consent was obtained from all individuals for the collection of samples and subsequent analysis. The protocols for the use of human PBMCs and animals were approved by the Institutional Review Board and the Institutional Animal Care and Use Committee on clinical and animal research of the University of the Ryukyus prior to initiation of the study. Strains of WKA/H, F344, SD rats were purchased from SLC (Shizuoka, Japan). NOG mice were purchased from the Central Institute of Experimental Animals (Kanagawa, Japan) and were kept in the specific-pathogen-free animal facilities of the Laboratory Animal Center, University of the Ryukyus. Mice were six to seven weeks old at the time of the intra-splenic transplantation of human PBMCs [20]. Fresh PBMCs were isolated from HTLV-I-negative normal donors by a Histopaque-1077 (Sigma) density gradient centrifugation.

*2.5. Isolation of Human T-Cells from Mouse Spleen*

Human CD4$^+$ T-cells were isolated from mouse spleen cells by positive immunoselection with the Dynal$^®$ CD4-positive isolation kit (Invitrogen), according to the manufacturer's protocol. In brief, mouse spleen cells were incubated with anti-CD4-coated beads for 30 min at 4 $°$C under gentle tilt rotation. Captured CD4$^+$ T-cells were collected with a magnet (Dynal MPC-S) and detached from beads with DETACHaBEAD CD4/CD8$^®$ (Invitrogen). Purity was >99% CD4$^+$ T-cells as determined by flow cytometry.

*2.6. Genomic DNA Extraction and Quantification of HTLV-I Proviral Load*

Genomic DNA was extracted by QIAamp kit (QIAGEN, Tokyo, Japan) according to the manufacturer's instructions. To examine the HTLV-I PVL, we carried out a quantitative PCR method using StepOnePlus (Applied Biosystems) with 100 ng of genomic DNA (roughly equivalent to $10^4$ cells) from PBMC samples as reported previously [21]. Based on the standard curve created by four known concentrations of template, the concentration of unknown samples was determined. Using β-actin as an internal control, the amount of HTLV-I proviral DNA was calculated by the following formula: copy number of HTLV-I tax per $1 \times 10^4$ PBMCs = [(copy number of tax) / (copy number of β-actin / 2)] $\times 10^4$. All samples were performed in triplicate.

*2.7. Flow Cytometry*

Before staining, live cells were Fc-blocked with 2 mg/mL pooled normal human IgG in FACS buffer (PBS containing 0.2% bovine serum albumin and 0.1% sodium azide) for 10 min on ice, and then incubated for 15 min at room temperature. After washing with FACS buffer, the cells were fixed in PBS containing 4% paraformaldehyde (Sigma) for 20 min at room temperature followed by permeabilization and washing in 0.5% saponin + 1% BSA (Sigma) containing FACS buffer. The cells were incubated with 0.1 μg/mL of HiLyte Fluor$^®$ 647-labeled anti-Tax mAb (clone Lt-4) for 20 min. Negative control cells were stained with HiLyte Fluor$^®$ 647 Lt-4 in the presence of 50 μg/mL of unlabelled Lt-4. Finally, the cells were washed twice and analyzed by standard flow cytometry using a FACSCalibur™ flow cytometer (BD) and FlowJo software (Tree Star).

## 3. Results

*3.1. Mother-to-Child Transfer of Passively Immunized LAT-27*

It has been demonstrated that rats are relatively permissive for *in vivo* infection with HTLV-I. Although rat T-cells transformed with HTLV-I produce no or little infectious HTLV-I virion, rats can be used to evaluate the efficacy of prophylactic vaccines. Thus, we examined whether LAT-27 could be transferred from mother to neonates without loss of protecting capacity against HTLV-I infection. Our preliminary studies showed that spleen cells of SD strain rats were more sensitive to HTLV-I infection *in vitro* than those of WKA/H and F344 strain rats. Thus, we chose SD rats for this *in vivo* study.

At first, pregnant SD rats were infused i.p. with 25 mg/head of either LAT-27 or isotype control mAb two times on −7 d and −2 d of delivery (*n* = 2). Two days after the birth, blood samples were collected from both mothers and newborns, and LAT-27 concentrations of their sera were quantitated using ELISA (Table 1).

**Table 1.** Mother-to-offspring transfer of LAT-27 in rats. Two pregnant SD rats were infused i.p. with 25 mg/head LAT-27 two times on −7 d and −2 d delivery. Two days after delivery, LAT-27 concentration of each serum was quantitated by ELISA.

| Pregnant Rat #1 | Serum LAT-27 conc. (μg/mL) | Pregnant Rat #2 | Serum LAT-27 conc. (μg/mL) |
|---|---|---|---|
| Mother | 102 | Mother | 51 |
| Offspring-1 | 204 | Offspring-1 | 102 |
| Offspring-2 | 204 | Offspring-2 | 51 |
| Offspring-3 | 102 | Offspring-3 | 102 |
| Offspring-4 | 102 | Offspring-4 | 102 |
| Offspring-5 | 204 | Offspring-5 | 51 |
| Offspring-6 | 102 | Offspring-6 | 51 |
| Offspring-7 | 204 | Offspring-7 | 51 |
| Offspring-8 | 204 | Offspring-8 | 102 |
| Offspring-9 | 102 | Offspring-9 | 102 |
| | | Offspring-10 | 51 |

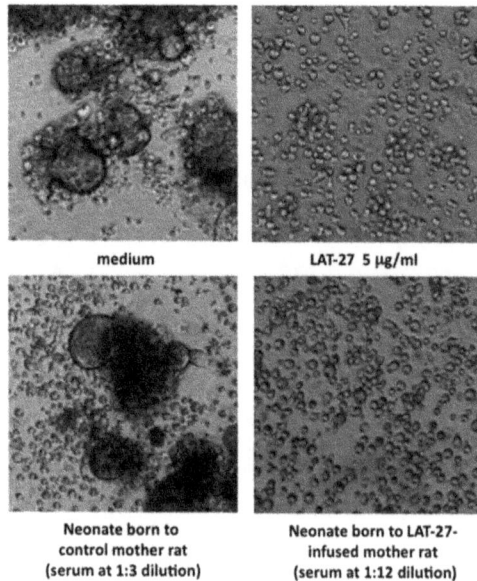

medium

LAT-27 5 μg/ml

Neonate born to control mother rat (serum at 1:3 dilution)

Neonate born to LAT-27-infused mother rat (serum at 1:12 dilution)

**Figure 1.** Neutralization of HTLV-I by mother-to-offspring-transferred LAT-27. Syncytium inhibition assay was carried out as described in Materials and Methods. ILT-M1 cells were treated with diluted serum or purified antibody in a flat-bottom 96-well micro-titer plate for 5 min, and then co-cultured with Jurkat cells. After cultivation for 16 h, syncytium formation was microscopically observed using an inverted microscope at magnification of 200×.

The data showed that a substantial dose of LAT-27 was transferred to the neonates from the mothers. It is of interest that the LAT-27 concentrations in newborn rat sera were equivalent to those in mother sera, suggesting an efficient transfer of LAT-27.

In order to confirm that the transferred LAT-27 in neonates retained neutralizing activity, we performed the syncytium inhibition assay (Figure 1).

Overnight co-culture of the HTLV-I-producing ILT-M1 cells and HTLV-I-negative Jurkat T-cells in media alone gave rise to the generation of a number of syncytia (Figure 1 upper left panel), which was completely inhibited by the addition of LAT-27 (5 µg/mL) (Figure 1 upper left panel). In contrast to that, the control serum obtained from a neonate born to an untreated normal mother did not inhibit the syncytium formation even at 1:3 dilution, thus the serum from a LAT-27-adminstered mother did neutralize HTLV-I infection at 1:12 dilutions. Taken together, these data indicated that LAT-27 could be transferred from mother to offspring without any loss of its neutralization property.

### 3.2. LAT-27 from Mothers Protects Newborns against HTLV-I Infection

Then, we addressed whether mother-to-child-transferred LAT-27 could protect newborn rats against HTLV-I infection. Pregnant SD rats received 25 mg/head LAT-27 two days before delivery, and their newborns were challenged i.p. with the mitomycin C-treated HTLV-I-producing cells on day 1 after birth. As controls, two newborn littermates of non-antibody-administered mothers were infected with HTLV-I before or after administration with 2 mg/head LAT-27. Three weeks after infection, rats were sacrificed and spleen cells were subjected to quantitative real-time PCR analysis for testing HTLV-I infection. Figure 2 showed that the rats born to the LAT-27-treated mother rat as well as the newborn rats directly administered LAT-27 after birth were negative for proviral DNA, suggesting that mother-to-child-transferred LAT-27 could work *in vivo*.

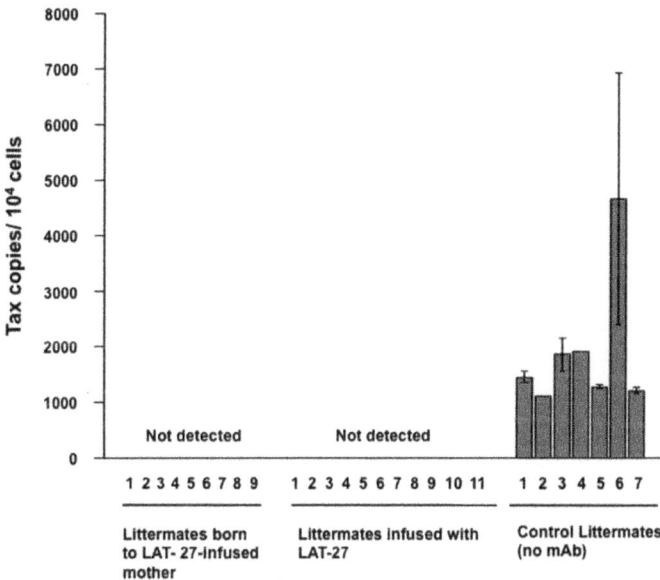

**Figure 2.** Protection of newborns from HTLV-I infection by mother-to-child-transferred LAT-27. Nine littermates from LAT-27 passively immunized mother rat (**left**), 11 littermates directly injected with LAT-27 (**center**) before infection, and seven naïve littermates (**right**) were i.p. injected with MMC-treated ILT-M1 cells. After three weeks, rats were tested for HTLV-I infection.

### 3.3. Retention of Humanized LAT-27 in NOG Mice

For future prophylactic and clinical utilization, we have succeeded in generating humanized LAT-27 (hu-LAT-27) as shown in Figure 3. hu-LAT-27 consists of a complementarity-determining region (CDR) of the original rat LAT-27 and human IgG1 backbone as illustrated in Figure 3A.

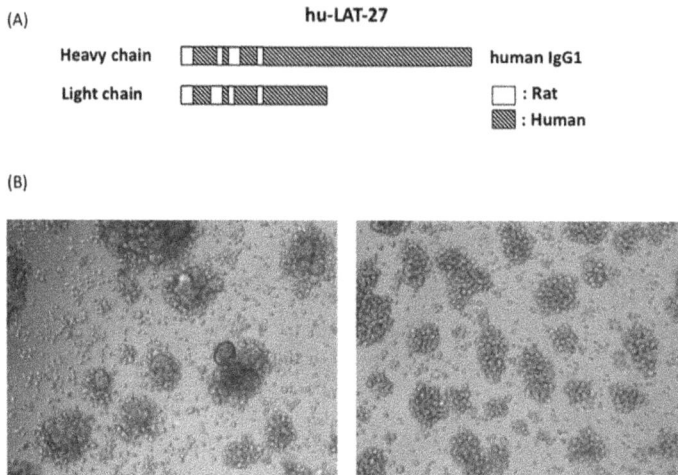

**Figure 3.** Schematic illustration of humanized LAT-27 and its neutralizing activity. Syncytium inhibition assay was carried out with either 200 µg/mL of normal human IgG (**left** of **B**) and 5 µg/mL of humanized LAT-27. After cultivation for 24 h, syncytium formation was microscopically observed using an inverted microscope at magnification of 100× (**right** of **B**).

hu-LAT-27 was produced in CHO cells. Due to a patent matter, the gene sequence of hu-LAT-27 will be reported elsewhere. The *in vitro* neutralization test showed that the minimum concentration of hu-LAT-27 for complete inhibition of syncytium formation was 5 µg/mL, which was comparable to the original rat LAT-27 (Figure 3B), demonstrating no decay in neutralizing ability after humanization. In order to examine whether hu-LAT-27 could work *in vivo*, we firstly tested the retention time of hu-LAT-27 in NOG mice. Mice were injected i.v. with 1 mg/head hu-LAT-27 (*n* = 3). Sera were collected daily for five days and the concentration of hu-LAT-27 was quantitated by ELISA as described in the Materials and Methods. The data presented in Table 2 indicated that serum levels of hu-LAT-27 gradually decreased day by day. The half-life of administrated LAT-27 was estimated to be approximately two days.

**Table 2.** Retention of humanized LAT-27 in NOG mice. Each of three NOG mice was infused i.v. with 1 mg/head hu-LAT-27, and the LAT-27 concentrations in their sera were measured daily by ELISA using gp46 peptide (180–204) as antigen and secondary anti-human IgG-HRP, in which purified hu-LAT-27 was used to draw a standard curve.

| Days after Injection | hu-LAT-27 Concentration in Serum (µg/mL) | | |
|:---:|:---:|:---:|:---:|
| | NOG#1 | NOG#2 | NOG#3 |
| 1 | 268 | 249 | 277 |
| 2 | 114 | 155 | 124 |
| 3 | 79 | 88 | 80 |
| 4 | 57 | 86 | 61 |
| 5 | 49 | 47 | 48 |

### 3.4. Protection of hu-PBL-NOG Mice against HTLV-I Infection by hu-LAT-27

Then, we tested the protective efficacy of hu-LAT-27 mAbs against *in vivo* infection with HTLV-I using our humanized mouse model as reported previously [18]. PBMCs from HTLV-I-negative donors were transplanted directly into the spleen of NOG mice together with the mitomycin-C-treated HTLV-I-producing cells. They were intra-peritoneally infused with either hu-LAT-27 or control human chimeric anti-CEA mAb before HTLV-I infection. Fourteen days after infection, these mice were sacrificed and fresh spleen cells were isolated, cultured for 24 h and tested for HTLV-I infection by flow cytometry. As shown in Figure 4A, human CD4+ T-cells from all mice immunized with the hu-LAT-27 before HTLV-I infection were negative for Tax antigen, suggesting complete protection against HTLV-I infection (lower panels). PCR testing of HTLV-I proviral loads supported this conclusion. The protective effect was hu-LAT-27-specific since the control anti-CEA did not confer protection against the infection (upper panels). When hu-LAT-27 was administered one day after HTLV-I infection, it did not protect mice against HTLV-I infection. These results suggested that hu-LAT-27 could be used for prophylaxis against HTLV-I infection *in vivo* when a sufficient amount of it was infused before infection with HTLV-I.

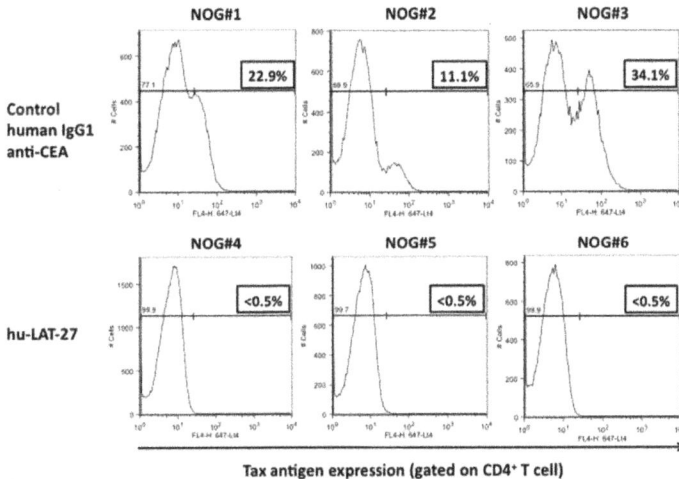

**Figure 4.** Protection of hu-PBL-NOG mice against HTLV-I by hu-LAT-27. The Mitomycin C (MMC)-treated HTLV-I-infected ILT-M1 ($1 \times 10^6$ cells) was mixed with freshly isolated PBMCs ($2 \times 10^6$ cells) from healthy donor in a final volume of 50 μL and then were directly injected into the spleens of NOG mice. MAbs (0.5 mg each: lower panels) or control IgG (0.5 mg each: upper panels) was inoculated i.v. one hour before cell transplantation. Fourteen days after cell transplantation, mice were sacrificed, blood was collected by cardiocentesis, and human lymphocytes were recovered from the spleen. The bisector indicated in all panels showed tax negative edge of histogram.

## 4. Discussion

The present study suggested a potential of monoclonal anti-gp46 neutralizing LAT-27 for passive immunization for prophylaxis against HTLV-I infection. The main mechanism for LAT-27-mediated protection of animals against HTLV-I *in vivo* is most likely neutralization via inducing conformational changes in the gp46 structure. In addition, as we have previously reported, LAT-27 may also inhibit HTLV-I infection by eliminating HTLV-I-producing cells via HTLV-1-specific ADCC in the presence of NK cells [17]. In the present study, we have chosen an excess dose of LAT-27 for infusion. Since the minimum HTLV-I neutralizing antibody concentration of LAT-27 *in vitro* is as low as 5 μg/mL, further studies are required to determine the minimum dose of LAT-27 for perfect *in vivo* protection against HTLV-I in each animal system.

Using a rat model, the present study first showed that LAT-27 could be transferred from dams to infant rats, and that the transferred LAT-27 was active for HTLV-I neutralization and protected the infants against intraperitoneal HTLV-I infection. It remains to be determined how long the antibodies were sustained in the littermates born to pregnant rats infused with LAT-27. In rats, it is known that maternal antibodies are transferred equally effectively to infants either via placenta or from the gut by breastfeeding [22]. It should be emphasized that administration of LAT-27 to pregnant rats did not show any obvious effects on delivery and growth of infants, showing the safety of this mAb in rats at least. It remains to be tested whether the transferred LAT-27 is protective against oral HTLV-I infection. Although it has been shown that rats are relatively susceptible to oral infection with HTLV-I [23], our preliminary studies failed to infect newborn (two-day-old) rats with HTLV-I by feeding cow milk serum containing live HTLV-I at least producing cells. In order to establish a good model for oral HTLV-I infection, choices of better rat strain, HTLV-I at least producing cells and timing of infection may be a key. Further studies are in progress.

For future clinical studies in humans, it is necessary to humanize the original LAT-27 (rat origin). In collaborative investigation with IBL Co., Ltd., we have succeeded in the generation of hu-LAT-27, and examined its function *in vitro* and *in vivo*. Similar to rat LAT-27, hu-LAT27 neutralized HTLV-I *in vitro* at a minimum concentration of 5 μg/mL (Figure 1), showing that there is no loss of neutralizing activity after humanization. Furthermore, hu-LAT-27 was capable of protecting NOG mice reconstituted with human PBMCs from intraperitoneal infection with HTLV-I (Figure 4). The half-life of hu-LAT-27 in NOG mice was roughly estimated two days. Since half-life of IgG *in vivo* is dependent on its affinity to neonatal FcR, such a short lifetime of hu-LAT-27 in NOG mice (Table 2) will be prolonged up to three weeks in humans, as in the cases of other humanized mAbs [24].

The present data suggest that hu-LAT-27 mAb will be useful in passive immunization against HTLV-I infection. The rationale of using these monoclonal antibodies instead of polyclonal antibodies obtained from healthy HTLV-I carriers for the prevention of HTLV-I infection is that, in contrast to HIV-1, the genome of HTLV-I is stable [11], and thus the LAT-27 reactive-neutralizing epitope (gp46 amino acid 191–196 of MT-2 virus) is broadly conserved among wild HTLV-I strains. Indeed, LAT-27 reacted all HTLV-I-producing cell lines tested, including naturally infected cell lines from ATL and HAM/TSP patients and asymptomatic carriers. So far, the potential of passive immunization with IgG from HTLV-I carriers has been show in rabbit [25,26], monkey [27] and humanized NOG mouse models [18]. Since polyclonal antibodies contain non-neutralizing antibodies, the efficacy of neutralization by neutralizing antibody could be interfered with by those non-neutralizing anti-gp46 antibodies. More importantly, since the total availability of anti-HTLV-I serum is limited, production capability in a large quantity of Good Manufacturing Practice-grade of HTLV-I neutralizing mAb may be a prerequisite for antibody-based prophylaxis against HTLV-I. Another choice, instead of hu-LAT-27, will be human mAbs with HTLV-I neutralizing capacity, as several human mAbs have been reported [28].

We anticipate that hu-LAT-27 mAb will be used as a safe passive vaccine against both horizontal and vertical infection with HTLV-I for people at high risk of HTLV-I infection, including adults and babies born to HTLV-I carriers. In addition, since organ transplantation to HTLV-I-negative recipients from HTLV-I-positive donors is another risk factor of HTLV-I infection and the development of HAM/TSP [29], pretreatment of recipients with hu-LAT-27 may be useful for the prevention of HTLV-I infection. It may be also possible that, for HTLV-I carriers whose anti-HTLV-I neutralizing antibody titers are low, administration of LAT-27 may function to inhibit further spreading of HTLV-I *in vivo*. Thus, it will be worthy to validate hu-LAT-27 in an animal model in more detail before starting a clinical study. In addition to this passive vaccine, in order to control HTLV-I infection all over the world, it is clear that an active vaccine which can elicit or boost anti-HTLV-I gp46 neutralizing antibody titer should be developed.

## 5. Conclusions

This study describes experimental results indicating a possible concept that passive immunization with our humanized anti-gp46 neutralizing mAb (hu-LAT-27) reduces the chance of HTLV-I infection of adults and babies at high risk. A combination of hu-LAT-27-based passive vaccination and yet-to-be-developed active vaccine might be potent in preventing the spread of HTLV-I infection around the world.

**Acknowledgments:** This work was supported by grants from the Ministry of Education, Culture, Sports, Science, and Technology and the Ministry of Health, Labor, and Welfare of Japan, and from Okinawa prefecture.

**Author Contributions:** Hideki Fujii, and Yuetsu Tanaka conceived and designed the study, performed research, wrote the manuscript. Mamoru Shimizu generated hu-LAT-27. Takuya Miyagi and Marie Kunihiro performed animal experiments. Reiko Tanaka and Yoshiaki Takahashi performed antibody research. Yuetsu Tanaka provided funding for this study. All authors contributed to the final version of the manuscript, read and approved it.

**Conflicts of Interest:** The authors declare no conflict of interest.

## References

1. Poiesz, B.J.; Ruscetti, F.W.; Gazdar, A.F.; Bunn, P.A.; Minna, J.D.; Gallo, R.C. Detection and isolation of type C retrovirus particles from fresh and cultured lymphocytes of a patient with cutaneous T-cell lymphoma. *Proc. Natl. Acad. Sci. USA* **1980**, *77*, 7415–7419. [CrossRef] [PubMed]
2. Yoshida, M.; Miyoshi, I.; Hinuma, Y. Isolation and characterization of retrovirus from cell lines of human adult T-cell leukemia and its implication in the disease. *Proc. Natl. Acad. Sci. USA* **1982**, *79*, 2031–2035. [CrossRef] [PubMed]
3. Hinuma, Y.; Nagata, K.; Hanaoka, M.; Nakai, M.; Matsumoto, T.; Kinoshita, K.I.; Shirakawa, S.; Miyoshi, I. Adult T-cell leukemia: Antigen in an ATL cell line and detection of antibodies to the antigen in human sera. *Proc. Natl. Acad. Sci. USA* **1981**, *78*, 6476–6480. [CrossRef] [PubMed]
4. Yoshida, M.; Seiki, M.; Yamaguchi, K.; Takatsuki, K. Monoclonal integration of human T-cell leukemia provirus in all primary tumors of adult T-cell leukemia suggests causative role of human T-cell leukemia virus in the disease. *Proc. Natl. Acad. Sci. USA* **1984**, *81*, 2534–2537. [CrossRef] [PubMed]
5. Gessain, A.; Barin, F.; Vernant, J.C.; Gout, O.; Maurs, L.; Calender, A.; de The, G. Antibodies to human T-lymphotropic virus type-I in patients with tropical spastic paraparesis. *Lancet* **1985**, *2*, 407–410. [CrossRef]
6. Osame, M.; Usuku, K.; Izumo, S.; Ijichi, N.; Amitani, H.; Igata, A.; Matsumoto, M.; Tara, M. HTLV-I associated myelopathy, a new clinical entity. *Lancet* **1986**, *1*, 1031–1032. [CrossRef]
7. Proietti, F.A.; Carneiro-Proietti, A.B.; Catalan-Soares, B.C.; Murphy, E.L. Global epidemiology of HTLV-I infection and associated diseases. *Oncogene* **2005**, *24*, 6058–6068. [CrossRef] [PubMed]
8. Igakura, T.; Stinchcombe, J.C.; Goon, P.K.; Taylor, G.P.; Weber, J.N.; Griffiths, G.M.; Tanaka, Y.; Osame, M.; Bangham, C.R. Spread of HTLV-I between lymphocytes by virus-induced polarization of the cytoskeleton. *Science* **2003**, *299*, 1713–1716. [CrossRef] [PubMed]
9. Kobe, B.; Center, R.J.; Kemp, B.E.; Poumbourios, P. Crystal structure of human T cell leukemia virus type 1 gp21 ectodomain crystallized as a maltose-binding protein chimera reveals structural evolution of retroviral transmembrane proteins. *Proc. Natl. Acad. Sci. USA* **1999**, *96*, 4319–4324. [CrossRef] [PubMed]
10. Jones, K.S.; Lambert, S.; Bouttier, M.; Bénit, L.; Ruscetti, F.W.; Hermine, O.; Pique, C. Molecular aspects of HTLV-1 entry: Functional domains of the HTLV-1 surface subunit (SU) and their relationships to the entry receptors. *Viruses* **2011**, *3*, 794–810. [CrossRef] [PubMed]
11. Gessain, A.; Gallo, R.C.; Franchini, G. Low degree of human T-cell leukemia/lymphoma virus type I genetic drift *in vivo* as a means of monitoring viral transmission and movement of ancient human populations. *J. Virol.* **1992**, *66*, 2288–2295. [PubMed]
12. Tanaka, Y.; Zeng, L.; Shiraki, H.; Shida, H.; Tozawa, H. Identification of a neutralization epitope on the envelope gp46 antigen of human T cell leukemia virus type I and induction of neutralizing antibody by peptide immunization. *J. Immunol.* **1991**, *147*, 354–360. [PubMed]
13. Tanaka, Y.; Tanaka, R.; Terada, E.; Koyanagi, Y.; Miyano-Kurosaki, N.; Yamamoto, N.; Baba, E.; Nakamura, M.; Shida, H. Induction of antibody responses that neutralize human T-cell leukemia virus type I infection *in vitro* and *in vivo* by peptide immunization. *J. Virol.* **1994**, *68*, 6323–6331. [PubMed]

14. Baba, E.; Nakamura, M.; Tanaka, Y.; Kuroki, M.; Itoyama, Y.; Nakano, S.; Niho, Y. Multiple neutralizing B-cell epitopes of human T-cell leukemia virus type 1 (HTLV-1) identified by human monoclonal antibodies. A basis for the design of an HTLV-1 peptide vaccine. *J. Immunol.* **1993**, *151*, 1013–1024. [PubMed]

15. Baba, E.; Nakamura, M.; Ohkuma, K.; Kira, J.; Tanaka, Y.; Nakano, S.; Niho, Y. A peptide-based human T cell leukemia virus type I vaccine containing T and B cell epitopes that induces high titers of neutralizing antibodies. *J. Immunol.* **1995**, *154*, 399–412. [PubMed]

16. Goncalves, D.U.; Proietti, F.A.; Ribas, J.G.; Araújo, M.G.; Pinheiro, S.R.; Guedes, A.C.; Carneiro-Proietti, A.B. Epidemiology, treatment, and prevention of human T-cell leukemia virus type 1-associated diseases. *Clin. Microbiol. Rev.* **2010**, *23*, 577–589. [CrossRef] [PubMed]

17. Tanaka, Y.; Takahashi, Y.; Tanaka, R.; Kodama, A.; Fujii, H.; Hasegawa, A.; Kannagi, M.; Ansari, A.A.; Saito, M. Elimination of human T cell leukemia virus type-1 (HTLV-1)-infected cells by neutralizing and ADCC-inducing antibodies against HTLV-1 envelope gp46. *AIDS Res. Hum. Retrovir.* **2014**, *30*, 542–552. [CrossRef] [PubMed]

18. Saito, M.; Tanaka, R.; Fujii, H.; Kodama, A.; Takahashi, Y.; Matsuzaki, T.; Takashima, H.; Tanaka, Y. The neutralizing function of the anti-HTLV-1 antibody is essential in preventing *in vivo* transmission of HTLV-1 to human T cells in NOD-SCID/γc null (NOG) mice. *Retrovirology* **2014**, *11*. [CrossRef]

19. Tanaka, Y.; Yasumoto, M.; Nyunoya, H.; Ogura, T.; Kikuchi, M.; Shimotohno, K.; Shiraki, H.; Kuroda, N.; Shida, H.; Tozawa, H. Generation and characterization of monoclonal antibodies against multiple epitopes on the C-terminal half of envelope gp46 of human T-cell leukemia virus type-I (HTLV-I). *Int. J. Cancer* **1990**, *46*, 675–681. [CrossRef] [PubMed]

20. Yoshida, A.; Tanaka, R.; Murakami, T.; Takahashi, Y.; Koyanagi, Y.; Nakamura, M.; Ito, M.; Yamamoto, N.; Tanaka, Y. Induction of protective immune responses against R5 human immunodeficiency virus type 1 (HIV-1) infection in hu-PBL-SCID mice by intrasplenic immunization with HIV-1-pulsed dendritic cells: Possible involvement of a novel factor of human CD4+ T-cell origin. *J. Virol.* **2003**, *77*, 8719–8728. [CrossRef] [PubMed]

21. Nagai, M.; Usuku, K.; Matsumoto, W.; Kodama, D.; Takenouchi, N.; Moritoyo, T.; Hashiguchi, S.; Ichinose, M.; Bangham, C.R.; Izumo, S.; *et al.* Analysis of HTLV-I proviral load in 202 HAM/TSP patients and 243 asymptomatic HTLV-I carriers: High proviral load strongly predisposes to HAM/TSP. *J. Neurovirol.* **1998**, *4*, 586–593. [CrossRef] [PubMed]

22. Grindstaff, J.L.; Brodie, E.D., 3rd; Ketterson, E.D. Immune function across generations: Integrating mechanism and evolutionary process in maternal antibody transmission. *Proc. Biol. Sci.* **2003**, *270*, 2309–2319. [CrossRef] [PubMed]

23. Komori, K.; Hasegawa, A.; Kurihara, K.; Honda, T.; Yokozeki, H.; Masuda, T.; Kannagi, M. Reduction of human T-cell leukemia virus type 1 (HTLV-1) proviral loads in rats orally infected with HTLV-1 by reimmunization with HTLV-1-infected cells. *J. Virol.* **2006**, *80*, 7375–7381. [CrossRef] [PubMed]

24. Lobo, E.D.; Hansen, R.J.; Balthasar, J.P. Antibody pharmacokinetics and pharmacodynamics. *J. Pharm. Sci.* **2004**, *93*, 2645–2668. [CrossRef] [PubMed]

25. Miyoshi, I.; Takehara, N.; Sawada, T.; Iwahara, Y.; Kataoka, R.; Yang, D.; Hoshino, H. Immunoglobulin prophylaxis against HTLV-I in a rabbit model. *Leukemia* **1992**, *6*, S24–S26.

26. Sawada, T.; Iwahara, Y.; Ishii, K.; Taguchi, H.; Hoshino, H.; Miyoshi, I. Immunoglobulin prophylaxis against milkborne transmission of human T cell leukemia virus type I in rabbits. *J. Infect. Dis.* **1991**, *164*, 1193–1196. [CrossRef] [PubMed]

27. Murata, N.; Hakoda, E.; Machida, H.; Ikezoe, T.; Sawada, T.; Hoshino, H.; Miyoshi, I. Prevention of human T cell lymphotropic virus type I infection in Japanese macaques by passive immunization. *Leukemia* **1996**, *10*, 1971–1974. [PubMed]

28. Kuroki, M.; Nakamura, M.; Itoyama, Y.; Tanaka, Y.; Shiraki, H.; Baba, E.; Esaki, T.; Tatsumoto, T.; Nagafuchi, S.; Nakano, S.; *et al.* Identification of new epitopes recognized by human monoclonal antibodies with neutralizing and antibody-dependent cellular cytotoxicity activities specific for human T cell leukemia virus type 1. *J. Immunol.* **1992**, *149*, 940–948. [PubMed]

29. Gövert, F.; Krumbholz, A.; Witt, K.; Hopfner, F.; Feldkamp, T.; Korn, K.; Knöll, A.; Jansen, O.; Deuschl, G.; Fickenscher, H. HTLV-1 associated myelopathy after renal transplantation. *J. Clin. Virol.* **2015**, *72*, 102–105. [CrossRef] [PubMed]

*viruses*

MDPI

Article

# Heat Shock Enhances the Expression of the Human T Cell Leukemia Virus Type-I (HTLV-I) Trans-Activator (Tax) Antigen in Human HTLV-I Infected Primary and Cultured T Cells

Marie Kunihiro [1,2], Hideki Fujii [1], Takuya Miyagi [1], Yoshiaki Takahashi [1], Reiko Tanaka [1], Takuya Fukushima [3], Aftab A. Ansari [4] and Yuetsu Tanaka [1,*]

[1]  Department of Immunology, Graduate School of Medicine, University of the Ryukyus, Okinawa 903-0215, Japan; m-kunihiro@emro.co.jp (M.K.); hfujii@med.u-ryukyu.ac.jp (H.F.); miya_skywalker2008@yahoo.co.jp (T.M.); ytakah3@med.u-ryukyu.ac.jp (Y.T.); reiko_tanaka@s5.dion.ne.jp (R.T.)
[2]  EM Research Organization Inc., Okinawa 901-2311, Japan
[3]  Laboratory of Hematoimmunology, School of Health Sciences, Faculty of Medicine, University of the Ryukyus, Okinawa 903-0215, Japan; fukutaku@med.u-ryukyu.ac.jp
[4]  Department of Pathology, Emory University School of Medicine, Atlanta, GA 30322, USA; pathaaa@emory.edu
*  Correspondence: yuetsu@s4.dion.ne.jp; Tel.: +81-398-895-3331

Academic Editor: Louis Mansky
Received: 21 April 2016; Accepted: 1 July 2016; Published: 11 July 2016

**Abstract:** The environmental factors that lead to the reactivation of human T cell leukemia virus type-1 (HTLV-I) in latently infected T cells in vivo remain unknown. It has been previously shown that heat shock (HS) is a potent inducer of HTLV-I viral protein expression in long-term cultured cell lines. However, the precise HTLV-I protein(s) and mechanisms by which HS induces its effect remain ill-defined. We initiated these studies by first monitoring the levels of the trans-activator (Tax) protein induced by exposure of the HTLV-I infected cell line to HS. HS treatment at 43 °C for 30 min for 24 h led to marked increases in the level of Tax antigen expression in all HTLV-I-infected T cell lines tested including a number of HTLV-I-naturally infected T cell lines. HS also increased the expression of functional HTLV-I envelope gp46 antigen, as shown by increased syncytium formation activity. Interestingly, the enhancing effect of HS was partially inhibited by the addition of the heat shock protein 70 (HSP70)-inhibitor pifithlin-μ (PFT). In contrast, the HSP 70-inducer zerumbone (ZER) enhanced Tax expression in the absence of HS. These data suggest that HSP 70 is at least partially involved in HS-mediated stimulation of Tax expression. As expected, HS resulted in enhanced expression of the Tax-inducible host antigens, such as CD83 and OX40. Finally, we confirmed that HS enhanced the levels of Tax and gp46 antigen expression in primary human CD4+ T cells isolated from HTLV-I-infected humanized NOD/SCID/γc null (NOG) mice and HTLV-I carriers. In summary, the data presented herein indicate that HS is one of the environmental factors involved in the reactivation of HTLV-I in vivo via enhanced Tax expression, which may favor HTLV-I expansion in vivo.

**Keywords:** HTLV-I; Tax; heat shock; adult T cell leukemia (ATL)

## 1. Introduction

The Human T cell leukemia virus type-I (HTLV-I) is the first human retrovirus that is etiologically associated with adult T cell leukemia (ATL) and HTLV-I associated myelopathy/tropical spastic paraparesis (HAM/TSP) [1–3]. HTLV-I is prevalent worldwide with foci of high prevalence in southwest Japan, the Caribbean islands, South America and parts of Central Africa, most of which are located in subtropical and tropical regions [4]. HTLV-I is transmitted through contact with bodily fluids containing infected cells most often either vertically from mother to child via breastfeeding or horizontally in adults [5]. The total rate of developing ATL or HAM/TSP is roughly estimated to be 5% among HTLV-I carriers [6]. Although the molecular basis for the development of these HTLV-I-related disorders is still unclear, high levels of HTLV-I proviral load (PVL), as shown by the number of proviral DNA copies per 100 cells, is suggested to be one of the risk factors for the diseases [7].

HTLV-I is dormant in vivo at least in peripheral blood or lymph nodes, so that freshly isolated lymphoid cells from HTLV-I infected individuals do not express detectable levels of mRNA or proteins of HTLV-I [8]. However, the continued presence of strong $CD8^+$ cytotoxic T lymphocyte (CTL) responses and readily detectable levels of antibodies specific for HTLV-I antigens in not only ATL and HAM/TSP patients but also asymptomatic HTLV-I carriers clearly indicate that repeated production of HTLV-I must occur in vivo. In accordance, once fresh peripheral blood mononuclear cells (PBMCs) of HTLV-I infected individuals are cultured in vitro for a short time, they begin to produce HTLV-I antigens. Little is known, however, about the mechanism(s) for HTLV-I activation in vivo. So far, several lines of in vitro studies have shown that HTLV-I is activated upon host cell activation by a variety of stimuli. These include lactoferrin [9], prostaglandin E2 (PGE2) [10], T cell activation agents such as anti-CD3 plus anti-CD28, anti-CD2 antibodies, mitogens including phytohaemagglutinin (PHA) and phorbol myristate acetate (PMA) [11–13], and agents that induce cellular stress-including HS and oxidation [14,15].

It is well established that heat shock proteins (HSPs) play roles not only in housekeeping functions by serving as molecular chaperones for cell survival but also in supporting gene expression of various DNA and RNA viruses [16]. According to the geographic bias of HTLV-I prevalence towards regions in subtropical and tropical areas and the finding that HS up-regulates HTLV-I synthesis in infected T cell lines in vitro [14,15], we hypothesized that prolonged exposure to strong sunlight may be one of the environmental factors for HTLV-I reactivation in vivo. This view is supported by the fact that infrared radiation that accounts for approximately 40% of the solar radiation energy that reaches the ground leads to the generation of heat and increases skin temperature at a level that induces HS [17]. Because HTLV-I production is dependent on its enhancer onco-protein, trans-activator (Tax), it may be possible that a thermal shock stimulates Tax expression at first followed by trans-activation of HTLV-I structural proteins and activation of Tax-inducible cellular proteins including OX40 and CD83 [18]. In accordance, it has been shown that HTLV-I Tax protein does associate with a number of HS proteins, such as HSP90 and HSP70 [19,20].

Using our library of anti-HTLV-I monoclonal antibodies (mAbs), and ATL patients' and HTLV-I carriers' blood samples, we determined the effect of HS in more detail by studying the expression of HTLV-I antigens in various HTLV-I-infected T cell lines and primary human $CD4^+$ T cells. We report herein that HS significantly enhanced the expression of Tax followed by enhanced expression of gp46 along with Tax-inducible host proteins. These results reported herein suggest a role for environmental heat stress on HTLV-I reactivation in vivo.

## 2. Materials and Methods

### 2.1. Reagents

The medium used throughout the studies consisted of RPMI 1640 medium (Sigma-Aldrich Inc., St. Louis, MO, USA), supplemented with 10% fetal calf serum, 100 U/mL penicillin and 100 µg/mL streptomycin (hereinafter called RPMI medium). The IL-2-dependent T cell lines utilized herein were maintained in vitro in RPMI medium containing 20 U/mL of recombinant human IL-2. Our mouse and rat monoclonal antibodies (mAbs) utilized in this study included mouse IgG3 anti-HTLV-I Tax (clone Lt-4) [21], rat IgG2a anti-HTLV-I gp46 (clone LAT-27) [22], and mouse anti-OX40 (clone B-7B5) [23]. These in-house mAbs were purified from the ascites fluids of CB.17-SCID mice by ammonium sulfate precipitation followed by gel filtration using Superdex G-200 (GE Healthcare,

Tokyo, Japan). Aliquots of these mAbs were labeled with fluorescein isothiocyanate (FITC), HyLite Fluor™ 488, HyLite Fluor™ 647 or HRP (Dojindo, Kumamoto, Japan) according to the manufacturer's instructions. Anti-human CD4 and CD83 mAb were purchased from Beckman Coulter, Inc. (Brea, CA, USA) and BioLegend (SanDiego, CA, USA), respectively. Anti- HSP70 antibody (clone: C92F3A-5) was purchased from StressMarq Biosciences Inc (Victoria, BC, Canada). Zerumbone (ZER) [24] and pifithlin-μ (PFT) [25] were purchased from Focus Biomolecules LLC (Plymouth Meeting, PA, USA) and Sigma-Aldrich. Inc, respectively. ZER and PFT were dissolved in DMSO and stored at $-20\ ^{\circ}$C until used. Mitomycin-C (MMC) was purchased from Kyowa Kirin (Tokyo, Japan) and used for the inactivation of live cells by incubation of cells at 50 μg/mL in RPMI medium at 37 $^{\circ}$C for 30 min. Arsenite ($As_2O_3$) and N-acetylcysteine (NAC) were purchased from Sigma-Aldrich, Inc. (St. Louis, MO, USA). Cell proliferation was evaluated using the Cell Counting Kit-8 (CCK-8) (Dojindo Laboratories, Kumamoto, Japan).

## 2.2. Cell Culture and Heat Shock (HS)

The IL-2-dependent T cell lines utilized herein included an ATL patient-derived CD4$^+$ ILT-H2 and the HAM/TSP patient-derived CD8$^+$ ILT-M1 [26]. Other CD4$^+$ T cell lines utilized were established in our laboratory from ATL patients and included ATL-026i, ATL-056i, ATL-083i, and the cell line YT/cM1 that was established from CD4$^+$ T cells isolated from a normal donor. The HTLV-I-unrelated T cell line that was used for the syncytium formation assay was the Jurkat cell line (ATCC, Rockville, MD, USA) [26]. An ATL-derived IL-2-independent and HTLV-I-producing B cell line, ATL-040 established in our laboratory was used for in vivo infection of NOD/SCID/γc null mice (NOG mice) with HTLV-I. The JPX-9 cell line in which HTLV-I Tax antigen can be induced by cultivation in the presence of 10 μM cadmium (Cd) was used to determine the effect of HS on the expressions of Tax-inducible CD83 and OX40 [18].

For HS treatment, cells in growth media in a volume of 1 mL in 15 mL plastic conical tubes (BD Biosciences, San Diego, CA, USA) were heated in a water bath at various temperatures for various times as indicated in the text. After heating, these cells were dispensed into 24-well culture plates (BD Biosciences, San Diego, CA, USA) and cultured at 37 $^{\circ}$C for 12~72 h. The protocol for the use of human PBMCs and animals were approved by the Human Institutional Review Board and the Institutional Animal Care and Use Committee on clinical and animal research of the University of the Ryukyus, prior to initiation of the present study.

## 2.3. Flow Cytometry (FCM) Analysis

The polychromatic phenotypic analysis of cells was performed as described previously [18,26]. Briefly, live cells were first Fc receptor-blocked with 1 mg/mL normal human IgG in FACS buffer (phosphate-buffered saline (PBS) containing 0.2% bovine serum albumin (BSA) and 0.1% sodium azide) for 15 min on ice. For cell surface staining, aliquots of these cells in a 96-well U-bottom plate were incubated with a panel of mAbs for 30 min on ice. For intra-cellular Tax staining, cells were fixed in 4% paraformaldehyde (PFA) for 5 min and washed in FACS buffer containing 1% BSA and 0.5% saponin, and then aliquots of these cells were stained with HyLite Fluor™ (AnaSpec Inc., Fremont, CA, USA) 647-labeled anti-Tax mAb (Lt-4) for 30 min on ice. Negative control cells were stained with fluorochrome-labeled Lt-4 in the presence of 100 μg/mL of unlabeled Lt-4 as shown previously [18]. Cells were washed and re-suspended in 1% PFA-FACS buffer, and analyzed using FACS Calibur (BD Biosciences, San Diego, CA, USA). Data was collected on a minimum of 100,000 events and analyzed using either the FlowJo (TreeStar, Inc., Ashland, OR, USA) or CellQuest software (BD Biosciences, Version 6.0).

## 2.4. Syncytium Formation Assay

The syncytium formation assay was performed as reported previously [26]. Briefly, YT/cM1 cells with or without HS treatment followed by in vitro culture for 24 h were co-cultured with an equal number of Jurkat cells in 96-well U-bottom plate at 37 °C. After 15 h, the cell cultures were transferred into a 96-well flat-bottomed well plate and syncytia were counted in 5 randomly selected fields of each well at a magnification of 100× and the mean values calculated.

*2.5. Enzyme-Linked Immunosorbent Assay (ELISA)*

Tax antigen concentration in whole cell lysates were determined using our in-house formulated and standardized enzyme-linked immunosorbent assay (ELISA) kit using a pair of anti-Tax mAbs that included the capture TAXY-7 mAb [27] and the detector HRP-labeled WATM-1 mAb [28]. Recombinant Tax protein (Proteintech, Rosemont, IL, USA) was used as a standard. The sensitivity of the assay was determined to be 0.5 ng/mL. Cell lysates were prepared by lysis of cells with a low-salt extraction buffer on ice for 30 min [29]. Protein concentration of each cell lysate sample was determined using Quick Start protein assay kit (Bio-Rad Laboratories, Hercules, CA, USA).

*2.6. Primary human CD4⁺ T Cells Infected with Human T Cell Leukemia Virus Type-I (HTLV-I)*

The protocol for infection of humanized mice with HTLV-I has been described previously [30]. Briefly, purified PBMCs ($4 \times 10^7$ cells/mouse) from HTLV-I-negative donors were transplanted into the peritoneal cavity of NOG mice together with MMC-treated HTLV-I producing ATL-040 cell line ($1 \times 10^7$ cells/mouse). Two weeks later, cells in the peritoneal lavage were collected, either untreated or heat-treated, and cultured for 1~2 days. The expression of Tax antigen in CD4⁺ human T cells was analyzed using flow cytometry (FCM) on day 1 and 2. In addition, whole heparinized peripheral blood of HTLV-I carriers was diluted 1:4 in RPMI medium, aliquoted, and an aliquot heat treated and another sham treated, and both cultured. The cells in the cultures were stained with anti-CD4, anti-Tax and anti-gp46 mAbs, and analyzed using FCM after hypotonic lysis of the red blood cells.

*2.7. Statistical Analysis*

Data were tested for statistical significance by either paired or unpaired Student's *t* test using Prism software (GraphPad Software, Version 4.03). Data from more than three-armed experiments were analyzed by one-way analysis of variance (ANOVA) with post hoc Holm test and Tukey test.

## 3. Results

*3.1. HS Up-Regulates the Expression of the HTLV-I Trans-Activator (Tax) Antigen*

At first, in order to determine whether HS affects the expression of Tax antigen in HTLV-I-infected T cells, we examined two IL-2-dependent CD4⁺ T cell lines generated from acute ATL patients, ATL-026i and ATL-056i. Aliquots of these cell lines were heated at various temperatures, 37, 39, 41, 43 and 45 °C for 30 min and cultured for 24 h. The intra-cellular expression of Tax and HSP70 antigens was analyzed by FCM. Figure 1a shows that while the frequencies of Tax-expressing cells increased by HS at 43 and 45 °C in the ATL-026i cell line, the ATL-056i cell line had a broader range from 39~45 °C for Tax expression. The enhanced expression of HSP70, a direct indicator of HS, was also observed by HS at 43 and 45 °C in the two cell lines. HS at 45 °C resulted in decreased cell viability as determined using a sensitive CCK-8 cell counting assay. Because the enhanced Tax expression reached a plateau by heating at 43 °C for 30 min, and that HSP70 expression was apparently enhanced at 43 °C, all subsequent studies were carried out with HS treatment at 43 °C.

Next, we determined the optimum exposure time for enhanced Tax expression. As shown in Figure 1b, incubation for 30 min was sufficient for the enhanced expression of both Tax and HSP70 with minimum cytotoxic effect. On the basis of these results, all subsequent studies were carried out using HS at 43 °C for 30 min. It is noteworthy that the MFI for Tax⁺ cells also slightly increased under HS at both bulk and single cell levels as shown in Supplemental Figures S1 and S2.

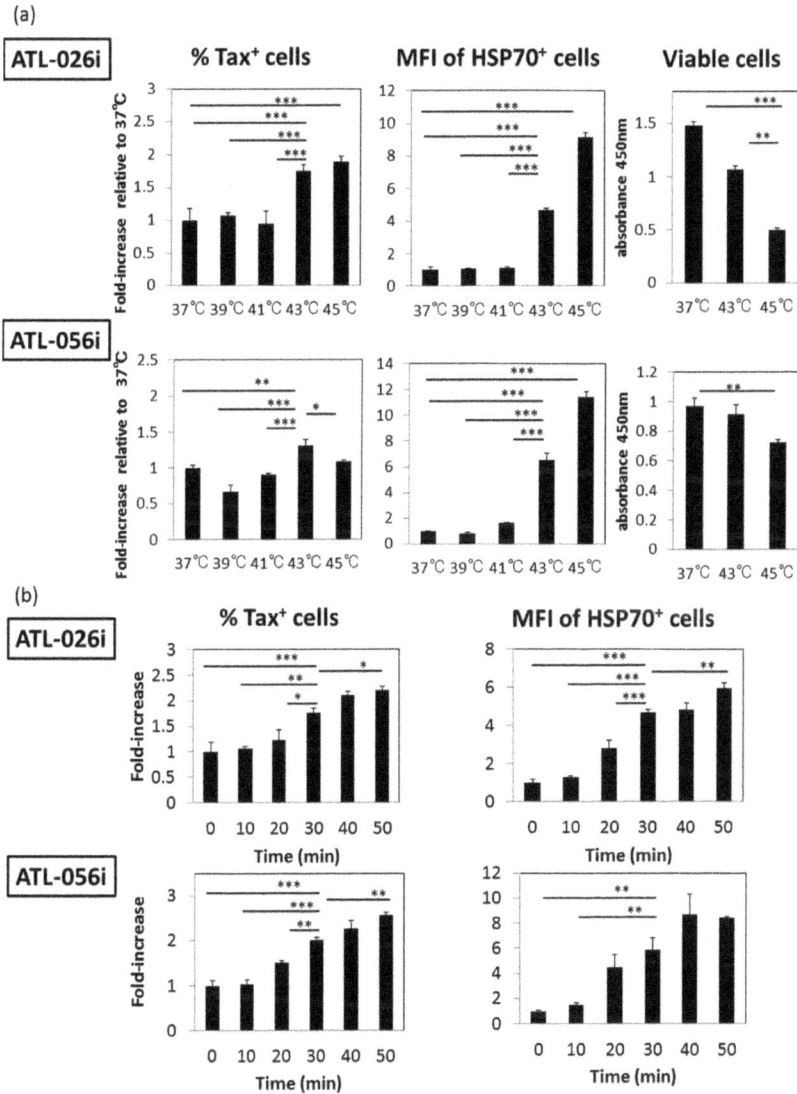

**Figure 1.** Effects of heat shock (HS) exposure on human T cell leukemia virus type-I (HTLV-I)-infected cell lines derived from acute adult T cell leukemia (ATL) patients: (**a**) Aliquots of ATL-026i and ATL-056i cells were incubated at various temperatures for 30 min and cultured for 24 h. The cells were then analyzed for the frequencies of trans-activator (Tax)$^+$ cells (left bar graphs) by flow cytometry (FCM) and the relative density (Mean Fluorescent Intensity, MFI) of heat shock protein 70 (HSP70) expression (middle bar graphs) and for cell viability using the CCK-8 cell counting kit (right bar graphs). (**b**) The kinetics of the up-regulation of Tax and HSP70 expression by the same two cell lines following exposure to 43 °C for various times is shown. The values denote the means ± SD. * $p < 0.05$, ** $p < 0.01$, *** $p < 0.001$.

*3.2. HS Increases the Total Amount of Tax Protein*

The intra-cellular localization of Tax has been shown to be altered in response to various forms of cellular stress, such as HS and ultra violet (UV) light, resulting in an increase in cytoplasmic Tax about 1~2 h after treatment and a decrease in Tax speckled structures [15], which might affect Tax detection by FCM. In order to confirm the enhancing effect of HS on Tax expression, we quantified the levels of total Tax protein in whole cell lysates by using our in-house Tax-specific ELISA. As shown in Figure 2, the levels of Tax protein increased significantly by exposure to HS in three distinct T cell lines including two ATL-derived CD4⁺ T cell lines and an in vitro- HTLV-I-immortalized CD4⁺ T cell line prepared from a normal donor (YT/cM1). These data thus confirm the fact that exposure to HS treatment increases the total amount of Tax antigen per culture during the 24 h culture period.

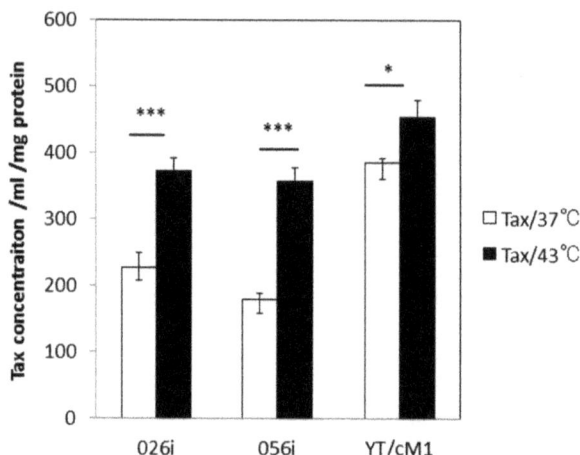

**Figure 2.** Effect of HS on total Tax protein expression: Aliquots of the ATL-derived cell lines (ATL-026i and ATL-056i) and the HTLV-I-immortalized CD4⁺ T cell line (YT/cM1) were incubated at either 37 °C or 43 °C for 30 min followed by incubation for 24 h in triplicates wells. The cells were then collected, washed, lysed in lysis buffer and the total protein concentrations were determined. Then amounts of total Tax concentration/mg total protein were quantitated using our Tax specific enzyme-linked immunosorbent assay ELISA. The values denote the means $\pm$ SD. * $p < 0.05$, *** $p < 0.001$.

*3.3. HS Up-Regulates a Functional Form of Envelope gp46.*

Next, we examined whether exposure to HS influences the expression of the HTLV-I structural envelope protein gp46 whose expression is known to be enhanced by Tax-mediated transactivation. As shown in Figure 3a,b, the up-regulation of gp46 antigen expression was apparent in three HTLV-I-infected T cell lines including the CD4⁺ and CD8⁺ T cell line (ILT-M1). In Figure 3a, it was obvious that gp46⁺ Tax⁻ cells also increased after HS. We speculate that the gp46 on Tax-negative cells may represent biofilm of HTLV-I particles produced from Tax-positive cells, or alternatively, gp46⁺ Tax⁻ cells may be at a resting phase after a productive infection phase. It is noteworthy that syncytium-forming capacity which is an indicator of HTLV-I infectivity, was enhanced significantly by the exposure of YT/cM1 cells to HS (Figure 3c,d), indicating that HS also enhanced the functional form of gp46 expression. In addition, production of gag p24 in the culture supernatant of YT/cM1 cells was also enhanced by HS (Supplemental Figure S3).

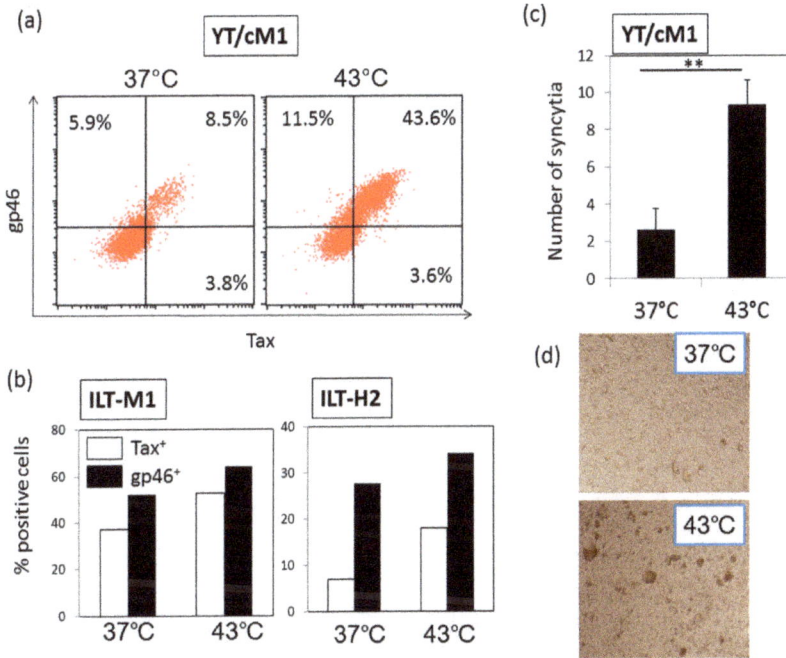

**Figure 3.** Enhancement of HTLV-I envelope gp46 expression by HS: Three HTLV-I-producing T cell lines CD4[+] YT/cM1, an HTLV-I associated myelopathy/tropical spastic paraparesis (HAM/TSP)-derived CD8[+] ILT-M1 and an ATL-derived CD4[+] ILT-H2, were either exposed to HS or mock treated and cultured for 24 h. (**a,b**) The levels of gp46 along with Tax antigen expression were determined by FCM. Data shown are representative of three independent experiments. (**c**) Syncytium formation capacity was compared between untreated and heat shocked YT/cM1 cells by a co-culture method using Jurkat cells as the indicator cell line. The numbers of syncytia were counted in triplicate cultures and the means $\pm$ SD are shown. ** $p < 0.01$. (**d**) Syncytia formed in each culture were microscopically observed using an inverted microscope at magnification of $100\times$.

### 3.4. Time Course Effects of HS

We examined the kinetics by which exposure to HS leads to optimal expression of HSP70, Tax and gp46 using two HTLV-I[+] T cell lines (ATL-026i and YT/cM1). As shown in Figure 4, there appeared to be a sequential increase in HSP70 expression peaking at during 12~24 h, followed by Tax and then finally gp46 antigen expression. The expression levels of these three antigens returned to control levels by ~72 h (unpublished data, [31]). The delay in enhanced gp46 expression in comparison to Tax antigen suggests that gp46 expression was dependent on Tax antigen expression.

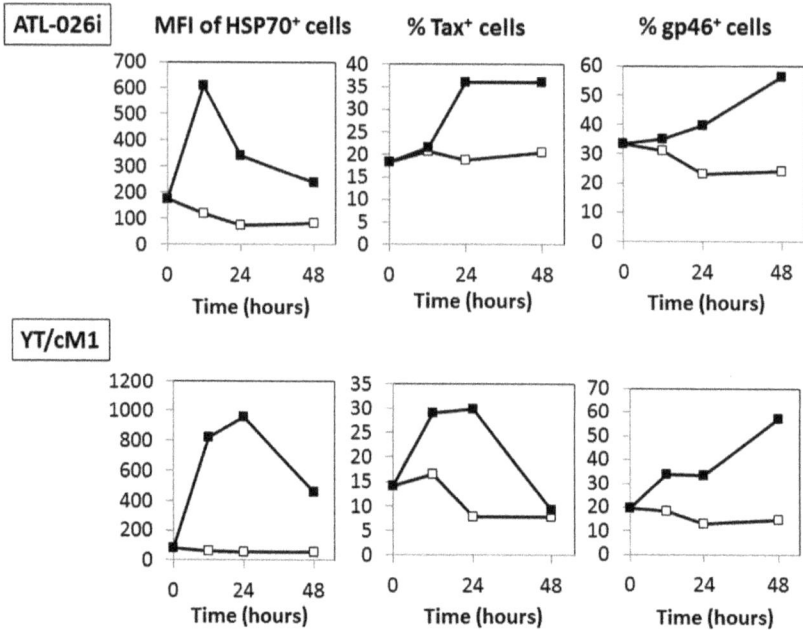

**Figure 4.** A comparison of time course of the expression among HSP70, Tax and gp46 antigens followed by exposure to HS: The ATL-026i (upper panel) and YT/cM1 (lower panel) cells were either exposed to HS (black square) or mock treated (open square), cultured for 12, 24 and 48 h, and then their phenotypes were analyzed by FCM. Data shown are representative of three independent experiments.

### 3.5. A Possible Involvement of Heat Shock Protein 70 (HSP70)

In attempts to test a possible involvement of HSP70 in the enhanced expression of Tax and gp46, we tested the effects of the HSP70-inducing chemical ZER and the HSP70 functional inhibitor PFT. The results of preliminary studies indicated that the optimal concentrations to be utilized for ZER was noted at 10 μM and PFT at 1 μM without any detectable effect on cell viability and cell growth. As shown in Figure 5a, incubation of cultures with ZER for 24 h significantly up-regulated both the frequencies of Tax+ and HSP70+ cells in the absence of HS. On the other hand, PFT treatment inhibited the increase in the frequencies of Tax+ cells by HS (Figure 5b). HSP70 expression itself was not altered by incubation with PFT since PFT is only a functional inhibitor of HSP70. These results indicate that HSP70 might be involved at least in part in the HS-induced enhancement of Tax antigen expression.

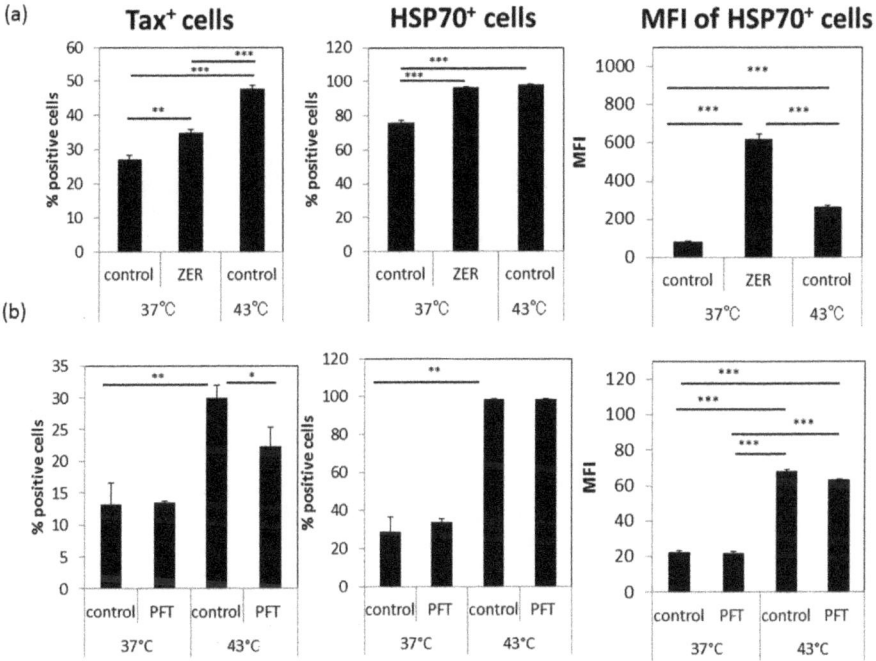

**Figure 5.** The effects of the HSP70 inducing and inhibiting agents on the expression of Tax antigen: Triplicate cultures of YT/cM1 cells were either exposed to HS or mock treated, and then cultured in the presence or absence of either (**a**) 10 μM zerumbone (ZER, HSP70 inducer) or (**b**) 1 μM pifithlin-μ (PFT, HSP70 functional inhibitor) for 24 h. The Tax and HSP70 expression was analyzed by FCM. The values denote the means ± SD. * $p < 0.05$, ** $p < 0.01$, *** $p < 0.001$.

### 3.6. Effect of HS on Tax-Inducible Host Antigens CD83 and OX40

Because Tax is known to up-regulate a variety of host antigens, some of which are associated with immortalization or transformation of HTLV-I-infected T cells, we examined the effect of HS on the expression of some of the Tax-inducible host antigens, including CD83 and OX40. As shown in Figure 6a, the levels of CD83 and OX40 expression were significantly up-regulated by exposure to HS in three HTLV-I-infected CD4$^+$ T cell lines. Representative dot blots are shown in Supplemental Figure S4. To demonstrate the involvement of Tax in these enhanced CD83 and OX40 antigen expressions, we examined the effect of HS exposure on JPX-9 cells which carry the Cd inducible Tax gene. Figure 6b showed that incubation of the JPX-9 cells in media containing Cd and HS exposure induced not only increased levels of Tax but also CD83 and OX40 on the cell surface. Thus, it was apparent that HS enhanced the expression of not only Tax antigen but also Tax-inducible host antigens.

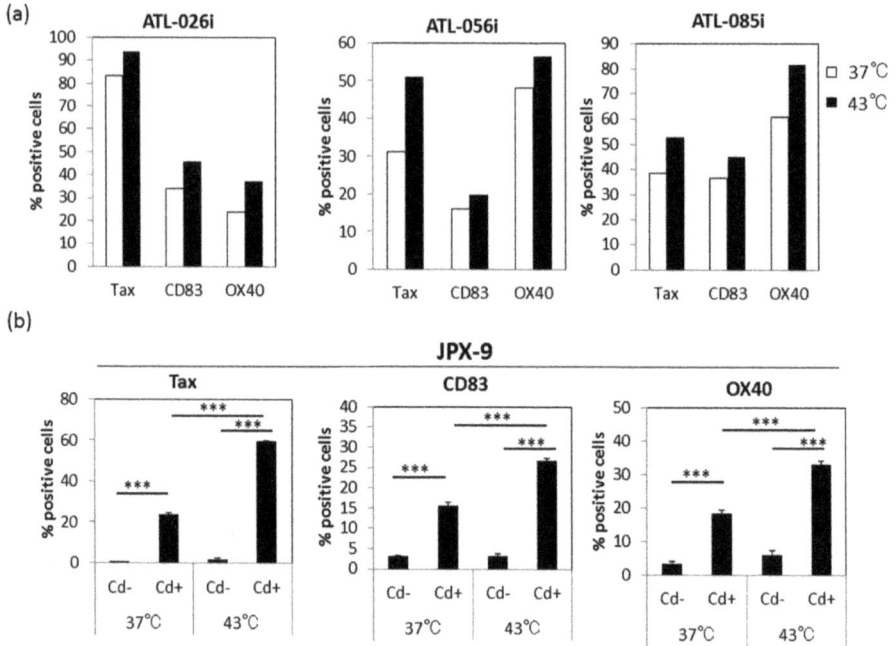

**Figure 6.** The up-regulation of the expression of Tax-inducible host antigens by exposure to HS: The ATL-derived CD4⁺ T cell lines (**a**) and JPX-9 cell line (**b**) were either exposed to HS or mock treated, and then cultured for 24 h in the presence or absence 1 μM Cd (Tax inducing agent in JPX-9 cells). The frequencies (%) of Tax antigen and Tax-inducible host antigens including CD83 and OX40 expressing cells were analyzed by FCM. Data shown are representative of three independent experiments. In (**b**), experiments were performed in triplicates, with incubation in media alone (Cd-) or media containing Cd (Cd+) and the frequencies of Tax, CD83 and OX40 cells determined. The means ± SD values are shown. *** $p < 0.001$.

### 3.7. Effect of HS Exposure on Primary Human CD4⁺ T Cells Infected with HTLV-I

Finally, we examined the effect of HS exposure on HTLV-I-infected primary human CD4⁺ T cells, including those obtained from HTLV-I-infected humanized NOG mice and HTLV-I carriers. As shown in Figure 7a, exposure to HS resulted in significant increase in the frequencies of Tax⁺ CD4⁺ T cells after in vitro culture for 24 h ($p < 0.05$), that returned to control levels by 48 h. Figure 7b shows that exposure to HS resulted in increase in the expression of Tax antigen in primary CD4⁺ T cells from three HTLV-I carriers ($p < 0.05$). Representative dot blots are shown in Supplemental Figure S5. Taken together, these data support the view that HTLV-I activation by exposure to HS is a general phenomenon rather than restricted to having effects on long-term cultured cell lines.

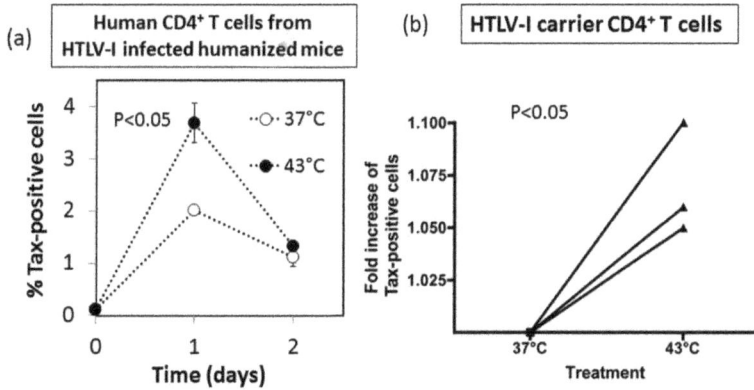

**Figure 7.** Effect of HS exposure on primary human CD4[+] T cells infected with HTLV-I: (a) NOD/SCID/γc null mice (NOG mice, $n = 3$) were humanized by transplantation with normal human peripheral blood mononuclear cells (PBMCs) and infected with HTLV-I (as described in Materials and Methods). After 2 weeks, cells within the peritoneal lavage were collected and either exposed to HS (black circle) or mock treated (open circle). These cells were cultured in triplicates and examined for the frequencies of Tax antigen expressing cells on day 0, 1 and 2. The values denote the means ± SE. * $p < 0.05$; (b) Aliquots of diluted whole blood from three different HTLV-I carriers were either exposed to HS (43 °C) or mock treated, and then cultured for 24 h. The gated population of CD4[+] T cells was analyzed for the expression of Tax and gp46 antigens by FCM.

## 4. Discussion

In the present study, we demonstrated that exposure to HS up-regulates the expression of HTLV-I Tax antigen during in vitro culture for 12~48 h along with induction of functional HTLV-I gp46 and the Tax-inducible host antigens CD83 and OX40 in not only human HTLV-I-infected T cell lines but also in primary human CD4[+] T cells. We conclude from these observations that HS is one of the environmental factors that can potentially activate integrated HTLV-I provirus in vitro in latently infected T cells, although in vivo studies are yet to be performed.

In the literature, Andrew et al. reported that heat stress enhances the expression of HTLV-I envelope and gag antigens in long-term cultured HTLV-I-transformed T cells lines, such as HUT-102 and MT-2 cells [14,32,33], and suggested that the enhancement is due to increased translation of HTLV-I structural proteins, but not changes in protein turnover. Since the synthesis of viral structural proteins of HTLV-I is dependent on its Tax protein, our present study suggested that Tax antigen synthesis precedes that of HTLV-I structural proteins. Indeed, results of the kinetic studies showed that the enhanced gp46 antigen expression came after enhanced Tax antigen expression (Figure 4). As expected, enhanced Tax antigen expression was accompanied by enhanced expression of some of Tax-inducible host antigens (Figure 6), indicating that enhanced Tax antigen molecules are functional. In addition, in line with the fact that HS is able to enhance HIV-1 Tat-mediated transactivation of the LTR independently of Tat expression levels, there might be an alternative mechanism for Tax-mediated enhancement of HTLV-I production upon HS.

The precise mechanism by which Tax antigen synthesis is up-regulated by exposure to HS is yet to be defined. On the basis of our preliminary data (data not shown) indicating that HS up-regulated total levels of Tax mRNA, it is possible that HS directly controls transcription. Thus, further studies are in progress to determine this possibility by measuring nuclear RNA levels for Tax RNA and genomic RNA in the cytoplasm. Since Tax operates by the activation of transcriptional factors such as the cAMP-response element-binding protein (CREB) for activation of not only viral but also host cellular genes, it can be speculated that CREB is also activated by exposure to HS in HTLV-I-infected cells.

The finding that ER stress is induced by exposure to HS [34] and in turn CREB is activated by ER stress in HeLa cells supports this view [35]. However, our preliminary FCM experiments showed that HS did not affect the levels of expression of phosphorylated forms of NF-kB and CREB molecules (Supplemental Figure S6).

As it is the case that the cellular onco-protein c-Myc protein is up-regulated following HS [36], rapid induction of Tax antigen after heat-stress may be an advantage of HTLV-I-infected cells for induction of the recovery process. In addition, enhanced Tax expression resulting in enhanced infectious HTLV-I viral production may also be an advantage for the spread of HTLV-I from heat-stressed cells. It is likely that HTLV-I infected cells after HS may become good targets for HTLV-I Tax-specific CTL [37,38] or ADCC by anti-gp46 antibodies plus NK cells [26]. However, CTL and NK cell activities are suppressed when exposed to temperatures >41 °C and 42 °C, respectively, [39,40]. We have confirmed that exposure to HS inactivates ADCC activity of PBMCs against HTLV-I-infected cells (unpublished data, [41]). Therefore, heat-stress may favor HTLV-I infection rather than immune-surveillance against HTLV-I at a local site of exposure to HS. Further studies to reveal molecular mechanisms underlying this HS-mediated HTLV-I activation are yet to be performed.

Another representative cellular stress that is critical for Tax expression is oxidative stress, such as $As_2O_3$. It has been utilized for ATL therapy as it induces the apoptosis of HTLV-I-infected cells [42]. The specificity of the effect of $As_2O_3$ on HTLV-I synthesis has been a subject of controversy. Thus, while Andrew et al. reported an enhancing effect [14,32,33], Nabeshi et al. in contrast showed a suppressing effect [43]. These discrepancies seem to be due to different experimental conditions. In our preliminary studies, we could not observe $As_2O_3$ mediated enhancement of Tax and gp46 expression in various conditions and various types of cell lines (Supplemental Figure S7). Thus, it is likely that HS stress and $As_2O_3$-mediated oxidative stress may utilize distinct pathways on Tax expression by HTLV-I infected cells.

The involvement of HSPs in viral infection has been reported for several viruses [44–46] including HTLV-I [19,47–50]. Our present data indicates that among the HSP family, HSP70 might be associated at least in part with HS-mediated enhancement of Tax antigen expression based on the data obtained with the use of the HSP70-inducer, ZER, and the HSP70 functional inhibitor PFT (Figure 5). The increased numbers of HSP70 molecules after exposure to HS may serve to function as molecular chaperon for Tax and the other HTLV-I antigen molecules together with the other constitutively expressed HSPs including HSP90 [49]. Further studies are in progress to determine what types of HSP are involved in the HS-mediated enhanced Tax expression.

Infrared sunlight is known to generate heat and for their ability to increase skin temperature depending on exposure dose and time. It has been reported that intra-dermal-temperature reaches up to 44 °C when irradiated with IR at 970 nm at 80 mW/$cm^2$ within 15 min [51]. Thus, it can be speculated that exposure to direct sunlight in subtropical or tropical areas might raise the skin temperature to above 43 °C resulting in HS of HTLV-I-infected cells. In addition, skin inflammation generated by exposure to strong solar UV radiation (sunburn) may cause the recruitment of HTLV-I-infected cells to the sites of inflammation. Furthermore, systemic physiological hyperthermia induced by either direct exposure to sunlight or an opportunistic infection (for example, infection with Strongyloides stercoralis) might also stimulate HTLV-I expression in vivo, as is the case for HIV-1 [52]. Although we failed to demonstrate that activation of HTLV-I-infected $CD4^+$ T cell lines via CD3 molecule further enhanced the effect of HS on Tax expression (Supplemental Figure S8), it is also possible that immune activation of fresh HTLV-I-infected T cells in vivo might synergize HS.

In this way, it is likely that, in subtropical and tropical areas, repeated skin exposure to strong sunlight for a prolonged time can cause HTLV-I reactivation in vivo.

## 5. Conclusions

In conclusion, results of the present study indicate that a mild exposure to heat shock may be one of the natural environmental factors that stimulate Tax antigen expression leading to not only reactivation of HTLV-I in latently infected T cells but also induction of expression of Tax-inducible host antigens, which may favor expansion of HTLV-I infection in vivo. It can be speculated that exposure to strong sunlight in subtropical and tropical areas is involved in the geographic bias of increased HTLV-I prevalence within these regions, and that, for HTLV-I carriers, precaution against exposure to strong sunlight for prolonged periods of time might be beneficial to prevent reactivation of HTLV-I.

**Supplementary Materials:** The following figures are available online at www.mdpi.com/1999-4915/8/7/191/s1. **Figure S1.** HS increased the MFI of Tax$^+$ cells. **Figure S2.** Correlation of Tax and HSP70 expression; **Figure S3.** HS increased p24 concentration; **Figure S4.** Correlation between intra-cellular Tax and cell surface expression of OX40 and CD83; **Figure S5.** HS upregulated expression of Tax and gp46 in fresh PBMC from careers; **Figure S6.** Effect of HS on the expression of NF-kB and CREB molecules; **Figure S7.** The failure of exposure to oxidative stress, As$_2$O$_3$, to induce increased expression of Tax and gp46; **Figure S8.** Failure to stimulate HTLV-I-infected cell lines with anti-CD3 mAb (OKT-3).

**Acknowledgments:** This work was supported by grants from the Ministry of Education, Culture, Sports, Science, and from the Project of Establishing Medical Research Base Networks against Infectious Diseases in Okinawa. We thank Takeshi Sairenji for beneficial discussions and comments.

**Author Contributions:** MK performed research and wrote the manuscript, HF and TM supported animal experiments, YTak processed human blood samples, RT generated mAbs, TF collected blood samples from HTLV-I-infected donors, and AAA and YT conceived and designed the study, YT also performed a part of research and provided funding for this study. All authors contributed to the final version of the manuscript, read and approved it.

**Conflicts of Interest:** The authors declare no conflict of interest.

## References

1. Hinuma, Y.; Nagata, K.; Hanaoka, M.; Nakai, M.; Matsumoto, T.; Kinoshita, K.I.; Shirakawa, S.; Miyoshi, I. Adult T-cell leukemia: Antigen in an ATL cell line and detection of antibodies to the antigen in human sera. *Proc. Natl. Acad. Sci. USA* **1981**, *78*, 6476–6480. [CrossRef] [PubMed]

2. Poiesz, B.J.; Ruscetti, F.W.; Reitz, M.S.; Kalyanaraman, V.S.; Gallo, R.C. Isolation of a new type C retrovirus (HTLV) in primary uncultured cells of a patient with sezary T-cell leukaemia. *Nature* **1981**, *294*, 268–271. [CrossRef] [PubMed]

3. Jacobson, S.; Raine, C.S.; Mingioli, E.S.; McFarlin, D.E. Isolation of an HTLV-1-like retrovirus from patients with tropical spastic paraparesis. *Nature* **1988**, *331*, 540–543. [CrossRef] [PubMed]

4. Gessain, A.; Cassar, O. Epidemiological Aspects and World Distribution of HTLV-1 Infection. *Front. Microbiol.* **2012**, *3*, 388. [CrossRef] [PubMed]

5. Watanabe, T. Current status of HTLV-1 infection. *Int. J. Hematol.* **2011**, *94*, 430–434. [CrossRef] [PubMed]

6. Arisawa, K.; Soda, M.; Endo, S.; Kurokawa, K.; Katamine, S.; Shimokawa, I.; Koba, T.; Takahashi, T.; Saito, H.; Doi, H.; Shirahama, S. Evaluation of adult T-cell leukemia/lymphoma incidence and its impact on non-Hodgkin lymphoma incidence in southwestern Japan. *Int. J. Cancer* **2000**, *85*, 319–324. [CrossRef]

7. Iwanaga, M.; Watanabe, T.; Utsunomiya, A.; Okayama, A.; Uchimaru, K.; Koh, K.R.; Ogata, M.; Kikuchi, H.; Sagara, Y.; Uozumi, K.; et al. Human T-cell leukemia virus type I (HTLV-1) proviral load and disease progression in asymptomatic HTLV-1 carriers: A nationwide prospective study in japan. *Blood* **2010**, *116*, 1211–1219. [CrossRef] [PubMed]

8. Kurihara, K.; Harashima, N.; Hanabuchi, S.; Masuda, M.; Utsunomiya, A.; Tanosaki, R.; Tomonaga, M.; Ohashi, T.; Hasegawa, A.; Masuda, T.; et al. Potential immunogenicity of adult T cell leukemia cells in vivo. *Int. J. Cancer* **2005**, *114*, 257–267. [CrossRef] [PubMed]

9. Moriuchi, M.; Moriuchi, H. A milk protein lactoferrin enhances human T cell leukemia virus type I and suppresses HIV-1 infection. *J. Immunol.* **2001**, *166*, 4231–4236. [CrossRef] [PubMed]

10. Moriuchi, M.; Inoue, H.; Moriuchi, H. Reciprocal interactions between human t-lymphotropic virus type 1 and prostaglandins: Implications for viral transmission. *J. Virol.* **2001**, *75*, 192–198. [CrossRef] [PubMed]

11. Guyot, D.J.; Newbound, G.C.; Lairmore, M.D. Signaling via the CD2 receptor enhances HTLV-1 replication in T lymphocytes. *Virology* **1997**, *234*, 123–129. [CrossRef] [PubMed]

12. Lin, H.C.; Dezzutti, C.S.; Lal, R.B.; Rabson, A.B. Activation of human T-cell leukemia virus type 1 TAX gene expression in chronically infected T cells. *J. Virol.* **1998**, *72*, 6264–6270. [PubMed]

13. Lin, H.C.; Hickey, M.; Hsu, L.; Medina, D.; Rabson, A.B. Activation of human T cell leukemia virus type 1 LTR promoter and cellular promoter elements by T cell receptor signaling and HTLV-1 TAX expression. *Virology* **2005**, *339*, 1–11. [CrossRef] [PubMed]

14. Andrews, J.M.; Oglesbee, M.J.; Trevino, A.V.; Guyot, D.J.; Newbound, G.C.; Lairmore, M.D. Enhanced human T-cell lymphotropic virus type-1 expression following induction of the cellular stress-response. *Virology* **1995**, *208*, 816–820. [CrossRef] [PubMed]

15. Gatza, M.; Marriott, S. Genotoxic stress and cellular stress alter the subcellular distribution of human T-cell leukemia virus type 1 TAX through a CRM1-dependent mechanism. *J. Virol.* **2006**, *80*, 6657–6668. [CrossRef] [PubMed]

16. Kim, M.Y.; Oglesbee, M. Virus-heat shock protein interaction and a novel axis for innate antiviral immunity. *Cells* **2012**, *1*, 646–666. [CrossRef] [PubMed]

17. Cho, S.; Shin, M.H.; Kim, Y.K.; Seo, J.-E.; Lee, Y.M.; Park, C.-H.; Chung, J.H. Effects of infrared radiation and heat on human skin aging in vivo. *J. Investig. Dermatol. Symp. Proc.* **2009**, *14*, 15–19. [CrossRef] [PubMed]

18. Tanaka, Y.; Mizuguchi, M.; Takahashi, Y.; Fujii, H.; Tanaka, R.; Fukushima, T.; Tomoyose, T.; Ansari, A.A.; Nakamura, M. Human T-cell leukemia virus type-1 TAX induces the expression of CD83 on T cells. *Retrovirology* **2015**, *12*, 56. [CrossRef] [PubMed]

19. Gao, L.; Harhaj, E.W. HSP90 protects the human T-cell leukemia virus type 1 (HTLV-1) tax oncoprotein from proteasomal degradation to support NF-kappa B activation and HTLV-1 replication. *J. Virol.* **2013**, *87*, 13640–13654. [CrossRef] [PubMed]

20. Cheng, H.; Cenciarelli, C.; Shao, Z.P.; Vidal, M.; Parks, W.P.; Pagano, M.; Cheng-Mayer, C. Human T cell leukemia virus type 1 TAX associates with a molecular chaperone complex containing HTID-1 and HSP70. *Curr. Biol.* **2001**, *11*, 1771–1775. [CrossRef]

21. Tanaka, Y.; Zeng, L.; Shiraki, H.; Shida, H.; Tozawa, H. Identification of a neutralization epitope on the envelope gp46 antigen of human T cell leukemia virus type I and induction of neutralizing antibody by peptide immunization. *J. Immunol.* **1991**, *147*, 354–360. [PubMed]

22. Takahashi, Y.; Tanaka, R.; Yamamoto, N.; Tanaka, Y. Enhancement of OX40-induced apoptosis by TNF coactivation in OX40-expressing T cell lines in vitro leading to decreased targets for HIV type 1 production. *AIDS Res. Hum. Retrovir.* **2008**, *24*, 423–435. [CrossRef] [PubMed]

23. Takahashi, Y.; Tanaka, Y.; Yamashita, A.; Koyanagi, Y.; Nakamura, M.; Yamamoto, N. OX40 stimulation by gp34/OX40 ligand enhances productive human immunodeficiency virus type 1 infection. *J. Virol.* **2001**, *75*, 6748–6757. [CrossRef] [PubMed]

24. Ohnishi, K.; Ohkura, S.; Nakahata, E.; Ishisaka, A.; Kawai, Y.; Terao, J.; Mori, T.; Ishii, T.; Nakayama, T.; Kioka, N.; et al. Non-specific protein modifications by a phytochemical induce heat shock response for self-defense. *PLoS ONE* **2013**, *8*, e58641. [CrossRef] [PubMed]

25. Goloudina, A.R.; Demidov, O.N.; Garrido, C. Inhibition of HSP70: A challenging anti-cancer strategy. *Cancer Lett.* **2012**, *325*, 117–124. [CrossRef] [PubMed]

26. Tanaka, Y.; Takahashi, Y.; Tanaka, R.; Kodama, A.; Fujii, H.; Hasegawa, A.; Kannagi, M.; Ansari, A.A.; Saito, M. Elimination of Human T cell Leukemia Virus Type-1-Infected Cells by Neutralizing and Antibody-Dependent Cellular Cytotoxicity-Inducing Antibodies Against Human T cell Leukemia Virus Type-1 Envelope gp46. *AIDS Res. Hum. Retrovir.* **2014**, *30*, 542–552. [CrossRef] [PubMed]

27. Tanaka, Y.; Yoshida, A.; Tozawa, H.; Shida, H.; Nyunoya, H.; Shimotohno, K. Production of a recombinant human T-cell leukemia virus type-1 trans-activator (TAX1) antigen and its utilization for generation of monoclonal antibodies against various epitopes on the TAX1 antigen. *Int. J. Cancer* **1991**, *48*, 623–630. [CrossRef] [PubMed]

28. Tanaka, Y.; Masuda, M.; Yoshida, A.; Shida, H.; Nyunoya, H.; Shimotohno, K.; Tozawa, H. An antigenic structure of the trans-activator protein encoded by human T-cell leukemia virus type-1 (HTLV-1), as defined by a panel of monoclonal antibodies. *AIDS Res. Hum. Retrovir.* **1992**, *8*, 227–235. [CrossRef] [PubMed]

29. Tanaka, Y.; Inoi, T.; Tozawa, H.; Yamamoto, N.; Hinuma, Y. A glycoprotein antigen detected with new monoclonal antibodies on the surface of human lymphocytes infected with human T-cell leukemia virus type-1 (HTLV-1). *Int. J. Cancer* **1985**, *36*, 549–555. [CrossRef] [PubMed]

30. Saito, M.; Tanaka, R.; Fujii, H.; Kodama, A.; Takahashi, Y.; Matsuzaki, T.; Takashima, H.; Tanaka, Y. The neutralizing function of the anti-HTLV-1 antibody is essential in preventing in vivo transmission of HTLV-1 to human T cells in NOD-SCID/γcnull (NOG) mice. *Retrovirology* **2014**, *11*, 74. [CrossRef] [PubMed]

31. Kunihiro, M.; Tanaka, Y. Transient upregulation of Tax and gp46 by heat shock. Unpublished.

32. Andrews, J.M.; Newbound, G.C.; Oglesbee, M.; Brady, J.N.; Lairmore, M.D. The cellular stress response enhances human T-cell lymphotropic virus type 1 basal gene expression through the core promotor region of the long terminal repeat. *J. Virol.* **1997**, *71*, 741–745. [PubMed]

33. Andrews, J.M.; Oglesbee, M.; Lairmore, M.D. The effect of the cellular stress response on human T-lymphotropic virus type I envelope protein expression. *J. Gen. Virol.* **1998**, *79*, 2905–2908. [CrossRef] [PubMed]

34. Kikuchi, D.; Tanimoto, K.; Nakayama, K. CREB is activated by ER stress and modulates the unfolded protein response by regulating the expression of IRE1α and PERK. *Biochem. Biophys. Res. Commun.* **2016**, *469*, 243–250. [CrossRef] [PubMed]

35. Li, L.; Tan, H.; Gu, Z.; Liu, Z.; Geng, Y.; Liu, Y.; Tong, H.; Tang, Y.; Qiu, J.; Su, L. Heat stress induces apoptosis through a Ca2+-mediated mitochondrial apoptotic pathway in human umbilical vein endothelial cells. *PLoS ONE* **2014**, *9*, e111083. [CrossRef] [PubMed]

36. Luscher, B.; Eisenman, R. c-Myc and c-Myb protein degradation: Effect of metabolic inhibitors and heat shock. *Mol. Cell. Biol.* **1988**, *8*, 2504–2512. [CrossRef] [PubMed]

37. Bangham, C.R.M.; Osame, M. Cellular immune response to HTLV-1. *Oncogene* **2005**, *24*, 6035–6046. [CrossRef] [PubMed]

38. Kannagi, M. Immunologic control of human T-cell leukemia virus type I and adult T-cell leukemia. *Int. J. Hematol.* **2007**, *86*, 113–117. [CrossRef] [PubMed]

39. Takahashi, A.; Torigoe, T.; Tamura, Y.; Kanaseki, T.; Tsukahara, T.; Sasaki, Y.; Kameshima, H.; Tsuruma, T.; Hirata, K.; Tokino, T.; et al. Heat shock enhances the expression of cytotoxic granule proteins and augments the activities of tumor-associated antigen-specific cytotoxic t lymphocytes. *Cell Stress Chaperones* **2012**, *17*, 757–763. [CrossRef] [PubMed]

40. Harada, H.; Murakami, T.; Tea, S.S.; Takeuchi, A.; Koga, T.; Okada, S.; Suico, M.A.; Shuto, T.; Kai, H. Heat shock suppresses human NK cell cytotoxicity via regulation of perforin. *Int. J. Hyperth.* **2007**, *13*, 657–665. [CrossRef] [PubMed]

41. Kunihiro, M.; Tanaka, Y. Heat shock inactivated ADCC activity of PBMC to HTLV-I infected cells. (unpublished; manuscript in preparation).

42. Jing, Y.K.; Dai, J.; Chalmers-Redman, R.M.E.; Tatton, W.G.; Waxman, S. Arsenic trioxide selectively induces acute promyelocytic leukemia cell apoptosis via a hydrogen peroxide-dependent pathway. *Blood* **1999**, *94*, 2102–2111. [PubMed]

43. Nabeshi, H.; Yoshikawa, T.; Kamada, H.; Shibata, H.; Sugita, T.; Abe, Y.; Nagano, K.; Nomura, T.; Minowa, K.; Yamashita, T.; et al. Arsenic trioxide inhibits human T cell-lymphotropic virus-1-induced syncytiums by down-regulating gp46. *Biol. Pharma. Bull.* **2009**, *32*, 1286–1288. [CrossRef]

44. Mutsvunguma, L.Z.; Moetlhoa, B.; Edkins, A.L.; Luke, G.A.; Blatch, G.L.; Knox, C. Theiler's murine encephalomyelitis virus infection induces a redistribution of heat shock proteins 70 and 90 in BHK-21 cells, and is inhibited by novobiocin and geldanamycin. *Cell Stress Chaperones* **2011**, *16*, 505–515. [CrossRef] [PubMed]

45. Khachatoorian, R.; Ganapathy, E.; Ahmadieh, Y.; Wheatley, N.; Sundberg, C.; Jung, C.L.; Arumugaswami, V.; Raychaudhuri, S.; Dasgupta, A.; French, S.W. The ns5a-binding heat shock proteins HSC70 and HSP70 play distinct roles in the hepatitis c viral life cycle. *Virology* **2014**, *454–455*, 118–127. [CrossRef] [PubMed]

46. Zhang, C.; Kang, K.; Ning, P.; Peng, Y.; Lin, Z.; Cui, H.; Cao, Z.; Wang, J.; Zhang, Y. Heat shock protein 70 is associated with CSFV NS5A protein and enhances viral rna replication. *Virology* **2015**, *482*, 9–18. [CrossRef] [PubMed]

47. Nagata, K.; Ide, Y.; Takagi, T.; Ohtani, K.; Aoshima, M.; Tozawa, H.; Nakamura, M.; Sugamura, K. Complex formation of human T-cell leukemia virus type I p40tax transactivator with cellular polypeptides. *J. Virol.* **1992**, *66*, 1040–1049. [PubMed]

48. Fang, D.; Haraguchi, Y.; Jinno, A.; Soda, Y.; Shimizu, N.; Hoshino, H. Heat shock cognate protein 70 is a cell fusion-enhancing factor but not an entry factor for human T-cell lymphotropic virus type I. *Biochem. Biophys. Res. Commun.* **1999**, *261*, 357–363. [CrossRef] [PubMed]

49. Kawakami, H.; Tomita, M.; Okudaira, T.; Ishikawa, C.; Matsuda, T.; Tanaka, Y.; Nakazato, T.; Taira, N.; Ohshiro, K.; Mori, N. Inhibition of heat shock protein-90 modulates multiple functions required for survival of human T-cell leukemia virus type I-infected T-cell lines and adult T-cell leukemia cells. *Int. J. Cancer* **2007**, *120*, 1811–1820. [CrossRef] [PubMed]

50. Fallouh, H.; Mahana, W. Antibody to heat shock protein 70 (HSP70) inhibits human T-cell lymphoptropic virus type I (HTLV-I) production by transformed rabbit T-cell lines. *Toxins* **2012**, *4*, 768–777. [CrossRef] [PubMed]

51. Barolet, D.; Christiaens, F.; Hamblin, M.R. Infrared and skin: Friend or foe. *J. Photochem. Photobiol. B* **2016**, *155*, 78–85. [CrossRef] [PubMed]

52. Roesch, F.; Meziane, O.; Kula, A.; Nisole, S.; Porrot, F.; Anderson, I.; Mammano, F.; Fassati, A.; Marcello, A.; Benkirane, M.; et al. Hyperthermia Stimulates HIV-1 Replication. *PLoS Pathog.* **2012**, *8*, e1002792. [CrossRef] [PubMed]

*Article*

# Analysis of the Prevalence of HTLV-1 Proviral DNA in Cervical Smears and Carcinomas from HIV Positive and Negative Kenyan Women

Xiaotong He [1,‡], Innocent O. Maranga [1,2,‡], Anthony W. Oliver [1], Peter Gichangi [2], Lynne Hampson [1,†] and Ian N. Hampson [1,*,†]

1   Viral Oncology Lab, University of Manchester, St Mary's Hospital, Manchester M13 9WL, UK;
    xiaotong.he@manchester.ac.uk (X.H.); dr.maranga@yahoo.com (I.O.M.);
    Anthony.W.Oliver@manchester.ac.uk (A.W.O.); lynne.hampson@manchester.ac.uk (L.H.)
2   Obstetrics and Gynaecology, University of Nairobi, Kenyatta National Hospital Nairobi, Nairobi 00202,
    Kenya; gichangip@yahoo.com
*   Correspondence: ian.n.hampson@manchester.ac.uk or Ian.hampson@manchester.ac.uk;
    Tel.: +44-(0)-161-701-6938
†   These authors contributed equally to this work.
‡   These authors contributed equally to this work.

Academic Editor: Louis M. Mansky
Received: 6 January 2016; Accepted: 26 August 2016; Published: 5 September 2016

**Abstract:** The oncogenic retrovirus human T-cell lymphotropic virus type 1 (HTLV-1) is endemic in some countries although its prevalence and relationship with other sexually transmitted infections in Sub-Saharan Africa is largely unknown. A novel endpoint PCR method was used to analyse the prevalence of HTLV-1 proviral DNA in genomic DNA extracted from liquid based cytology (LBC) cervical smears and invasive cervical carcinomas (ICCs) obtained from human immunodeficiency virus-positive (HIV+ve) and HIV-negative (HIV−ve) Kenyan women. Patient sociodemographic details were recorded by structured questionnaire and these data analysed with respect to HIV status, human papillomavirus (HPV) type (Papilocheck®) and cytology. This showed 22/113 (19.5%) of LBC's from HIV+ve patients were positive for HTLV-1 compared to 4/111 (3.6%) of those from HIV−ve women ($p = 0.0002$; odds ratio (OR) = 6.42 (2.07–26.56)). Only 1/37 (2.7%) of HIV+ve and none of the 44 HIV−ve ICC samples were positive for HTLV-1. There was also a significant correlation between HTLV-1 infection, numbers of sexual partners ($p < 0.05$) and smoking ($p < 0.01$). Using this unique method, these data suggest an unexpectedly high prevalence of HTLV-1 DNA in HIV+ve women in this geographical location. However, the low level of HTLV-1 detected in HIV+ve ICC samples was unexpected and the reasons for this are unclear.

**Keywords:** human immunodeficiency virus (HIV); human T-cell lymphotropic virus type 1 (HTLV-1); human papilloma virus (HPV); retrovirus; liquid based cytology (LBC); invasive cervical cancer (ICC); proviral DNA; PCR

---

## 1. Introduction

Human T-cell lymphotropic virus type 1 (HTLV-1) was the first pathogenic human retrovirus to be discovered and is known to be associated with various mild to severe pathologies resulting from chronic lifelong infection [1–4]. It is the aetiological agent for adult-T-cell leukaemia (ATL) and HTLV-1-associated myelopathy/tropical spastic paraparesis (HAM/TSP) with the former often proving fatal whilst the latter causes significant debilitating morbidity. Other associated diseases include arthropathy, HTLV-1 associated infective dermatitis, uveitis and polymyositis [5–7]. The cumulative risk of a HTLV-1 carrier developing ATL has been estimated at between 2.5% and 5% although a latency period of 50–70 years is typical [8,9].

Although HTLV-1 is reported to currently infect between 10 and 25 million people globally, the true prevalence in different geographical regions of the world is largely unknown [4,7,10,11]. The 2012 review of global HTLV-1 prevalence by Gessain and Cassar [11] summarises the most recent data and contains comprehensive details of patient numbers, subgroups studied and outcomes. Current estimates rely on serological screening of blood donors and pregnant women which often excludes high-risk groups such as human immunodeficiency virus (HIV) positive individuals and ethnic minorities [4,7]. HTLV-1 infection is endemic in many parts of the world including southern Japan, Equatorial Africa, Central and South America and in immigrant descendants of people from these regions [12]. In a recent review of global HTLV-1 prevalence, the highest prevalence was found in the Japanese islands with 36.4%. Other reports in Africa are between 6.6% and 8.5% in Gabon and 1.05% in Guinea whereas elsewhere, such as in the Caribbean islands, it is 6% [4,7,13]. The prevalence in female sex workers has ranged from 3.2% in Congo, to 5.7% in Fukuoka (Japan) and from 8.7% to 21.8% in Callao (Peru) [14]. A study done in São Paulo (Brazil) on HIV positive intravenous drug addicts (IVDAs) showed a HTLV-1 prevalence of 15.3% [15].

Both HIV and HTLV-1 share similar risk factors since they are both transmitted through sexual contact, blood transfusion, sharing of needles among intravenous drug users and vertical transmission from mother to child via breast-feeding, [8,10,16,17]. With regard to breast-feeding, transmission of HTLV-1 via this modality is known to be related to duration, proviral load and antibody titre in addition to other less well defined factors [18]. The development of neoplasia in patients with HIV and acquired immune deficiency syndrome (HIV/AIDS) is generally attributed to failure of immune surveillance and associated co-infections with oncogenic viruses such as human herpes virus 8 (HHV8) and Kaposi sarcoma, Epstein–Barr virus (EBV) with non-Hodgkins lymphoma (NHL), and human papilloma virus (HPV) with invasive cervical carcinomas (ICC) [19–22].

During the past 20 years, HIV type 1 (HIV-1)/HTLV-1 co-infection has emerged as a global health problem with increasing numbers of cases in South America and Africa [8]. Although both viruses have a tropism for CD4+ T helper cells, the overall effect of virus-virus interactions on their related pathologies is still controversial [8]. Although Africa is considered to be a large reservoir for HTLV-1 infection, there is a paucity of prevalence data from Sub-Saharan Africa and especially East Africa [7,11]. Based on screening pregnant women, in Western Africa Nigeria has been reported to have the highest prevalence of 5.5% followed by Zaire in Central Africa at 4.6% where subtype B is the most common [11]. It is thus significant that screening for HTLV-1 during pregnancy and before blood transfusion is not routine in most Sub-Saharan African countries [10,23]. Furthermore, it is also very likely that other higher-risk groups, such as intravenous drug addicts (IVDA) and sex workers, will have a much higher prevalence of the virus. As mentioned the study of IVDA's in São Paulo found 15.3% prevalence [15] which is much higher than the 0.1% prevalence observed in pregnant women in the same location [11]. Since there is, as yet, neither an effective vaccine against HTLV-1 nor a cure for its associated pathologies, the virus has the potential to impose a significant social and financial burden [10,24] in this area of the world. Moreover, given that Sub-Saharan Africa is also home to two thirds of the global HIV pandemic, it is important to assess how these two retroviruses associate in this population and how this may contribute to the pathology of different diseases.

Several previous studies have attempted to link HTLV-1 infection with the aetiology of ICC and have produced inconsistent findings. For example, a study carried out in Yucatan (Mexico) where HTLV-1 prevalence is low but the incidence of ICC is high, found no statistical difference between these two groups [25]. However, a larger study carried out in Japan found the prevalence to be higher in ICC patients younger than 59 years. Furthermore, increased numbers of HTLV-1 infections were also found in women from all age groups who developed vaginal carcinomas (VC) when compared to age-matched healthy controls [26]. The conclusion from this study was that HTLV-1 infection may promote cervical carcinogenesis and may also affect the prognosis of ICC or VC. These results were consistent with a study carried out on Jamaican women where a higher prevalence of HTLV-1 was found in cervical intraepithelial neoplasia 3/invasive cervical cancer (CIN3/ICC) patients when compared to controls [27]. However, these data were not supported in a more recent report [28].

In light of these previous observations, the objective of the current study was to evaluate the prevalence of HTLV-1 proviral DNA extracted from liquid based cytology (LBC) specimens from HIV positive and negative patients attending specialist referral clinics in Kenyatta National Hospital in Nairobi, Kenya. The same analysis was carried out on DNA extracted from biopsies obtained from patients with ICC in the same hospital.

## 2. Materials and Methods

### 2.1. Study Population and Sample Collection

A cross-sectional study was carried out among women attending Kenyatta National Hospital (KNH), Nairobi, Kenya. Consecutive female patients attending a Specialist HIV Clinic and Family Planning Clinic were recruited between April 2008 and February 2009. Women aged between 21 and 52 years were included, while those who had prior destructive procedures for cervical disease and/or hysterectomy were excluded. A total of 224 patients were recruited, including 111 HIV negative (median age = 35 years, range = 21–52) and 113 HIV positive (median age = 35 years, range = 21–52) with 56 of these in receipt of highly active antiretroviral therapy (HAART). After undergoing voluntary counselling and testing for HIV, a structured questionnaire was administered and a blood sample was collected. During the same period, 37 HIV positive and 40 HIV negative women newly diagnosed with ICC at KNH were also randomly recruited. They underwent examination under anaesthesia (EUA), a biopsy was taken which was then formalin-fixed and paraffin-embedded (FFPE). FFPE samples were used for histopathology and DNA extractions.

This study was approved by the Kenyatta National Hospital's Ethics and Research Board (ERB), the University of Nairobi (No KNH-ERC/01/4988) and the University of Manchester, Oldham Research Ethics Committee amendment 5 project 07/Q1405/14. All patients gave written informed consent.

### 2.2. LBC Samples

All patients were examined with a speculum and cervical samples collected using a cervex brush, which was stirred into a vial of PreservCyt®transport solution (ThinPrep®Pap test, Hologic Inc., Bedford, MA, USA). Cytology slides were prepared using a Cellspin Cytocentrifuge, (Tharmac, Waldsolms, Germany) and were stained with Papanicolaou (pap) stain. All pap stained slides were independently examined by two different pathologists and the Bethesda 2001 criteria were used for slide interpretation [29].

### 2.3. HIV Testing

HIV testing was done using Determine® test kit (Abbot Pharmaceuticals, Chicago, IL, USA), and, if positive, this was confirmed by Uni-Gold® (Trinity Biotech Plc, Bray, Ireland).

### 2.4. Extraction of Genomic DNA

All residual PreserveCyt®material was used for automated DNA extraction using BioRobot® M48 (Qiagen, Hilden, Germany) as described by the manufacturer. Approximately $4 \times 10$ µm FFPE ICC sections were used for DNA extraction using the Qiagen Qiacube®(Qiagen, Crawley, West Sussex, UK) according to the manufacturer's instructions.

### 2.5. Papillocheck® HPV Genotyping

DNA was quantified using a Nanodrop UV spectrophotometer (Thermo Fisher, Altrincham, Cheshire, UK) and used for HPV genotyping with the PapilloCheck® test (Greiner Bio-One, Stonehouse, Gloucestershire, UK) as per the manufacturer's instructions. This identifies 24 different HPV genotypes:

six low-risk HPV types (LR: 6, 11, 40, 42, 43, and 44/55) and 18 high-risk HPV types (HR: 16, 18, 31, 33, 35, 39, 45, 51, 52, 53, 56, 58, 59, 66, 68, 70, 73, and 82).

*2.6. PCR*

GAPDH PCR was performed to assess the quality of input DNA using a 50 µL reaction mixture containing 2 µL of each DNA sample (50 ng), 2.5 units of BioTaq DNA polymerase (Bioline Ltd., London, UK), 0.2 mM dNTPs, and 0.2 µM of each primer in 10 mM Tris-HCl pH 8.3, 50 mM KCl and 2.5 mM MgCl$_2$. HTLV-1 Tax PCRs were optimized using a multiplex PCR kit (Qiagen) as recommended by the manufacturer. All reactions were carried out in duplicate, using a Veriti™ Thermal Cycler (Applied Biosystems, Paisley, UK) with the conditions and primers indicated in Table 1. The sensitivity of the method was tested by adding a dilution series of genomic DNA extracted from Tax transduced JPX-9 cells [30] to 50 ng of genomic DNA from a HTLV-1 negative patient per reaction. PCR products were separated by 1.5%–2.5% agarose gel electrophoresis, stained with ethidium bromide and examined under UV.

**Table 1.** Primers and PCR conditions used.

| Primer | Sequence | Conditions | Amplimer Size (bp) |
|---|---|---|---|
| GAPDH-F<br>GAPDH-R | 5′-CATTGACCTCAACTACATGGT-3′<br>5′-TCGCTCCTGGAAGATGGTGAT-3′ | 94 °C × 5 min; 33 cycles:<br>94°C × 25 s; 53 °C × 25 s;<br>72 °C × 25 s; 72°C × 7 min | 130 |
| HTLV-1 Tax-F<br>HTLV-1 Tax-R | 5′-CACCTGTCCAGAGCATCAGA-3′<br>5′-TCTGGAAAAGACAGGGTTGG-3′ | 95 °C × 15 min; 45 cycles:<br>94°C × 30 s; 57 °C × 30s;<br>75 °C × 30 s; 75°C × 7 min | 264 |

*2.7. DNA Sequencing*

The gel-separated DNA bands were excised, isolated, purified and sequenced using an ABI BigDye Cycle sequencing kit (Applied Biosystems, Warrington, UK) as indicated by the manufacturer and a 3100 ABI sequencer (Applied Biosystems, Warrington, UK).

*2.8. Statistical Analysis*

The data were captured in an Access database and exported to SPSS version 16.0 for analysis after cleaning and validation. Statistical tests of significance were done using Pearson's chi-square with Yates correction or with Fisher's test if the expected frequencies were less than five. Odds ratio (OR), adjusted OR (AOR) and the 95% confidence intervals (CI) were used to measure strengths of associations. Further, cross-tabulations were also used to assess the distribution between HTLV-1 and HIV status of the participants, and the 95% CI around the prevalence of virus infection was computed using R program version 3.2.2 package binom. A *p*-value of less than 0.05 was considered statistically significant.

## 3. Results

*3.1. DNA Quality*

As indicated, PCR of glyceraldehyde 3-phosphate dehydrogenase (GAPDH) with GAPDH specific primers was used to assess the quality and integrity of all DNA samples prior to these being used for unknown test PCR's with Tax specific primers as illustrated in Figures S1 and S2.

## 3.2. PCR Detection of Proviral HTLV-1 Tax DNA

As illustrated in Figure 1, this simple hot-start, end-point PCR method specifically detected a single 264 bp HTLV-1 Tax DNA amplimer from as little as 0.1 fg of input genomic DNA from the JPX-9 cell line.

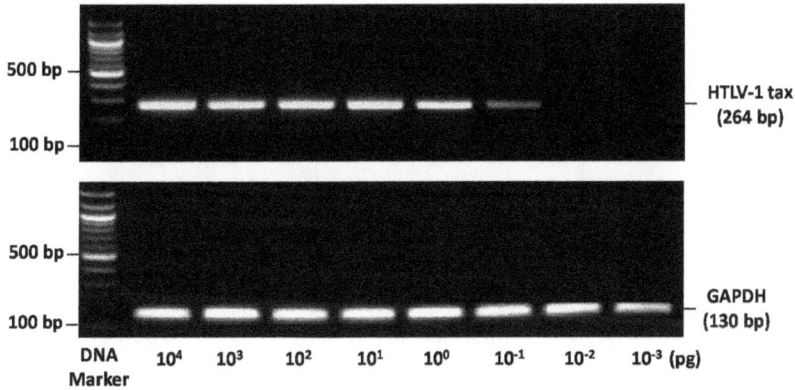

**Figure 1.** Sensitivity of human T-cell lymphotropic virus type 1 (HTLV-1) Tax PCR detection. End-point PCR method able to specifically detect a single HTLV-1 Tax DNA amplimer from 0.1 fg of input genomic DNA (DNA from the human JPX9 cell line was used as a positive control for Tax). GAPDH: glyceraldehyde 3-phosphate dehydrogenase.

The specificity of the single Tax amplimer product was confirmed by excision and DNA sequencing of the 264-bp PCR product and the identity verified by the use of National Center for Biotechnology Information (NCBI) Basic Local Alignment Search Tool (BLAST) (See Supplementary Data S1, Figure S4 and Supplementary Data S2 (BLAST)).

## 3.3. Analysis of the Association between HTLV-1 Infection and HIV Infection in LBC DNA

Genomic DNA extracted from 113 HIV positive and 111 HIV negative LBC samples was analysed for the presence of HTLV-1 Tax proviral DNA. As shown in Figure S1, a single HTLV-1 Tax amplimer was successfully amplified from 26 out of 224 of the LBC samples giving a total infection rate of 11.6% (95% confidence interval (CI): 7.7%–16.5%). Most significantly, it was found that 22 of the HTLV-1 positive samples were also HIV positive producing an infection rate of 19.5% (95% CI: 12.6%–27.9%) in the 113 HIV positive patients.

The HTLV-1 prevalence was four out of 111 (3.6%) (95% CI: 0.9%–8.9%) in HIV negative women, which indicates that the frequency of HTLV-1 infection is >5 times higher in HIV positive than negative patients, indicating that HTLV-1 infection is significantly associated with a HIV positive status ($p < 0.01$).

## 3.4. Analysis of the Association of HTLV-1 Infection with Abnormal Cervical Cytology

Using the same LBC samples, we have previously shown that HIV infection was associated with abnormal cervical cytology [31]. In light of this, HTLV-1 infection was assessed in relation to our previous cervical cytology findings where low-grade and high-grade squamous intraepithelial lesion (LSIL and HSIL) and abnormal squamous cells of undetermined significance (ASCUS) were analysed independently. In contrast to HIV infection, there was no statistically significant evidence for an association between HTLV-1 and abnormal cervical cytology ($p = 0.231$ using Pearson's chi-square with Yates correction). Nevertheless, it was noted that all the HTLV-1 infections were in normal/LSIL and none were found in HSIL/ASCUS (See Figure S4).

*3.5. Analysis of the Association between HTLV-1 Infection, Stage of HIV and Use of Highly Active Antiretroviral Therapy (HAART)*

According to World Health Organization (WHO) clinical staging system, the 113 HIV positives were classified as stage I–IV, consisting of stage I in 24, II in 16, III in 38 and IV in 35 patients. There was no statistical significance between HTLV-1 positive and HIV clinical stages. Additionally, HTLV-1 positivity was not associated with increasing advancement of HIV disease or use of HAART ($p > 0.05$) (See Table S3). WHO guidelines for HIV staging can be found on the AIDS Education and Training Center (AETC) website [32].

*3.6. Analysis of the Association between HTLV-1 Infection and HPV in LBC DNAs*

As described above, 24 subtypes of HPV, including six low-risk (LR) and 18 high-risk (HR) were identified in LBC specimens using the Papillocheck® system. Of the samples analysed, one sample had an invalid result, while 121 (55%, 95%CI: 48.2%–61.7%) had at least one HPV subtype with the highest number of genotypes detected in any one individual being seven, covering 107 HR (88.4%, 95% CI: 81.3%–93.5%) and 14 LR (11.6%, 95% CI: 6.5%–18.7%) HPV subtypes. None of the HPV subtypes covered showed any association with HTLV-1 apart from HPV type 53 where there was a significant positive association ($p < 0.05$) (Figure 2 and Table 2).

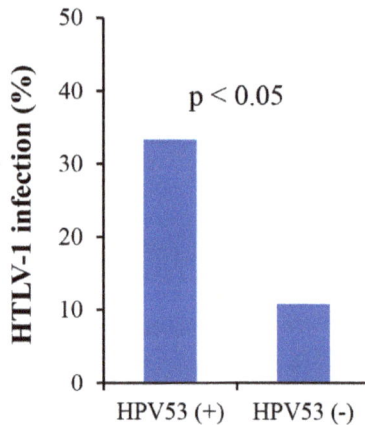

**Figure 2.** Association of HTLV-1 and human papillomavirus (HPV) type 53 infections. HPV type 53 showed a statistically significant association with HTLV-1 infection (p <0.05). None of the other HPV sub-types showed a positive association.

**Table 2.** Analysis of the association between HTLV-1 and HPV infections.

|  | HPV (+) | HPV (–) |
|---|---|---|
| **HTLV-1 (+)** | 13 | 13 |
| **HTLV-1 (–)** | 108 | 89 |
| *p*-value | > 0.05 | |

*3.7. Analysis of the Relationship between HTLV-1 Infection and Patient Age*

LBC material for this study was obtained from women with an age range of 21 to 52 years and median or mean ages of 35 or 35.2 years, respectively. The median age of HTLV-1 negatives was 35 and HTLV-1 positives 31 years with no significant difference observed between the median age of HTLV-1 positives with respect to either HPV or HIV status (Figure 3). Within HTLV-1 positives, no significant difference was found between infection and any of the age groups.

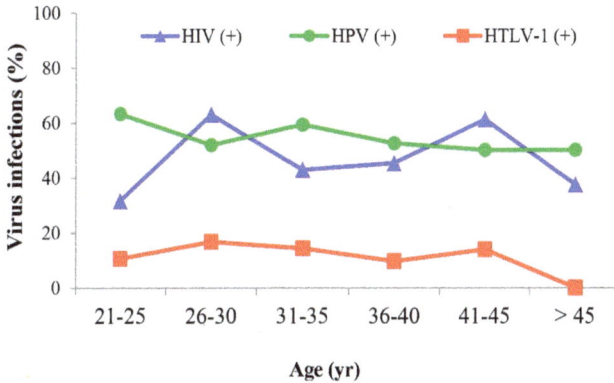

**Figure 3.** Relationship between HTLV-1, HPV, human immunodeficiency virus (HIV) infections and age. There was no significant difference observed between the median age of HTLV-1 positives with respect to either HPV or HIV status. However, there was a trend towards a lower HTLV-1 prevalence among women >45 years old when compared to women positive for HPV and HIV.

Among the 77 ICC study subjects, the median and mean ages were 42 and 44 years, respectively, with a range of 27–68 years. Any association between HTLV-1 infection and age could not be calculated since only one ICC sample was positive.

*3.8. Analysis of the Relationship between HTLV-1 Infection and Patient Sociodemographics*

Analysis of the relationship between HTLV-1 infection and patient's life style shown in Figure 4 indicated that the rate of HTLV-1 infection in the smoking group was five times higher than that in the non-smoking group. Women with increased numbers of marriages and sex partners were also significantly more likely to have HTLV-1 infections confirming that sexual transmission is an important means of HTLV-1 infection. No significant associations between HTLV-1 infections were found with religion, educational level, occupation, socio-economic status, contraceptive use, intravenous drug use, blood transfusion, genital herpes and either rural or urban residency (see Figure S5).

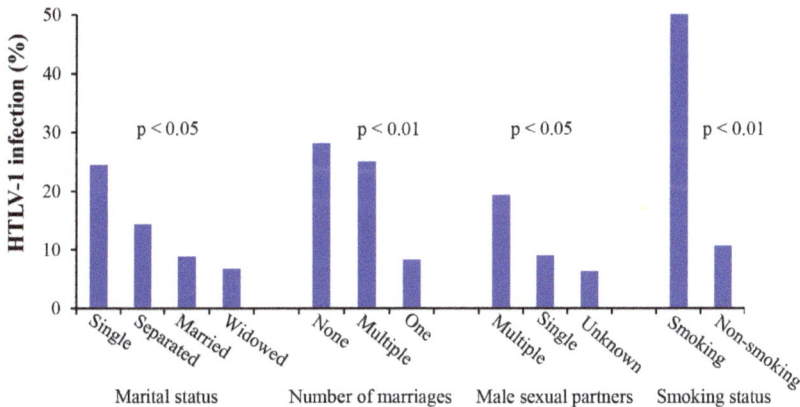

**Figure 4.** Analysis of the relationship between HTLV-1 infection and life styles factors. The rate of HTLV-1 infection in the smoking group was five times higher than that in non-smoking group. Women with increased numbers of marriages and sexual partners were significantly more likely to have HTLV-1 infections confirming that sexual contact is an important means of HTLV-1 transmission.

*3.9. Analysis of HTLV-1 DNA in ICC DNA's*

HTLV-1 DNA was only detected in one out of 77 ICC biopsies, of which 37 were HIV positive (see Figure S2). This equates to an overall infection rate of 1.3% (95% CI: 0.033%–7%), which is significantly lower than the 12.1% (95% CI: 8.1%–17%) found in DNA from LBC material. However, if the HTLV-1 infections found in LBC and ICC material from just HIV positive women are compared, this equates to 19.5% (95% CI: 12.6%–27.9%) versus 2.7% (95% CI: 0.068%–14%), respectively. Thus, HIV positive women with ICCs have approximately seven-fold less numbers of HTLV-1 infections than would be predicted from the rate observed in LBC material from HIV positive women.

## 4. Discussion

Although HTLV-1 was discovered 35 years ago, information on its prevalence and associated risk factors in Sub-Saharan Africa countries have remained poorly defined. HIV, HTLV-1 and HPV can share the same sexual route of transmission and, to our knowledge, this is the first study carried out in Kenya to analyse the prevalence of HTLV-1 in LBC cervical specimens and formalin fixed and paraffin-embedded (FFPE) ICC biopsies from HIV positive and negative women. The most significant observation was the high prevalence of HTLV-1 proviral DNA found in LBC samples from HIV positive women, consistent with previous work carried out in Guinea-Bissau, showing that HIV infection is a potential risk factor for acquiring HTLV-1 [33]. Moreover, the seven-fold lower incidence of HTLV-1 DNA found in ICCs from HIV positive women begs the question, what is the explanation for this? Clearly, it could be related to differences in the type of sample and extraction procedures used for LBC's and ICCs although there are other possibilities. For example, women who are positive for both HTLV-1 and HIV have increased AIDS related mortality, which could prevent them from living long enough to develop ICC. However, it is still not established that concomitant HTLV-1/HIV infection does promote progression to AIDS [8,34].

A limitation of the current study is the lack of corroborating serological data on HTLV-1 prevalence. Although the PCR method is sensitive, it is very clear this may underestimate the true prevalence since it depends on the inclusion of sufficient numbers of infected cells in extracted material. Indeed, this could also provide an alternative explanation for the lower detection of HTLV-1 DNA found in ICC samples. However, another possibility is that HTLV-1 could suppress HPV mediated cervical carcinogenesis in HIV positive women, which raises the question: is there a potential mechanism for such an effect? Both HIV and HTLV-1 infect CD4+ cells and it has been previously shown that elevated numbers of CD4+ cells are associated with HPV-related dysplasia [35–37]. Moreover, persistent HPV infections are known to suppress T-helper type 1 (Th-1) and promote a T-helper type 2 (Th-2) responses, which results in defective interferon (IFN) production. This, in turn promotes the development of HPV related neoplasia [38]. Thus, if co-infection with HTLV-1 and HIV produced a shift from Th-2 to a Th-1 response, this could potentially inhibit cervical carcinogenesis and explain the reduced prevalence of HTLV-1 in ICCs from HIV positive women. Indeed, it is highly significant that the work of Abrahão et al. has shown that concomitant HTLV-1/HIV infection augments production of the Th-1 cytokines interleukin-2 (IL-2) and IFN above that of single HTLV-1 or HIV infected individuals which is entirely consistent with this hypothesis [39]. It is also curious that none of the 22 HPV subtypes tested, apart from HPV 53, showed any association with HTLV-1. The importance of this is that, in an earlier report, we demonstrated that type 53 was significantly more common in LBC's from HIV positive than HIV negative women although it was never detected in any of the ICCs examined [31].

With regard to other cancers, a protective role for HTLV-1 is not without precedent. Hirata et al. evaluated the relationship between HTLV-1 infection and the occurrence of several types of cancers namely biliary tract, pancreatic, esophageal, gastric, colorectal, liver, and lung cancers using logistic regression analysis adjusted for age and sex [40]. Curiously, the HTLV-1 infection rate found in gastric cancer patients was significantly lower than in controls ($p$ = 0.01) but no associations were observed with any of the other cancers studied.

Another important question raised by the current study is that, given the high prevalence of HTLV-1 in HIV positive Kenyan women, is there a correspondingly higher incidence of HTLV-1 related pathologies in these women as observed in Japan [41]? However, this may not be apparent as the average 59 years life expectancy in Kenya is much shorter than the 83 years in Japan. Since HTLV-1 related adult-T-cell leukaemia occurs after a long latency period, typically spanning 5–7 decades, it is thus highly likely that infected individuals in Sub-Saharan Africa will succumb to other illnesses prior to the development of HTLV-1 related pathologies. Indeed the median age for LBC and ICC subjects in the current study was 35 and 42 years, respectively.

Other noteworthy finding from our study was the significant association between HTLV-1 infection and increased numbers of direct and indirect sexual contacts in addition to smoking. Given that smoking is known to impair the function of cervical Langerhans cells, it is possible that the increased incidence of HTLV-1 in smokers could be related to impaired local immunity which contributes to increased rates of sexual transmission of the virus [42].

In summary, it is known that the influence of HIV infection on the incidence of HPV related ICC seems to vary depending on the geographical location of the study [43]. For example, work carried out in European locations shows an increased incidence of ICC in HIV positive women [43–45], whereas this has not been observed in other, mainly African, studies [46,47]. In light of our findings, it is tempting to speculate that the observed differences in ICC incidence rates found in HIV positive women in different geographical locations may be, at least in part, due to variations in the prevalence of HTLV-1 infection since this is much more common in Sub-Saharan Africa than Europe. Clearly, larger studies are needed in order to improve the statistical significance of this finding, which could also be supplemented with analysis of Th-1 and Th-2 cytokine profiles.

As discussed, there is a paucity of epidemiological data on HTLV-1 in Sub-Saharan Africa, which also bears the heaviest global HIV burden. The HTLV-1 prevalence found in LBC samples from HIV positive women (20.4%) suggests this could be higher than anticipated reinforcing the finding that Kenya may be endemic for HTLV-1 and also confirming the association between HTLV-1 and HIV. Clearly, there is an urgent need to conduct larger population based studies and to institute public health measures to curb the continued spread of HTLV-1. For example, addressing the issue of vertical transmission through breast-feeding (by considering safe replacement feeds for those whose mothers are carriers), increased condom use to curb sexual transmission and provision of universal HTLV-1 screening on all blood donations, tissue and organ donors. Currently, routine HTLV-1 screening is not carried out in most African countries including Kenya—a gap which may be promoting the spread of this virus. Furthermore, there is no curative treatment for HTLV-1 or its associated pathologies and an effective vaccine is equally unavailable which puts a heavy social and financial burden on sufferers, their families and the healthcare systems. Thus, public health interventions aimed at counselling and educating high-risk individuals and the public are of critical significance [10].

**Supplementary Materials:** The following are available online at www.mdpi.com/1999-4915/8/9/245/s1, Figure S1: PCR Analysis of HTLV-1 Tax in 220 LBC DNA's, Figure S2: PCR Analysis of HTLV-1 Tax in 77 ICC DNA's, Figure S3: HTLV-1 prevalence and cervical cytology in HIV+ve and −ve patients, Figure S4: Nested PCR analysis of the Tax 181 pb amplimer in the 26 HTLV-1 positive LBC's, Table S1: Relationship between HTLV-1 and WHO HIV staging and HAART, Table S2: HTLV-1 status with respect to patient sociodemographics, Supplementary_Data_S2: NCBI BLAST of nested Tax amplimer sequences shown in supplementary Figure S4.

**Acknowledgments:** The work described was funded by charitable donations from the Humane Research Trust, The Janice Cholerton Post Graduate Support Fund/East Africa Medical Trust (EAMT), The Caring Cancer Trust, and The Cancer Prevention Research Trust. Post graduate student I.O.M. was part-funded by the International Atomic Energy Agency and Wellbeing of Women. The funders had no role in study design, data collection and analysis, decision to publish, or preparation of the manuscript.

**Author Contributions:** I.O.M. co-wrote the manuscript and carried out this study as part of his PhD; X.H. developed the PCR detection method, carried out sequencing and BLAST analysis and co-wrote the manuscript; A.W.O. co-wrote the manuscript; P.G. co-wrote the manuscript; L.H. had the idea and critically reviewed the manuscript; I.N.H. had the idea and co-wrote the manuscript.

**Conflicts of Interest:** The authors declare no conflicts of interest.

## References

1.  Moens, B.; Lopez, G.; Adaui, V.; Gonzalez, E.; Kerremans, L.; Clark, D.; Verdonck, K.; Gotuzzo, E.; Vanham, G.; Cassar, O.; et al. Development and validation of a multiplex real-time PCR assay for simultaneous genotyping and human T-lymphotropic virus type 1, 2, and 3 proviral load determination. *J. Clin. Microbiol.* **2009**, *47*, 3682–3691. [CrossRef] [PubMed]

2.  Gallo, R.C. Human retroviruses after 20 years: A perspective from the past and prospects for their future control. *Immunol. Rev.* **2002**, *185*, 236–265. [CrossRef] [PubMed]

3.  Poiesz, B.J.; Ruscetti, F.W.; Gazdar, A.F.; Bunn, P.A.; Minna, J.D.; Gallo, R.C. Detection and isolation of type C retrovirus particles from fresh and cultured lymphocytes of a patient with cutaneous T-cell lymphoma. *Proc. Natl. Acad. Sci. USA* **1980**, *77*, 7415–7419. [CrossRef] [PubMed]

4.  Hlela, C.; Shepperd, S.; Khumalo, N.P.; Taylor, G.P. The prevalence of human T-cell lymphotropic virus type 1 in the general population is unknown. *AIDS Rev.* **2009**, *11*, 205–214. [PubMed]

5.  LaGrenade, L.; Hanchard, B.; Fletcher, V.; Cranston, B.; Blattner, W. Infective dermatitis of Jamaican children: A marker for HTLV-I infection. *Lancet* **1990**, *336*, 1345–1347. [PubMed]

6.  Mochizuki, M.; Watanabe, T.; Yamaguchi, K.; Takatsuki, K.; Yoshimura, K.; Shirao, M.; Nakashima, S.; Mori, S.; Araki, S.; Miyata, N. HTLV-I uveitis: A distinct clinical entity caused by HTLV-I. *Jpn. J. Cancer Res.* **1992**, *83*, 236–239. [CrossRef] [PubMed]

7.  Verdonck, K.; Gonzalez, E.; Van Dooren, S.; Vandamme, A.M.; Vanham, G.; Gotuzzo, E. Human T-lymphotropic virus 1: Recent knowledge about an ancient infection. *Lancet Infect. Dis.* **2007**, *7*, 266–281. [CrossRef]

8.  Casoli, C.; Pilotti, E.; Bertazzoni, U. Molecular and cellular interactions of HIV-1/HTLV coinfection and impact on AIDS progression. *AIDS Rev.* **2007**, *9*, 140–149. [PubMed]

9.  Tanaka, G.; Okayama, A.; Watanabe, T.; Aizawa, S.; Stuver, S.; Mueller, N.; Hsieh, C.C.; Tsubouchi, H. The clonal expansion of human T lymphotropic virus type 1-infected T cells: A comparison between seroconverters and long-term carriers. *J. Infect. Dis.* **2005**, *191*, 1140–1147. [CrossRef] [PubMed]

10. Proietti, F.A.; Carneiro-Proietti, A.B.; Catalan-Soares, B.C.; Murphy, E.L. Global epidemiology of HTLV-I infection and associated diseases. *Oncogene* **2005**, *24*, 6058–6068. [CrossRef] [PubMed]

11. Gessain, A.; Cassar, O. Epidemiological Aspects and World Distribution of HTLV-1 Infection. *Front. Microbiol.* **2012**, *3*, 388. [CrossRef] [PubMed]

12. Gallo, R.C. Human retroviruses in the second decade: A personal perspective. *Nat. Med.* **1995**, *1*, 753–759. [CrossRef] [PubMed]

13. Holmgren, B.; da Silva, Z.; Larsen, O.; Vastrup, P.; Andersson, S.; Aaby, P. Dual infections with HIV-1, HIV-2 and HTLV-I are more common in older women than in men in Guinea-Bissau. *AIDS* **2003**, *17*, 241–253. [CrossRef] [PubMed]

14. Wignall, F.S.; Hyams, K.C.; Phillips, I.A.; Escamilla, J.; Tejada, A.; Li, O.; Lopez, F.; Chauca, G.; Sanchez, S.; Roberts, C.R. Sexual transmission of human T-lymphotropic virus type I in Peruvian prostitutes. *J. Med. Virol.* **1992**, *38*, 44–48. [CrossRef] [PubMed]

15. De Araujo, A.C.; Casseb, J.S.; Neitzert, E.; de Souza, M.L.; Mammano, F.; Del Mistro, A.; De Rossi, A.; Chieco-Bianchi, L. HTLV-I and HTLV-II infections among HIV-1 seropositive patients in Sao Paulo, Brazil. *Eur. J. Epidemiol.* **1994**, *10*, 165–171. [CrossRef] [PubMed]

16. Manns, A.; Hisada, M.; La Grenade, L. Human T-lymphotropic virus type I infection. *Lancet* **1999**, *353*, 1951–1958. [CrossRef]

17. Feigal, E.; Murphy, E.; Vranizan, K.; Bacchetti, P.; Chaisson, R.; Drummond, J.E.; Blattner, W.; McGrath, M.; Greenspan, J.; Moss, A. Human T cell lymphotropic virus types I and II in intravenous drug users in San Francisco: Risk factors associated with seropositivity. *J. Infect. Dis.* **1991**, *164*, 36–42. [CrossRef] [PubMed]

18. Percher, F.; Jeannin, P.; Martin-Latil, S.; Gessain, A.; Afonso, P.V.; Vidy-Roche, A.; Ceccaldi, P.E. Mother-to-Child Transmission of HTLV-1 Epidemiological Aspects, Mechanisms and Determinants of Mother-to-Child Transmission. *Viruses* **2016**, *8*, 40. [CrossRef] [PubMed]

19. Ambroziak, J.A.; Blackbourn, D.J.; Herndier, B.G.; Glogau, R.G.; Gullett, J.H.; McDonald, A.R.; Lennette, E.T.; Levy, J.A. Herpes-like sequences in HIV-infected and uninfected Kaposi's sarcoma patients. *Science* **1995**, *268*, 582–583. [CrossRef] [PubMed]

20. Schalling, M.; Ekman, M.; Kaaya, E.E.; Linde, A.; Biberfeld, P. A role for a new herpes virus (KSHV) in different forms of Kaposi's sarcoma. *Nat. Med.* **1995**, *1*, 707–708. [CrossRef] [PubMed]

21. Aoki, Y.; Tosato, G. Neoplastic conditions in the context of HIV-1 infection. *Curr. HIV Res.* **2004**, *2*, 343–349. [CrossRef] [PubMed]

22. Clarke, B.; Chetty, R. Postmodern cancer: The role of human immunodeficiency virus in uterine cervical cancer. *Mol. Pathol.* **2002**, *55*, 19–24. [CrossRef] [PubMed]

23. Mbanya, D.N.; Takam, D.; Ndumbe, P.M. Serological findings amongst first-time blood donors in Yaounde, Cameroon: Is safe donation a reality or a myth? *Transfus. Med.* **2003**, *13*, 267–273. [CrossRef] [PubMed]

24. Carneiro-Proietti, A.B.; Catalan-Soares, B.C.; Castro-Costa, C.M.; Murphy, E.L.; Sabino, E.C.; Hisada, M.; Galvao-Castro, B.; Alcantara, L.C.; Remondegui, C.; Verdonck, K.; et al. HTLV in the Americas: Challenges and perspectives. *Rev. Panam. Salud Publica* **2006**, *19*, 44–53. [CrossRef] [PubMed]

25. Góngora-Biachi, R.; González-Martínez, P.; Castro-Sansores, C.; Bastarrachea-Ortiz, J. Infection with HTLV virus type I-II in patients with cervico-uterine cancer in the Yucatan peninsula Mexico. *J.Ginecol. Obstet. Mex.* **1997**, *65*, 141–144.

26. Miyazaki, K.; Yamaguchi, K.; Tohya, T.; Ohba, T.; Takatsuki, K.; Okamura, H. Human T-cell leukemia virus type I infection as an oncogenic and prognostic risk factor in cervical and vaginal carcinoma. *Obstet. Gynecol.* **1991**, *77*, 107–110. [CrossRef]

27. Strickler, H.D.; Rattray, C.; Escoffery, C.; Manns, A.; Schiffman, M.H.; Brown, C.; Cranston, B.; Hanchard, B.; Palefsky, J.M.; Blattner, W.A. Human T-cell lymphotropic virus type I and severe neoplasia of the cervix in Jamaica. *Int. J. Cancer* **1995**, *61*, 23–26. [CrossRef] [PubMed]

28. Castle, P.E.; Escoffery, C.; Schachter, J.; Rattray, C.; Schiffman, M.; Moncada, J.; Sugai, K.; Brown, C.; Cranston, B.; Hanchard, B.; et al. Chlamydia trachomatis, herpes simplex virus 2, and human T-cell lymphotrophic virus type 1 are not associated with grade of cervical neoplasia in Jamaican colposcopy patients. *Sex. Transm. Dis.* **2003**, *30*, 575–580. [CrossRef] [PubMed]

29. Raab, S.S.; Hart, A.R.; D'Antonio, J.A.; Grzybicki, D.M. Clinical perception of disease probability associated with Bethesda System diagnoses. *Am. J. Clin. Pathol.* **2001**, *115*, 681–688. [CrossRef] [PubMed]

30. Nagata, K.; Ohtani, K.; Nakamura, M.; Sugamura, K. Activation of endogenous c-fos proto-oncogene expression by human T-cell leukemia virus type I-encoded p40tax protein in the human T-cell line, Jurkat. *J. Virol.* **1989**, *63*, 3220–3226. [PubMed]

31. Maranga, I.O.; Hampson, L.; Oliver, A.W.; He, X.; Gichangi, P.; Rana, F.; Opiyo, A.; Hampson, I.N. HIV Infection Alters the Spectrum of HPV Subtypes Found in Cervical Smears and Carcinomas from Kenyan Women. *Open Virol. J.* **2013**, *7*, 19–27. [CrossRef] [PubMed]

32. AETC HIV Classification: CDC and WHO Staging Systems. Available online: http://aidsetc.org/guide/hiv-classification-cdc-and-who-staging-systems (accessed on April 2014).

33. Van Tienen, C.; van der Loeff, M.F.; Peterson, I.; Cotten, M.; Holmgren, B.; Andersson, S.; Vincent, T.; Sarge-Njie, R.; Rowland-Jones, S.; Jaye, A.; et al. HTLV-1 in rural Guinea-Bissau: Prevalence, incidence and a continued association with HIV between 1990 and 2007. *Retrovirology* **2010**, *7*, 322–332. [CrossRef] [PubMed]

34. Brites, C.; Sampalo, J.; Oliveira, A. HIV/human T-cell lymphotropic virus coinfection revisited: Impact on AIDS progression. *AIDS Rev.* **2009**, *11*, 8–16. [PubMed]

35. Abdel-Hady, E.S.; Martin-Hirsch, P.; Duggan-Keen, M.; Stern, P.L.; Moore, J.V.; Corbitt, G.; Kitchener, H.C.; Hampson, I.N. Immunological and viral factors associated with the response of vulval intraepithelial neoplasia to photodynamic therapy. *Cancer Res.* **2001**, *61*, 192–196. [PubMed]

36. Al-Saleh, W.; Giannini, S.L.; Jacobs, N.; Moutschen, M.; Doyen, J.; Boniver, J.; Delvenne, P. Correlation of T-helper secretory differentiation and types of antigen-presenting cells in squamous intraepithelial lesions of the uterine cervix. *J. Pathol.* **1998**, *184*, 283–290. [CrossRef]

37. Bais, A.G.; Beckmann, I.; Lindemans, J.; Ewing, P.C.; Meijer, C.J.L.M.; Snijders, P.J.F.; Helmerhorst, T.J.M. A shift to a peripheral Th2-type cytokine pattern during the carcinogenesis of cervical cancer becomes manifest in CIN III lesions. *J. Clin. Pathol.* **2005**, *58*, 1096–1100. [CrossRef] [PubMed]

38. Scott, M.; Nakagawa, M.; Moscicki, A.B. Cell-mediated response to human papillomavirus infection. *Clin. Diagn. Lab. Immunol.* **2001**, *8*, 209–220. [CrossRef] [PubMed]

39. Abrahao, M.H.; Lima, R.G.; Netto, E.; Brites, C. Short communication: Human lymphotropic virus type 1 coinfection modulates the synthesis of cytokines by peripheral blood mononuclear cells from HIV type 1-infected individuals. *AIDS Res. Hum. Retrovir.* **2012**, *28*, 806–808. [CrossRef] [PubMed]

40. Hirata, T.; Nakamoto, M.; Nakamura, M.; Kinjo, N.; Hokama, A.; Kinjo, F.; Fujita, J. Low prevalence of human T cell lymphotropic virus type 1 infection in patients with gastric cancer. *J. Gastroenterol. Hepatol.* **2007**, *22*, 2238–2241. [CrossRef] [PubMed]

41. World Health Organization. Health statistics and health information systems: Life tables for WHO Member States; Part III Global health indicators. In *World Health Statistics 2012*; WHO Press: Geneva, Switzerland, 2012.

42. Campaner, A.B.; Nadais, R.F.; Galvao, M.A. The effect of cigarette smoking on cervical langerhans cells and T and B lymphocytes in normal uterine cervix epithelium. *Int. J. Gynecol. Pathol.* **2009**, *28*, 549–553. [CrossRef] [PubMed]

43. Dal Maso, L.; Serraino, D.; Franceschi, S. Epidemiology of AIDS-related tumours in developed and developing countries. *Eur. J. Cancer* **2001**, *37*, 1188–1201. [CrossRef]

44. Franceschi, S.; Dal Maso, L.; Arniani, S.; Lo Re, A.; Barchielli, A.; Milandri, C.; Simonato, L.; Vercelli, M.; Zanetti, R.; Rezza, G. Linkage of AIDS and cancer registries in Italy. *Int. J. Cancer* **1998**, *75*, 831–834. [CrossRef]

45. Franceschi, S.; Dal Maso, L.; Pezzotti, P.; Polesel, J.; Braga, C.; Piselli, P.; Serraino, D.; Tagliabue, G.; Federico, M.; Ferretti, S.; et al. Incidence of AIDS-defining cancers after AIDS diagnosis among people with AIDS in Italy, 1986–1998. *J. Acquir. Immune Defic. Syndr.* **2003**, *34*, 84–90. [CrossRef] [PubMed]

46. Odida, M.; Sandin, S.; Mirembe, F.; Kleter, B.; Quint, W.; Weiderpass, E. HPV types, HIV and invasive cervical carcinoma risk in Kampala, Uganda: A case-control study. *Infect. Agents Cancer* **2011**, *6*, 8. [CrossRef] [PubMed]

47. Ter Meulen, J.; Eberhardt, H.C.; Luande, J.; Mgaya, H.N.; Chang-Claude, J.; Mtiro, H.; Mhina, M.; Kashaija, P.; Ockert, S.; Yu, X.; et al. Human papillomavirus (HPV) infection, HIV infection and cervical cancer in Tanzania, east Africa. *Int. J. Cancer* **1992**, *51*, 515–521. [CrossRef] [PubMed]

**viruses**

**MDPI**

*Article*

# Human T-Lymphotropic Virus Type I (HTLV-1) Infection among Iranian Blood Donors: First Case-Control Study on the Risk Factors

**Mohammad Reza Hedayati-Moghaddam [1],\*, Farahnaz Tehranian [2,3] and Maryam Bayati [2,3]**

[1] Research Center for HIV/AIDS, HTLV and Viral Hepatitis, Iranian Academic Center for Education, Culture and Research (ACECR), Mashhad Branch, University Campus, Azadi Sq., Mashhad 91775-1376, Iran

[2] Blood Transfusion Research Center, High Institute for Research and Education in Transfusion Medicine, Tehran 14665-1157, Iran; farahnaz_tehranian@yahoo.com (F.T.); bayati.mar@gmail.com (M.B.)

[3] Razavi Khorasan Blood Transfusion Center, Mashhad 91379-13119, Iran

\* Correspondence: drhedayati@acecr.ac.ir or drhedayati@yahoo.com; Tel.: +98-51-38821533; Fax: +98-51-38810177

Academic Editor: Louis M. Mansky

Received: 20 June 2015 ; Accepted: 29 October 2015 ; Published: 4 November 2015

**Abstract:** Human T-cell lymphotropic virus type 1 (HTLV-1) infection is an endemic condition in Northeast Iran and, as such, identification of risk factors associated with the infection in this region seems to be a necessity. All the possible risk factors for HTLV-1 seropositivity among first-time blood donors were evaluated in Mashhad, Iran, during the period of 2011–2012. Blood donation volunteers were interviewed for demographic data, medical history, and behavioral characteristics and the frequencies of risk factors were compared between HTLV-1 positive (case) and HTLV-1 negative (control) donors. The data was analyzed using Chi square and *t*-tests. Logistic regression analysis was performed to identify independent risk factors for the infection. Assessments were carried out on 246 cases aged 17–60 and 776 controls aged 17–59, who were matched based on their ages, gender, and date and center of donation. Logistic analysis showed low income (OR = 1.53, $p$ = 0.035), low educational level (OR = 1.64, $p$ = 0.049), being born in the cities of either Mashhad (OR = 2.47, $p$ = 0.001) or Neyshabour (OR = 4.30, $p$ < 0001), and a history of blood transfusion (OR = 3.17, $p$ = 0.007) or non-IV drug abuse (OR = 3.77, $p$ < 0.0001) were significant predictors for infection with HTLV-1. Lack of variability or small sample size could be reasons of failure to detect some well-known risk factors for HTLV-1 infection, such as prolonged breastfeeding and sexual promiscuity. Pre-donation screening of possible risk factors for transfusion-transmissible infections should also be considered as an important issue, however, a revision of the screening criteria such as a history of transfusion for more than one year prior to donation is strongly recommended.

**Keywords:** HTLV-1 infection; risk factors; blood donors; Mashhad; Iran

## 1. Introduction

Human T-cell lymphotropic virus type 1 (HTLV-1) was the first human retrovirus to have been discovered in 1980 [1]. Approximately 5 to 10 million people are infected with HTLV-1 worldwide [2]. HTLV-1 infection is observed throughout all parts of the world; however, Southwestern Japan, Caribbean Basin, South America, and Central Africa have been identified as being endemic regions for the virus [2]. In addition to these regions, the virus is known to be endemic in Northeast Iran especially in the cities of Mashhad and Neyshabour [3,4]. HTLV-1 is the etiological agent for adult T-cell leukemia (ATL) and HTLV-1-associated myelopathy/tropical spastic paraparesis (HAM/TSP) [2,3]. Despite this, more than 95% of infected individuals remain as asymptomatic carriers for the duration of their lives [5].

The infection can be transmitted through the transfusion of contaminated blood or blood products, unprotected sexual contact, sharing of contaminated syringes and other instruments, or via transmission from mother to child [2,6]. Risk of sero-conversion followed by transfusion is estimated to be between 40%–60% [6]. Receiving red blood cells, platelet, and whole blood compared to plasma products is also thought to be associated with a higher risk of transmission [7]; however, this risk decreases after freezing the blood for more than two weeks [6,7]. Routine screenings of blood volunteers for HTLV antibodies, along with the exclusion of high risk individuals, have contributed in reducing virus transmission through blood transfusion in endemic areas [7]. In this context, several preventive measures have been implemented by the Iranian Blood Transfusion Organization as a means to guarantee blood safety in the country and these include recording medical histories and physical examinations of the volunteers by trained physicians, encouraging people to donate blood on a regular basis, exclusion of remunerated or family replacement donations, exclusion of individuals with histories of possible risk factors such as bloodletting, tattooing, high risk sexual contact, drug abuse, *etc.* and screening for blood-borne agents including HBV, HCV, HIV, HTLV, and *Treponema pallidum* with sensitive and accurate assays [8]. The first study of Iranian blood donors in 1994 showed the prevalence of HTLV-1 infection to be 0.29%; 1.97% among Mashhadi blood donors, and 0%–0.5% in other cities [9]. After that, all donated bloods are screened for HTLV-1 and HTLV-2 antibodies in some provinces from Northeastern Iran [8]. However, the prevalence of HTLV-1 infection is still considerable among the general population (2.12%) and blood donors (0.45%) of Mashhad [3,10]. Moreover, remarkable prevalence of the infection has been reported among both blood donors and frequently blood recipients from other regions of the country as well [11–14].

To our knowledge, no survey has been performed to determine the risk factors for HTLV-1 infection in Iran. Therefore, in the current study the frequency of associated risk factors for HTLV-1 infection was investigated among blood donors in Mashhad, Northeastern Iran. Identification of individuals with these factors and excluding them from blood donation would result in a reduction of transfusion-transmitted cases of the infection.

## 2. Materials and Methods

This case-control epidemiological study was conducted among first-time blood volunteers who had been referred to blood transfusion centers of Mashhad, Iran between September 2011 and August 2013. A total of 54,436 individuals donated blood, of which 321 individuals (0.59%; 95% CI: 0.53%–0.66%) had HTLV-1 infection based on screening and subsequent confirmatory test results. The cases included 316 blood donation volunteers from Mashhad city with confirmed HTLV-1 seropositivity and the controls were selected randomly from Mashhadi donors who had shown no reactivity for HTLV-1 antibodies in screening tests. Four controls were individually matched to each case on the basis of their ages ($\pm 2$ years), gender, and date and center of donation. ELISA method (EIAgen HTLV I-II Ab Kit, Adaltis S.r.l., Rome, Italy) was used as a primary detection tool of HTLV-1 antibodies and the positive results were confirmed via a Western blot analysis (MP-Diagnostics HTLV-Blot 2.4, MP Biomedicals Asia Pacific Pte. Ltd., Singapore, Singapore). All blood donors were routinely visited by the physician of the blood transfusion center before donation and checked for the presence of any blood-borne infections such as HBV, HCV, HIV, HTLV, and *Treponema pallidum*. Blood samples from the control group showed no reactivity to the abovementioned transfusion-transmitted agents.

Participants were interviewed by trained research assistants using a questionnaire on demographic and socio-economic characteristics, such as age, gender, marital status, education, income, birth place, family size, weight and height, duration of breastfeeding, medical histories including sexually-transmitted infections (STIs), blood transfusion, hospitalization, surgery, invasive diagnostic tests (endoscopy, *etc.*), dentistry procedures with bleeding (tooth extraction, gum surgery, root canal treatment, fixed dental prosthesis), suturing, acupuncture, needlestick injuries (in health settings or beauty salons), history of risky behaviors such as tattooing, cupping, tatbir (*Qama Zani*, the act of striking the head with a sword or knife until blood gushes out as a ritual), body piercing, unsafe sexual contact, drug abuse, and imprisonment.

This study was approved by The Research and Technology Deputy of Iranian Academic Center for Education, Culture and Research (ACECR) (Number: 2033-10) with regard to ethical issues. All participants in the study did so voluntarily, and signed a written consent.

All statistical analyses were performed by SPSS version 16 (Chicago, IL, USA) using a Chi square test and *t*-test. Logistic regression analysis was also performed to identify independent risk factors for HTLV-1 infection. Statistical significance was assumed for *p* values of less than 0.05.

## 3. Results

### 3.1. Demographic Characteristics

From 316 cases, 246 (77.8%) individuals aged 17–60 years had been referred for consulting and were subsequently interviewed. Moreover, 1241 controls were invited to participate in the study, of which 776 (62.5%) persons aged 17–59 years, agreed to be interviewed. As Table 1 shows, age distributions of both cases and controls were not significantly different between studied individuals and non-participants (*p* = 0.990 and *p* = 0.478, respectively). Among the case group, male to female ratio in participants was considerably less than that in non-participants (*p* = 0.004), albeit, no differences in the ratio was observed between referred and non-referred individuals from the control group (*p* = 0.329).

**Table 1.** Age and sex distributions of studied samples and non-participants from both cases and controls.

| | Cases | | | Controls | | |
|---|---|---|---|---|---|---|
| Variable | Studied Samples (*n* = 246) No (%) | Non-Participants (*n* = 70) No (%) | *p* Value † | Studied Samples (*n* = 776) No (%) | Non-Participants (*n* = 465) No (%) | *p* Value † |
| Age (years) | | | 0.990 | | | 0.478 |
| <30 | 52 (21.1) | 15 (21.4) | | 193 (24.9) | 132 (28.4) | |
| 30–39 | 68 (27.6) | 18 (25.7) | | 203 (26.2) | 124 (26.7) | |
| 40–49 | 77 (31.3) | 24 (34.3) | | 237 (30.5) | 133 (28.6) | |
| ≥50 | 49 (19.9) | 13 (18.6) | | 143 (18.4) | 76 (16.3) | |
| Gender | | | 0.004 | | | 0.329 |
| Male | 181 (73.6) | 64 (91.4) | | 613 (79.0) | 378 (81.3) | |
| Female | 65 (26.4) | 6 (8.6) | | 163 (21.0) | 87 (18.7) | |

† Chi square test.

**Table 2.** Demographic and socio-economic features associated with HTLV-1 infection in blood donors.

| Variable | HTLV-1-Positive (*n* = 246) No (%) | HTLV-1-Negative (*n* = 776) No (%) | *p* Value † |
|---|---|---|---|
| Age (years) | 39.1 ± 10.6 * | 38.3 ± 10.6 * | 0.25 |
| Gender | | | |
| Male | 181 (73.6) | 613 (79) | 0.08 |
| Female | 65 (26.4) | 163 (21) | |
| Marital status | | | |
| Single | 24 (9.8) | 98 (12.6) | |
| Married | 213 (86.9) | 667 (86) | 0.09 |
| Divorced | 8 (3.3) | 11 (1.4) | |

Table 2. *Cont.*

| Variable | HTLV-1-Positive (*n* = 246) No (%) | HTLV-1-Negative (*n* = 776) No (%) | *p* Value † |
|---|---|---|---|
| **Remarriage** | | | |
| Yes | 19 (8.7) | 25 (3.7) | |
| No | 200 (91.3) | 652 (96.3) | 0.003 |
| **Education** | | | |
| Illiterate | 4 (1.6) | 15 (1.9) | |
| Primary school (1–5 years) | 67 (27.2) | 105 (13.6) | |
| Secondary and high school (6–12 years) | 113 (45.9) | 406 (52.6) | <0.0001 |
| Academic education | 62 (25.2) | 250 (32.3) | |
| **Income per month (Million Rials)** | | | |
| <5 | 100 (40.8) | 213 (27.6) | |
| 5–9.9 | 113 (46.1) | 401 (51.9) | |
| 10–19.9 | 27 (11) | 131 (17) | 0.001 |
| ⩾20 | 5 (2) | 27 (3.5) | |
| **Birth place** | | | |
| Mashhad (Khorasan Razavi province) | 163 (66.3) | 473 (61) | |
| Nyshabour (Khorasan Razavi province) | 20 (8.1) | 28 (3.6) | |
| Other cities of Khorasan Razavi province | 43 (17.5) | 144 (18.6) | <0.0001 |
| Other provinces | 20 (8.1) | 131 (16.9) | |
| **Family size** | 3.7 (1.2) * | 3.8 (1.3) * | 0.20 |
| **BMI** | | | |
| <25 | 79 (33.2) | 220 (39.5) | |
| 25–30 | 108 (45.4) | 339 (47) | 0.73 |
| ⩾30 | 51 (21.4) | 163 (22.6) | |
| **Duration of breastfeeding** | | | |
| <6 months | 8 (5.2) | 47 (7.9) | |
| ⩾6 moths | 144 (94.7) | 547 (92.1) | 0.26 |

BMI: Body mass index; * mean ± SD; † *t*-test for age and family size, Chi square test for other variables.

Demographic and socio-economic characteristics of donors are shown in Table 2. No significant difference was observed in the mean age and gender between case and control groups (*p* = 0.25 and *p* = 0.08, respectively). Cases had higher frequency of remarriage compared to controls (*p* = 0.003). The proportion of individuals with primary education or less in the case group (28.8%) was higher than those observed in the control group (15.5%; *p* < 0.0001). Furthermore, patients with the infection had significantly lower incomes compared to the controls (*p* = 0.001). According to Table 2, frequency of subjects who were born in the cities of Mashhad and Neyshabour in the case group was more than those in controls (*p* < 0.0001). In addition, no significant difference was found in the marital status (*p* = 0.09), body mass index (*p* = 0.73), family size (*p* = 0.21), and duration of breastfeeding (*p* = 0.26) between the both groups.

*3.2. Medical Conditions*

Medical histories of blood donors which were assumed to be associated with the risk of HTLV-1 infection are shown in Table 3. A history of blood or blood products transfusion in the case group was higher than that in controls (*p* = 0.005). Fifteen donors, including six cases (3.7%) and nine controls (1.5%), reported blood reception in the province of Khorasan Razavi (nearly all in Mashhad city) after the routine screening for HTLV-1 infection was started in this region (*p* = 0.07).

In addition, history of STIs was very low in both case and control groups (2.4% and 1.8%, respectively; *p* = 0.53). On the other hand, no significant differences were found in the history of hospitalization, surgery, acupuncture, suturing, invasive dental treatment, invasive diagnostic procedures, and needlestick between both groups (Table 3). Three cases of HTLV-1-HBV co-infection and one case of HTLV-1-HCV co-infection were detected but no association with HIV and *Treponema pallidum* was observed in the case group.

**Table 3.** Medical histories of blood donors associated with HTLV-1 infection.

| Variable | HTLV-1-Positive (*n* = 246) No (%) | HTLV-1-Negative (*n* = 776) No (%) | *p* Value † |
|---|---|---|---|
| **Hospitalization** | | | |
| Yes | 128 (51.2) | 461 (57.4) | 0.10 |
| No | 122 (48.8) | 342 (42.6) | |
| **Surgery** | | | |
| Yes | 98 (40.8) | 327 (42.5) | 0.88 |
| No | 142 (59.2) | 444 (57.6) | |
| **Blood or blood products transfusion** | | | |
| Yes | 13 (5.3) | 15 (1.9) | 0.005 |
| No | 233 (94.7) | 761 (98.1) | |
| **Invasive diagnostic procedure** | | | |
| Yes | 18 (7.3) | 57 (7.3) | 0.99 |
| No | 227 (92.7) | 719 (92.7) | |
| **Invasive dental treatment** | | | |
| Yes | 188 (76.4) | 550 (71.2) | 0.11 |
| No | 58 (23.6) | 222 (28.8) | |
| **Acupuncture** | | | |
| Yes | 2 (0.8) | 17 (2.2) | 0.16 |
| No | 244 (99.2) | 759 (98.7) | |
| **Suturing** | | | |
| Yes | 82 (33.6) | 245 (31.8) | 0.54 |
| No | 162 (66.4) | 527 (68.3) | |
| **Needlestick** | | | |
| Yes | 13 (5.3) | 37 (4.8) | 0.74 |
| No | 233 (94.7) | 739 (95.2) | |
| **STIs** | | | |
| Yes | 6 (2.4) | 14 (1.8) | 0.53 |
| No | 240 (97.6) | 761 (98.2) | |

† Chi square test; STIs: sexually-transmitted infections.

### 3.3. Risky Behaviors

Possible risky behaviors related to HTLV-1 infection are presented in Table 4. Frequency of cupping, piercing, tattooing, and imprisonment history did not differ between the groups; with only two cases (0.8%) and three controls (0.4%) having a history of *tatbir*. On the other hand, a history of drug abuse in the case group was significantly higher than those found in the controls (*p* < 0.0001). However, none of the subjects had stated a history of injecting drug use (IDU). Moreover, history of pre- and extra-marital sexual contact and number of lifetime sexual partners in both groups were similar (*p* = 0.17 and *p* = 0.28). On the other hand, among 22 cases with a history of pre- or extra-marital sex, 4.5% had one partner, 31.8% had two partners, and 63.6% had at least three partners. In controls with a same history, 43.6%, 27.3%, and 29.1% of 55 donors had one, two, and three or more partners, respectively (*p* = 0.002).

**Table 4.** Risky behaviors associated with HTLV-1 infection among blood donors.

| Variable | HTLV-1-Positive (*n* = 246) No (%) | HTLV-1-Negative (*n* = 776) No (%) | *p* Value † |
|---|---|---|---|
| **Cupping** | | | |
| Yes | 85 (34.7) | 275 (35.4) | |
| No | 160 (65.3) | 501 (64.6) | 0.89 |
| **Tattooing** | | | |
| Yes | 16 (6.5) | 35 (4.5) | |
| No | 230 (93.5) | 741 (95.5) | 0.21 |
| **Piercing** | | | |
| Yes | 47 (19.2) | 155 (20.1) | |
| No | 198 (80.8) | 616 (79.9) | 0.70 |
| **Drug abuse** | | | |
| Yes | 23 (9.3) | 15 (1.9) | |
| No | 223 (90.7) | 761 (98.1) | <0.0001 |
| **Pre- and extra-marital sex** | | | |
| Yes | 25 (10.2) | 58 (7.5) | |
| No | 219 (89.8) | 715 (92.5) | 0.17 |
| **Number of life time sexual partners** | | | |
| 0 | 22 (9.2) | 86 (11.2) | |
| 1 | 196 (81.7) | 634 (82.3) | 0.28 |
| ≥2 | 22 (9.2) | 50 (6.5) | |
| **Imprisonment** | | | |
| Yes | 9 (3.7) | 25 (3.2) | |
| No | 237 (96.3) | 751 (96.8) | 0.73 |

† Chi square test.

*3.4. Regression Analysis of Risk Factors Related to HTLV-1 Infection*

All variables with a significant relation to HTLV-1 infection in univariate analysis were entered into the logistic regression model. As Table 5 shows, significant associations were found between the infection and low educational levels (OR = 1.64, 95% CI: 1.04–2.69), low income (OR = 1.53, 95% CI: 1.03–2.26), birth place (OR = 4.30, 95% CI: 1.91–9.67 for those born in the city of Neyshabour and OR = 2.47, 95% CI: 1.42–4.33 for Mashhad city), a history of blood transfusion (OR = 3.17, 95% CI: 1.37–7.33), and drug abuse (OR = 3.77, 95% CI: 1.79–7.55).

**Table 5.** Results from logistic regression analysis for HTLV-1 associated risk factors in blood donors.

| Variable | OR | 95% CI for OR | *p*-Value |
|---|---|---|---|
| **Remarriage** | | | |
| Yes | 1.50 | 0.75–2.99 | 0.252 |
| No | 1.0 | | |
| **Education level** | | | |
| Illiterate or primary school (0–5 years) | 1.64 | 1.04–2.69 | 0.049 |
| Secondary and high school (6–12 years) | 0.73 | 0.43–1.27 | 0.266 |
| Academic education | 1.0 | | |
| **Monthly income (Million Rials)** † | | | |
| <5 | 1.53 | 1.03–2.26 | 0.035 |
| 5–9.9 | 1.53 | 0.49–4.73 | 0.461 |
| 10–19.9 | 1.13 | 0.35–3.71 | 0.838 |
| ⩾20 | 1.0 | | |
| **Birth City** | | | |
| Mashhad (Khorasan Razavi province) | 2.47 | 1.42–4.23 | 0.001 |
| Neyshabour (Khorasan Razavi province) | 4.30 | 1.91–9.67 | <0.0001 |
| Other cities of Khorasan Razavi province | 1.82 | 0.96–3.45 | 0.067 |
| Cities of other provinces | 1.0 | | |
| **History of blood transfusion** | | | |
| Yes | 3.17 | 1.37–7.33 | 0.007 |
| No | 1.0 | | . |
| **History of drug abuse** | | | |
| Yes | 3.77 | 1.79–7.55 | <0.0001 |
| No | 1.0 | | |

OR: Odd ratio; CI: Confidence Interval. † One million Rials was approximately equal to 30 USD in the time of study.

## 4. Discussion

This study was conducted to identify possible risk factors related to HTLV-1 infection among first-time donors who referred to blood centers in Mashhad, Iran. Findings showed that the main risk factors associated with HTLV-1 infection were low educational levels, low income, being born in the cities of Mashhad and Neyshabour, and histories of blood transfusion and drug abuse.

In a study conducted on Australian blood donors during 2000–2006, the prevalence of HTLV-1 seropositivity was three per 100,000 donors, of which 5% had at least one unidentified risk factor at the time of blood donation. In general, no dominant risk factors emerged, but the major identified factors related to the infection were as follows: the country of birth and parental ethnicity (24%), sex with individual from overseas (23%), at-risk household contacts (19%), tattooing or piercing (8%), surgery or endoscopy (8%), blood product transfusion (8%), and other blood contacts, such as needle sticks (4%) [15].

In the current study, being born in the cities of Mashhad or Neyshabour showed a strong association with infection of HTLV-1. Very high prevalence of the infection in the general population of these cities had been previously reported [3,4]. However, studies conducted in other cities of Iran, had shown that HTLV-1 prevalence was not considerable [13,16].

In the present study, the case group had a significant lower education and income levels compared to the controls. It seems that low education and income levels among these people would have reduced their access to health information and may increase the prevalence of risk factors related to HTLV-1 infection, such as risky sexual behaviors or contact with contaminated blood through tattooing or piercing among these individuals. In another study on the general population of Mashhad, HTLV-1 infection was not associated with income, but the highest prevalence of the infection was observed among illiterates. Nevertheless, regression analysis showed no significant association between educational level and the infection [3]. Similarly, Custer *et al.* and Dourado *et al.* could not find any significant association between literacy and income with the infection in blood donors [17,18]. Conversely, in a case-control study by Rouet *et al.*, low educational level was identified to be a main risk factor for HTLV-1 infection among blood donors in French West Indies [19]. Furthermore, Blas *et al.* demonstrated that low literacy was indeed an important risk factor for HTLV-1 infection [20].

Breastfeeding, especially longer than six months, is one of the HTLV-1 transmission routes [21]. The current study, showed high frequencies of breastfeeding for ⩾ 6 months in both groups. However prolonged breastfeeding was slightly higher in the case group compared to the controls, but this difference was not statistically significant. Lack of variability about prolonged breastfeeding might be a reason that we could not find a significant association between mode of feeding in infancy period and prevalence of the infection. Similarly, Blas *et al.* could not find a significant association between the infection and breastfeeding due to very high prevalence of breastfeeding in both HTLV-1 seropositive and seronegative individuals [20]. On the other hand, breastfeeding was reported as one of the important risk factors for HTLV-1infection among blood donors from Taiwan with an odds ratio of 4.4 [22].

As expected, this study showed that HTLV-1 infection is significantly associated with a history of blood reception. Nevertheless, there was no significant difference between the study groups if the history was limited to reception in Khorasan Razavi blood centers after 1994 when routine screening for HTLV-1 infection began in this region. A history of blood transfusion during the last year is one of the deferral criteria for blood donations in Iran and all of the blood donors had not received any blood for at least one year before the blood donation date. In a survey in Neyshabour, the proportion of the infection among individuals with a history of the transfusion was four times higher than others [4]. Additionally, in several studies on blood donors from Taiwan, Canada and Brazil a history of blood transfusion increased the risk of infection by a factor of nine to ten times [22–24]. On the other hand, HTLV-1 infection was not significantly associated with transfusion among healthy blood donors from Nigeria [25] and USA [17].

Based on the results of the present study, the risk of infection was not increased in individuals with a history of surgery, hospitalization, needlestick, cupping, or tattooing. Similarly, in the study conducted on the general population of Mashhad, HTLV-1 infection was not more prevalent among individuals with a history of these variables on the regression analysis [3]. In another survey among blood donors from Taiwan, a history of surgery was also not associated with the infection [22].

This study unexpectedly identified non-IDU as a risk factor for HTLV-1 infection. The frequency of non-IDU in the case group was significantly higher than in controls (9% and 2%, respectively). Likewise, Soares *et al.* showed that HTLV-1 seropositive cases in Brazilian blood donors more often used non-intravenous illegal drugs (OR = 3.3), while no significant difference was observed between HTLV-1 seropositive and seronegative groups for the frequency of intravenous drugs use [24]. In Iran, people who use opium orally or by inhalation (except heroin and crack) could donate their blood but are permanently excluded from donation if they report at least one IDU or use sharp tools for substance inhalation. It seems that the other risk factors which are not declared by the participants with a history of non-IDU or not evaluated by the researchers might have been involved in the infection of HTLV and other blood-borne viruses [26].

A history of STIs during one last year is a deferral factor for blood donation in Iran, and in the current study a history of STIs among blood donors was expectedly very low as a result. A history of extramarital sexual contact was similar among cases and controls, however, among this subgroup, cases had more lifetime sexual partner than controls. In a study conducted on injecting drug users in Mashhad prison, no association was found between HTLV-1 infection and a history of STIs and the frequency of multiple sexual partners did not differ between the HTLV-1 positive and negative groups [27]. In contrast, several studies have reported that HTLV-1 infection is associated with a history of STIs or with having multiple sexual partners [23,28,29]. Rouet *et al.* demonstrated that a history of STIs and Chlamydia seropositivity are indeed predictive factors for HTLV-1 infection among blood donors from Guadeloupe, French West Indies [19]. In addition, Melbye *et al.* showed significant correlation between having more than three partners and HTLV-1 infection among females in Guinea-Bissau, West Africa. However, they could not find a significant correlation between a history of gonorrhea and genital ulcer with HTLV-1 infection [30].

## 5. Conclusions

In summary, a history of blood transfusion, being born in the cities of Mashhad and Neyshabour, having low education and income levels, and non-IDU were significantly associated to HTLV-1 infection in first-time donors from Mashhad, Iran. Failure to detect other well-known risk factors linked to HTLV-1 infection such as breastfeeding, sexual promiscuity, and IDU could have been due, in part, to a lack of variability or the small sample size used. Various blood safety programs, such as pre-donation screening and deferral policies for blood donors with a history of possible risk factors for transfusion-transmissible infections, might also have contributed to the reduction of blood donation volunteers with these risk factors in Iran. However, a revision of the screening criteria such as a history of transfusion for more than one year prior to donation is strongly recommended.

**Acknowledgments:** This study was financially supported by Deputy for Research and Technology of Iranian Academic Center for Education, Culture and Research (ACECR). We would like to thank the blood donors who participated in this study. Also, we appreciate Razavi Khorasan Blood Transfusion Center for their great co-operation. Many thanks to Dr. Aram Meshkini, Dr. Narjes Sahebzadeh, Dr. Taraneh Honarparvar, Dr. Seyed Sana Davoodi Moghaddam and Dr. Mohsen Tadayyon for their kind cooperation and assistance.

**Author Contributions:** Acquisition of data and design: Mohamad Reza Hedayati-Moghadam, Farahnaz Tehranian, and Maryam Bayati; Drafting of manuscript: Mohamad Reza Hedayati-Moghadam and Farahnaz Tehranian; Data analysis: Mohamad Reza Hedayati-Moghadam.

**Conflicts of Interest:** The authors declare no conflict of interest.

## References

1. Goon, P.K.; Igakura, T.; Hanon, E.; Mosley, A.J.; Barfield, A.; Barnard, A.L.; Kaftantzi, L.; Tanaka, Y.; Taylor, G.P.; Weber, J.N.; *et al.* Human T cell lymphotropic virus type I (HTLV-I)-specific CD4+ T cells: Immunodominance hierarchy and preferential infection with HTLV-I. *J. Immunol.* **2004**, *172*, 1735–1743. [CrossRef] [PubMed]
2. Gessain, A.; Cassar, O.; Sohgandi, L.; Azarpazhooh, M.R.; Rezaee, S.A.; Farid, R.; *et al.* Epidemiological aspects and world distribution of HTLV-1 infection. *Front. Microbiol.* **2012**, *3*. [CrossRef] [PubMed]
3. Rafatpanah, H.; Hedayati-Moghaddam, M.R.; Fathimoghadam, F.; Bidkhori, H.R.; Shamsian, S.K.; Ahmadi, S. High prevalence of HTLV-I infection in Mashhad, Northeast Iran: A population-based seroepidemiology survey. *J. Clin. Virol.* **2011**, *52*, 172–176. [CrossRef] [PubMed]
4. Hedayati-Moghaddam, M.R.; Fathimoghadam, F.; Eftekharzadeh Mashhadi, I.; Soghandi, L.; Bidkhori, H.R. Epidemiology of HTLV-1 in Neyshabour, Northeast of Iran. *Iran. Red Crescent Med. J.* **2011**, *13*, 424–427. [PubMed]
5. Matsuura, E.; Yamano, Y.; Jacobson, S. Neuroimmunity of HTLV-I Infection. *J. Neuroimmune Pharmacol.* **2010**, *5*, 310–325. [CrossRef] [PubMed]
6. Verdonck, K.; Gonzalez, E.; van Dooren, S.; Vandamme, A.M.; Vanham, G.; Gotuzzo, E. Human T-lymphotropic virus 1: Recent knowledge about an ancient infection. *Lancet Infect. Dis.* **2007**, *7*, 266–281. [CrossRef]
7. Proietti, F.A.; Carneiro-Proietti, A.B.; Catalan-Soares, B.C.; Murphy, E.L. Global epidemiology of HTLV-I infection and associated diseases. *Oncogene* **2005**, *24*, 6058–6068. [CrossRef] [PubMed]
8. Abolghasemi, H.; Maghsudlu, M.; Kafi-Abad, S.A.; Cheraghali, A. Introduction to Iranian blood transfusion organization and blood safety in Iran. *Iran. J. Public Health* **2009**, *38*, 82–87.
9. Rezvan, H.; Ahmadi, J.; Farhadi, M. A cluster of HTLV-1 infection in northeastern of Iran. *Transfus. Today* **1996**, *7*, 8–9.
10. Tarhini, M.; Kchour, G.; Zanjani, D.S.; Rafatpanah, H.; Otrock, Z.K.; Bazarbachi, A.; Farid, R. Declining tendency of human T-cell leukaemia virus type I carrier rates among blood donors in Mashhad, Iran. *Pathology* **2009**, *41*, 498–499. [CrossRef] [PubMed]

11. Karimi, A.; Nafici, M.; Imani, R. Comparison of human T-cell leukemia virus type-1 (HTLV-1) seroprevalence in high risk patients (thalassemia and hemodialysis) and healthy individuals from Charmahal-Bakhtiari Province, Iran. *Kuwait Med. J.* **2007**, *39*, 259–261.

12. Khameneh, Z.R.; Baradaran, M.; Sepehrvand, N. Survey of the seroprovalence of HTLV I/II in hemodialysis patients and blood donors in Urmia. *Saudi J. Kidney Dis. Transpl.* **2008**, *19*, 838–841. [PubMed]

13. Ghaffari, J.; Kowsarian, M.; Mahdavi, M.R.; Shahi, K.V.; Rafatpanah, H.; Tafreshian, A.R. Prevalence of HTLV-1 infection in patients with thalassemia major in Mazandaran, north of Iran. *Jundishapur J. Microbiol.* **2013**, *6*, 57–60. [CrossRef]

14. Moradi, A.; Mansurian, A.; Ahmadi, A.; Ghaemi, E.; Kalavi, K.; Marjani, A.; Sanei Moghaddam, E. Prevalence of HTLV-1 antibody among major thalassemic patients in Gorgan (South East of Caspian Sea). *J. Appl. Sci.* **2008**, *8*, 391–393. [CrossRef]

15. Polizzotto, M.N.; Wood, E.M.; Ingham, H.; Keller, A.J. Reducing the risk of transfusion-transmissible viral infection through blood donor selection: The Australian experience 2000 through 2006. *Transfusion* **2008**, *48*, 55–63. [CrossRef] [PubMed]

16. Kalavi, K.; Moradi, A.; Tabarraei, A. Population-based seroprevalence of HTLV-I infection in Golestan province, South East of Caspian Sea, Iran. *Iran. J. Basic Med. Sci.* **2013**, *16*, 225–228. [PubMed]

17. Custer, B.; Kessler, D.; Vahidnia, F.; Leparc, G.; Krysztof, D.E.; Shaz, B.; Kamel, H.; Glynn, S.; Dodd, R.Y.; Stramer, S.L.; *et al.* Risk factors for retrovirus and hepatitis virus infections in accepted blood donors. *Transfusion* **2015**, *55*, 1098–1107. [CrossRef] [PubMed]

18. Dourado, I.; Alcantara, L.C.; Barreto, M.L.; da Gloria Teixeira, M.; Galvao-Castro, B. HTLV-I in the general population of Salvador, Brazil: A city with African ethnic and sociodemographic characteristics. *J. Acquir. Immune Defic. Syndr.* **2003**, *34*, 527–531. [CrossRef] [PubMed]

19. Rouet, F.; Herrmann-Storck, C.; Courouble, G.; Deloumeaux, J.; Madani, D.; Strobel, M. A case-control study of risk factors associated with human T-cell lymphotrophic virus type-I seropositivity in blood donors from Guadeloupe, French West Indies. *Vox Sang.* **2002**, *82*, 61–66. [CrossRef] [PubMed]

20. Blas, M.M.; Alva, I.E.; Garcia, P.J.; Carcamo, C.; Montano, S.M.; Mori, N.; Muñante, R.; Joseph, R.Z. High prevalence of human T-lymphotropic virus infection in indigenous women from the peruvian Amazon. *PLoS ONE* **2013**, *8*, e73978. [CrossRef] [PubMed]

21. Van Tienen, C.; Jakobsen, M.; van der Loeff, M.S. Stopping breastfeeding to prevent vertical transmission of HTLV-1 in resource-poor settings: Beneficial or harmful? *Arch. Gynecol. Obstet.* **2012**, *286*, 255–256. [CrossRef] [PubMed]

22. Lu, S.C.; Kao, C.L.; Chin, L.T.; Chen, J.W.; Yang, C.M.; Chang, A.C.; Chen, B.H. Intrafamilial transmission and risk assessment of HTLV-I among blood donors in southern Taiwan. *Kaohsiung J. Med. Sci.* **2001**, *17*, 126–132. [PubMed]

23. O'Brien, S.F.; Goldman, M.; Scalia, V.; Yi, Q.L.; Fan, W.; Xi, G.; Dines, I.R.; Fearon, M.A. The epidemiology of human T-cell lymphotropic virus types I and II in Canadian blood donors. *Transfus. Med.* **2013**, *23*, 358–366. [CrossRef] [PubMed]

24. Soares, B.C.; Proietti, A.B.; Proietti, F.A. HTLV-I/II and blood donors: Determinants associated with seropositivity in a low risk population. *Rev. Saude Publica* **2003**, *37*, 470–476. [CrossRef] [PubMed]

25. Durojaiye, I.; Akinbami, A.; Dosunmu, A.; Ajibola, S.; Adediran, A.; Uche, E.; Oshinaike, O.; Odesanya, M.; Dada, A.; Okunoye, O.; *et al.* Seroprevalence of human T lymphotropic virus antibodies among healthy blood donors at a tertiary centre in Lagos, Nigeria. *Pan Afr. Med. J.* **2014**, *17*. [CrossRef] [PubMed]

26. Galea, S.; Nandi, A.; Vlahov, D. The social epidemiology of substance use. *Epidemiol. Rev.* **2004**, *26*, 36–52. [CrossRef] [PubMed]

27. Rowhani-Rahbar, A.; Tabatabaee-Yazdi, A.; Panahi, M. Prevalence of common blood-borne infections among imprisoned injection drug users in Mashhad, North-East of Iran. *Arch. Iran. Med.* **2004**, *7*, 190–194.

28. Schreiber, G.B.; Murphy, E.L.; Horton, J.A.; Wright, D.J.; Garfein, R.; Chien, H.C.; Nass, C.C. Risk factors for human T-cell lymphotropic virus types I and II (HTLV-I and -II) in blood donors: The Retrovirus Epidemiology Donor Study. NHLBI Retrovirus Epidemiology Donor Study. *J. Acquir. Immune Defic. Syndr. Hum. Retrovirol.* **1997**, *14*, 263–271. [CrossRef] [PubMed]

29. Ansaldi, F.; Comar, M.; D'Agaro, P.; Grainfenberghi, S.; Caimi, L.; Gargiulo, F.; Bruzzone, B.; Gasparini, R.; Icardi, G.; Perandin, F.; *et al.* Seroprevalence of HTLV-I and HTLV-II infection among immigrants in northern Italy. *Eur. J. Epidemiol.* **2003**, *18*, 583–588. [CrossRef] [PubMed]

30. Melbye, M.; Poulsen, A.G.; Gallo, D.; Pedersen, J.B.; Biggar, R.J.; Larsen, O.; Dias, F.; Aaby, P. HTLV-1 infection in a population-based cohort of older persons in Guinea-Bissau, West Africa: Risk factors and impact on survival. *Int. J. Cancer* **1998**, *76*, 293–298. [CrossRef]

*viruses*

MDPI

*Article*

# Low Proviral Load is Associated with Indeterminate Western Blot Patterns in Human T-Cell Lymphotropic Virus Type 1 Infected Individuals: Could Punctual Mutations be Related?

Camila Cánepa *, Jimena Salido, Matías Ruggieri, Sindy Fraile, Gabriela Pataccini, Carolina Berini [†] and Mirna Biglione [†]

Instituto de Investigaciones Biomédicas en Retrovirus y SIDA, UBA-CONICET, Paraguay 2155, piso 11, C1121ABG, CABA, Argentina; canepa.camila@gmail.com (C.C.); jimenasalido@gmail.com (J.S.); ruggierimatias50@gmail.com (M.R.); sfraile77@gmail.com (S.F.); gaby.pataccini@hotmail.com (G.P.); cberini@fmed.uba.ar (C.B.); mbiglione@fmed.uba.ar (M.B.)

* Correspondence: canepa.camila@gmail.com; Tel.: +54-11-4508-3689 (ext. 125); Fax: +54-11-4508-3705
† These authors contributed equally to this work.

Academic Editor: Louis M. Mansky
Received: 28 May 2015 ; Accepted: 22 October 2015 ; Published: 28 October 2015

**Abstract:** Background: indeterminate Western blot (WB) patterns are a major concern for diagnosis of human T-cell lymphotropic virus type 1 (HTLV-1) infection, even in non-endemic areas. Objectives: (a) to define the prevalence of indeterminate WB among different populations from Argentina; (b) to evaluate if low proviral load (PVL) is associated with indeterminate WB profiles; and (c) to describe mutations in LTR and *tax* sequence of these cases. Results: Among 2031 samples, 294 were reactive by screening. Of them, 48 (16.3%) were WB indeterminate and of those 15 (31.3%) were PCR+. Quantitative real-time PCR (qPCR) was performed to 52 HTLV-1+ samples, classified as Group 1 (G1): 25 WB+ samples from individuals with pathologies; Group 2 (G2): 18 WB+ samples from asymptomatic carriers (AC); and Group 3 (G3): 9 seroindeterminate samples from AC. Median PVL was 4.78, 2.38, and 0.15 HTLV-1 copies/100 PBMCs, respectively; a significant difference ($p$=0.003) was observed. Age and sex were associated with PVL in G1 and G2, respectively. Mutations in the distal and central regions of Tax Responsive Elements (TRE) 1 and 2 of G3 were observed, though not associated with PVL.The 8403A>G mutation of the distal region, previously related to high PVL, was absent in G3 but present in 50% of WB+ samples ($p$ = 0.03). Conclusions: indeterminate WB results confirmed later as HTLV-1 positive may be associated with low PVL levels. Mutations in LTR and *tax* are described; their functional relevance remains to be determined.

**Keywords:** HTLV-1/2; proviral load; Western blot; indeterminate; mutations

## 1. Introduction

Human T-cell lymphotropicvirus type 1 and 2 (HTLV-1/2) are distributed worldwide. HTLV-1 infects an estimated 20 million people in the world and is considered the etiologic agent of adult T-cell leukemia/lymphoma (ATLL), HTLV-associated myelopathy/tropical spastic paraparesis (HAM/TSP), and HTLV-1 uveitis [1]. HTLV-1 presents foci of endemicity in the Caribbean, Southeastern Japan, sub-Saharan Africa, the Middle East, and areas of South America, while HTLV-2 is naturally endemic in natives from Africa and aborigines of the Americas [1,2]. Concerning phylogeny, seven subtypes have been identified within HTLV-1: cosmopolitan (a), Central African (b and d), Melanesian (c), a variant from Zaire (e), one from Gabon (f), and one from Cameroon (g). The cosmopolitan subtype, disseminated worldwide, is composed of five subgroups: transcontinental (A), Japanese (B), West African (C), North African (D), and Black Peruvian (E) [3–5]. In Argentina,

HTLV-1 cosmopolitan subtype transcontinental subgroup A is the major subgroup detected in the endemic area of thenorthwest as well as in blood donors, pregnant women, and different at-risk populations in non-endemic regions [6,7].

Mandatory screening for HTLV-1/2 in blood banks, which includes detection by an enzyme immunoassay (EIA) or particle agglutination (PA), has been implemented in many countries so far. According to the current algorithm, a serological confirmation, usually by Western blot (WB), should be performed after reactive screening results [8]. However, despite improvements made in the WB assay specificity over the past years, HTLV-indeterminate WB results continue to be frequent in blood donors, mainly in inter-tropical areas, posing a major challenge for routine diagnosis worldwide [9–11].

It has been observed that the use of screening tests with low specificity significantly increases the number of indeterminate WB results that are later confirmed negative for the infection by molecular techniques [12]. Other possible explanations include cross-reactivity against other retroviruses or microbial agents, as occurs with *Plasmodium falciparum* in Central Africa, Indonesia, and the Philippines [13–15]. Regarding seroindeterminate cases later confirmed positive for the infection, several hypotheses have been proposed such as the presence of defective virus or low copy numbers of prototypic HTLV-1/2 that could be yielding a light antibody response [16]. Punctual mutations in key viral genes could be another alternative. Netto *et al.* have reported an association between G232A in the Tax Responsive Element (TRE) 1 and an increase in PVL levels [17]. Furthermore, it has been observed that non-synonymous mutations of the HTLV-1 *tax* gene could display markedly attenuated abilities to transactivate the provirus [18].

Over the last decade, a sensitive and specific nested polymerase chain reaction (n-PCR) assay able to confirm HTLV-1/2 infections in individuals with an indeterminate profile or HTLV positive but not typeable results by WB became an important tool for diagnosis [19,20]. Years later, quantitation of HTLV-1 proviral load (PVL) by quantitative real-time PCR (qPCR) was implemented for the follow-up of patients with associated pathologies worldwide [21–23]. Recently, both qPCR and multiplex (mqPCR) have been proposed as molecular testing for the confirmation of HTLV-1/2 diagnosis, aimed to address the issue of indeterminate results. However, according to reported data, these techniques still show sensibility problems [24,25].

As a consequence of frequent indeterminate WB results leading to difficulties in interpretation and counseling in our country, this study aims to (i) define the prevalence and banding profile frequency of cases with indeterminate results by WB among different populations from Argentina; (ii) evaluate whether a low PVL in HTLV-1 positive individuals is one of the causes of these results; and (iii) identify the presence of punctual mutations, both in Long Terminal Repeats (LTR) and *tax* regions, of indeterminate cases.

## 2. Results

### 2.1. Prevalence Studies

Prevalence of WB indeterminate results corresponding to samples from four different populations of Argentina is shown in Table 1. Three of them, Men who have Sex with Men (MSM), Injecting Drug Users (IDUs), and Female Sex Workers (FSW), belong to a previous epidemiological study [26]. The remaining one, the HTLV Diagnosis and Confirmation population (HDC), is composed of individuals referred from blood banks or hospitals to our Institute. The global methodology, including the number of samples tested at each step, is illustrated in Figure S1. The total number of WB indeterminate samples (IS) and, of those, the ones that were later confirmed as HTLV-1 or HTLV-2 positive by molecular techniques, are also shown in Table 1.

Table 1. Prevalence of HTLV-1/2 infection and frequency of WB indeterminate patterns in four populations of Argentina. MSM: Men who have Sex with Men; IDU: Injecting Drug Users; FSW: Female Sex Workers; HDC: samples received at a Reference Institute for HTLV Diagnosis and Confirmation (HDC) from blood banks or hospitals of Argentina. ELISA: enzyme-linked immunosorbent assay. PA: particle agglutination.

| | Reactive by PA or ELISA | Indeterminate samples (IS) n (%) | IS Confirmed HTLV-1+ by n-PCR n (%) | IS Confirmed HTLV-2+ by n-PCR n (%) | Total HTLV-1 Prevalence % (n/N) | Total HTLV-2 Prevalence % (n/N) |
|---|---|---|---|---|---|---|
| MSM (N=667) | 26 | 11 (1.65) | 3 (27.28) | 0 (0) | 0.45% (3/667) [c] | 0% (0/667)[c] |
| IDU (N=173) | 36 | 4 (2.31) | 4 (100) | 2 [b] (100) | 4.62% (8/173) [c] | 15.6% (27/173) [c] |
| FSW (N=613) | 25 | 3 (2.12) | 3 (23.10) | 0 (0) | 1.46% (9/613) [c] | 0.2% (1/613) [c] |
| HDC (N=578) | 207 | 30 (5.19) | 3 (15.79) [a] | 2 (10.53)[a] | 18.8% (109/578) | 5.36% (31/578) |
| Total | 294 | 48 (16.33) | 13 (35.13) | 4 (10.81) | 6.35% (129/2031) | 2.90% (59/2031) |

[a] Out of 30 seroindeterminate samples, only 19 could be tested by molecular techniques, as no DNA was available for the other 11. [b] These two samples were HTLV-1/2 co-infected. [c] Data reported by Berini *et al.* 2007 [26].

The different WB indeterminate banding patterns are detailed in Table 2 for all samples from the HDC population (*n*=30), including positive and negative ones by n-PCR. The banding patterns corresponding to the other three populations were previously described by Berini *et al.* 2007 [26].

Table 2. Description of WB indeterminate patterns for positive and negative samples by n-PCR among 578 samples received at a Reference Institute for HTLV Diagnosis and Confirmation (HDC) from blood banks or hospitals of Argentina.

| WB Indeterminate Banding Pattern | N | HTLV-1/2 Negative | HTLV-1/2 Positive | Not Performed |
|---|---|---|---|---|
| GD21 | 6 | 3 | 1 | 2 |
| GD21 + others | 7 | 3 | 3 | 1 |
| rgp46-1 and/or 2 | 4 | 1 | 1 | 2 |
| p19 | 2 | 2 | 0 | 0 |
| p19 + p24 | 4 | 2 | 0 | 2 |
| p19 + others | 1 | 0 | 0 | 1 |
| HGIP | 6 | 3 | 0 | 3 |
| Total | 30 | 14 | 5 | 11 |

The following experiments were performed in nine out of 13 seroindeterminate confirmed HTLV-1 positive cases found among the four studied populations, as samples from the other four were scarce. Figure 1 shows the banding pattern in each case.

## 2.2. Performance of the qPCR

The qPCR quantitation limit was 34 albumin copies/reaction and three *pol* copies/reaction. Samples with seroindeterminate results were run together, and an additional dilution was added to *pol* standard curve, as lower Threshold Cycle (Ct) values were expected. Acceptance criteria were accomplished and linearity was maintained ($R^2 > 0.99$). The intra-assay coefficient of variation (CV) was directly proportional to viral load levels: 14% at high load (>10 HTLV-1 copies/100 cells), 9% at medium load (1–10 HTLV-1 copies/100 cells), and 7% at low load (<1 HTLV-1 copy/100 cells); inter-assay CV at high loads was 24%. Of the 52 samples, PVL was detected in 51 and successfully quantified in 44 of them, including 25 pathology cases that were not in treatment at the time of the sample extraction (G1), 15 samples from asymptomatic carriers (AC) (G2), and four from seroindeterminate cases (G3), also AC. PVL was detected but not quantified in three G2 and four G3 samples, as the acceptance criteria was not met because of variable Ct values for *pol* gene. This viral gene was not detected in one of the G3 samples, although it could be amplified by n-PCR.

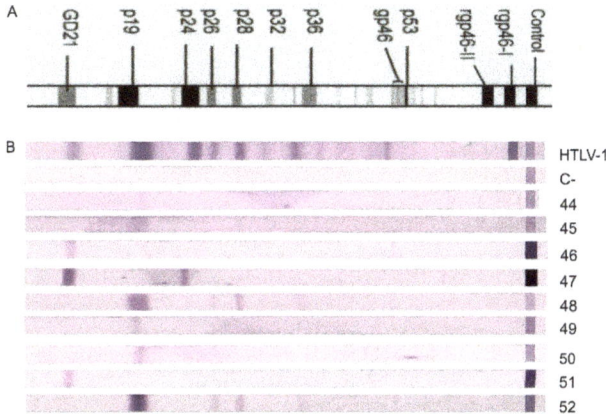

**Figure 1.** Western blot patterns of indeterminate cases confirmed HTLV-1 positive by n-PCR. Seroreactivity pattern using the MPD HTLV Blot 2.4 kit, which contains a recombinant GD21 (common for HTLV-1 and HTLV-2) and two synthetic peptides (rpg46-I and rpg46-II), specific either for HTLV-1 or HTLV-2. "HTLV-1": HTLV-1 positive control. "C-": negative control. 44–52: banding profile for each of the nine seroindeterminate cases analyzed.

### 2.3. PVL Values

Individual PVL values are shown in Table 4. Median PVL values and standard errors were 4.78 (2.4), 2.38 (1.02), and 0.15 (0.07) HTLV-1 copies/100 PBMCs for G1, G2, and G3, respectively; a significant difference could be observed between the three groups ($p = 0.003$), as shown in Figure 2a. The difference was also significant ($p = 0.005$) when samples with seropositive WB results were analyzed as a single group [G1 + G2]. As shown in Figure 2b, no significant difference ($p = 0.07$) was observed when samples from acute leukemia (ATL; $n = 4$) were compared with HAM/TSP ones ($n = 21$).

### 2.4. PVL Distribution

Regarding WB indeterminate samples, all successfully detected PVLs (8/9) were lower than 1 HTLV-1 copy/100 cells. Concerning G1 and G2 samples, an overlap in the range of PVL values was observed (Table 3). In three cases, viral loads below 1 HTLV-1 copy/100 cells were observed in individuals with pathology (12%).

**Table 3.** Individual proviral load values (PVL) distribution by groups (G). G1: positive samples by Western blot (WB) from individuals with pathology; G2: positive samples by WB from asymptomatic carriers (AC); and G3: indeterminate samples by WB from AC. PVLs are expressed as HTLV-1 copies/100 PBMCs.

| PVL Range | G1 ($n$=25) | G2 ($n$=18) | G3 ($n$=8) |
|-----------|-------------|-------------|------------|
| <1 | 12% | 44.4% | 100% |
| 1–10 | 52% | 44.4% | - |
| >10 | 36% | 11.2% | - |

**Table 4.** Age, gender, and individual proviral load values (PVL) of cases confirmed as HTLV-1 positive by nested PCR (n-PCR). Samples were classified as Group 1: positive samples by WB from individuals with pathology that are not on treatment ($n = 25$), Group 2: positive samples by WB from asymptomatic carriers (AC) ($n = 18$), and Group 3: indeterminate samples by WB from AC ($n = 9$). Codes for 14 LTR and/or *tax* sequences are detailed in brackets. Indeterminate patterns for G3 are also described in brackets. PVLs are expressed as HTLV-1 copies/100 PBMCs.

| Sample N° (Sequence Code) | Group | Age | Gender | PVL |
|---|---|---|---|---|
| 1 | 1- Leukemia | 53 | M | 33.9768 |
| 2 | 1- Leukemia | 66 | F | 40.1995 |
| 3 (ATL1) | 1- Lymphoma | 48 | F | 1.2974 |
| 4 (ATL2) | 1- Leukemia | 67 | F | 12.4920 |
| 5 | 1- HAM/TSP | 43 | F | 0.7081 |
| 6 | 1- HAM/TSP | 14 | F | 8.9051 |
| 7 | 1- HAM/TSP | 27 | F | 3.1244 |
| 8 | 1- HAM/TSP | 38 | F | 1.6633 |
| 9 | 1- HAM/TSP | 52 | F | 8.5679 |
| 10 | 1- HAM/TSP | 65 | M | 5.3882 |
| 11 | 1- HAM/TSP | 37 | F | 1.2808 |
| 12 | 1- HAM/TSP | 51 | F | 4.7829 |
| 13 (Neu28) | 1- HAM/TSP | 42 | F | 13.0850 |
| 14 (Neu14) | 1- HAM/TSP | 39 | F | 1.0119 |
| 15 | 1- HAM/TSP | 52 | M | 35.0971 |
| 16 | 1- HAM/TSP | 59 | F | 12.8610 |
| 17 | 1- HAM/TSP | 50 | M | 1.3508 |
| 18 | 1- HAM/TSP | 56 | F | 29.5394 |
| 19 | 1- HAM/TSP | NA | M | 0.1183 |
| 20 | 1- HAM/TSP | 26 | F | 3.1234 |
| 21 | 1- HAM/TSP | 71 | F | 10.4542 |
| 22 | 1- HAM/TSP | 35 | M | 0.5227 |
| 23 | 1- HAM/TSP | 67 | F | 1.6661 |
| 24 | 1- HAM/TSP | 52 | M | 15.5252 |
| 25 | 1- HAM/TSP | 49 | M | 4.0326 |
| 26 | 2 | 50 | M | 12.4340 |
| 27 | 2 | 64 | M | 1.8929 |
| 28 | 2 | 50 | M | 0.0832 |
| 29 | 2 | 46 | F | 1.2681 |
| 30 (ASYAR3) | 2 | 35 | M | 4.6143 |
| 31 (ASYAR2) | 2 | 26 | F | 0.7502 |
| 32 | 2 | 47 | F | 0.2476 |
| 33 | 2 | 25 | M | 8.4668 |
| 34 | 2 | 33 | M | 2.3861 |
| 35 | 2 | 39 | F | 2.8778 |
| 36 (ASYAR1) | 2 | 47 | F | 0.4813 |
| 37 | 2 | 52 | F | <3 copies/ reaction |
| 38 | 2 | NA | F | <3 copies/ reaction |
| 39 | 2 | 38 | F | 0.1832 |
| 40 (BDAR20) | 2 | 38 | M | 3.9461 |
| 41 | 2 | 57 | M | 5.9663 |
| 42 | 2 | NA | M | <3 copies/ reaction |
| 43 | 2 | 46 | M | 10.5358 |
| 44 | 3 (p19) | 59 | M | 0.0013 |
| 45 | 3 (p19) | 24 | M | 0.3365 |
| 46 (BDAR21) | 3 (GD21) | 28 | M | 0.1493 |
| 47 (BDAR18) | 3 (p24, GD21) | 52 | M | 0.1452 |
| 48 (FSW8) | 3 (HGIP) | 32 | F | <3 copies/ reaction |
| 49 (FSW9) | 3 (p19) | 52 | F | <3 copies/ reaction |
| 50 (FSW7) | 3 (p19) | 25 | F | <3 copies/ reaction |
| 51 (BDAR19) | 3 (GD21) | 31 | M | <3 copies/ reaction |
| 52 | 3 (HGIP) | 21 | M | *pol* not detected |

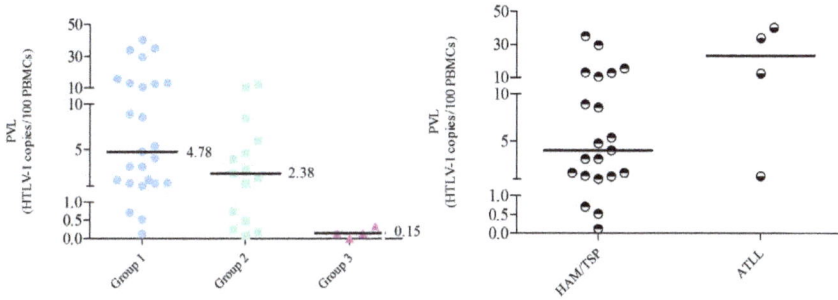

**Figure 2.** Individual proviral load values (PVLs). PVLs are expressed as HTLV-1 copies/100 PBMCs. (**a**) Samples are classified as Group 1: positive samples by Western blot (WB) from individuals with pathology that are not on treatment (*n* = 25), Group 2: positive samples by WB from asymptomatic carriers (AC) (*n* = 15), and Group 3: indeterminate samples by WB from AC (*n* = 4); PVL values are shown. A significant difference is observed between the three groups (p = 0.003) (GraphPad Prism V5). (**b**) PVLs for samples of Group 1, classified by disease: ATLL (*n* = 4) or HAM/TSP (*n* = 21) are shown, p = 0.07.

### 2.5. PVL Association with Gender, Age, and Optical Density

A moderate correlation (S=0.56) between PVL values and age at the moment of the sample extraction was observed in G1, which included patients with pathology who were not in treatment. Meanwhile, there was a significant association between PVL and gender in G2 (*p* = 0.01) (Figure 3). No correlation was observed in G3. Significantly lower optical density values were observed in most of the seroindeterminatesamples (*n* = 6/7), when compared with plasma samples from G1 and G2 selected randomly.

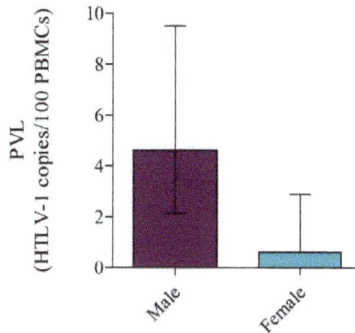

**Figure 3.** Median individual proviral loads (PVL) of G2 samples distrubuted by gender. A significant association (p =0.01) is observed between PVL and gender in Western blot (WB) positive samples from asymptomatic carriers (AC) distributed as: Males (*n* = 9) and Females (*n* = 6). Median PVL values were 4.61 in males and 0.61 in females. Median values together with interquartile ranges are shown.

### 2.6. HTLV-1 Phylogeny

A rooted neighbor-joining (NJ) tree of 134 HTLV-1 strains based upon a 458-bp fragment of the LTR region was performed, including 14 new strains from Argentina (G1: *n*=4; G2: *n* = 4; G3: *n* = 6); three G3 sequences could not be obtained. Cosmopolitan subtype HTLV-1a was clearly separated from HTLV-1 subtypes b, c, d, e, f, and g, with a bootstrap value of 69%. Within the cosmopolitan subtype, five subgroups were identified as previously described [4]. Although all of them were consistently separated by NJ, only two subgroups, West African/Caribbean subgroup C and North African subgroup D were well supported, showing higher bootstrap values than 75% (90% and 96%, respectively). Thirteen strains clustered within the cosmopolitan subtype transcontinental subgroup A, while the remaining one (N° 13, Neu28) from G1 clustered among sequences from Brazil and Peru previously reported by our group as Divergent Strains (Figure S2). Of these, five strains (ASYAR2, N°31; BDAR20, N°40; ASYAR3, N°30; ATL2, N°4; and Neu14, N°14) grouped with sequences from the Big Latin American cluster and three sequences (BDAR21, N°46; ATL1, N°3; and BDAR19, N°51) from the Small Latin American cluster. All eight sequences from the Latin American clusters grouped with Amerindian strains and other populations from Argentina previously reported by our group. One of the G3 strains (FSW7, N°50) grouped in the South African cluster along with one Argentine blood donor (BD1) and one Peruvian female sex worker (FSW1) residing in Argentina. The remaining four (BDAR18, N°47; FSW8, N°48; FSW9, N°49; and ASYAR1, N°36) clustered within the transcontinental group, near the other Argentine sequences previously reported by our group but not in any specific cluster.

### 2.7. Sequence Analysis

LTR and *tax* genes could be amplified in eight and four of all the seroindeterminate cases analyzed, respectively. Of those, six LTR and two *tax* sequences could be obtained with high quality, despite having repeated the procedure several times and assayed different DNA amounts with the remaining samples. A total of 25 mutations were detected among LTR sequences from G3 samples, of which seven were linked to geographical subtypes. Four mutations were observed in more than one sequence and the remaining 14 in only one, as detailed in Table 5. Of them, one mutation in the TRE-1 domain was present in the distal (dr) and one in the central (cr) region of the LTR. Regarding the TRE-2, two punctual mutations (8522T>C, 8545G>A) were observed in sample N° 50, while the other three were present in most Argentine sequences. While not found in any of the LTR sequences from G3, a mutation was observed in the distal region of TRE-1 (8403A>G) in 50% of the remaining sequences, which corresponded to samples with positive HTLV-1 results by WB and a significant difference ($p = 0.03$) between these frequencies was established. As for sequences from patients with pathology, eight (2 ATL and 6 HAM/TSP) out of 15 had this same mutation, thus maintaining the significance ($p = 0.04$).

*Tax* sequences showed 12 punctual mutations at the nucleotide level; of them, five were present in most of the Argentine sequences. Through the analysis of non-synonymous (NS) mutations in functional domains of the Tax protein, four amino acidic mutations were detected, two of them corresponding to geographical subtypes (A221V, S304N). The remaining mutations (H43P and M154V) were detected in sequences N° 46 and 47, respectively; H43P is located within the Nuclear Localization Signal domain, the Zn Finger, and CREB activation by Tax, while M154V is in the NF-κB activation region and the Dimerization Domain.

**Table 5.** Punctual mutations detected in LTR (U-3, R, and U-5) and *tax* gene sequences of indeterminate Western Blot samples confirmed as HTLV-1 positive by nested PCR. Six LTR and two *tax* sequences from seroindeterminate samples were analyzed, together with four sequences from HTLV-1 patients with pathology, four from HTLV-1 asymptomatic carriers, and 44 sequences from Argentina, previously obtained by our group and available in Gene Bank. Geographical: mutation linked to geographical subtypes; dr: distal region; cr: central region.

| Punctual Mutation | Sequence N° | Region | Punctual Mutation | Sequence N° | Region |
|---|---|---|---|---|---|
| 8295G>A | 46- 48 | LTR; U-3 | 8718C>T | 51 | LTR; R |
| 8367C>A | Geographical | LTR; U-3 | 8779T>C | 47 | LTR; R |
| 8381G>A | 48 | LTR; U-3 | 8822G>A | 46 | LTR; R |
| 8391G>A | 46 | LTR; U-3 | 8828A>G | 47 | LTR; R |
| 8392G>A | 46 | LTR; U-3 | 8912T>C | 46- 47 | LTR; U-5 |
| 8420C>T | 49 | LTR; U-3; TRE-1; dr | 8955G>A | 47 | LTR; U-5 |
| 8428_8429insA | Geographical | LTR; U-3; TRE-1 | 7383C>T | 46 | *tax* |
| 8446G>A | Geographical | LTR; U-3; TRE-1 | 7398C>T | 46- 47 | *tax* |
| 8471G>T | 50 | LTR; U-3; TRE-1; cr | 7401C>T | Geographical | *tax* |
| 8509A>G | Geographical | LTR; U-3; TRE-2 | 7431G>A | 47 | *tax* |
| 8509_8511delA | Geographical | LTR; U-3; TRE-2 | 7448A>C | 46 | *tax* |
| 8522T>C | 50 | LTR; U-3; TRE-2 | 7780A>G | 47 | *tax* |
| 8545G>A | 50 | LTR; U-3; TRE-2 | 7914T>C | Geographical | *tax* |
| 8546T>C | Geographical | LTR; U-3; TRE-2 | 7920C>T | Geographical | *tax* |
| 8606C>G | Geographical | LTR; U-3 | 7933C>T | 47 | *tax* |
| 8606C>A | 49- 51 | LTR; U-3 | 7982C>T | Geographical | *tax* |
| 8632G>A | 46 | LTR; U-3 | 8001A>G | 46 | *tax* |
| 8655G>T | 46 | LTR; U-3 | 8231G>A | Geographical | *tax* |
| 8665C>T | 46- 48 | LTR; R | | | |

## 3. Materials and Methods

### 3.1. Samples

A retrospective cross-sectional study was carried out on four different populations. Of those, Injecting Drug Users (IDU; $n = 173$), Female Sex Workers (FSW; $n = 613$), and Men who Have Sex with Men (MSM; $n = 667$) have been previously recruited and studied by our group; the corresponding HTLV-1/2 prevalence was reported in Berini *et al.* [26]. The fourth population included individuals with previous serological reactive results for HTLV-1/2 and/or with symptoms of HTLV-1 associated pathologies, whose samples have been sent from blood banks and/or hospitals from all over Argentina to our Institute for HTLV diagnosis or confirmation (HDC) between January 2011 and January 2014. The prevalence of the former population is reported here. All participants provided a written informed consent and the project was approved by the Institutional Review Board of Nexo Civil Association, Argentina. Samples confirmed as HTLV-1 positive by molecular techniques were classified in three different groups (G). G1 consisted of all the positive samples by WB from individuals with pathology who were not on treatment received for HDC between 2011 and 2014: 21 HAM/TSP patients and 4 Acute Leukemia Lymphoma (ATLL) patients. G2 included 18 WB positive samples from asymptomatic carriers (AC), who were chosen randomly for this study. G3 included 9 indeterminate samples by WB, also from AC. Serological and socio-demographic information of all nine G3 individuals is described in Supplementary Table 1.

### 3.2. Diagnostic Algorithm

Antibody screening for HTLV-1/2 was performed by particle agglutination technique (SERODIA-HTLV-I, Fujirebio, Tokyo, Japan) and/or by enzyme-linked immunosorbent assay (ELISA) (Murex HTLV-I+II, Abbott Laboratories Argentina, Buenos Aires, Argentina or HTLV I&II Ab v. ULTRA, Dia.Pro, Milan, Italy). Reactive samples were subjected to WB confirmation (HTLV blot 2.4, Genelabs Diagnostics, Science Park, Singapore). A WB was scored as HTLV-1 or HTLV-2 positive, untypeable, indeterminate, or negative according to the manufacturer's criteria.

For molecular confirmation of indeterminate or HTLV-positive samples by WB, DNA was extracted from peripheral blood mononuclear cells (PBMCs) by column extraction (ADN PuriPrep-S kit, Highway®, Inbio, Tandil, Argentina) and analyzed with "in-house" n-PCR for HTLV-1 and 2 *pol* and *tax* regions as previously described [19,20]. PCR was considered positive when amplicons from at least one amplification reaction were clearly detectable following agarose gel analysis [11]. Samples with non-reactive results by PA or ELISA were also further confirmed by n-PCR in order to avoid misdiagnosis in patients on retroviral treatment (because of other infections) and immune-compromised patients.

### 3.3. DNA Quantitation

Absolute quantitation of PVL was performed by real-time SYBR Green PCR, using an ABI Prism 7500Prism System (Applied Biosystems, Foster City, CA, USA).The HTLV-1 *pol* gene was amplified using 5 µL DNA, 12.5 µL SYBR Green PCR Master Mix (Applied Biosystems), and 200 nM of each primer (SK110-1: 5′-CCCTACAATCCAACCAGCTCAG-3′ and SK111-1: 5′-GTGGTGAAGCTGCCATCGGGTTTT-3′). PCR amplification of the albumin gene (ALB-S: 5′-GCTGTCATCTCTTGTGGGCTGT-3′ and ALB-AS: 5′-AAACTCATGGG AGCTGCTGGTT-3′) was performed as a separate reaction, as an endogenous reference to avoid variation due to differences in either the PBMC number or the DNA extraction method used. Cycle conditions were the following: 2 min at 50 °C and 10 min at 95 °C followed by 40 cycles of 15 s at 95 °C and 1 min at 65 °C. Melting curves were performed after the end of the amplification cycles to validate the specificity of the amplified products. Standard curves were generated using 10-fold serial dilutions of DNA from MT2 cells ($10^4$–$10^0$), and normalized to two copies of the HTLV-1 *pol* gene and two copies of the cellular albumin gene per MT2 cell [27]. All standard dilutions, controls, and individual samples were run in triplicate for both HTLV-1 and albumin DNA quantitation. Standard curves were accepted when slopes were between –3.10 and –3.74 and the $R^2$ was >0.99 [28]. The accuracy of the diagnostic test was assessed by measuring intra-assay and inter-assay variability. Intra-assay variability was evaluated by calculating the coefficient of variation (CV) of three viral load replicates from three DNA samples in three different ranges defined as low (<1 HTLV-1 copy/100 cells), medium (1–10 HTLV-1 copies/100 cells), and high (>10 HTLV-1 copies/100 cells). Inter-assay variability was calculated by measuring the CV for a high PVL sample in three independent runs. CV rather than standard deviations were used as they are not affected by the PVL absolute value. HTLV-1 proviral load was reported as [(*pol* average copy number)/(albumin average copy number/2)]*100 and expressed as the number of HTLV-1 copies/100 cells.

### 3.4. Molecular Analysis

Indeterminate samples were subjected to hemi-nPCR, aimed at amplifying LTR and *tax* genes. Amplification of the 3′ LTR region was performed using 8200LA (5′-CTCACACGGCCTCATACAGTACTC-3′) and R2 (5′-GTGCTATAGGATGGGCTGTCGC-3′) as outer primers and 3VINT (5′-GAACGCRACTCAACCGGCRYGGATGG-3′) and 3LTRf (5′-TCCCCATTTCTCTATTTTTAACG-3′) as inner primers (528 bp, ATK-1 genome position 8196–8699). Amplification of the *tax* region was carried out with outer primers HFL75 (5′-GCTATAGTCTCCTCCCCCTGC-3′) and 3VINT (5′-GAACGCRACTCAACCGGCRYGGATGG-3′) and inner primers TaxF (5′-ATGGCCCACTTCCCAGGGTT-3′) and TaxR (5′-TCAGACTTCTGTTTCTCGGA-3′), specific for HTLV-1. A slight modification was made to the protocols described elsewhere, in order to amplify *tax* and LTR genes in seroindeterminate samples by enhancing the DNA amount [29]. Direct sequencing reactions were done using a Big Dye Terminator 3.1 Cycle Sequencing RR-100 (Applied Biosystems). Sequences were generated on a 3500xL Genetic Analyzer AB/HITACHI according to the manufacturer's instructions. Sequences were edited manually (Sequencher 4.8) and then aligned using Clustal W (BioEdit 7.0.4.1 sequence alignment editor). For the sequence analysis of both genes, the ATK-1 genome was included as a reference prototype sequence and four samples from G1 together with four samples from G2 included as controls. Furthermore, 50 LTR sequences were also added, all of them corresponding to samples from Argentina and available in PubMed. MEGA software v. 5.2.2 was used for translation of *tax* gene sequences, based on the conventional genetic code.

To construct a comprehensive phylogenetic dataset, 14 of the LTR sequences (six seroindeterminate samples, four from HTLV-1 patients with pathology, and four from HTLV-1 AC) (See 5.5 "Accession numbers") were aligned along with 120 HTLV-1 reference strains obtained from the GenBank database, preferentially chosen because they were either from Argentina or from neighboring countries with high migration rates to Argentina. The Mel 5 reference strain (Melanesian origin, subtype c) was used as an outgroup. Once aligned, the dataset consisted of 458 bp corresponding to the 3′ LTR region. The phylogenetic analysis was performed by neighbor joining (NJ) using MEGA 5.2 and the tree topology was visualized with TreeView (http://taxonomy.zoology.gla.ac.uk/rod/treeview.html) [30].

*3.5. Accession Numbers*

Accession numbers corresponding to all new sequences mentioned here are detailed below. ASYAR2LTR: KT633516; BDAR20LTR: KT633517; FSW7LTR: KT633518; ASYAR3LTR: KT633519; BDAR19LTR: KT633520; FSW8LTR: KT633521; Neu14LTR: KT633522; BDAR18LTR: KT633523; BDAR21LTR: KT633524; FSW9LTR: KT633525; ASYAR1LTR: KT633526; ATL1tax: KT633527; Neu2tax: KT633528; ASYAR2tax: KT633529; BDAR18tax: KT633530; ATL2tax: KT633531; ASYAR1tax: KT633532; ASYAR3tax: KT633533; BDAR20tax: KT633534; Neu14tax: KT633535; BDAR21tax: KT633536. The following numbers correspond to the sequences obtained from GenBank for the molecular analysis. MEL5 (L02534); Efe1 (Y17014), ITIS (Z32527), PH236 (L76307); 2810YI (AY818432); Lib2 (Y17017); pyg19 (L76310); HS35 (DI3784), FrGu1 (AY324785), BO (U12804), Pr52 (U12806), Pr144 (U12807), Bl1.Peru (Y16481), RKl4.Peru (AF054627), BCl2.1 (U32557), H5 (M37299), Ni1-3.Peru (Y16484, Y16487, Y16485), ATL-YS (U19949), ATK-1 (J02029), MT4.LB ( Z31661), Br4 (AY324788), Bl3.Peru (Y16483), Neu13 (EU622623), Neu10 (EU622620), MT2 (L03562), 73RM (M81248), Ar11 (AY324777), FSW6 (EU622605), MSM2 (EU622609), BD7 (EU622588), Sur229-30 (AY374468, AY374466), BD3 (EU622586), BD10 (EU622590), TBH1-3 (L76026, L76025, L76034), FSW1 (EU622600), BD1 (EU622584), BD8 (EU622589), Gya468 (AY374459), Gya813 (AY374462), SurHM22 (AY374467), Gya542 (AY374460), BOI (L36905), BD13 (EU622593), BRRJ136.96 (DQ323759), KUW1-2 (L42253, L42255), IRN2 (U87261), CH26 (D23690), Abl.A (U87264), BCl1.2 (U32552), BRRP445 (DQ323755), BRRJ276.95 (DQ323750), BRRJ56.00 (DQ323754), BRRJ53.97 (DQ323753), Neu4 (EU622615), PW2 (EU622625), Neu5 (EU622616), MSM4 (EU622611), IDU4 (EU622608), Neu3 (EU622614), BD2 (EU622585), CAM (AF063819), BRRJMDP (DQ323751), ARGSOT (AF007755), AMA (X88871), CMC (X88872), TBH4 (L76028), BRRP495(DQ323755), FSW4 (EU622603), Me3.Peru (Y16480), Ar55 (AY324782), FSW5 (EU622604), Sur1597 (AY374465), Gya572 (AY374461), Neu1 (EU622612), FSW2 (EU622601), JCP (X88875), BRRJFA (DQ323757), Me1.Peru (Y16478), MASU (X88877), FCR (X88873), BRRJ86.97 (DQ323760), MAQS (X88876), Ar5 (AY324783), Qu2.Peru (Y16476), Me2.Peru (Y16479), Qu3.Peru (Y16477), J37 (FJ751855), BD16 (EU622596), IDU1-3 (EU622598, EU622606, EU622607), BD15 (EU622595), BD14 (EU622594), BD12 (EU622592), Neu11 (EU622621), BD11 (EU622591), J77 (FJ758161), J43 (FJ751856), BD4 (EU622587), J68 (FJ751858), J20 (FJ751854), Neu12 (EU622622), Ar49 (AY324793), Ar15 (AY324778), PW1 (EU622624), FSW3 (EU622602), Qu1.Peru (Y16475), Neu7 (EU622618), ARGDOU (AF007751), Neu8 (EU622619), BD17 (EU622597), J47 (FJ751857), MSM3 (EU622610), Neu2 (EU622613).

*3.6. Statistical Analysis*

Data analysis was performed using the Kruskal–Wallis non-parametric method; when two groups were compared, the Mann–Whitney–Wilcoxon test was used. To evaluate the presence of association between PVL and age of individuals, correlation was determined based on Spearman coefficient (S). GraphPad Prism (version 6.03) software was applied and significant differences were defined as $p < 0.05$.

## 4. Discussion

HTLV indeterminate WB patterns have been reported worldwide, although they are more frequent in tropical areas [31,32]. Most of these reports refer to blood donors, and seroindeterminate frequency varies according to HTLV-1/2 endemicity (*i.e.*, the geographical area studied) [27,33]. In this study, we report for the first time in our country the prevalence of indeterminate WB results among at-risk populations: 2.31%, 2.12%, and 1.65% for IDUs, FSW, and MSM, respectively. Moreover, the prevalence of indeterminate WB patterns (5.19%) in the HDC population recruited in our Institute was higher than the one reported in our country for blood donors (0.1%), as expected due to a biased population [27]. On the other hand, it must be considered that screening in our laboratory has always been performed with the most efficient assays available in the country. Concerning the indeterminate WB banding patterns in the HDC population (both positive and negative cases by n-PCR), GD21 (alone or with other bands) and HGIP were the most frequent. Several studies have demonstrated the presence of HGIP patterns mostly among blood donors, and it has been suggested that generally these are not caused by HTLV-1 infection [32]. Nonetheless, as reported previously by our group, two HGIP cases corresponded to samples that turned out positive for HTLV-1 infection (one blood donor and one IDU) [10]. Regarding the "N pattern" recently described by Filippone *et al.*, it was not observed neither in the HDC population, nor in at-risk populations (pattern described in Berini *et al* [26]) [11].

There are few reports suggesting that low HTLV-1/2 PVL could cause indeterminate WB patterns in samples from infected individuals [16,34]. Regarding HTLV-1, low PVL levels for these cases were suggested, especially in the ones in which no PVL could be successfully quantified, although no direct comparisons with seropositive patients were performed [35–37]. In this study, PVL values were compared between two groups with positive WB results: G1 (individuals with pathology without retroviral treatment) and G2 (asymptomatic carriers: AC). A significant difference was observed between them, in line with previously published data [21,22]. Furthermore, a third group was included consisting of nine samples from AC with seroindeterminate WB, of which four PVLs could be determined. Even when considering the small sample size of G3, a significant difference was observed between these three groups. Although a seroconversion could not be discarded in the indeterminate cases, these data demonstrate that in some cases indeterminate WB results could be associated with low HTLV-1 PVL; a low viral replication rate may consequently trigger a weak immune response and low concentrations of anti-HTLV-1 antibodies.

Regarding age and gender, it was determined that these variables were associated with PVL levels among G1 and G2, respectively. While Vakili *et al.* observed no significant association between PVL in HAM/TSP patients and healthy carriers with age and gender, it has been reported that for both HTLV-1 and 2 infections, women have lower PVL levels than men, consistent with our results [38–40]. In contrast, Hisada *et al.* showed no gender differences in PVL [41]. Thus, further studies should be performed in order to clarify this issue. Regarding serological status, Manns *et al.* reported in 1999 that the anti-HTLV-1 antibody titer had a positive correlation with PVL levels, and years later, Akimoto *et al.* confirmed these results [42,43]. In this study, and similarly to what has been reported by Filippone *et al.*, the optical density values obtained were lower for most of the seroindeterminate profile plasma samples when compared to HTLV-1 WB-positive ones [11].

As for phylogeny, all 14 new strains described in this study belonged to the cosmopolitan subtype and most of them classified within the transcontinental subgroup A (one sample classified as divergent). Half of the G3 sequences did not group in any specific cluster, while two grouped in the small Latin American cluster (BDAR19, BDAR21) and one in the South African cluster (FSW7), together with other sequences previously reported by our group (FSW1 and BD1) [44]. These data confirm our previous publication concerning the presence of HTLV-1 transcontinental and African strains circulating in Argentina, although most of the infected individuals in our study were not of black origin, supporting the hypothesis of multiple introductions of HTLV-1 of the cosmopolitan subtype in the New World [5].

Even though some of the observed mutations in LTR and *tax* genes were linked to geographical subtypes, others were further analyzed in order to establish their relevance, as it is well known that

*Viruses* **2015**, *7*, 5643–5658

during replication and transcription the LTR/Tax system is extremely important. The promoter region LTR responds to the transactivation mediated by the Tax protein, which directly interacts by binding the DNA or indirectly by binding cellular transcription factors, in the Tax regulation elements known as TRE-1 and TRE-2 [45]. Although mutations in sequences from G3 were observed both in the distal and central regions of TRE-1, no significant associations were established given the number of mutated sequences. The same is valid for the TRE-2 region, which can also mediate transactivation by Tax-1 [46,47], considering that it contains binding sites for a large number of transcription factors, including AP-2, HNF-3, Ets family members, NFκB, and Sp1 [48]. Interestingly, we have not observed the 8403A>G mutation (distal region of TRE-1) in any of the G3 sequences, although it was present in 50% of the remaining sequences, all of them positive by WB. Therefore, a significant association was established between the presence of this mutation and seropositivity. Particularly, 53% of sequences from patients with pathology were mutated in base 8403, consistent with a report from Brazil, in which this mutation was significantly associated with high PVL values [17]. Regarding the Tax protein, non-synonymous mutations in the CREB activation domain as well as in the NF-κB and Zn Finger activation domains were detected. Previous data indicate that mutations in these regions displayed markedly attenuated abilities to transactivate the provirus and to reduce the ability to induce nuclear expression of NF-κB [18]. Whether the presence of mutations observed in this study could explain a diminished transactivation activity of Tax protein and therefore a low PVL still remains to be determined. Only functional studies would indicate their possible impact on indeterminate WB profiles.

Considering diagnosis, indeterminate WB results cannot be avoided without an improvement of serological commercial kits aimed at enhancing sensibility and specificity. Thus, and taking into account the high prevalence of these seroindeterminate cases worldwide [10,11,49], the usefulness of serological confirmation is questionable, highlighting the difficulties in interpretation and counseling. Seroindeterminate results represent a big challenge for health professionals, especially in those countries with endemic areas and no national programs for controlling HTLV infection. Furthermore, WB kits are far more expensive than n-PCR, being a relevant factor for the healthcare system. Costa and Thorstensson *et al.* have recommended different strategies for reducing costs and improving the accuracy of the diagnosis [24,50]. While both proposed two EIAs for screening, Costa recommends qPCR to confirm the infection. Even though qPCR is actually the standard method for PVL quantitation, other technologies have also been introduced. Recently, an mq-PCR testing algorithm for the diagnosis of HTLV-1/2 infection has been proposed [25]. Nevertheless, in some serologically confirmed positive cases, PVL could not be detected, especially in HTLV-2 samples [25,51]. In that context, Brunetto *et al.* reported the utility of digital droplet PCR (ddPCR) in the quantitation of HTLV-1 PVL [52]. They postulate that, even though both methods show a strong correlation and similar performance, ddPCR exhibits lower inter and intra-assay variability as it is based on a Poisson algorithm for quantitation of genes instead of the standard curve used in qPCR [52]. On the other hand, another study showed a higher sensitivity for qPCR compared to ddPCR when detecting cytomegalovirus in clinical samples [53]. Based on these data, more studies should be performed in order to establish the most efficient methodology for HTLV-1/2 PVL quantitation. Furthermore, a consensus regarding qPCR data interpretation and analysis for HTLV-1/2 PVL quantitation, as well as a universal expression unit, should be achieved in order to avoid confusion and misunderstanding. Besides, most HTLV-1/2 endemic areas correspond to developing countries and access to qPCR equipment is not always possible.

Therefore, we propose to re-evaluate the diagnostic algorithm, considering molecular confirmation by n-PCR for reactive samples, instead of qPCR, right after the combination of two screening tests. This alternative would avoid serological confirmation by WB, the most expensive stage of the diagnosis algorithm, until a better confirmation technique is available and standardized. Further studies with significant panels including HTLV-1, HTLV-2, and indeterminate samples should be carried out in order to establish whether it is time saving, effective, and less expensive.

## 5. Conclusions

This study describes the prevalence of indeterminate WB patterns among different populations from Argentina and demonstrates that in some cases these profiles may be associated with low HTLV-1 PVL. Mutations in LTR and *tax* have been described among both indeterminate and positive HTLV-1 cases, highlighting 8403A>G in the distal region of TRE-1, already related to high PVL. Still, the functional relevance of these mutations remains to be determined.

**Acknowledgments:** The authors would like to thank Dr. Federico Remes Lenicov for his counseling as well as Mirta Villa, Claudio Gomez, and Ricardo Casime from the *Instituto de Investigaciones Biomédicas en Retrovirus y SIDA* (INBIRS), UBA-CONICET, part of the patient assistance service. . We would also like to highlight Williams Pedrozo's commitment with regard to sample collection in Misiones province and, lastly, Lic. Sergio Mazzini for his assistance with English revision.

**Author Contributions:** Mirna Biglione, Carolina Berini, and Camila Cánepa conceived and designed the study. Camila Cánepa and Jimena Salido carried out sample selection, extraction and quantitation of PVL, as well as n-PCR confirmation and sequencing, together with Matías Ruggieri's collaboration. Carolina Berini advised and helped with the sequencing and corresponding analysis. Camila Cánepa performed the statistical analysis. Mirna Biglione, Carolina Berini, Camila Cánepa, and Jimena Salido drafted the manuscript. All authors contributed to the final version of the manuscript, read and approved it.

**Conflicts of Interest:** The authors declare no conflict of interest.

## References

1. Gessain, A.; Cassar, O. Epidemiological Aspects and World Distribution of HTLV-1 Infection. *Front. Microbiol.* **2012**, *3*. [CrossRef] [PubMed]
2. Roucoux, D.F.; Murphy, E.L. The epidemiology and disease outcomes of human T-lymphotropic virus type II. *AIDS Rev.* **2004**, *6*, 144–154. [PubMed]
3. Gessain, A.; Mahieux, R. Epidemiology, origin and genetic diversity of HTLV-1 retrovirus and STLV-1 simian affiliated retrovirus. *Bull. Soc. Pathol. Exot.* **2000**, *93*, 163–171. [PubMed]
4. Vidal, A.U.; Gessain, A.; Yoshida, M.; Tekaia, F.; Garin, B.; Guillemain, B.; Schulz, T.; Farid, R.; Thé, G. Phylogenetic classification of human T cell leukaemia/lymphoma virus type I genotypes in five major molecular and geographical subtypes. *J. Gen. Virol.* **1994**, *75*, 3655–3666. [CrossRef] [PubMed]
5. Van Dooren, S.; Gotuzzo, E.; Salemi, M.; Watts, D.; Audenaert, E.; Duwe, S.; Ellerbrok, H.; Grassmann, R.; Hagelberg, E.; Desmyter, J.; *et al.* Evidence for a post-Columbian introduction of human T-cell lymphotropic virus in Latin America. *J. Gen. Virol.* **1998**, *79*, 2695–2708. [CrossRef] [PubMed]
6. Biglione, M.M.; Astarloa, L.; Salomón, H.E. Referent HTLV-I/II Argentina Group. High prevalence of HTLV-I and HTLV-II among blood donors in Argentina: A South American health concern. *AIDS Res. Hum. Retrovir.* **2005**, *21*, 1–4. [CrossRef] [PubMed]
7. Gastaldello, R.; Hall, W.W.; Gallego, S. Seroepidemiology of HTLV-I/II in Argentina: An overview. *J. Acquir. Immune Defic. Syndr.* **2004**, *35*, 301–308. [CrossRef] [PubMed]
8. Licensure of Screening Tests for Antibody to Human T-Lymphotropic Virus Type 1. *MMWR* **1988**, *37*, 736–747.
9. Khabbaz, R.F.; Heneine, W.; Grindon, A.; Hartley, T.M.; Shulman, G.; Kaplan, J. Indeterminate HTLV serologic results in U.S. blood donors: Are they due to HTLV-I or HTLV-II? *J. Acquir. Immune Defic. Syndr.* **1992**, *5*, 400–404. [PubMed]
10. Berini, C.A.; Eirin, M.E.; Pando, M.A.; Biglione, M.M. Human T-cell lymphotropic virus types I and II (HTLV-I and -II) infection among seroindeterminate cases in Argentina. *J. Med. Virol.* **2007**, *79*, 69–73. [CrossRef] [PubMed]
11. Filippone, C.; Bassot, S.; Betsem, E.; Tortevoye, P.; Guillotte, M.; Mercereau-Puijalon, O.; Plancoulaine, S.; Calattini, S.; Gessain, A. A new and frequent human T-cell leukemia virus indeterminate Western blot pattern: Epidemiological determinants and PCR results in central African inhabitants. *J. Clin. Microbiol.* **2012**, *50*, 1663–1672. [CrossRef] [PubMed]
12. Prince, H.E.; Gross, M. Impact of initial screening for human T-cell lymphotropic virus (HTLV) antibodies on efficiency of HTLV Western blotting. *Clin. Diagn. Lab. Immunol.* **2001**, *8*. [CrossRef]

13. Mahieux, R.; Horal, P.; Mauclère, P.; Mercereau-Puijalon, O.; Guillotte, M.; Meertens, L.; Murphy, E.; Gessain, A. Human T-cell lymphotropic virus type 1 gag indeterminate western blot patterns in Central Africa: Relationship to Plasmodium falciparum infection. *J. Clin. Microbiol.* **2000**, *38*, 4049–4057. [PubMed]

14. Porter, K.R.; Liang, L.; Long, J.W.; Bangs, M.J.; Anthony, R.; Andersen, E.M.; Hayes, C.G. Evidence for anti-Plasmodium falciparumantibodies that cross-react with human T-lymphotropic virus type I proteins in a population in Irian Jaya, Indonesia. *Clin. Diagn. Lab Immunol.* **1994**, *1*, 11–15. [PubMed]

15. Hayes, C.G.; Burans, J.P.; Oberst, R.B. Antibodies to human T lymphotropic virus type I in a population from the Philippines: Evidence for cross-reactivity with Plasmodium falciparum. *J. Infect. Dis.* **1991**, *163*, 257–262. [CrossRef] [PubMed]

16. Abrams, A.; Akahata, Y.; Jacobson, S. The prevalence and significance of HTLV-I/II seroindeterminate Western blot patterns. *Viruses* **2011**, *3*, 1320–1331. [CrossRef] [PubMed]

17. Netto, E.C.; Brites, C. Characteristics of Chronic Pain and Its Impact on Quality of Life of Patients with HTLV-1-associated Myelopathy/Tropical Spastic Paraparesis (HAM/TSP). *Clin. J. Pain.* **2011**, *27*, 131–135. [CrossRef] [PubMed]

18. Smith, M.R.; Smith, M.R. Identification of HTLV-I tax trans-activator mutants exhibiting novel transcriptional phenotypes. *Genes Dev.* **1990**, *4*, 1875–1885.

19. Heneine, W.; Khabbaz, R.F.; Lal, R.B.; Kaplan, J.E. Sensitive and specific polymerase chain reaction assays for diagnosis of human T-cell lymphotropic virus type I (HTLV-I) and HTLV-II infections in HTLV-I/II-seropositive individuals. *J. Clin. Microbiol.* **1992**, *30*, 1605–1607. [PubMed]

20. Tuke, P.W.; Luton, P.; Garson, J.A. Differential diagnosis of HTLV-I and HTLV-II infections by restriction enzyme analysis of 'nested' PCR products. *J. Virol. Methods* **1992**, *40*, 163–173. [CrossRef]

21. Furtado, M.; Andrade, R.G.; Romanelli, L.C.; Ribeiro, M.A.; Ribas, J.G.; Torres, E.B.; Barbosa-Stancioli, E.F.; Proietti, A.B.; Martins, M.L. Monitoring the HTLV-1 proviral load in the peripheral blood of asymptomatic carriers and patients with HTLV-associated myelopathy/tropical spastic paraparesis from a Brazilian cohort: ROC curve analysis to establish the threshold for risk disease. *J. Med. Virol.* **2012**, *84*, 664–671. [CrossRef] [PubMed]

22. Nagai, M.; Usuku, K.; Matsumoto, W.; Kodama, D.; Takenouchi, N.; Moritoyo, T.; Hashiguchi, S.; Ichinose, M.; Bangham, C.R.; Izumo, S.; *et al.* Analysis of HTLV-I proviral load in 202 HAM/TSP patients and 243 asymptomatic HTLV-I carriers: High proviral load strongly predisposes to HAM/TSP. *J. Neurovirol.* **1998**, *4*, 586–593. [CrossRef] [PubMed]

23. Olindo, S.; Lézin, A.; Cabre, P.; Merle, H.; Saint-Vil, M.; Edimonana-Kaptue, M.; Signate, A.; Césaire, R.; Smadja, D. HTLV-1 proviral load in peripheral blood mononuclear cells quantified in 100 HAM/TSP patients: A marker of disease progression. *J. Neurol. Sci.* **2005**, *237*, 53–59. [CrossRef] [PubMed]

24. Costa, E.A.; Magri, M.C.; Caterino-de-Araujo, A. The best algorithm to confirm the diagnosis of HTLV-1 and HTLV-2 in at-risk individuals from São Paulo, Brazil. *J. Virol. Methods* **2011**, *173*, 280–286. [CrossRef] [PubMed]

25. Waters, A.; Oliveira, A.L.; Coughlan, S.; de Venecia, C.; Schor, D.; Leite, A.C.; Araújo, A.Q.; Hall, W.W. Multiplex real-time PCR for the detection and quantitation of HTLV-1 and HTLV-2 proviral load: Addressing the issue of indeterminate HTLV results. *J. Clin. Virol.* **2011**, *52*, 38–44. [CrossRef] [PubMed]

26. Berini, C.A.; Pando, M.A.; Bautista, C.T.; Eirin, M.E.; Martinez-Peralta, L.; Weissenbacher, M.; Avila, M.M.; Biglione, M.M. HTLV-1/2 among high-risk groups in Argentina: Molecular diagnosis and prevalence of different sexual transmitted infections. *J. Med.Virol.* **2007**, *79*, 1914–1920. [CrossRef] [PubMed]

27. Mangano, A.M.; Remesar, M.; del Pozo, A.; Sen, L. Human T lymphotropic virus types I and II proviral sequences in Argentinian blood donors with indeterminate Western blot patterns. *J. Med. Virol.* **2004**, *74*, 323–327. [CrossRef] [PubMed]

28. Bustin, S.A.; Benes, V.; Garson, J.A.; Hellemans, J.; Huggett, J.; Kubista, M.; Mueller, R.; Nolan, T.; Pfaffl, M.W.; Shipley, G.L.; *et al.* The MIQE guidelines: Minimum information for publication of quantitative real-time PCR experiments. *Clin. Chem.* **2009**, *55*, 611–622. [CrossRef] [PubMed]

29. Eirin, M.E. Epidemiología molecular del virus linfotrópico T-humano tipo 1 (HTLV-1) en Argentina: Análisis étnico-geográfico y variabilidad viral. Doctoral Thesis, Facultad de Ciencias Exactas y Naturales. Universidad de Buenos Aires, CABA, Argentina, 2011.

30. Gascuel, O. BIONJ: An improved version of the NJ algorithm based on a simple model of sequence data. *Mol. Biol. Evol.* **1997**, *14*, 685–695. [CrossRef] [PubMed]

31. Cesaire, R.; Bera, O.; Maier, H.; Martial, J.; Ouka, M.; Kerob-Bauchet, B.; Ould Amar, A.K.; Vernant, J.C. Seroindeterminate patterns and seroconversions to human T-lymphotropic virus type I positivity in blood donors from Martinique, French West Indies. *Transfusion* **1999**, *39*, 1145–1149. [CrossRef] [PubMed]

32. Rouet, F.; Meertens, L.; Courouble, G.; Herrmann-Storck, C.; Pabingui, R.; Chancerel, B.; Abid, A.; Strobel, M.; Mauclere, P.; Gessain, A. Serological, epidemiological, and molecular differences between human T-cell lymphotropic virus Type 1 (HTLV-1)-seropositive healthy carriers and persons with HTLV-I Gag indeterminate Western blot patterns from the Caribbean. *J. Clin. Microbiol.* **2001**, *39*, 1247–1253. [CrossRef] [PubMed]

33. The HTLV European Research Network. Seroepidemiology of the human T-cell leukaemia/lymphoma viruses in Europe. *J. Acquir. Immune Defic. Syndr. Hum. Retrovirol.* **1996**, *13*, 68–77.

34. Olah, I.; Fukumori, L.M.; Smid, J.; de Oliveira, A.C.; Duarte, A.J.; Casseb, J. Neither molecular diversity of the envelope, immunosuppression status, nor proviral load causes indeterminate HTLV western blot profiles in samples from human T-cell lymphotropic virus type 2 (HTLV-2)-infected individuals. *J. Med. Virol.* **2010**, *82*, 837–842. [CrossRef] [PubMed]

35. Yao, K.; Hisada, M.; Maloney, E.; Yoshihisa, Y.; Hanchard, B.; Wilks, R.; Rios, M.; Jacobson, S. Human T Lymphotropic Virus Types I and II Western Blot Seroindeterminate Status and Its Association with Exposure to Prototype HTLV-I. *J. Infect. Dis.* **2006**, *193*, 427–437. [CrossRef] [PubMed]

36. Mangano, A.; Altamirano, N.; Remesar, M.; Bouzas, M.B.; Aulicino, P.; Zapiola, I.; DelPozo, A.; Sen, L. HTLV-I proviral load in Argentinean subjects with indeterminate western blot patterns. *Retrovirology* **2011**, *8* (Suppl. 1). [CrossRef]

37. Demontis, M.A.; Hilburn, S.; Taylor, G.P. Human T cell lymphotropic virus type 1 viral load variability and long-term trends in asymptomatic carriers and in patients with human T cell lymphotropic virus type 1-related diseases. *AIDS Res. Hum. Retrovir.* **2013**, *29*, 359–364. [CrossRef] [PubMed]

38. Vakili, R.; Sabet, F.; Aahmadi, S.; Boostani, R.; Rafatpanah, H.; Shamsian, A.; Rahim Rezaee, S.A. Human T-lymphotropic Virus Type I (HTLV-I) Proviral Load and Clinical Features in Iranian HAM/TSP Patients: Comparison of HTLV-I Proviral Load in HAM/TSP Patients. *Iran J. Basic Med. Sci.* **2013**, *16*, 268–272. [PubMed]

39. Hodson, A.; Laydon, D.; Bain, B.J.; Fields, P.A.; Taylor, G.P. Pre-morbid human T-lymphotropicvirus type I proviral load, rather than percentage of abnormal lymphocytes, is associated with an increased risk of aggressive adult T-cell leukemia/lymphoma. *Haematologica* **2013**, *98*, 385–388. [CrossRef] [PubMed]

40. Montanheiro, P.; Olah, I.; Fukumori, L.M.I.; Smid, J.; Penalva de Oliveira, A.C.; Kanzaki, L.I.B.; Fonseca, L.A.; Duarte, A.J.; Casseb, J. Low DNA HTLV-2 proviral load among women in Sao Paulo City. *Virus Res.* **2008**, *135*, 22–25. [CrossRef] [PubMed]

41. Hisada, M.; Miley, W.J.; Biggar, R.J. Provirus load is lower in human T lymphotropic virus (HTLV)-II carriers than in HTLV-I carriers: A key difference in viral pathogenesis? *J. Infect. Dis.* **2005**, *191*, 1383–1385. [CrossRef] [PubMed]

42. Manns, A.; Miley, W.J.; Wilks, R.J.; Morgan, O.; Hanchard, B.; Wharfe, G.; Cranston, B.; Maloney, E.; Welles, S.; Blattner, W.A.; *et al.* Quantitative Proviral DNA and Antibody Levels in the Natural History of HTLV-I Infection. *J. Infect. Dis.* **1999**, *180*, 1487–1493. [CrossRef] [PubMed]

43. Akimoto, M.; Kozako, T.; Sawada, T.; Matsushita, K.; Ozaki, A.; Hamada, H.; Kawada, H.; Yoshimitsu, M.; Tokunaga, M.; Haraguchi, K.; *et al.* Anti-HTLV-1 tax antibody and tax-specific cytotoxic T lymphocyte are associated with a reduction in HTLV-1 proviral load in asymptomatic carriers. *J. Med.Virol.* **2007**, *79*, 977–986. [CrossRef] [PubMed]

44. Eirin, M.E.; Dilernia, D.A.; Berini, C.A.; Jones, L.R.; Pando, M.A.; Biglione, M.M. Divergent strains of human T-lymphotropic virus type 1 (HTLV-1) within the Cosmopolitan subtype in Argentina. *AIDS Res. Hum.Retrovir.* **2008**, *24*, 1237–1244. [CrossRef] [PubMed]

45. Bosselut, R.; Lim, F.; Romond, P.C.; Frampton, J.; Brady, J.; Ghysdael, J. Myb protein binds to multiple sites in the human T cell lymphotropic virus type 1 long terminal repeat and transactivates LTR-mediated expression. *Virology* **1992**, *186*, 764–769. [CrossRef]

46. Marriott, S.J.; Boros, I.; Duvall, J.F.; Brady, J.N. Indirect binding of human T-cell leukemia virus type I tax1 to a responsive element in the viral long terminal repeat. *Mol. Cell. Biol.* **1989**, *9*, 4152–4160. [CrossRef] [PubMed]

47. Numata, N.; Ohtani, K.; Niki, M.; Nakamura, M.; Sugamura, K. Synergism between two distinct elements of the HTLV-I enhancer during activation by the trans-activator of HTLV-I. *New Biol.* **1991**, *3*, 896–906. [PubMed]

48. Datta, S.; Kothari, N. H.; Fan, H. *In vivo* genomic footprinting of the human T-cell leukemia virus type 1 (HTLV-1) long terminal repeat enhancer sequences in HTLV-1-infected human T-cell lines with different levels of Tax I activity. *J. Virol.* **2000**, *74*, 8277–8285. [CrossRef] [PubMed]

49. Costa, J.M.; Segurado, A.C. Molecular evidence of human T-cell lymphotropic virus types 1 and 2 (HTLV-1 and HTLV-2) infections in HTLV seroindeterminate individuals from São Paulo, Brazil. *J. Clin. Virol.* **2009**, *44*, 185–189. [CrossRef] [PubMed]

50. Thorstensson, R.; Albert, J.; Andersson, S. Strategies for diagnosis of HTLV-I and –II. *Transfusion* **2002**, *42*, 780–791. [CrossRef] [PubMed]

51. Busch, M.P.; Switzer, W.M.; Murphy, E.L.; Thomson, R.; Heneine, W. Absence of evidence of infection with divergent primate T-lymphotropic viruses in United States blood donors who have seroindeterminate HTLV test results. *Transfusion* **2000**, *40*, 443–449. [CrossRef] [PubMed]

52. Brunetto, G; Massoud, G.; Leibovitch, E.C.; Caruso, B.; Johnson, K.; Ohayon, J.; Fenton, K.; Cortese, I.; Jacobson, S. Digital droplet PCR (ddPCR) for the precise quantification of human T-lymphotropic virus 1 proviral loads in peripheral blood and cerebrospinal fluid of HAM/TSP patients and identification of viral mutations. *J. Neurovirol.* **2014**, *20*, 341–351.

53. Hayden, R.T.; Gu, Z.; Abdul-Ali, D.; Shi, L.; Pounds, S.; Caliendo, A.M. Comparison of droplet digital PCR to real-time PCR for quantitative detection of cytomegalovirus. *J. Clin. Microbiol.* **2013**, *51*, 540–546. [CrossRef] [PubMed]

MDPI AG

St. Alban-Anlage 66

4052 Basel, Switzerland

Tel. +41 61 683 77 34

Fax +41 61 302 89 18

http://www.mdpi.com

*Viruses* Editorial Office

E-mail: viruses@mdpi.com

http://www.mdpi.com/journal/viruses